D1453558

INDUSTRIAL DESIGN OF PLASTICS PRODUCTS

INDUSTRIAL DESIGN OF PLASTICS PRODUCTS

M. Joseph Gordon, Jr.

WILEY-INTERSCIENCE
A JOHN WILEY & SONS, INC., PUBLICATION

This book is printed on acid-free paper. ♾

Published by John Wiley & Sons, Inc., Hoboken, New Jersey.

Published simultaneously in Canada.

Library of Congress Cataloging-in-Publication Data:

Gordon, Joseph (M. Joseph)
Industrial design of plastics products / M. Gordon Joseph, Jr.
p. cm.
"A Wiley-Interscience publication"
ISBN 0-471-23151-7 (cloth : alk. paper)
1. Plastics. 2. Engineering design. I. Title.

TA455.P5 G596 2002
620.1'923–dc21 2002028801

Printed in the United States of America.
10 9 8 7 6 5 4 3 2 1

CONTENTS

PREFACE

Plastic product design relies on the same engineering formulas and procedures used for the design of metal and other materials for today's products. The use of plastics as a building material requires a more in-depth knowledge of how the material responds to forces and the environment it is subjected to in service. Therefore, to produce acceptable products, the plastic product designer must answer the same metal design product questions and must analyze specific material design questions specific for plastic materials. Plastic material property areas to be considered are creep, temperature extremes, time under load, impact forces and method of loading the part, and how the material will react under the product end use environment.

Metal products have these same concerns but to a lesser degree. Metals that can be replaced by plastic, for certain applications, are typically well within their acceptable material physical property values that are not affected in the same manner as are plastic materials. This discrepancy in understanding the difference usage of a plastic versus a metal is almost totally replaced by the design and property information available from the major plastic material suppliers: Hoechst Celanese, DuPont, BASF, and other companies that supply answers for how their materials will perform. This information is supplied in the form of material data property graphs, charts, and tables for products tested at end-use conditions. However, the data may not always contain the specific information that the designer requires to complete the product analysis in plastic. Plastic product designers must therefore know what information they need to correctly design the product and also make the correct material selection from a material list of over 30 generic plastic resin families. The specific resin grades in these families can total over 100,000 specific types of products.

This text attempts to address the specific areas of knowledge that a plastic designer must have to correctly design a product in plastic. This is accomplished by combining material, environmental properties with structural design requirements using the standard design formulas. By applying the data for plastic materials from material suppliers, the products end use objectives are obtained.

The designer must consider in the analysis the correct plastic design and manufacturing principles for the materials considered for the product. Failure to follow the plastic practical design and manufacturing principles can result in a product that is well designed but difficult or too costly to manufacture. Each plastic material has its individual manufacturing limitations that the design team must consider. Correctly designing a plastic product involves a greater understanding of the materials variables than designing with metal. Plastic material physical properties vary more as temperature varies from 32 to 150°F where creep and physical property variance are the major considerations. Selection of a plastic material involves the design team considering the product's end-use temperature range, creep under load, and where forces are concentrated at specific locations that can cause problems for the product.

Design and manufacture of plastic products using injection molding involves mold and processing variables that the design team must consider for manufacture. The design team must know the material, manufacturing, and processing variables during the products design phase. After the part is designed, the design team must continue to maintain control over the mold's design and the manufacturing process. The product must be designed for the method of manufacture with the correct mold built to produce the product. The mold must be designed for good material flow, dimensional and temperature control, and the reduction of molded-in material stress to obtain the required part dimensions. This also involves selecting the correct size of injection molding machine with preferably a closed-loop continuous-feedback process control system for monitoring and controlling the manufacturing process.

Successful plastic product design involves using the design team method. Early in the program the company must build a team of personnel for design and material; mold and production personnel must then determine what material, mold, and process are best to manufacture the product. Teaming will ensure the greatest degree of success for the product to meet all requirements. A team leader is selected by the team members, who then pool their knowledge and experience to consider the products requirements. This includes the design, mold, manufacture, and product costs. To assist them, design, material, and manufacturing checklists were developed to answer the multitude of questions that arise in producing a good product in plastic. Then, using the standard product design equations and procedures the product will be developed to meet the customer's end-use requirement and specifications for form, fit, and function.

Plastics, as a distinct family of materials, are very versatile and adaptive manufacturing materials. There are no other family of materials that will support loads; resist impacts; allow self-assembly among parts, having one part perform many varied functions such as springs, wear surfaces, support platform, and ratchet mechanisms; and provide chemical and electrical protection plus color coding and/or decorating on its surface. Of course, the correct generic plastic, within the plastics family, must be chosen if all these requirements were required of a single part. And yes, there are plastic materials to satisfy these requirements and then even more. How this will be accomplished in a step-by-step procedure for the selection of the correct material, the analysis of the parts end-use function to meet agency, customer, and consumer safety requirements will be developed in the following chapters.

An area often overlooked by the designer, as some may feel it is not their responsibility, but it is, is how the part will be manufactured. This includes final piece part cost, assembly method, combining of part functions, decorating, packaging, part tolerances, manufacturing, and tooling considerations, which affect part shrinkage, and final dimensions and the final part quality requirements.

The designer is part of a team of specialists who are involved with how the part is manufactures and the parts design will have a major influence on how the part is made and what quality and part tolerances are attainable. The design team should be included in all discussions with the mold or tool designer, the material supplier and the manufacturer, either in house or at a custom injection molder.

To design only the part and leave the rest up to others is not the correct procedure for plastic part design. There are too many variables and critical areas to consider before a part is released for manufacture.

I have seen too many programs begin with a good design turn into a manufacturing, assembly, testing certification, or end-use requirement disaster. This need not have happened if the right checks and balances had been in place, all possible questions answered and all parties involved meeting regularly to discuss the programs progress. The design team method assures that this will always happen.

Beginning with Chapter 3, the designer is committed to a type of manufacture—in this book injection molding—for the product. The following chapters then take the designer, mold designer, production manager, material supplier, and quality control group to task to produce the part or product to meet anticipated part costs, quality requirements, and customer requirements and satisfaction.

The design team must never forget who the customer is because without customer satisfaction, the whole program is doomed. All the named parties, including management, who must provide the assets and support to make the program successful, must agree on workable procedures to make the program successful and profitable. The customer is also an integral part of the program to ensure that their needs are met in performance, cost, timing, and quality for the part's end-use requirements. This includes the simplest of parts as a single gear in a subassembly, to the finished assembled mechanism made up of many plastic and/or metal components that must function in the chosen environment for an anticipated length of time or part life.

Plastic part design does not entail merely looking through a material supplier's handbook and copying their recommended section thickness, boss diameters, and so on. Plastic part design encompasses all these plus selecting a material that will accomplish the requirement, at a manufacturing and finished piece price cost to provide a service to the customer and yield the supplier a profit. Often a more expensive plastic, in cents per cubic inch, can or may not be chosen that will yield more versatility in design, faster manufacture, and equal customer satisfaction versus a less expensive plastic.

All plastic materials must be judged on their ability to provid to the essential ingredients into the design and cost equation. When a designer finds one material that can do the job, keep it. But don't forget to explore other new resins and alloys that can provide the same and even more benefits at possible lower cost, manufacturing, and product capabilities.

ACKNOWLEDGMENTS

I want to thank my wife, Joyce, for her love and continued support during the writing of my second book. I also want to acknowledge the support received from Dr. Edward Immergut, retired, of Hanser-Gardner Publishing Company and others for their comments and insight in the review of the material for this book. I also thank them for assisting and recommending additional design and material information to be included to help companies ensure that the products they design and manufacture will continue to be the best they are capable of producing. This, in turn, will assist in ensuring that their customers requirements and specifications will always be met and/or exceeded.

The prevention of design, manufacturing, and service related product problems are the key objectives of this text. When the information presented is followed only positive manufacturing and quality products should result.

July 2002

M. JOSEPH GORDON, JR.
MSME

INTRODUCTION: PLASTIC PART DESIGN AND DEVELOPMENT

The design and development of any part begins with the selection of a design team. Members are selected from specific departments of the company, material suppliers, and possibly the customer's representative. These members assist the designer in determining the product end-use requirements plus form, fit, and function. The design team is responsible for communicating the requirements of their department into the products design for manufacture.

There are too many variables that impact the success of a program, and answers must be determined early in the program for the product to satisfy the requirements of the customers and the final users of the product. This information is developed using the experience and knowledge of the team members. Checklists are also used to answer the important and necessary product requirement questions early in the program to ensure that all design, liability, and quality information is available. This method of developing new products and converting existing products to plastic tie all member departments of a company into one team to ensure that success is achieved. The customer's requirements, specifications, and level of tolerances must also be known. Quality is never a question, as the manufacturer is responsible for ensuring that all products meet the needs of the customers and their manufacturing specifications. Using this information, the designer can now begin the initial design selection process involving part structural analysis, material selection, method of manufacture, and cost analysis and can consider any secondary operations such as assembly, decoration, packaging, and shipping, leading toward a finished product. Designers have found plastics, a distinct family of materials, to be very versatile, economical, and adaptive in manufacturing. Plastics are proven materials of construction that are used for the design and manufacture of multifunction products in all major consumer and industrial markets. No other family of materials have prove as universal a building material as plastics. Plastics support loads, withstand severe temperature variations in the ever-changing environment, resist impacts, and allow self–assembly among parts, with one part performing many varied functions such as springs, wear surfaces, support platform, and ratchet mechanisms, and providing chemical and electrical protection to their users. These products are easily decorated and can impart information on their surfaces to become even more cost-saving, valuable, and useful to their customers.

The selection of the correct generic plastic or compound is easy when the products requirements are known. Plastic materials are designed to accomplish most requirements except those in the most rigorous applications with extreme temperatures or high forces, and plastics are usualy less costly than metals. It is the responsibility of the designer and the design team to run the design analysis and select the material and process to make the program successful and profitable for the supplier and customers. This will be accomplished in an ordered procedure, by determining the part's end-use, agency and code requirements, or customer testing by using the appropriate design criteria and principles for the selection of the product's design layout and material.

Also considered is the design of the part for combining and reducing the number of parts, ease of assembly, decoration considerations, and manufacturability. The part must be designed for zero variation and failure as required by the customer's specifications. This includes developing, early in the design program, piece part cost estimates, assembly method, combining of part functions, decorating, packaging, realistic part tolerances, and most importantly, manufacturing and tooling considerations. The parts design geometry, material, tooling, manufacturing process, assembly, decorating, packaging, and shipping has an effect of the part's final dimensions and the final part quality that can be attained.

The designer is only one member of a team of specialists who are involved in determining how the product is manufactured. The parts design will have a major influence on how the product is made and the part tolerances attainable, to ensure that the products quality is never in jeopardy.

The design team is composed of representatives from management, sales, engineering, manufacturing, and quality and specialists outside the company such as the mold or tool designer, and the material and support services suppliers, for assembly, decorating, and packaging as required. To just design the part and leave the rest of how it is to be manufactured and completed to others is not the world-class procedure for plastic part design. There are too many variables and critical areas to consider before a part is released for manufacture.

Too many programs that begin with a good design can turn into a manufacturing, assembly, testing certification, quality, or end-use requirement disaster. It need not happen if the right checks and balances were in place, all possible questions answered, and all participants meeting regularly to discuss the program progress. The design team method is the best insurance that this will always happen.

Beginning with Chapter 3, the design information is specific to injection molding for the product. The following chapters then take the product designer, mold designer, production manager, material supplier, and quality control group to task to develop and manufacture the product to meet anticipated product costs, quality requirements, and customer requirements and satisfaction.

The design team must never forget who the customer is because without customer satisfaction the whole program is doomed. All the design team members, including management, who must provide the assets and support to make the program successful, must agree on workable procedures and solutions to any problems to make the program successful and profitable. The customer is also an integral part of the program to ensure that their needs are met for the parts performance, cost, scheduling, and quality. This includes the simplest of parts as a single gear in a subassembly, to the finished assembled mechanism made up of many plastic and/or metal components that must function in the chosen environment for an anticipated part life.

Plastic product design is not just looking through a material supplier's handbook, selecting a material, and copying their recommended section thickness, boss diameters, and so on. Plastic product design encompasses many disciplines that must be coordinated to accomplish the product's performance parameters at a manufacturing and finished piece price cost to provide a quality product to the customer and yield a profit for the supplier.

STANDARD PLASTIC TERMINOLOGY per ASTM D1600-92

Abbreviation	Term
ABA	Acrylonitrile-butadiene-acrylate
ABS	Acrylonitrile-butadiene-styrene plastics
ACPES	Acrylonitrile-chlorinated polyethylene-styrene
AEPDM	Acrylonitrile/ethylene-propylenediene/styrene
AES	Acrylonitrile-ethylene-styrene
AMBA	Acrylonitrile-methyl acrylate-acrylonitrile-butadiene rubber
AMMA	Acrylonitrile/methyl methacrylate
ASA	Acrylonitrile-styrene-acrylate
ARP	Aromatic polyester
CMC	Carboxymethyl cellulose
CS	Casein
CA	Cellulose acetate
CAB	Cellulose acetate butyrate
CAP	Cellulose acetate propionate
CN	Cellulose nitrate
CE	Cellulose plastics, general
CP	Cellulose propionate
CTA	Cellulose triacetate
CPE	Chlorinated polyethylene
CPVC	Chlorinated poly(vinyl chloride)
CF	Cresol-formaldehyde
EP	Epoxy, epoxide
EC	Ethyl cellulose
EEA	Ethylene/ethyl acrylate
EMA	Ethylene/methacrylic acid
EPM	Ethylene-propylene polymer
EPD	Ethylene-propylenediene
ETFE	Ethylene-tetrafluoroethylene copolymer
EVAL	Ethylene-vinyl alcohol
EVA	Ethylene/vinyl acetate
FF	Furan formaldehyde
HDPE	High-density polyethylene plastics
IPS	Impact-resistant polystyrene
LLDPE	Linear low-density polyethylene plastics

STANDARD PLASTIC TERMINOLOGY per ASTM D1600-92

Abbreviation	Term
LMDPE	Linear medium-density polyethylene plastics
LCP	Liquid crystal polymer
LDPE	Low-density polyethylene plastics
MDPE	Medium-density polyethylene plastics
MBS	Methacrylate-butadiene-styrene
MF	Melamine-formaldehyde resin
MPF	Melamine/phenol-formaldehyde resin
MC	Methyl cellulose
PA	Nylon (see also Polyamide)
PFA	Perfluoro(alkoxy alkane)
FEP	Perfluoro(ethylenepropylene) copolymer
PF	Phenol-formaldehyde resin
PFF	Phenol-furfural resin
PAA	Poly(acrylic acid)
PAN	Polacrylonitrile
PADC	Poly(allyl diglycol carabonate)
PMS	Poly(α-methylstyrene)
PA	Polyamid (nylon)
PAI	Polyamideimide
PARA	Polyaryl amide
PAE	Polyarylether
PAEK	Polyaryletherketone
PASU	Polyarylsulfone
PBAN	Polybutadiene-acrylonitrile
PBS	Polybutadiene-styrene
PB	Polybutene-1
PBA	Poly(butyl acrylate)
PBT	Poly(butylene terephthalate)
PC	Polycarbonate
PDAP	Poly(diallyl phthalate)
PAK	Polyester alkyd
PAUR	Poly(ester urethane)
PEK	Polyether ketone
PEUR	Poly(ether urethane)
PEBA	Polyether block amide
PEEK	Polyetheretherketone
PEI	Poly(etherimide)
PES	Poly(ether sulfone)
PE	Polyethylene
PEO	Poly(ethylene oxide)

STANDARD PLASTIC TERMINOLOGY per ASTM D1600-92

Abbreviation	Term
PET	Poly(ethylene terephthalate)
PETG	Poly(ethylene terephthalate) glycol comonomer
PI	Polyimide
PISU	Polyimidesulfone
PIB	Polyisobutylene
PMCA	Poly(methyl-α-chloroacrylate)
PMMA	Poly(methyl methacrylate)
PMP	Poly(4-methylpentene-1)
PCTFE	Polychlorotrifluoroethylene
POM	Polyoxymethylene, polyacetal
PPE	Poly(phenylene ether)
PPO	Poly(phenylene oxide [deprecated term and acronym; see preferred term Poly(phenylene ether) and acronym]
PPS	Poly(phenylene sulfide)
PPSU	Poly(phenylene sulfone)
PPA	Polyphthalamide
PP	Polypropylene
PPOX	Poly(propylene oxide)
PS	Polystyrene
PSU	Polysulfone
PTFE	Polytetrafluoroethylene
PUR	Polyurethane
PVK	Polyvinylcarbazole
PVP	Polyvinylpyrrolidone
PVAC	Poly(vinyl acetate)
PVAL	Poly(vinyl alcohol)
PVB	Poly(vinyl butyral)
PVC	Poly(vinyl chloride)
PVCA	Poly(vinyl chloride acetate)
PVF	Poly(vinyl fluoride)
PVFM	Poly(vinyl formal)
PVDC	Poly(vinylidene chloride)
PVDF	Poly(vinylidene fluoride)
SP	Saturated polyester plastic
SI	Silicone plastics
SAN	Styrene-acrylonitrile plastic
SB	Styrene-butadiene plastic
S/MA	Styrene-maleic anhydride plastic
SMS	Styrene/α-methylstyrene plastic
SRP	Styrene-rubber plastics

STANDARD PLASTIC TERMINOLOGY per ASTM D1600-92

Abbreviation	Term
TPEL	Thermoplastic elastomer
TEEE	Thermoplastic elastomer, ether–ester
TEO	Thermoplastic elastomer, olefinic
PEBA	Thermoplastic elastomer polyether block amide
TES	Thermoplastic elastomer, styrenic
TPES	Thermoplastic polyester
ARP	Copolyester [oiky(artktereogtgakate)]
PAT	Polyarylate [poly(arylterephthalate)]liquid crystal polymer
TPUR	Thermoplastic polyurethane
TSUR	Thermoset polyurethane
UHMWPE	Ultra-high-molecular-weight polyethylene
UP	Unsaturated polyester
UF	Urea formaldehyde resin
VCEMA	Vinyl chloride-ethylene-methyl acrylate resin
VCEV	Vinyl chloride-ethylene-vinyl acetate resin
VCE	Vinyl chloride-ethylene resin
VCMA	Vinyl chloride-methyl acrylate resin
VCMMA	Vinyl chloride-methyl methacrylate resin
VCOA	Vinyl chloride-octyl acrylate resin
VCVAC	Vinyl chloride-vinyl acetate resin
VCVDC	Vinyl chloride-vinylidene chloride resin

Note: For blends and alloys, the standard terminology is the sum of the abbreviated terms of the two comoponents. For example, poly(ethylene terephthalate) plus rubber = PET + RBR where an usspecified rubber is used, and acrylonitrile-butadiene-styrene plus polyamide = ABS + PA.

CHAPTER 1

PLASTIC PROGRAM ANALYSIS AND DEVELOPMENT

Thermoplastics for replacement of metals and for new products began essentially in the late 1930s, and their development and use for products has proliferated ever since. The first thermoplastics, other than cellulosics, were developed primarily for the clothing industry as "man-made fibers" from varying feed sources, primarily the chemical and petroleum industry, as they are carbon chain molecules. With the increase in fiber production there resulted a large fiber scrap problem. The first plastics were developed as fibers, and their primary use was for clothing. During manufacture there was scrap, and a method to reuse or recover the fibers was initiated. This was a costly waste since reclaiming this waste was not feasible at that time, which involved chemically breaking the fiber scrap down to their basic starting chemical elements. From studies undertaken by the chemical companies, a method to reuse this essentially virgin plastic was needed. The plastic injection molding industry was developed from these needs and studies to reuse the materials.

Engineers and chemists learned that these thermoplastics, after being reextruded and chopped into pellets, could be reheated, softened, made to flow and under high pressure, and injected into a mold to manufacture a new product with good physical, mechanical, chemical, and electrical properties. As time and product demands grew, the scrap fibers were not always suitable for the more rigorous demands required of plastic products. From this need, virgin plastic resin grades were developed to meet the specific needs of industry in the automotive, consumer, and industrial electrical areas. Virgin resins were required to guarantee the plastic products physical properties, color, and processing would be consistent, from lot to lot.

New product demands of industry to reduce weight and cost and to improve product performance drove the chemical industry to continue to develop new plastic resins. The existing and new modified plastics were designed to satisfy new and more demanding requirements that current resins could not meet. From this simple beginning has developed the plastic injection molding industry with over 30 basic generic families of plastics resins in use today. Adding in the thermoset and thermoplastic elastomers and their many alloy grades, the product designer today has over 11,000 grades and types of materials from which to select for their products. There are over 5000 grades of engineering thermoplastics available, and the list keeps growing. See Appendixes A for a list of these generic resins and their basic chemical structure. Thermoplastics are developed from a carbon chain molecule primarily from crude oil that is broken down into many components. The base resins properties can be enhanced when they are mixed and/or alloyed with other plastics, fillers, reinforcements, modifiers, plasticizer, color, and processing aids to develop more unique and distinctive materials to fulfill growing and specific product applications.

Each generic family of resins from acetal, acrylic, cellulosic, fluoroplastics, nylon, the polyolefin families, and styrene to the vinyl resins is unique and fulfills specific product and user applications. These resins can replace existing materials such as metals, wood, and ceramics, to create new products to satisfy, at a competitive price, many nonplastic materials. To consider the world today without plastics is almost too inconceivable to imagine.

For good product development there are many design, cost, and material considerations to consider. The basic design question is "What determines the material to be selected for an application, a new or a redesigned product?" Today product designers have a wide menu of materials to select from, and their selection is based on their experience, creativity, and product performance requirements. Also to be considered are manufacturability and cost factors as well as what the customer will accept and use. Plastics today fulfill and satisfy the most rigorous customer requirements and their performance objectives at a lower cost than did the previous materials. The majority of product designers are better trained in the use of metals, and not plastics, for product design. More colleges and universities are adding and teaching courses in the use of plastic materials. The designer and engineer are taught the concepts of good machine and structure design, but enough emphasis is never placed on plastic material selection and what their unique properties can add to products, except in specific curricula. This text attempts to fill the gap between the training phase, college, and design school, and the knowledge and practice of what is required to develop a good plastic product design. This includes defining and meeting customer requirements, with the design and manufacturing team assisting the designer in selecting the correct plastic, using end-use material test data, and designing and selecting efficient tooling and controllable and repeatable manufacturing process procedures to manufacture the product by injection molding. Program success also includes the ability to meet cost and manufacturing schedules and ensuring that the product will perform as required in its end-use environment.

The design team, consisting of designer, tool engineer, production specialist, material supplier, quality assurance, and the customer, must be taught what is required to perform these tasks. This is accomplished by education and training to identify and gather the information using checklists, developed in this text (see Appendix A), and applying the data gathered to meet the product industry or consumer requirements.

Many factors, people, and situations influence the selection of a material for a specific part. These may include what the competition is using, material supplier recommendations, past experience, and demand and needs required for the application.

The design team must also be aware that plastics may not always work for all products. However, as designers in industry are learning, there are now very few applications, except in high heat, high sustained load, or severe chemical situations, where a plastic material is not suitable. It is the design team's responsibility to know the product's total requirements, including the product's design parameters and material composition to determine the precision of its manufacture. The designer and team members are responsible for the product's successful manufacture. The design of the tooling to produce products to required tolerances is critical to eliminate product failures. Too often the product's requirements for the tools design are not communicated to the tool designer, and problems occur during tool and production startup. Therefore, early selection of the correct personnel for the design team can eliminate many design and manufacturing problems. The selection of the design team is accomplished early in the program, as is training of all team members in the product's requirements and end-use functions to ensure that they are met before beginning production.

SELECTING THE DESIGN TEAM FOR SUCCESS

The total quality management (TQM) principles for any program begin with the principle of "managing for design success." Involved are seven key principles for successful design that can lead to a successful completion of a program.

1. Design the product from the outside in. Ensure that the customer's use of the product is key to all product development, not the technology.
2. Surprise the user with extra benefits. Delivering extra features in a product above the customer's expectations creates product and customer loyalty. This increases the probability of creating a really "hot" or unique featured product.
3. Deep-partner by teaming and consulting with all your companies business units. This will unleash the power and knowledge they posses for the design and manufacturing creativity to define and develop new products. New and enhanced products will assist in the growth of the company and your customers.
4. Partner widely. The design team must partner with personnel in their company and their suppliers to seek their input and knowledge of improving product design and to improve the quality of manufacture. Evaluate any potential liability or product warranty problems using the warranty checklist.
5. Define the product up front. Screen out early in the program designs that do not meet requirements. It is critical to get the right product for the market before committing to tooling and manufacture. Using the product development checklist will ensure that the essential answers and documentation is gathered to evaluate a new program.

6. Get physical fast. Prototype early for visualization of the product. Quick feedback results from team review reduces the time of the product development cycle.
7. Design for manufacturability. Always design for quality, cost, and delivery (QCD) parameters. Manufacturing criteria are as important to a successful product as aesthetics, economics, and function.

Following these seven principles, the design team can work toward selecting and completing a successful design program within the time and cost schedule established for their customers.

USING CHECKLISTS TO DEVELOP PRODUCT REQUIREMENTS

Designing a new or redesigning an existing metal product in plastic requires basic steps and procedures to successfully compete in today's world-class markets. The process is too complicated for a single person, the designer, to successfully complete. The design process requires a team approach utilizing internal company personnel plus material, tooling, equipment, and services and possibly outside knowledgeable consultants who can also be your suppliers. There are too many variables—designs, materials, molding, and processing—for a single, plastic expert, to know and supply answers to make all the correct decisions.

Areas benefiting from the use of plastics in product design are

1. Weight reduction
2. Consolidation of parts and functions
3. Reduction in total per part cost
4. Greater design freedom
5. Reduction of secondary finishing and assembly operations
6. Availability of resins to fit multiple and specific applications
7. Ability to withstand high temperatures up to 500°F and chemical, corrosive environments with enhanced electrical performance properties
8. Decorating and information features molded in or on the part
9. Product ability to function in all or most environmental conditions

DEFINING THE DESIGN TEAM

The design team may consist of two separate groups: (1) the primary team, composed of company department representatives, and (2) the secondary team, made up of suppliers and/or consultants supplying information not readily available from the primary team members.

The company's internal design team typically consists of four to six permanent members, with others available as required who will work together until the program is completed. A team leader or program manager is selected to guide the design program team and provide continuity between members. Team members will come from design, engineering, production, research and development, sales, marketing, management, finance, purchasing, and quality control. The team's input is used to keep the program on schedule by answering the array of technical, material, tooling, production cost, and scheduling questions that will develop during the program. Team members and their responsibility are shown in Figure 1.1 and are used as required during the program.

The secondary team will consist of material, tooling, molding, assembly, and decorating equipment suppliers and possibly outside consultants who supply, when requested, their expertise to the program. If the product is directed at a specific market segment, experts in this field should be included along with outside industrial design experts when required. The primary team directs all activities and defines design, performance, manufacturing and cost requirements with each team member supplying their specific input as required for the program as shown in Figure 1.2. At the start of a program, each team member should be assigned specific responsibilities in areas in which they can draw on their own experience and contacts with outside suppliers. The secondary team will provide information requested by the primary design team members. This will avoid confusion and duplication and ensure that representatives will be available with the required information when necessary to keep the program on schedule. Your out-of-company consultants and suppliers will find this less confusing, and you will have the desired information when needed.

The secondary team members will provide in-depth assistance and information such as specific resin property information, code and specific product design assistance and reviews, prototype testing, fabrication, quality checkpoints, and tests, including preproduction evaluations. They can review the product design and perform sophisticated engineering analysis, train personnel in manufacturing, assembly, and decorating equipment operation. Also, they can ensure the mold, production equipment, material, and design of the product are individually suited to each other to manufacture a cost-effective product through a successful development program. Figure 1.3 is a total quality process control flowchart showing the program's overall development. The initial phase of the product's design program, after member selection, begins by gathering information for the product using the product development and design checklists in Appendix A (checklists 7 and 2, respectively). These checklists guide the development of the product for the entire program from sales, design, manufacture, and quality requirements and guide the design team toward the best selection of material candidates, manufacturing methods, and quality assurance for all the product components. The next step in the program involves evaluating the product's specific requirements for the design that includes material selection, tooling for manufacture, quality requirements, production method, assembly, decorating, and packaging. This stage of information

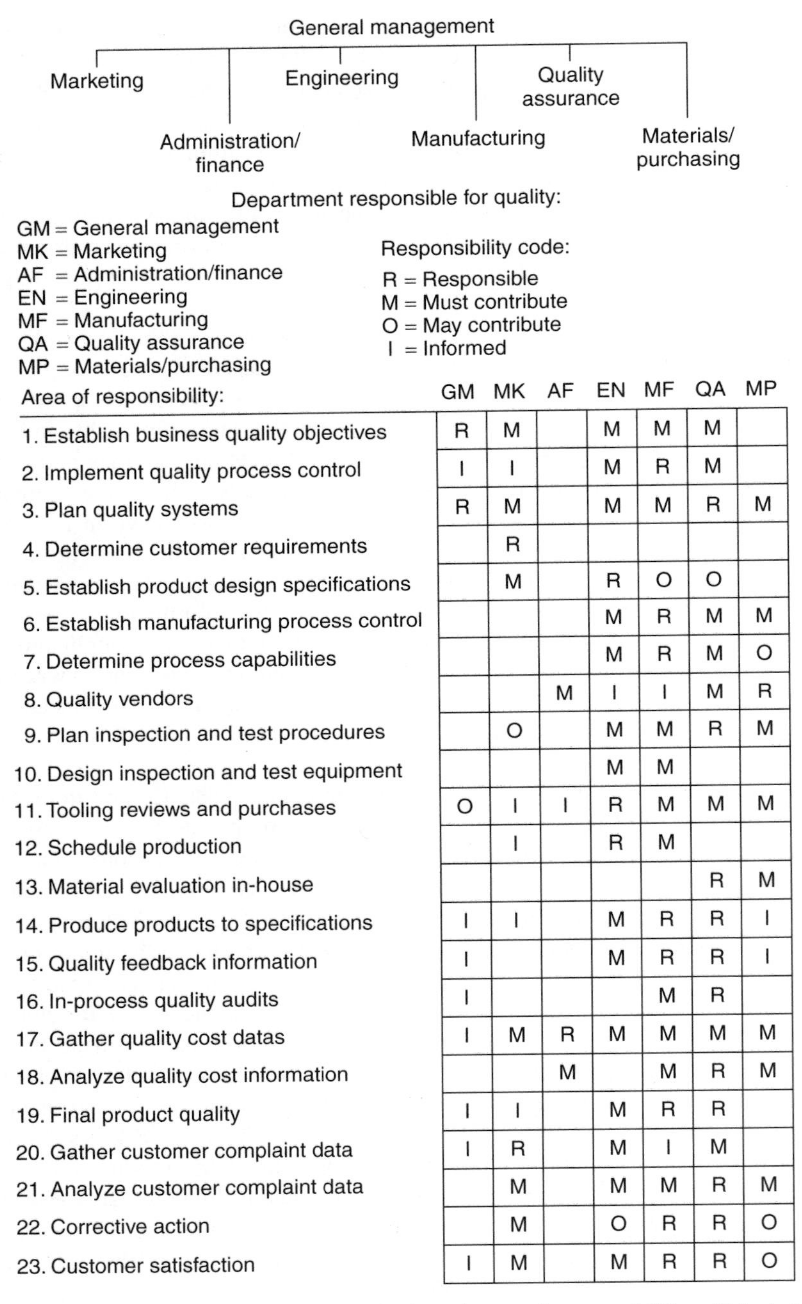

Area of responsibility:	GM	MK	AF	EN	MF	QA	MP
1. Establish business quality objectives	R	M		M	M	M	
2. Implement quality process control	I	I		M	R	M	
3. Plan quality systems	R	M		M	M	R	M
4. Determine customer requirements		R					
5. Establish product design specifications		M		R	O	O	
6. Establish manufacturing process control				M	R	M	M
7. Determine process capabilities				M	R	M	O
8. Quality vendors			M	I	I	M	R
9. Plan inspection and test procedures		O		M	M	R	M
10. Design inspection and test equipment				M	M		
11. Tooling reviews and purchases	O	I	I	R	M	M	M
12. Schedule production		I		R	M		
13. Material evaluation in-house						R	M
14. Produce products to specifications	I	I		M	R	R	I
15. Quality feedback information	I			M	R	R	I
16. In-process quality audits	I				M	R	
17. Gather quality cost datas	I	M	R	M	M	M	M
18. Analyze quality cost information			M		M	R	M
19. Final product quality	I	I		M	R	R	
20. Gather customer complaint data	I	R		M	I	M	
21. Analyze customer complaint data		M		M	M	R	M
22. Corrective action		M		O	R	R	O
23. Customer satisfaction	I	M		M	R	R	O

Figure 1.1 Company departmental organization chart and responsibilities. (Adapted from Ref. 1.)

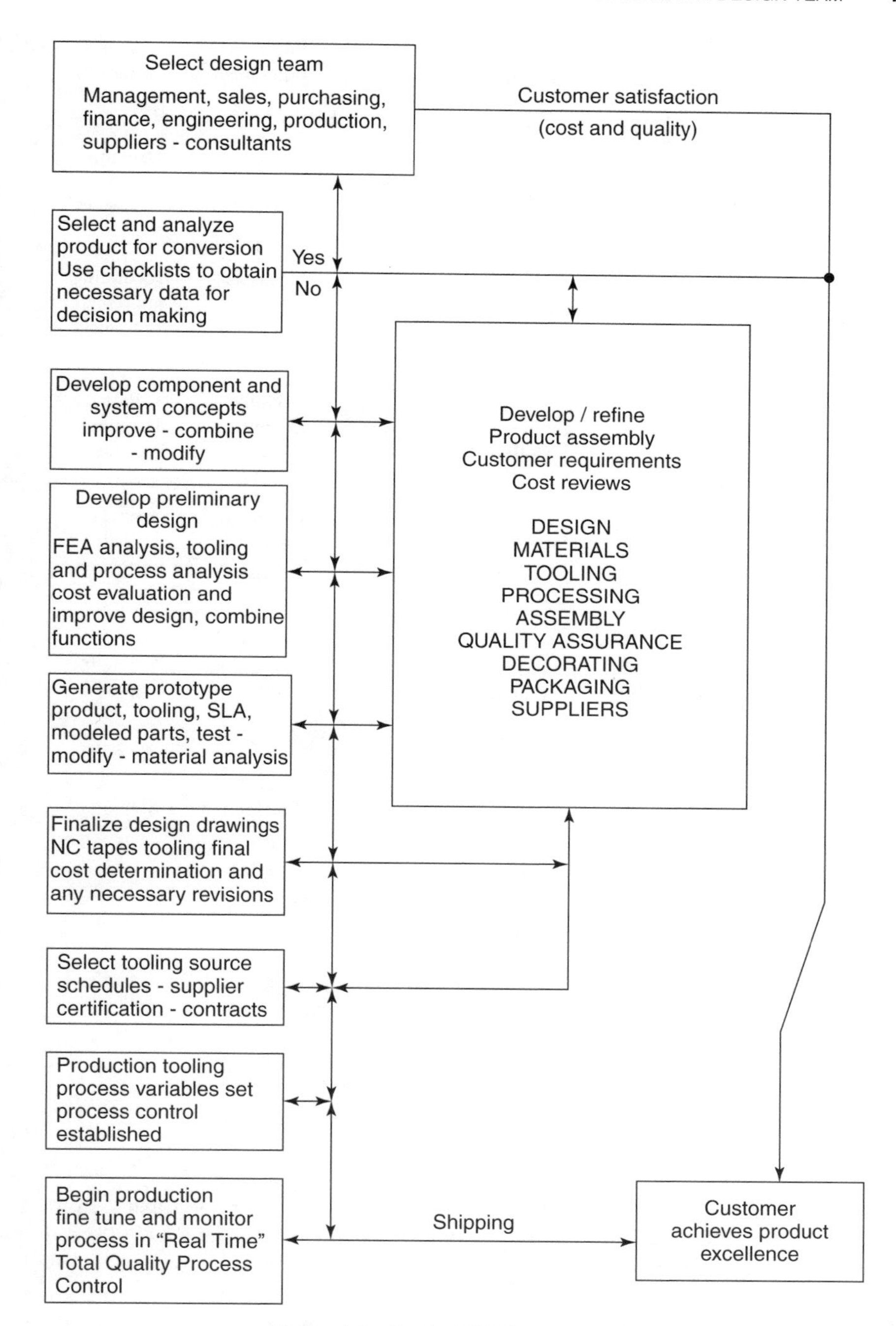

Figure 1.2 Product development path.

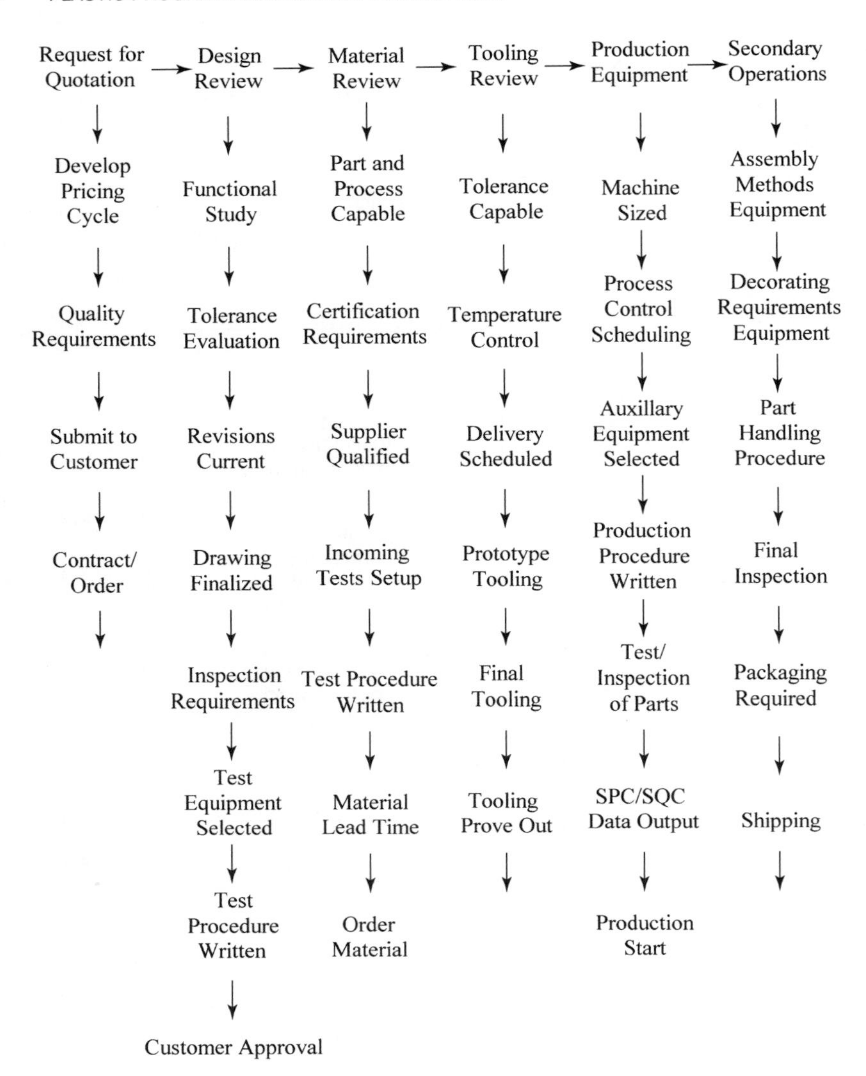

Figure 1.3 Total quality process control flowchart.

gathering and documenting is very important, as it is part of the company's procedures used for attaining and maintaining quality certification as an ISO 9000 and/or QS 9000 product supplier. These and other checklists shown in Appendix A will be discussed to assist the design team in developing and documenting information. The checklists are used to develop information for new and existing products if records were never completed as a form of reverse engineering for product record documentation.

DESIGN AND DEVELOPMENT SCHEDULING

The design and development schedule needs to be established early in a program with milestones and key decision points established for the products development program. The schedule must allocate sufficient time to detail system requirements and select variable design concepts to arrive at the optimum design and cost parameters for the finished product. If insufficient time and detail are not now spent, it could lead to an inferior product resulting in costly delays in the program schedule. The schedule, developed at the beginning of the program, should consist of the items listed on the design, development, and program scheduling checklists in Appendix A (checklists 2, 7, and 8, respectively).

The designer, with input and support from the design team, can now generate an initial product or system design with product layout drawings based on information from the design checklist. The design team can then evaluate the product design combining multipart functions when possible and selecting one or more design layouts for continued analysis.

From a preliminary design layout the products initial cost analysis can be developed. The analysis evaluates the cost of the products preliminary design; its type of manufacture, including assembly and decorating, and potential material candidates. This analysis can eliminate uneconomical design concepts and lead the team in developing product improvements, benefits, and cost reduction by evaluating the product variables and estimated costs. Developed in the analysis are anticipated tooling price, estimated manufacturing cycles, material usage, scrap loss, and assembly, decoration, and packaging cost estimates with anticipated efficiencies with included overhead cost factors.

When the initial design layout cost analysis is completed, the products sale and profit margins can be developed. Ensure that key customer and economic factors are considered to the product success in the marketplace. Usually at this stage in the program the product's final design is selected, the schedule is finalized, and any remaining program design and fabrication factors can now be better estimated and finalized for the product. These include tool design and type, including method of manufacture, decorating requirements, type of mechanical assembly, and packaging to point of use or sale.

The product's development continues with a prototype product produced that will give the team information on the product's suitability for the application. When possible, the prototype is tested as realistically as possible to actual end-use conditions for a sufficient length of time to prove acceptability.

After prototype acceptance, the production tool drawings can be finalized. If a prototype tool is built, preproduction samples are produced and evaluated and product and tooling adjustments made if necessary. At this stage of the program, all questions concerning the product should be known and answered so that production can begin. A typical application development flowchart is shown in Figure 1.4 to illustrate the major tasks and steps in bringing a successful program to market. The remainder of this book develops the information that will assist the design team and their partners in producing a customer acceptable product.

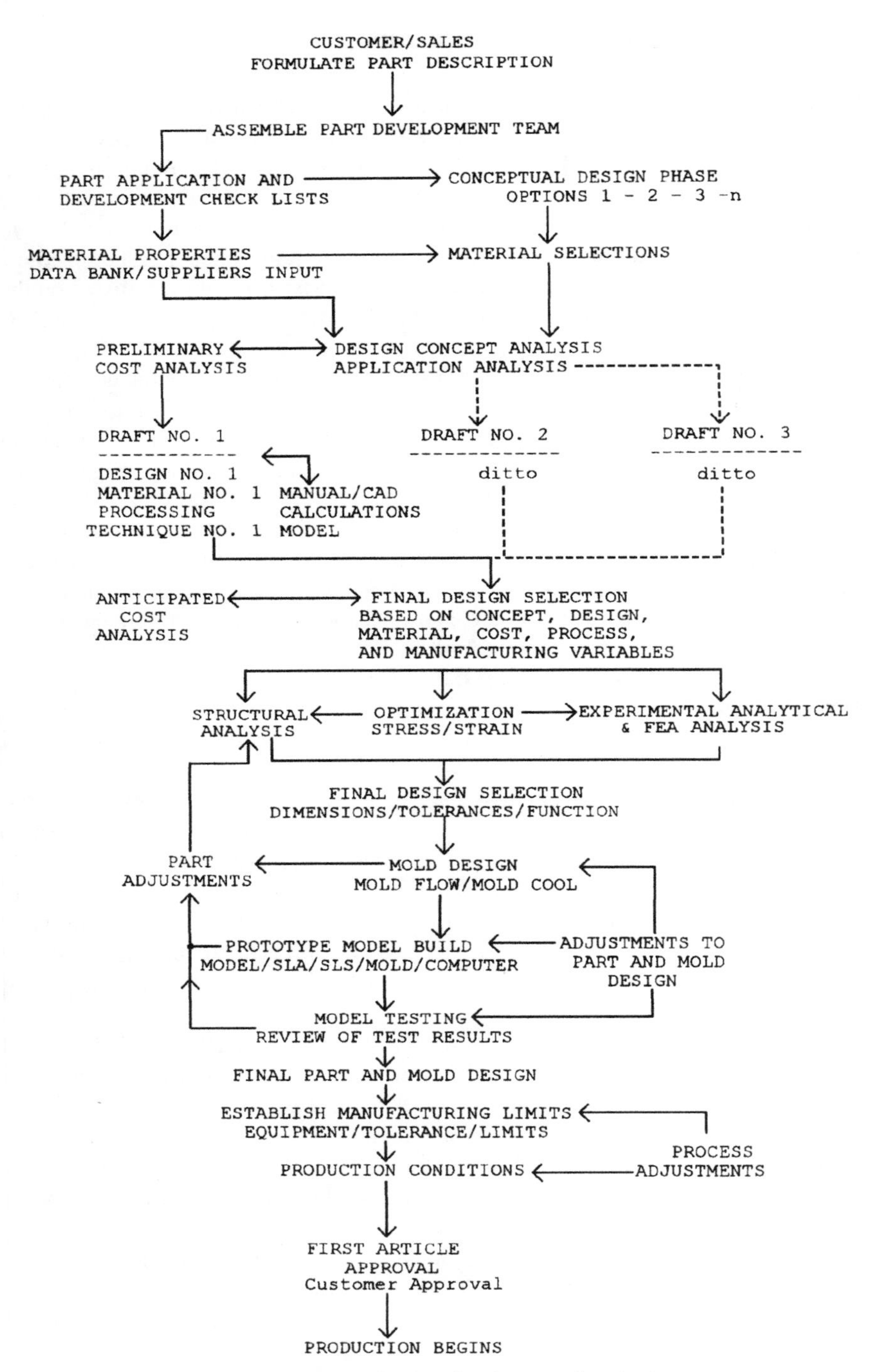

Figure 1.4 Application development flowchart.

COMPANY COMMITMENT TO PLASTICS

What is required for a company to make a product material change to plastics? Many companies are now considering this change, and each have their own reasons. One company considered changing because their current machines making their aluminum parts are almost worn out. The cost to replace them is extensive, and their product design is 20 years behind their major competitor, who is using acetal plastic for the same part. The company must now redesign their products. The reason is cost-driven, and if they want to stay competitive, they must make the change. Until now there was no urgent need to change. Their current machines kept their manufacturing costs in control, and their market share was not in jeopardy. But now they must change, be trained quickly, and learn all they can about plastics to make the conversion. It will not be easy, but they can make the transition with the right people and support. For a company to remain dormant just because their market share is strong is a poor reason not to consider what plastics can offer them in the future. Each company must develop a long-range, 5-year minimum forecast, for future products and increased sales. Each company must evaluate their future needs on the basis of their current product mix and manufacturing capabilities and project it into the future to ensure that it meets their long-term financial goals. All products need improvement to stay competitive, and plastics can provide the materials for a company to evaluate their future product's mix and markets answering the following questions for consideration for converting to plastics.

PRODUCT CONVERSION TO PLASTIC

The following questions require answers for conversion to plastic:

1. What is my competition doing?
2. What are my long-term company and product goals?
3. Will my manufacturing costs meet these goals?
4. Will my products compete in cost and customer acceptance?
5. How can I modify, improve, and reduce costs?
6. Will my manufacturing equipment meet these goals?
7. Can I improve my product, lower costs, and gain greater market share?
8. Will plastics give me this edge?
9. What are my costs both in the plant and outside?
10. Do I have the expertise to do it with current personnel?
11. What will it take to make me more competitive?
12. What agency and codes now affect my product that require evaluation?
13. What cost, testing, certification, and time requirements are involved?
14. Who do I contact, machinery, material suppliers, and/or consultants?

15. What are the costs?
16. How do I begin?
17. When do I begin?
18. Is ISO 9000-2000 and/or QS9000 required with six-sigma process (6σ) control to enter markets?

This last question is very important to current original equipment manufacturers (OEMs) and their part suppliers. Becoming an ISO 9000-2000– and/or QS 9000–certified product design, manufacture, and service certified supplier could open up new markets, or close doors for company's products.

Europe and the world have adopted the latest update ISO 9000-2000 certification program to assist in guaranteeing their customers that lot-to-lot production of products are defect-free and will meet their requirements and specifications. The latest revision to ISO 9000-2000 adds 34 new requirements that must now be met for companies certified to ISO 9000-1984, and new companies now becoming certified to ISO.

All of these questions—and there are others to be considered—need to be answered to make a good business decision to convert over to plastics. Companies now using plastics have answered these questions. Some were well directed and guided into plastics replacing metal parts, while others have struggled and paid in lost programs and market shares. The majority of the successful companies used the resources and expertise of their plastic resin suppliers and outside consultants, and by hiring experienced personnel to make the transition.

Resin suppliers are and were the main driving force toward conversion to plastics. They assisted their customers in product conversion to plastics by developing new and better plastics to meet their customers' part requirements. They also assisted in design programs and followed through in recommending and assisting in manufacture, assembly, and decorating areas. Without the Honeywell (formerly Allied Signal), BASF, Dow, DuPont, Hoechst Celanese, Monsanto, and many other reputable resin suppliers, these changes would not have occurred as fast or as successful.

For a company to commit or move into plastics requires a strong commitment to change in type of personnel hiring and business philosophy. Most design personnel come from a metals background. This is basically what today's universities are producing. Only a few specially dedicated universities are graduating engineers experienced in designing with plastic resins. Many of these engineers learn their trade, so to speak, on the job. There are so many varied plastic resins available, that no specific person can know them all and be competent in designing products with them. Therefore, these engineers and designers rely on their material suppliers for the design and product data on their materials. The material supplier design handbooks provide only the basic information needed for the design of new products. Therefore, designers must access their data files to select the correct material values, safety factors, and supplier design recommendations when beginning a new program.

To be successful in using plastics, a company should understand what is required of their personnel—a team effort—in providing the assets and direction to make it happen. This means that their employees and management need to know and understand the entire plastic process from design to consumer satisfaction and safety. Whether they do all the manufacture in house or use outside suppliers to fabricate their parts, each link in the product's program relies on the preceding work or operation performed.

Beginning with sales, where the ideas are generated from consumer needs and wants, the idea(s) flow to the design department. The designer confers with sales and often the customers to firm up their needs and requirements. The designer then selects materials to perform the task and consults with the material suppliers who furnish product data to meet the parts requirements. Often design suggestions can aid the designer in their initial and often final material selection. During the design stage it is necessary to have meetings with purchasing, manufacture, material suppliers, sales, and the customer. This involves joint meetings to answer technical and manufacturing questions and resolve any disputes that may occur to keep the program on schedule. It is management's responsibility to ensure that all these operations occur and the final result is customer satisfaction and a profit.

Too often each department within a company operates independently and unknowingly creates problems for the program without realizing how their actions impacted the project. Each department in a company has their own stated objectives to make a profit for the company. If they are not supplied and assisted by their other sister departments and able to express their own ideas for the program, they may be unable to meet their objectives. Therefore, their input and involvement in the program are essential for it to be a success. This is why the "design team" method is the correct method to use for having a "design for success" program. ISO 9000-2000 and QS 9000 with "real time" process control:

American industry is advancing toward product *excellence* through the use of "total quality process control" procedures as six-sigma capabilities. Manufacturers are implementing ISO/QS 9000 standards for establishing their product base for excellence in product quality. The evaluation, documentation, training, and investigative analysis of their products and production systems are a monumental task. It requires many personnel and hours of work to ensure that their design and manufacturing systems are capable of consistently producing quality products.

Understanding what is required to become ISO/QS 9000–certified is the first task. The second, implementing the standards with the required documentation and procedures, is more difficult. Establishing product, material, and product manufacturing specifications is almost impossible until the product's actual requirements and manufacturing tolerances are determined and known. In many cases this information will have to be established. Initially, during a product's design phase, the products materials were selected to meet the product form, fit, and functional needs and customer specifications and any required agency certification requirements.

Before ISO/QS 9000, the manufacturing process usually required minimal documentation as it was manufactured using the companies "standard operation procedures." ISO/QS 9000 requires documentation and procedures that ensure that the products produced will meet their customers' quality manufacturing specifications in real time. "Real time" means that the production process will consistently produce quality parts by exercising control over the manufacturing system variables by using "total quality process control" methods. The current method of accessing product quality after manufacture is by inspecting a lot of parts, hours, or days after production. This has not proved to be a reliable quality program. It is an accepted but outdated production method relying totally on manufacturing to consistently product quality products. Their task is to make products that will meet contracted, acceptable quality level (AQL) requirements. However, AQL sampling and inspection allow defects, although the tighter the AQL limit, the fewer defects per production lot are permitted. AQL inspection does not allow for any irregularities such as material variability, equipment not in calibration, mold wear, temperature variations, or process control variation. AQL inspection only sets limits on what is an acceptable product as determined by the inspection process. It can take hours to check a production lot's quality, and if defects are too high, the entire lot is rejected. If accepted, the lot may still have a small percentage of defects, which may be discovered only later (resulting in more costly rejected finished products).

Quality, therefore, must begin at the start of a manufacturing process. Materials, products, production equipment, and manufacturing specifications must be established to always produce quality parts in real time. This begins with defining the product requirements, designing the product to meet customer and manufacturing specifications, specifying materials within required limits, and qualifying vendors and suppliers to meet these specifications. Then the supplier's manufacturing equipment and personnel must be capable of making the products consistently to these specifications.

Knowing your equipment, personnel, and the manufacturing process in depth will assist in keeping in the manufacturing cycle in process control. Knowing when it varies by establishing procedures to monitor key elements during manufacture will assist in maintaining process control. Unfortunately, many companies have the capability to do this, but do not! It is discovered later when the inspection shows a high defect rate, indicating that the manufacturing process was out of control. The use of statistical process control techniques can be effectively used to assist in analyzing where quality processes must be controlled. This begins at the material design stage and continues through manufacturing so that any problem can be corrected and eliminated. These process control items will be used to ensure that control of the manufacturing process is maintained by monitoring and using the output data in real-time manufacturing process analysis.

SIX-SIGMA PROCESS CONTROL

Six-sigma process control is a new quality tool used to bring this technology into the manufacturing area. Dedicated cost-saving and quality assurance saving projects, stipulated to yield a minimum of $175,000.00 in savings, are analyzed by dedicated teams and defects are reduced by 20,000 times from three-sigma limits from 67,000 to 3.4 defects per million (dpm).

The significant characteristics influencing a product output can be improved when known. The process involves

Identification of the process
Understanding the variables
Optimization of the variables and process
Maintaining them at optimum and acceptable levels

To become six-sigma-compliant is expensive. The rewards are extensive and yield superior cost savings, products, reliability, and customer loyalty. Training is extensive and lengthy, with trained employees earning the title of "black belt" or "green belt." The employees are trained to implement the changes and are selected for their outstanding abilities to contribute to this program. Training consists of a one-week special course over a period of 4 months plus working four to six projects per year with their company producing bottom-line savings of $175,000.00 per project. Many companies are now searching for these superior trained personnel to implement similar programs, even though they're savings are not within this dollar range. Six-sigma success relies on the company's financial and technical support for implementation of improvement plans while company personnel and disseminate the six-sigma tools and methods in their six-sigma dedicated improvement teams. The average cost to train a certified black-belt trainer is $35,000 to $40,000 plus their time and the team members working on projects full time.

Excellent results are attainable with companies such as Honeywell, General Electric, Monsanto, Dow, and others attaining "best in class" status obtaining reduced costs, improved cycle time, and defect elimination and maintaining these results with increased customer satisfaction and increased profits. Many smaller companies, including suppliers to these major companies, now operate under six-sigma control. The major companies to whom they supply products trained their product supplier's black belts. This ensures the parent company that products delivered are up to the highest standard of quality and that they manufacture their final products for their customers. They also reap the financial rewards of their suppliers reaching six-sigma control in their operations, which more than covers the expense of training the companies black-belt candidates. This is the basis for the ideal quality system. It is to ensure quality products are consistently produced in real time with support documentation to meet customer requirements. Thousands of dollars and hours will be spent on becoming ISO/QS 9000 and six-sigma-certified/qualified, and the effort must be placed in the right areas to always keep the manufacturing system in control.

Control of incoming material and the manufacturing equipment processing variables must be established and any variability brought under control during production:

Is this easy? No!
Can it be done? Yes!

To do this requires identifying the key material, product, and process control variables. The data generated must be available to the operators during manufacture, as they must control the manufacturing system within the established control limits.

To do this, material, auxiliary, and main production equipment must be evaluated for their capability to stay in control within established product manufacturing limits. The machine's operation variables, pressure, speeds, temperature, cycle time, and other parameters must be specified and controlled to consistently produce quality products. This can be done, and the implementation is a key part in the *total quality process control system.* Operators must know their equipment and product's quality control specifications and be able to verify that they are within the control limits in real-time manufacturing. Assistance in these areas resides within the quality control department, equipment and material suppliers, and maintenance and engineering personnel, including upper management.

Management is key in assuring that the assets, personnel, time, and support are in place for their company to become a quality leader. Engineering, purchasing, manufacturing, and sales will then assist in determining, based on ISO 9000 standards, what is required to consistently produce a quality product in real time using total quality process control techniques. Results from using these techniques and procedures can be

1. Real-time quality manufactured product
2. Improved equipment efficiency and output
3. Control of manufacturing equipment and their operation
4. Improved profitability by reducing manufacturing costs
5. Increased customer base after becoming ISO/QS 9000–certified
6. Use of checklists for the development of data and information for products and manufacture

Total quality process control is not a single system; it is the only system for always producing quality products in *real time* to meet the demands of the world-class marketplace of the twenty-first century.

REFERENCE

1. *Subject—Quality Policy,* GE Semiconductor Products Dept., Instructions 3.16 4/26/68 MGF 217, p. 1 of 3.

CHAPTER 2

CHECKLISTS FOR PRODUCT DESIGN, DEVELOPMENT, AND MANUFACTURE

PROGRAM DEVELOPMENT CHECKLISTS

Companies considering plastic for a new or existing product have many questions to answer. Your customers, either directly or as a supplier to original equipment manufacturers (OEMs), want a quality product to meet their customers' expectations and requirements with a reasonable profit margin. This requires a product designed using a material that will not create any manufacturing, assembly, decorating, or future end-use liability problems for them and/or their customers.

Customers also require a contract with their suppliers to provide the service, knowledge, quality, and manufacturing expertise to meet their product's program schedule. Then, if a problem develops, a source for the solution to their problem is available and documented for all to follow. This information is described in detail for product, material, and manufacturing to provide answers to their questions. This is easily documented when made in house or by outside product suppliers when they are using ISO/QS 9000–certified procedures. ISO/QS 9000 requires material and process documentation to verify that their product and production quality requirements are in control. Automotive and other large consumer and industry product suppliers now require this documentation as a normal business procedure.

To ensure that all the required information is developed and documented, the use of checklists has gained in acceptance and importance. The "checklists" ask the questions required for developing the product and ensuring that manufacture results in only defect-free products. They are not replacements for ISO/QS 9000. Checklists are used to support the development of required information for any program. Checklists documenting this information (see Appendix A) are

1. Assembly
2. Product design
3. Decorating
4. Materials
5. Manufacturing
6. Packaging
7. Product development
8. Program scheduling
9. Price estimating
10. Purchasing
11. Quality
12. Sales contract
13. Warranty problems
14. Engineering change request
15. Supplier quality self-survey
16. Supplier survey report
17. Mold design and tooling requirements and materials

The checklists assist the design team in gathering required information in a controlled, organized, repeatable, and easily accessible database. The information used to integrate the programs requirements into a computer-integrated manufacturing (CIM) system. This allows the design team to access needed information to accomplish the product development program. The checklist answers to the specific questions enhance in the program's success and profitability. Checklists are used to ensure that all questions concerning a product's design or redesign through manufacture have been anticipated and answered.

Design team members use these checklists to gather information and to ensure that it is as accurate and complete as possible for their area of responsibility. Too often a critical piece of information is forgotten at the start of a program that later adds cost, schedule delays, or failure of the product. A proven preventive action program used by Lockheed Martin for the launch of a space shuttle is a checklist. The checklist must be completed before each space shuttle launch covering over 1 million potential problem areas in the launch sequence. Checklists assist in ensuring that nothing is overlooked. Many people balk at using checklists, but they do their job very effectively as a quality checkpoint for each step in a program or process.

The Lockheed checklist was developed by creatively thinking through each step of the launch program. Each step in the launch sequence was analyzed and documented with checkpoints to evaluate information and system operations. This also includes any potential part of the system where a problem could jeopardize the success of the launch. Implementing a preventive-action plan consists of developing a list of key thought-generating questions or "words" that open up the technical mind to a simple question such as "What if?" When all the "What if"s are answered, a final release point is created to ensure that all the questions are answered and that all personnel are satisfied that the launch system is correct before proceeding to the next step in the launch sequence.

No checklist will ever be totally complete, and as new questions and ways of doing business occur, new questions requiring answers should be added to the checklist database.

Product Design, Development, and Scheduling Checklists

The initial programs to use are the *design, development, and program scheduling checklists,* which begin by analyzing what information is needed to initiate and complete a plastic product program. All questions may not apply, and others may have to be expanded to obtain more specific and detailed answers. Suppliers can modify the specific checklist to suit their business operations. The questions are intended to ensure that the information needed is developed early in the program so that the answers will be timely and can be used in the program's total analysis.

Product Development

The product development checklist asks the specific questions needed to determine whether a market exists for a product, volume potential, pricing required, design, manufacture, customer, and liability requirements, plus testing, quality, assembly, decoration, and packaging to ship the product to market. Also, it develops checkpoints or milestone points for the analysis of continuing or dropping the program. A critical-path schedule and timeline can be developed to track the program. It is basically the master plan checklist for the entire program. The following checklists supplement it.

Sales Contract

The sales contract checklist outlines all areas required for working with the customers, subcontractors, and manufacturers, including any outside assemblers, decorators, and vendors for material subassemblies and services. It addresses the questions and vendors needed for materials, parts, and supplier services along with having approved sources for these materials. The checklist addresses the customer, the product design, testing, quality, price, and manufacturing with documentation along with contract terms, scheduling, and delivery of product within the terms of the sales agreement and contract. The sales contract can eliminate problems, assign responsibility, and save time and money if a problem ever occurs.

Product Design

The product design checklist is used to define the product end-use requirements and the customer's expectations. It is used for assisting in the product design and selection of the best and most economical materials based on the applied forces the product will experience and where they act on the product.

How accurately and completely this initial information is developed will determine the success of the program. Therefore, as many product requirements as possible must be obtained from all personnel involved in the program. The design team's input from sales to shipping and supplier/vendor support can be used for developing the information needed to assist the design team in developing a product to meet the customer's requirements and specifications. The design checklist also assists in addressing issues in the manufacturing cycle such as the mold, and how to manufacture, assemble, and decorate the product. When the product is in the final design stage with a material selected for manufacture, this will allow the mold, assembly, decoration, and manufacturing questions to be completed. The design checklist is extensive and designed to answer questions for many different manufacturing methods. Therefore, only those questions pertaining to the specific manufacturing method for the product need to be answered.

Engineering Change

A very important item to discuss is the use of an engineering change request (ECR) form and checklist. The ECR form, included as checklist 14 in Appendix A, is used to communicate any product, process, or design change initiated by the customer or engineering, manufacturing, or quality assurance pertaining to the product.

The respective component of the product or product is identified, specific drawing or document specified, and the change detailed and the reason specified. The document is then routed, individually or electronically, at one time to the respective departments involved with the product change. A specified response time to the author is included on all ECRs to ensure that it is acted on in a timely manner by each manager. All managers will discuss the change with their department personnel and compile a list of any significant change effects that it may create plus the cost impact to the program. Should the change affect how a supporting department manages the product, the manager and other personnel must decide what is best for the product, program, and customer. Then and only then should they sign off their department's approval on the change request. The department initiating the change request has the final approval for ensuring that the change request is implemented or determining the reasons why it is not implemented. The change request, after all department managers have approved the change, is then filed, becoming a part of the product file. Engineering change request procedures should be documented in the company's quality assurance program, and all personnel involved should be knowledgeable in the procedure for ensuring that the change request is implemented within the timeframe required to keep the program on schedule and within cost.

Materials

The materials checklist is involved with selecting the primary material of construction for the product. Questions concerning product requirements address issues such as agency certification, suitable supplier resins, testing, color, packaging, and purchasing or customer furnishing for the product's manufacture. It is helpful in purchasing, manufacture, and quality to ask and inform vendors what is necessary and required for the program's success.

If the product's manufacture is to be contracted out to a custom molder rather than manufactured in house, it will provide the information needed to specify, purchase, and ensure that the customer requirements are met for meeting production and delivery requirements and contract terms and pricing.

Purchasing

The purchasing checklist covers the areas for the purchase, certification, vendor approval, and certification of plastic resins, and finished products with their tolerance and specifications required to manufacture a quality product.

Vendor Survey

The vendor survey checklist is the document the company can use in evaluating a supplier's ability to be a quality supplier of parts or services. It is a detailed checklist in asking and evaluating the vendor's quality, manufacturing system, documentation, and personnel. It can also be used to evaluate ISO/QS 9000–certified suppliers who can supply similar goods and/or services. It is a comprehensive evaluation and rating system that generates a numerical score on the vendor's capability to become a certified supplier.

Customer Tooling and Tooling Design Requirements (Mold Design)

These checklists are specific for the manufacturing method of injection molding. Each industry needs to develop their own respective tooling checklist. Suppliers of specific manufacturing equipment can assist in tooling checklists for their specific industries.

Three separate tooling (the term *molding* or *mold* is often substituted for *tooling*) issues are addressed when referring to tooling design checklists. Appendix A includes mold/tooling checklists that ask questions regarding the molds requirements for the product. Questions concerning the mold design, product cavity layout, materials of construction, cooling, venting, gating, runner, and cavity operating system of the production mold can now be designed. The checklist can also be used for developing the requirements for a prototype mold. Each separate checklist assists the design team and customer in determining the quality required for the mold and exploring areas in the molds design, materials, tolerances, and quality for the finished product manufacture. The goal is to manufacture a quality product with a mold, on a machine and with equipment that is capable of producing defect-free parts, cycle to cycle for the anticipated life of the product.

Pricing—Determining Piece Part and Final Product Manufacturing Cost

Determining the estimated initial and final product cost is very important for any new product program. With a wide range of plastic materials to chose from, the lowest-priced plastic, based on a per pound selling price, may not always be the most economical. There are other key material variables that should be considered besides, price per pound when determining a product's final cost and selling price. A plastic material's physical and processing properties versus price can have an impact on the final design of the product. This is why the design team works as a unit to determine how the product will be manufactured and with what material.

Variables affecting a product's final cost but, not listed in any specific order of importance, are

1. End-use requirements
2. Design
3. Material
4. Manufacture
   a. Mold
   b. Processing equipment
   c. Cycle time
5. Assembly
6. Decoration
7. Packaging
8. Shipping

The product's total cost and individual piece part cost are determined, if multiple parts are based on these eight variables. The product end-use function and requirements must be met and will be the major factor in material selection and overall product cost. Initially several plastic materials may be likely candidates. Each material will influence the design of the part, such as section thickness, ability to combine multiple functions, specific gravity, part weight, cost per pound, and strength factors.

Once the plastic material is selected, other manufacturing items must be considered: number of cavities, mold materials of construction, runner length and flow length, cavity part tolerances, and estimated cycle time, including the sizing and selection of the injection molding machine and auxiliary support equipment. All of these items influence the design of the mold and manufacturing cycle to produce the required number of parts within the estimated product cost and to meet the manufacturing schedule. Selection of the injection-molding machine must be based on determining, within an acceptable hourly manufacturing rate, the ability to hold the mold and produce the product. The machine must have adequate melt and clamp capacity and preferably computerized process control.

The product's design must also consider the assembly and any decoration or information on the product. How the product is packaged and is delivered to the customer in usable condition involves packaging and part protection using reusable or one-way containers. It is important that the product always arrive on time at the customer location. How this is accomplished depends on the preplanning for the manufacturing schedule. The ordering of parts and material for manufacture must include calculation of sufficient lead time so that all parts or products come together on time for final assembly and packaging. This involves planning and exact timing from approved suppliers to ensure that all material parts are available at the precise time needed to manufacture the product. Many manufacturing operations now operate on "just in time" (JIT) delivery to reduce the cost of inventory on hand and to ensure a rapid turnaround of inventory and consumable materials, the cost of inventory dollars.

The cost of quality control is an overhead expense, usually accounted for in the injection-molding machine's hourly operation rate but if not, it must be included as an additional required cost. Charts and graphs have been developed to assist the designer team in determining anticipated part cost for a program. A critical-path control chart is used to ensure that all areas of the program stay on schedule to attain and meet cost and manufacturing objectives.

Cost Calculation for Plastic Product Development. Plastics are the most economical of building and consumer product materials available. Their versatility provides many design options and coupled with their low cost to manufacture products, they are the product of choice in today's marketplace. It is almost unimaginable to consider what our style of living and environment would be like without plastics. The cost of plastics today in cents per cubic inch has never been lower, based on using metals. Plastics cannot do everything a metal can do, but the newer plastic alloys are rapidly narrowing the gap in many conventional metal product markets.

The design team is responsible for the product's final cost, which includes both the design and manufacture of the product. If a product is not designed correctly for plastic in function, material, and end-use properties, it may fail. Therefore, it is the design team's responsibility to meet these requirements and to ensure that it can be manufactured to meet cost goals. The designer must never sacrifice quality of performance for price objectives. There are many families of plastic materials that can be chosen to meet end-use performance characteristics and anticipated cost goals. The design of a plastic product is just 20% of the finished piece product cost. This could be higher depending on the product complexity and time to design and research information required for the product.

Estimating Product Cost. A product's cost is based on several factors including the material for the product and what requirements it must meet and whether it is the most economical material that can meet these requirements. These are questions that the checklists can assist in answering, and in many product situations more than one material candidate can meet the end-use requirements for the product. The difference is the price per pound and specific gravity or material weight per product plus the molding cycle issues that must be addressed. Some resins will mold or cycle much faster than others and during the design have a thinner section thickness due to higher physical properties than other resins that may be considered. Therefore, the product's material selection is based on end-use part requirements. If more than one material candidate is selected, the design team should consider each material's physical properties to reduce the parts section thickness. Part volume and weight is evaluated before final material selection to consider the cost of the product's material in dollars per pound and cents per cubic inch. Some plastic materials are more expensive and have a higher specific gravity than others but have higher physical properties, which allow the designer to use a thinner product section thickness. This yields a lower part weight and faster molding cycle time, resulting in an overall lower piece part cost. The material candidates must be evaluated for piece part cost during the parts design by comparing key material physical properties to the product's structural needs and required part section thickness. For example, using a material's modulus of elasticity (a measure of the material's rigidity and strength) as a ratio for comparison of a product section thickness, the following comparison can be made:

$$\text{The ratio of material property section strength to a change in material thickness} = \frac{E_A}{E_B}$$

where E_A is the flexural modulus of material A and E_B is the flexural modulus of material B. The ratio of section strength E_A/E_B may allow a possible savings in material section thickness due to higher physical property strength.

Product part cost is determined on more than material cost. Initially the product end-use requirements must be met, but there are more factors to consider in developing the product's final piece part cost. The design of the product includes required section thickness for strength plus considering the combination of multipart product functions into the same part, thus reducing assembly operations and additional cost of parts in the final assembly. This can include molded-in mounting surfaces and snap/press-fit assembly, eliminating screws and assembly time. Using the material's properties to cut and reduce assembly costs will make it possible to assemble the product less expensively.

In determining manufacturing costs, you must consider the molds size and cost, number of product cavities, cooling requirements, and product tolerances. Manufacturing also considers cycle time, processing equipment, melt and clamp capacity, and process control, as well as information on decorating, including avoiding secondary finishing operations. Do not overlook provision of product protection after molding. Consider packing parts in reusable dunnage to transport and protect the product in your plant. Also, consider how the product is packaged, by hand or automatically at the manufacturing site.

Scheduling for manufacture is very important, just in time or built ahead of demand and inventoried at the plant for later delivery, which involves storage costs. In scheduling, always consider realistic manufacturing time with mold change cost, plus matching of colored parts and postmold shrinkage before assembly. Evaluate same-time manufacture for all colored plastic parts from the same lot of material and finally the cost of material purchased in one time period. Never forget the cost of quality, which is an overhead expense best spent controlling the manufacturing operation and addressing defects in a timely manner. Charts and graphs have been developed to assist the design team in determining the anticipated piece part cost for a product program that will now be developed.

The product's final cost will not be determined until actual production begins, but the initial cost calculations will show if the program is within anticipated cost goals. The following piece part cost estimation is developed which should be within 2 to 5% of the products final price.

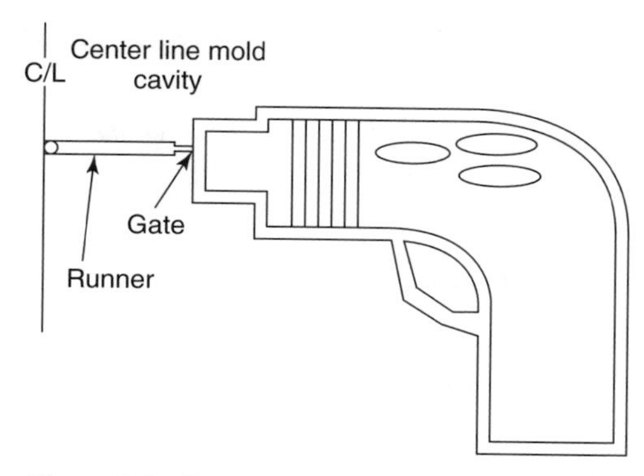

Figure 2.1 Power drill housing: two-cavity mold.

Estimating Product Cost: Case Study of Product Cost Analysis. A preliminary power tool housing for an electric hand drill has been designed as shown in Figure 2.1, less internal support ribbing and attachment bosses. Two materials are evaluated and found suitable for physical property requirements and are to be evaluated for individual cost. The original design material strength calculations were performed using material B's physical properties, a 20% GR (glass-reinforced) ABS and material A, a 15% GR polycarbonate impact grade resin.

Material A	MaterialB
15% GR PC	20% GR ABS
Specific gravity = 1.25	Specific gravity = 1.22
Elasticity modulus $E = 870 \times 10^3$	Elasticity modulus $E = 750 \times 10^3$
Price = \$1.95/lb	Price = \$1.75/lb
Thickness = ?	Thickness = 0.125

The part volume was calculated at 8 in.^{3} (cubic inches) from an existing metal die cast housing to be replaced by plastic. Material A is a viable candidate with similar shrinkage but a higher price per pound with a slightly longer molding cycle. The physical property to be used in the evaluation is the individual materials section modulus (E). The question to be answered is whether a higher modulus material will permit a thinner section thickness, resulting in less material for the finished product and lower product cost. The yearly estimated product volume is 250,000 drill housings.

Comparison of calculated product thickness for material A to reduced section thickness for material B yields

$$\Delta(\text{thickness } E_A \text{ to } E_B) = 3\sqrt{\frac{E_A}{E_B}} = 3\sqrt{870{,}000/750{,}000} = 3\sqrt{1.16}$$

$$\Delta(\text{thickness } E_A \text{ to } E_B) = 1.051 \qquad \text{(thinner section for material A)}$$

$$\text{Thickness A} = \frac{\text{thickness B}}{1.051} = \frac{0.125}{1.051} = 0.119\text{-in. thick section.}$$

where adjusting part A material volume (V) = 8 in.^{3}/1.051 in. = 7.61 in.

Calculating product weight (W) versus specific gravity (Sg) and part volume yields:

$$\frac{W_A}{W_B} = \frac{Sg_A}{Sg_B}$$

$$W_A = \frac{W_B\, Sg_A}{Sg_B}$$

$$W_A = 7.61\text{ in.}^3\,(1.22)(0.0361) = 0.335\text{ lb}$$

$$W_B = 8\text{ in.}^3\,(1.25)(0.0361) = 0.361\text{ lb}$$

$$\Delta(\text{part weight}) = \frac{W_B}{W_A} = \frac{0.361}{0.335} = 1.077\,(\text{material A is lighter})$$

If the weight of part B is known, use the specific-gravity relationship to calculate the weight of part A:

$$\text{Part weight } W_A = \frac{[W_B\, Sg_A]}{Sg_B} = (\text{oz} \times 16\,\text{oz/lb}) = \text{lb}$$

If part weight must be calculated, use the following analysis:

$$\text{Part weight } W = \text{part volume}\,(V)\,[\text{material specific gravity (Sg)}]\,(0.0361)$$

According to the data, the product using material A is lighter:

$$W_A = 0.335\text{ lb} \quad \text{vs.} \quad W = 0.361\text{ lb}$$

Calculating material part cost yields

$$\text{Part cost} = (W)\,\text{material cost in \$/lb}$$

$$\text{Part cost material A} = 0.335 \times \$1.95 = \$0.653$$

$$\text{Part cost material B} = 0.361 \times \$1.75 = \$0.632$$

$$\Delta\left(\text{part cost}\,\frac{\text{B}}{\text{A}}\right) = \frac{0.632}{0.653} = 0.96\,(\text{material B is lower-cost})$$

To complete the comparison, the molding cycle time for each material must be considered and estimated.

Estimating Molding Cycle Time. The part's cycle must now be estimated and is determined by two factors: (1) the time required to close, inject the plastic, and open and eject the part; and, (2) the time required for the part to cool sufficiently to be ejected without any distortion; this time variable is determined by the thickness of the part section. The cycle must not be limited by the runner system for ejecting the part after packout and cooling. Runner thickness must be designed to efficiently supply the melt to each cavity with minimum pressure drops in the system, and this is the responsibility of the mold designer to size correctly. Also, remember that the crystalline or engineering resins will usually require a smaller runner system, in contrast to the amorphous resins, which have slightly higher flow characteristics. The engineering resins will set up faster, which will reduce cycle time. Filled or reinforced resins will mold faster, set up more quickly, and be stiffer, allowing them to be ejected earlier than nonreinforced resins. The following formula is used to estimate cycle time (CT) (The material supplier will also be an excellent reference source for cycle times for their resins):

$$\text{(Estimated) cycle time: CT} = 8 + T(200)$$

$$8 = \text{mold-open factor}$$

$$T = \text{controlling wall thickness in inches}$$

$$200 = \text{cooling factor in seconds}$$

The mold-open factor may have to be adjusted up or down according to machine speed, automatic operation, mold type, and whether any internal cams or unscrewing actions are required and the use of molded-in insets. Typically, automatic operations, high-speed machines, and hot-runner molds will require less time. Large parts, operator-assisted part removal at ejection, and hand-loaded inserts require longer cycles.

If robots are used, cycle time may increase slightly (a few seconds), but part quality is markedly improved, due to uniform cycle times.

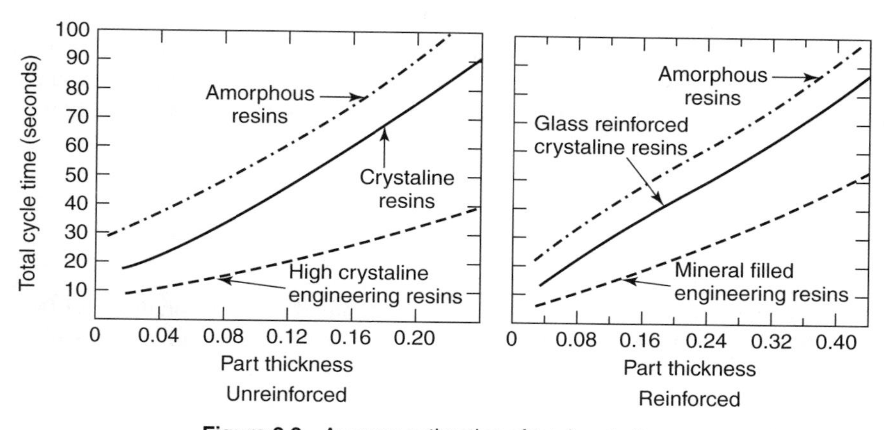

Figure 2.2 Average estimation of total cycle time.

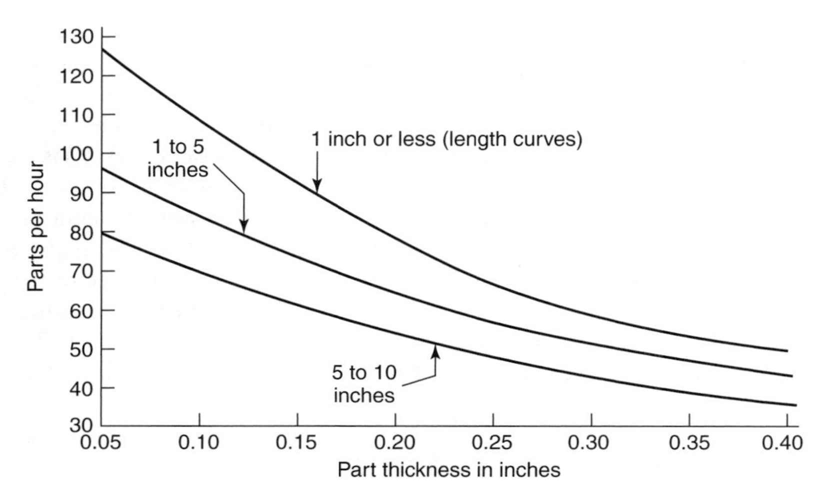

Figure 2.3 Estimated production per hour: length versus thickness.

Using the graphs in Figure 2.2, we can estimate amorphous, crystalline, and engineering resins cycle time. There are a wide range of cycle time values for resin combinations. Typically the lower the modulus of the resin and the higher its viscosity, the longer the cycle time. The following examples demonstrate how molding cycle time (CT) for each material must be calculated or estimated:

$$\mathrm{CT} = 8 + \mathrm{T}(200) \qquad (T = \text{part section thickness})$$

$$\mathrm{CT_A} = 8 + (0.119)(200) = 32 \text{ second (s)}$$

$$\mathrm{CT_B} = 8 + (0.125)(200) = 33 \text{ s}$$

Since all parts are not equal, flow length and section thickness must be considered in determining estimated cycle time. Figure 2.3 relates part thickness to part length and can be used for estimation if the cycle time is compatible with the parts geometry. Therefore, for part B with a 0.125-in.-thick and a 10-in.-long flow path in the mold, the parts per hour are

$$\mathrm{CT_B} = \frac{3600}{68} = 53 \text{ s}$$

Therefore, the estimated cycle time of 32–33 seconds is too low and does not factor in the flow length. If the part were 1 in. or less, the curve and formula would agree:

$$\mathrm{CT_A} = \frac{3600}{106} = 34 \text{ s}$$

In this example the longer cycle times must be used to correctly estimate cycle time based on material flow length and section thickness:

$$CT_A = \frac{3600}{70} = 52\,\text{s}$$

Determining the number of product mold cavities involves a number of factors, including the annual product volume, monthly product requirements, part size and injection-molding machine size available for the mold to fit, and later melt generating capacity. When using a custom molder, the production schedule should be adjusted to select a machine to run the product to meet the monthly product requirements. This should be discussed with the custom molder to ensure that the injection-molding machine used is compatible with the product's quality. If necessary, this should be confirmed by contract before release for production.

Estimating the Number of Mold Cavities. Number of cavities (NC) = (annual number parts) (CT cycle in seconds $\times 10^{-7}$) where three shifts per day, 6 days/week, 95% yield, and 80% utility are assumed. Always round off NC to the next closest even number. This number must provide the required volume of parts for the schedule, where the sequence of mold cavities is 2–4–8–16–32, and so on, for a balanced pressure drop in all cavities.

If the part size is very large or requires core pull in the plane of the cavity or in the number of square inches of product surface area, only a single-, two- or four-cavity mold should be built. The tolerances for the finished part have a definite effect on the selection of the number of mold cavities. If part tolerances are very critical, as for a tool housing, the number of cavities selected must ensure that the product tolerances are met for each cavity. Basically, the lower the number of cavities, the tighter the tolerance control. Each cavity will be slightly different, and the fewer variables involved cavity to cavity, more likely the molder is to obtain cycle-to-cycle product repeatability. Estimation of a molds size and the number of cavities must also factor in the molding machine's size to support and hold the mold. If the estimated sales volume requires the molding cycle to produce more parts in a short period of time, multiple molds may be required.

Determining the number of product cavities in the mold is based on product size, complexity of mold operations, temperature control of the cavities, and dimensional tolerances. The tool housing example uses a two-cavity mold, based on part configuration and layout in a balance mold runner configuration producing one complete tool housing per cycle, one right half and left half. Monthly sales volume is estimated according to the projected annual sales of 250,000 drills. Monthly manufacturing volume is $\frac{250{,}000}{12}$, which equals 20,834 housings.

The number of mold cavities must be determined as follows:

$$\text{Number of cavities (NC)} = (\text{annual number parts})\,(\text{CT cycle in s} \times 10^{-7})^{*}$$

$$NC_A = (250{,}000)(52\,\text{s} \times 10^{-7}) = 1.3$$

$$NC_B = (250{,}000)(53\,\text{s} \times 10^{-7}) = 1.325$$

The number of cavities (NC) is based on required monthly product volume. The mold will consist of two cavities, with, one cavity for each half of the housing:

Product production per hour (PH) for each material:

$$PH = \frac{NC}{CT} \times 3600$$

Products per hour	Runtime
$PH_A = \frac{1}{52} \times 3600 = 69.23$	$\frac{20{,}834}{69.23} = 301$ h/120 h/week = 2.5 weeks
$PH_B = \frac{1}{53} \times 3600 = 67.9$	$\frac{20{,}834}{67.9} = 307$ h/120 h/week = 2.56 weeks

Production time requires a minimum of $2\frac{1}{2}$ weeks based on 3-shift operation of 5 days/3 shifts per day per week or 120 h/week plus an initial setup time. If the product is molded in house, the injection-molding machine must meet the requirements and be available, or the product must be outsourced to an injection molder. How you schedule production can also have an affect on assembly dimensions. All plastic materials have some postmold shrinkage and a determination should be made in advance exactly what product dimensions are critical or required for incoming quality inspection and assembly. The longer the products are stored before assembly, the more dimensional variance that can occur. Therefore, tolerances on the product should account for this factor to avoid rejection of products due to out-of-tolerance dimensions before assembly.

Products molded in colors that have to match at assembly should be manufactured, whenever possible, from the same lot of material and in the same time period, as color can vary over time as a result of environmental changes, moisture pickup, and other factors such as processing conditions and melt temperature. This is always best when matching parts for a product that has been designed; the products are manufactured by the same molder, at the same or near timeframe using the same lot of plastic material. This is desired so that the products can be compared for the same matching color and fit when assembled in the plant product. Addressing these factors early in the program can reduce or eliminate aesthetic problems during or after assembly for show products.

General Mold Size. The mold must first fit between the tie bars of the injection-molding machine and not extend beyond the machine platen used to mount and support the mold. Also any internal mold operation that extend beyond the mold's primary dimensions as core pulls and unscrewing equipment, attached to the mold's outside surfaces, must fit in between or in and around the machine's free space.

1. Mold width and height are a minimum of $\frac{1}{2}$ in. wider per side than the part cavity dimensions and typically 1–2 in. for structural support strength.

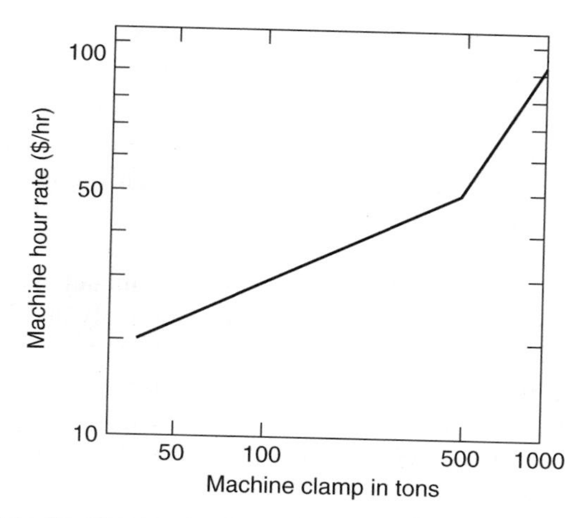

Figure 2.4 Machine hourly rate per ton of clamp. (Adapted from Ref. 2).

2. Stack height (height of mold closed) is 2.5 times the depth of the part cavity plus 4 in. This allows for the part depth plus 2 in. of steel safety stock for support of the base of the cavity, plus a minimum of a 1-in. plate on the core, an ejector stroke equal to part depth, and a 1-in.-thick ejector, plus a minimum of 1-in. clamp plates on the platens. There are exceptions for three-plate and hot-runner molds requiring greater distances. This also provides sufficient steel for the cooling line routing around the cavity for temperature and dimensional control.

Estimation of Injection-Molding Machine Size and Capability. Determining the size of the injection-molding machine to produce the product is the next consideration. The size of the machine determines the hourly rate for manufacturing cost calculation. The machine's hourly rate is based on clamp tonnage as shown in the example in Figure 2.4. Selection of the molding machine is usually made by manufacturing and is done in several ways: what is available, whether it will hold the mold and produce required melt capacity, the clamp type (toggle or hydraulic), the mold clamping pressure (tons of pressure to hold the mold closed), and process control capability. The injection-molding machine selected must meet or exceed the manufacturing requirements for the mold, product cycle time, and the tolerances required by the customer. Never select a molding machine incapable of meeting these requirements, as product quality will suffer.

Selecting the injection-molding machine to manufacture the product is determined by the following items:

1. Mold size
2. Number of cavities
3. Melt and shot capacity

4. Type of screw in the barrel
5. Mold clamp pressure capability

The mold size and cavity number will have already been determined as described in the preceding section and are used to calculate the remaining information for the sizing of the injection-molding machine.

Melt Generation and Shot Capacity. *Shot size* is the amount of resin required to fill the mold cavity, runner, and sprue for each cycle. Optimally the shot size should be 25–85% of the machine's rated melt generation capacity. Some molders have selected molding machines with inadequate melt capacity that is insufficient for the mold shot size required for filling the mold and producing parts. This is not recommended as the extra energy imparted by the screw can create excessive heat in the melt and degrade the plastic resin and possibly the product. Also, the reverse is true if the machine's melt capacity is too great, and the plastic resin is subjected to a longer residence time in the barrel. If the resonance time in the barrel is over 5 min, for heat-sensitive resins, the additional heat can cause degradation in physical properties and affect color quality. Each situation results in overheating the resin in the machine barrel in an attempt to obtain the required amount of melt for the cavity. Purchase of a new barrel, sized to meet the product's melt requirements is recommended, to ensure that the machine will have good-quality melt.

A method used to estimate shot weight is shown in Figure 2.5. The volumes of the runner and the sprue must also be added to the part volume weight to correctly estimate shot weight requirements. In selection of the injection-molding machine size for a program, use 1–3 shot weights of plastic resin melt capacity created in the barrel. For example, a 6–9-oz melt capacity barrel machine would be selected to mold parts with a shot weight of 3 oz. This would leave 1–2 resin shot weights of melt in varying molten stages in the barrel being readied for injection at all times. The residence or holdup time in the injection machine's barrel should not exceed 5 min. With a longer

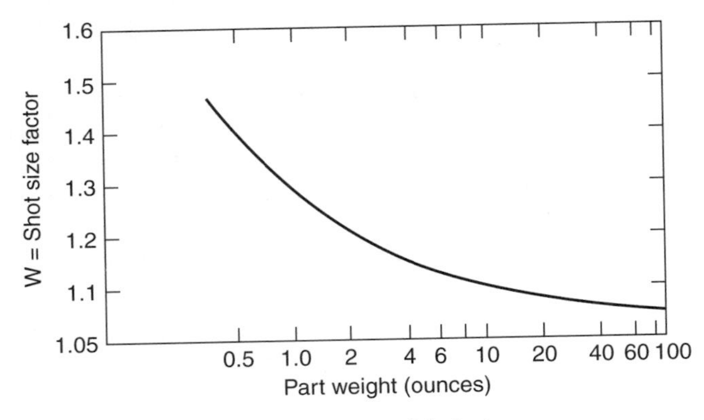

Figure 2.5 Shot weight factor.

residence time, the elevated processing temperatures could cause degradation of the resins. If for any reason the machine is not on cycle and the holdup time exceeds 5 min, the screw should be allowed to turn, slowly purging the material from the barrel. This will ensure that the material is not degraded.

A method for determining the machine size for shot weight and melting capacity is as follows, where shot weight volume and machine requirements are compared. Using the same example for the tool housing, the analysis proceeds as follows. Assume that product B (20% GR ABS) has a part weight of 0.361 lb times 2 cavities per mold, which equals 0.722 lb or 11.55 oz of material plus 3 oz for sprue and runner, bringing the total shot weight to 15 oz.

A machine in the 260-tonnage range has an average shot weight capacity of 35 oz of polystyrene. Polystyrene is used as the melt gauge capacity for all injection-molding machines when melt capacity is estimated. Other resins' melt capacity must be adjusted to that of polystyrene to arrive at the injection-molding machine's melt capacity for a particular resin. The resin used (20% GR ABS) requires a higher melt capacity than does polystyrene, so only 80% of the machine's rated melt capacity or 28 oz is used. The shot volume weight in ounces falls within the 25–85% shot size range or 53% of the barrel and the screw's capability to produce the required melt per cycle.

This range of melt capacity will allow the machine to prepare the plastic resin correctly, without degrading it, for every cycle. The lower clamp tonnage machine will also have a lower cost-per-hour operating rate. Clamp and shot weights are important considerations, and when used within these specifications, the products should meet specifications when all other variables are controlled.

Molding Machine Screw Type. The type of screw in the injection-molding machine is seldom considered in molding machine selection. The type of screw (see Fig. 2.6)—general-purpose, high-compression, nylon screw—and the length/diameter ratio (*L*/*D*) and diameter of the screw are seldom understood by many manufacturing personnel. The type of screw design used in a molding machine is very critical for certain resins, such as heat-sensitive, high-temperature-melting, crystalline, amorphous, reinforced, and general-purpose resins. The screw generates the shear heat in the barrel as the material is compressed in the screw flights as it is conveyed down the heated barrel. Always check with your material supplier as to the screw type they recommend for their resins and the compression ratio recommended for the screw.

Determining Molding Machine Clamp Requirements. The injection-molding machine must be selected to have sufficient clamp pressure to keep the mold clamped tight during the high injection pressure cycle. This pressure must be held until the plastic resin melt has solidified in the mold cavity and at the gate. If not frozen at the gate area, the mold can depressurize, known as "breathing" and occurring at the gate. This will cause molten resin to flow back into the runner system, resulting in a loss of the product's critical dimensions.

A molding machine's clamp pressure is calculated on the basis of the projected surface area of the mold cavity surface, including the projected area of the sprue and runner system. The type of machine clamp selected—hydraulic or toggle—should be

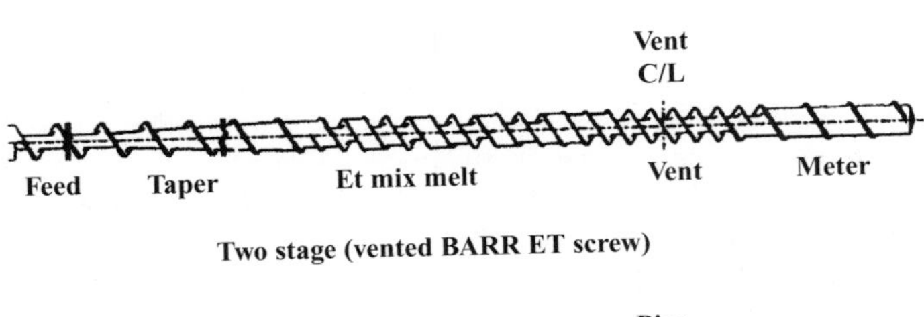

Pins

Feed

Taper

Metering

Pin type mixing screw

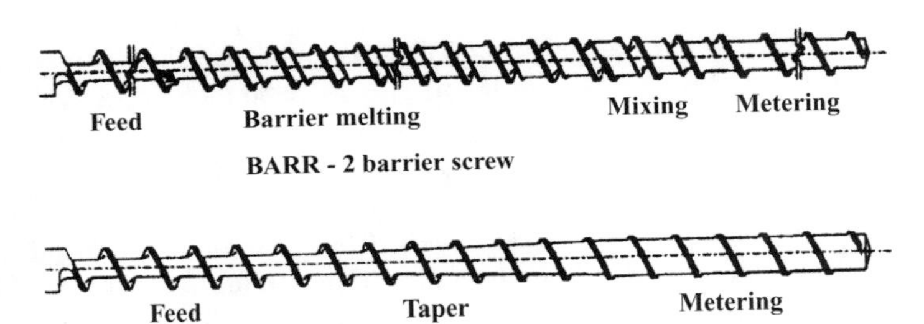

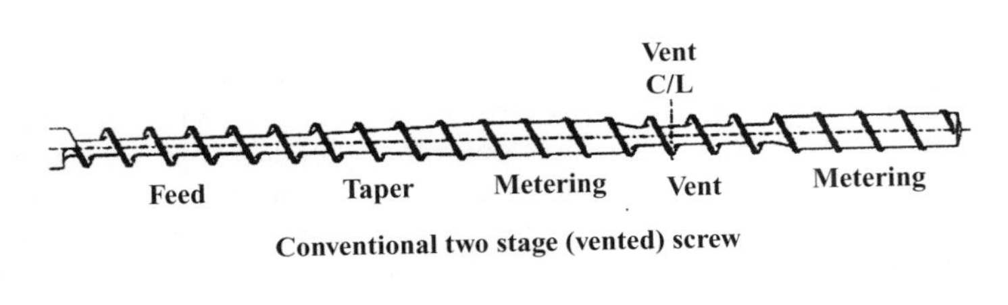

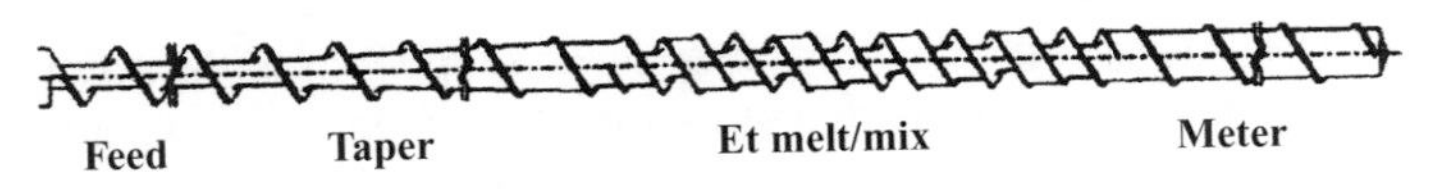

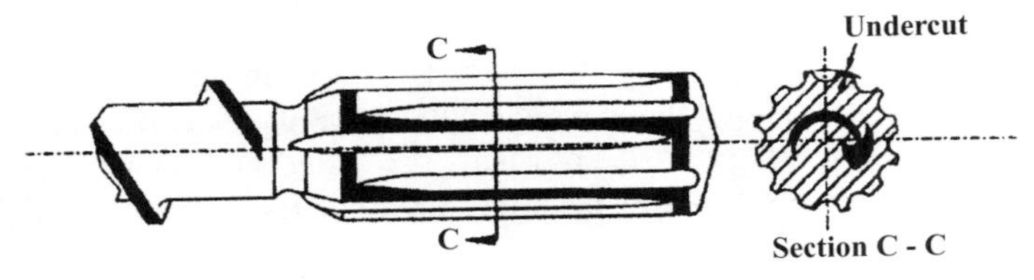

Figure 2.6 Screw designs. (Courtesy of Robert Barr, Inc., Virginia Beach, VA.)

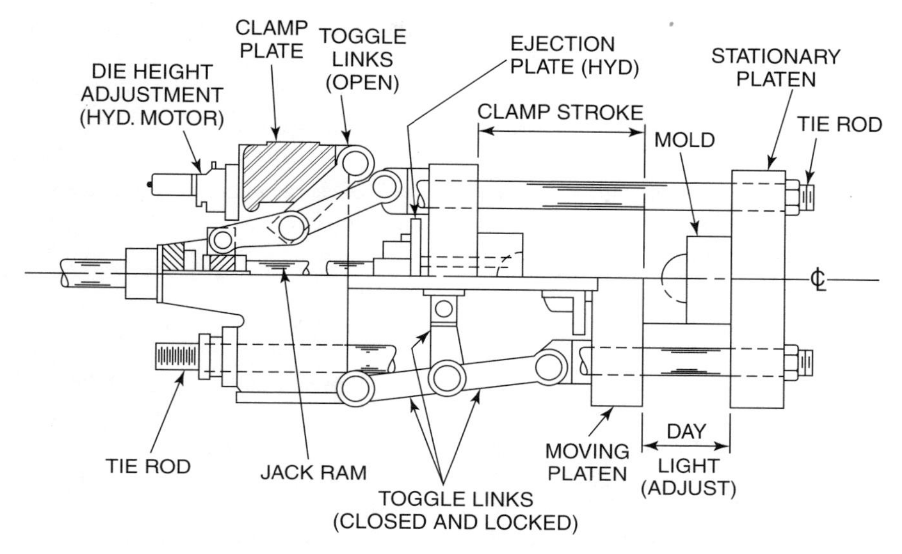

Figure 2.7 Toggle molding machine. (Adapted from Ref. 4).

considered when cycle time is estimated. Toggle machines (Fig. 2.7) with overcenter clamping design have a slightly longer opening and closing operation times. The hydraulic operated molding machine clamping operation does not have this time delay in operation. Hydraulic clamping pressure is attained once the mold closes and the hydraulic pressure pump applies the required holding pressure in a smooth operation. The molding machine's necessary clamp force (CF) in tons per square inch for a mold cavity's projected surface area as required for typical resins is listed in Table 2.1. However, depending on injection speed and mold filling speed, the clamp pressure may have to be increased to prevent the mold breathing, opening slightly during injection, which can flash the mold and part. Also, thin ($\leq$0.050 in.) wall section parts may require higher clamp pressure to overcome the higher fill pressure required to fill the section.

Remember that the longer and thinner the flow path and more fluid in the resin's melt properties, the higher the mold clamping force may be required to keep the mold

Table 2.1 Resin mold clamping requirements

Material	Weight (tons) of Clamp per Square Inch of Part and Runner Surface Area
PE/PP	$1\frac{1}{2}$
ABS/Styrene	2–3
PC/nylon	2–5
Polysulfone	3–5

tightly closed during injection and packing of the mold cavity. Mold clamping force is determined as follows:

$$CF_{machine} = \text{projected part area} + \text{runner and sprue area} \times MR$$

where $CF_{machine}$ is the machine's clamp force and MR is the material required for the resin clamp force to keep the mold closed tight in tons per square inch of projected surface area. Simplified, this is

$$CF = (\text{projected part area} + \text{runner and sprue area}) \times (\text{MR for CF, tons/in.}^2)$$

Injection-molding machines are available in many sizes of melt capacity, barrel and screw diameter, and clamp capacity tonnage to select for manufacture of a variety of product sizes. When calculating a machine's mold clamping requirements, the machine's clamping tonnage should be at least 20% greater than the clamp force calculated to hold the mold halves tightly closed. If insufficient clamp is selected, the mold can open during the injection cycle, depressurizing the mold, flashing the part, and causing out-of-tolerance product dimensions. Therefore, the ideal injection machine to selected must meet the following three criteria; (1) mold size to fit between the tie bars of the machine and within the platen dimensions, (2) shot and melt capacity to fill the mold with sufficient reserve of melt for the next cycle but not enough volume in the barrel to exceed the shot capacity by 3 times during the molding cycle, and (3) sufficient mold clamp pressure to keep the mold tightly closed during injection and mold cavity packout.

Machine Hourly Rate. When determining the economics for selecting the size of the injection-molding machine, the machine's hourly operating rate is a major factor in estimating and establishing part cost. Each size of machine has an established hourly rate of operation based on its size and overhead cost determined by the company's overall operating cost. Machine cost, like inflation, keeps increasing annually and is usually very competitive among custom molders. Average injection molding machine hourly costs in U.S. dollars (MC) for the second quarter of 2001 are shown in Table 2.2, developed by a survey of custom injection molders, and can be approximated by the following equation:

$$MC = 0.075\,(CF) + 12$$

This MC equation gives a good estimate of the average machine hourly costs and must be adjusted annually for inflation and operating expenses, power, and maintenance, as examples.

Machine Setup Charges. The initial product price quote for a program includes many factors, such as machine properties, resin, volume of parts, mold cost, operating overhead, profit, and the method used to amortize these items over the life of

Table 2.2 Average machine hourly operation rates for custom-injection molders with operator and profit dollars per hour (average) reported for second quarter 2001

Machine Tonnage Range	<50	50–99	100–299	300–499	500–749	750–999	1000–1499	1500–1999
Northeast	35.11	37.26	41.76	48.31	55.86	71.88	87.50	90.00
Southeast	30.27	37.33	41.05	48.80	70.31	132.00	—	—
North central	32.78	34.59	40.55	48.65	65.98	79.18	93.81	159.34
South central	24.20	28.80	36.43	44.51	62.92	75.96	98.91	132.51
West	35.57	38.98	43.59	48.03	61.23	80.93	102.00	207.00
National average[a]	32.73	35.71	41.01	48.05	62.74	82.65	94.18	145.95

[a]Data weighted geographically according to Plastic Technology's Manufacturing Census.

Source: http://www.plasticstechnology.com/articlews/hrates.html; "Your Business Hourly Rate Survey—October 2001" (online article).

Correction factors for Table 2.2

Press Tonnage & Operator	Without Profit (%)	Without Operator (%)	Without Either (%)
<100	12.2	16.6	30.3
100–299	8.2	12.0	24.1
300–499	7.5	10.1	20.5
500–749	5.4	9.7	17.3
750–999	5.9	11.9	21.3
1000+	2.1	11.3	15.5

Deduct these amounts (cumulative national averages over several surveys) from the values in Table 2.2 for rates without profit, operator, or both.

the product. Setup (SU) charges are determined by the press size and the time for installing the mold. There are essentially two methods used to establish SU charges: the conventional and/or the quick mold change (QMC) method.

The conventional method is standard operating procedure within the plant with the changing of a mold taking hours, depending on many variable factors, including urgency, mold size, complexity of the mold operations, auxiliary equipment support availability, experience of the technicians, plus the material in the machine. Changing out the material involves purging the barrel and may require the screw to be pulled, cleaned, or changed depending on the new material's melting and processing requirements.

The QMC method stages the mold adapted for quick mold change with the mold and machine specifically adapted for rapid mold change in minutes. QMC requires the mold to be preheated to operating temperature and ready to run once it is connected in the machine. Also, the material in the barrel will be the same or at least from the same generic family, so minimal purging is necessary to ensure that the barrel is clean and cross-contamination of resins is eliminated once the mold is in place. This is where production scheduling can save considerable time and expense, if job planing has factored in all variables in the production schedule.

The MC charge is a recurring cost each time the product is produced. Typical setup charges for different size molding machines are shown in Table 2.3.

Table 2.3 Machine setup charge

Press Size (tons)	Charge (U.S. $)
25–125	150.00
125–500	250.00
≥600	300.00–500.00

Setup time includes installing the mold, and purge and cleanout of the injection molding machine barrel and material hopper for the last resin run and will vary with the last material used and the experience of the setup team.

Changing a mold may take only 15–30 min using QMC techniques, or it may require several hours for conventional changeover of the mold. The production run size is based on the customer order quantity and the most economical scheduling to produce the product. These cost factors are considered when estimating and quoting the original program. Usually when longer production runs are scheduled, the product price can be decreased because there are fewer setup charges and greater efficiency of operation.

The production run size should be predicated on economic order quantities (EOQs) for scheduling and minimizing product storage and inventory costs. Mold cost in the initial part cost estimate will not be known exactly but can be estimated and, when finalized, used in the final product cost calculations. This cost, which can be many thousands of dollars, is usually amortized over 1 or 2 years of production. The cost of tooling will be predicated on material, tolerances, and volume of parts and required product quality. One important area to consider is the mold quality. The mold must be built to produce the required quantity and quality of products to meet the customer's requirements. It should also be built for ease of replacing worn components in the mold's working parts and cavity areas, the gate block, and other high wear areas subjected to the incoming resin flow.

Calculating Product Manufacturing Cost. Using the information developed the products manufacturing costs can now be calculated. This is based on the same example already discussed. Consider a power tool housing two separate plastic resins, A (15% GR PC) and B (20% GR ABS), in a two-cavity mold. One cycle will produce one tool housing. The clamp force (CF) required is based on a 2-cavity mold with a projected surface area, including sprue and runner system using resin A as shown in Figure 2.1. The required clamping force is 4 tons/in.^{2} to keep the mold closed during injection and packing. The mold's projected surface area for product and runner system is 72 in.^{2}:

$$\text{CF} = \text{in.}^2 \times \text{tons CF for material selected}$$

A 4-ton/in.^{2} clamp pressure was selected for glass-reinforced polycarbonate:

$$\text{CF}_\text{A} = (72)(4) = 288 \text{ tons of clamp force}$$

For a safety factor use 1.2 times the calculated CF, which equals 345 tons of clamp force required. The closest machine in this range is a 360-ton machine. This can vary with the company or the custom molder's available machine size. But a 360-ton machine will be adequate. Material B requires only 3 tons/in.^{2} of clamp force, so the clamp required is 216 tons:

$$CF_B = (72)(3) = 216\,\text{tons of clamp force} \times 1.2 = 250\,\text{tons}$$

With these conditions established, the machine cost (MC) comparison continues:

$$MC = 0.075(CF) + 12$$

$$MC_A = 0.075(360) + 12 = 40.12\$/\text{h}$$

$$MC_B = 0.075(250) + 12 = 31.50\$/\text{h}$$

Final Product Cost Projection. Assume a setup charge based on a press size of 250- and 360-ton clamp capacity yielding a setup charge of $250.00 per production run for each machine.

Final product manufacturing costs are calculated using the piece part estimating form shown in Figure 2.8 for materials A and B and total cost (processing) are calculated as shown in Figure 2.9. In the example, material B, 20% GR ABS is very economical, $0.972 versus GR PC, at $1.155 per housing. GR ABS requires a smaller molding machine with a lower machine hourly rate. Also, the difference in molding cycle time is too minimal for GR PC to be advantageous in this example. One possibility in this example would be to evaluate an engineering material as a glass-reinforced nylon material. This material will mold faster than either material evaluated and with a high modulus of elasticity, and may result in a thinner section thickness. The nylon may also offer impact properties lacking in the primary material candidates. Only evaluation of the material will show if there is a chance.

Using the part cost estimating form shows the cost differences between materials and any possible saving between the material choices for the part. The other factors can then be calculated and summarized in tabular form to show the total anticipated part or program cost. The mold cost is typically amortized over the first year or two of product production to obtain the true part cost relationship. An important item to remember is that the lowest quote is not always the most economical. If the molder is not familiar with running the material selected for the part or is unable to obtain the part tolerances consistently, then a more experienced molder with higher costs may be the better choice. Many OEMs qualify their sub–product suppliers, custom molders, and their own in-house molding capability for all their molds before selecting their custom molders to quote on their product program. Many custom molders are now ISO/QS 9000–certified if they want to participate in the OEM's product supplier programs. Always ensure that your molding operation, in house or through custom molders, are qualified and approved manufacturers for your company.

CUSTOMER: DATE:
ADDRESS:
CONTACT: PHONE: FAX: E-MAIL:

PART NAME: JOB NUMBER:
DRAWING NO.:

PIECE PART COST ESTIMATING PER 1000 PARTS

A. MATERIAL	:	:	:	:
B. RESIN COST ($/LB)	:	:	:	:
C. SPECIFIC GRAVITY (Sg)	:	:	:	:
D. PART WEIGHT (lbs)	:	:	:	:
E. PART WEIGHT (D x 1000)	:	:	:	:
F. MATERIAL COST (B x E)/0.95	:	:	:	:
G. CYCLE TIME (CT)	:	:	:	:
H. NUMBER OF CAVITIES (NC) *a	:	:	:	:
I. PARTS/HOUR (H/G x 3600)	:	:	:	:
J. CAVITY AREA (PROJECTED) *b	:	:	:	:
K.CLAMP FORCE *c (CF) TONS x (J x MATERIAL FACTOR)	:	:	:	:
L. SHOT WEIGHT (oz) (D x H x W *d x 16 oz/lb)	:	:	:	:
M. MACHINE HOUR COST (RATE x (MC *e)	:	:	:	:
N. PROCESSING COST ($/1000 PARTS) M/I x 1000	:	:	:	:
O. ADJUSTED PROCESSING COSTS *f [N/(0.95)(0.80)]	:	:	:	:
TOTAL COST (PROCESSING PER 1000 PARTS	:	:	:	:

*a Assumed three shifts/day, 6 days/week (*f), one years production produced
*b Projected cavity area & runner/sprue, mold cavity in square inches x number of cavities, plus runner and sprue area of mold surface in square inches
*c 80% to 20% maximum shot weight of resin, use material clamp factor to estimate tons of clamp required
*d Use reference chart for shot weight Figure A.
*e Use machine hour rate chart Figure B, adjust for current machine rates
*f Assumes 95% yield and 80% utility of molding process

Figure 2.8 Piece part estimating worksheet.

CUSTOMER: Quality Plastics
ADDRESS: 3319 Your Street
CONTACT: Mr. Powell
PART NAME: Drill housing
DRAWING NO.: 1345D-0

DATE: 11/15/01

PHONE: 727 555-1212 FAX: ditto E-MAIL: qplastics.com
JOB NUMBER: 1234321

PIECE PART COST ESTIMATING PER 1000 PARTS

	Trial 1	Trial 2	Trial 3	Trial 4
A. MATERIAL	:Matl. A	Matl. B	:	::
B. RESIN COST ($/LB)	:1.95	1.75 :	:	:
C. SPECIFIC GRAVITY (Sg)	:1.25	1.22 :	:	:
D. PART WEIGHT (lbs)	:0.352	0.361 :	:	:
E. PART WEIGHT (D x 1000)	:352	361 :	:	:
F. MATERIAL COST (B x E)/0.95	:772.53	665.00	:	:
G. CYCLE TIME (CT)	:52	53 :	:	:
H. NUMBER OF CAVITIES (NC) *a	:2	2 :	:	:
I. PARTS/HOUR (H/G x 3600)	:138	135 :	:	:
J. CAVITY AREA (PROJECTED) *b	:72	72 :	:	:
K.CLAMP FORCE *c (CF) TONS x (J x MATERIAL FACTOR)	:360	250 :	:	:
L. SHOT WEIGHT (oz) (D x H x W *d x 16 oz/lb)	:13	13 :	:	:
M. MACHINE HOUR COST (RATE x (MC *e)	:40.12	31.50 :	:	:
N. PROCESSING COST ($/1000 PARTS) M/I x 1000	:290.72	233.33 :	:	:
O. ADJUSTED PROCESSING COSTS *f [N/(0.95)(0.80)]	:382.52	307.02 :	:	:
TOTAL COST (PROCESSING PER 1000 PARTS	:$1155.06	$972.02	:	

*a Assumed three shifts/day, 6 days/week (*f), one years production produced
*b Projected cavity area & runner/sprue, mold cavity in square inches x number of cavities, plus runner and sprue area of mold surface in square inches
*c 80% to 20% maximum shot weight of resin, use material clamp factor to estimate tons of clamp required
*d Use reference chart for shot weight Figure A.
*e Use machine hour rate chart Figure B, adjust for current machine rates
*f Assumes 95% yield and 80% utility of molding process

Figure 2.9 Estimation final product price.

PROGRAM SCHEDULING FOR MANUFACTURE

The scheduling checklist includes the action items for keeping the program on schedule. Beginning at the start of the program and covering all events and steps in the development of the product program. For the success of a program, events must happen and be completed within a specified time period. Program events must be established to gather, approve, and order parts and services to keep the program on schedule and avoid problems and delays to ensure that all the necessary information is available, in a timely manner, to make meaningful and correct decisions.

The actual schedule can be put into a software program database to suit the user, usually in a software program, Lotus 1-2-3 style, for easy use and the ability to modify it as necessary.

MANUFACTURING

The checklist for manufacture of the program product begins after the design phase is complete. It is used to (1) assist in the design and selection of the mold to produce the product and (2) ensure that the injection-molding machine, auxiliaries, and process is capable of producing the product in real time, defect-free, cycle-to-cycle. The mold is one of the most important single items in the manufacturing process. The quality of the part is controlled by the mold's dimensional and temperature ability to reproduce the part cycle to cycle.

The next important item is the manufacturing process. The injection-molding machine should have a process control unit for the repeatable operation of the machine cycle to cycle. A closed-loop/continuous-feedback process control system is necessary for continuous monitoring of the machine variables. The system should also be able to make minor cycle-to-cycle adjustments to control the molding process. This will ensure that the plastic melt is delivered to the mold at the same temperature, volume, and pressure in a repeatable cycle to cycle. Manufacturing is responsible for ensuring that their equipment, molding machine, auxiliaries, and any secondary machinery control operations are always within specification to guarantee that the product produced meets manufacture and customer requirements.

Even correctly designed products can fail in manufacture or service if the mold and the manufacturing process do not continually meet quality standards. The designer can assist the mold builder and designer by referencing critical product dimensions from only one reference point or part surface, and notes this on their drawing. Also, depending on the number of mold cavities, only one or two critical dimensions may be able to be held by the mold and process. The designer should not use typical metal tolerances on the part and specify realistic and attainable tolerances that are critical to the product function. The mold designer will then be able to control these dimensions using irregular part surfaces such as contour changes, cutouts, bosses, ribs, and walls, to ensure that the final product dimensions are held.

The Society of the Plastics Industry, Inc. (SPI) has developed, for generic plastics, recommended part tolerance guidelines. Designers and molders can use these in determining the degree of tolerances anticipated under commercial and fine tolerance molding conditions. This information is available in their publication *Standards and Practices of Plastics Molders*. Material suppliers also provide shrinkage and tolerance information for their specific resins and compounds.

The type of mold selected for the part will depend on a number of factors including part volume, tolerances required, part appearance, material, and injection-molding machine size. The injection-molding machine designer can work with the tool designer to ensure the tool steel selection and cooling layout for the part to control dimensions in the mold will produce the part to required specifications. These questions are addressed in the mold design checklists.

After the highest-quality mold is built for the parts requirements, the injection-molding production department assumes responsibility for the product's manufacture. The design team is still responsible until the manufacturing group produces products consistently, cycle to cycle. The team may then work with production to improve cycle time and product quality for the manufacturing cycle, such as six-sigma capabilities for long molding runs.

Total quality process control monitoring of the manufacturing process continues today per ISO/QS 9000 requirements. ISO procedures can assist in ensuring the manufacturing process always stays in total quality process control and is continuously monitored in real time, not simply by measuring parts after hundreds have been produced many hours after manufacture.

QUALITY CONTROL

Use of the quality control checklist can assist in ensuring that the quality of the manufactured product meets the customer's contract requirements. The requirements of every product will vary with customer expectations and specifications. The quality of the product must meet these specifications with the engineering, production, and quality departments working together to ensure that the product, process, mold, material, and processing equipment will produce zero-defect parts, cycle to cycle.

Prior to manufacture the design team should implement process control procedures for all business and manufacturing operations in the company. This is the task of the manufacturing engineers' and business department managers' designated implementers. Once these are in place and operating correctly, the quality department will write the failure mode and effects analysis (FMEA) for these business and manufacturing operations to identify potential problem areas and resolve methods to eliminate them prior to manufacture. The FMEA is also used to ensure that all manufacturing steps and quality checkpoints are established to verify that the manufacturing process is in control and able to meet the product's requirements and specifications.

The design team requires that first-article approved products will be the same as the final production product. The design team can assist in establishing procedures and methods to monitor production output to assist production in keeping the manufacturing process in control. Should problems occur, the design team will assist in

Table 2.4 Quality improvement methods

Program Name or Description	Worker Involvement	Specialist Oriented	Group	Individual	Procedure	Work Methods	Quality	Prod Design	Morale Enhancement	Motovation
Qual circles	X		X		X	X	X		X	X
Zero defect	X			X			X			
Employee										
Suggestion	X			X		X	X	X	X	X
Work										
Simplify	X		X		X	X				X
Quality of worklife	X		X			X	X	X	X	X
Scanion plan	X		X			X				X
VE/VA IE work study		X	X		X			X		
		X		X	X	X				
QA/QC		X		X	X		X			
Organizational development	X	X	X	X	X				X	X
Fishbone	X		X	X	X	X	X	X	X	X
SPC	X			X	X		X			X
DOE	X		X	X		X	X	X	X	X
CP/CpK		X		X	X		X		X	X
FMEA		X		X	X	X	X	X	X	
PAP	X		X		X	X	X	X		X
PPAP	X		X		X	X	X	X		X
QFD		X		X	X	X	X		X	X
GMP	X		X		X	X	X		X	X

Source: Adapted from Ref. 5.

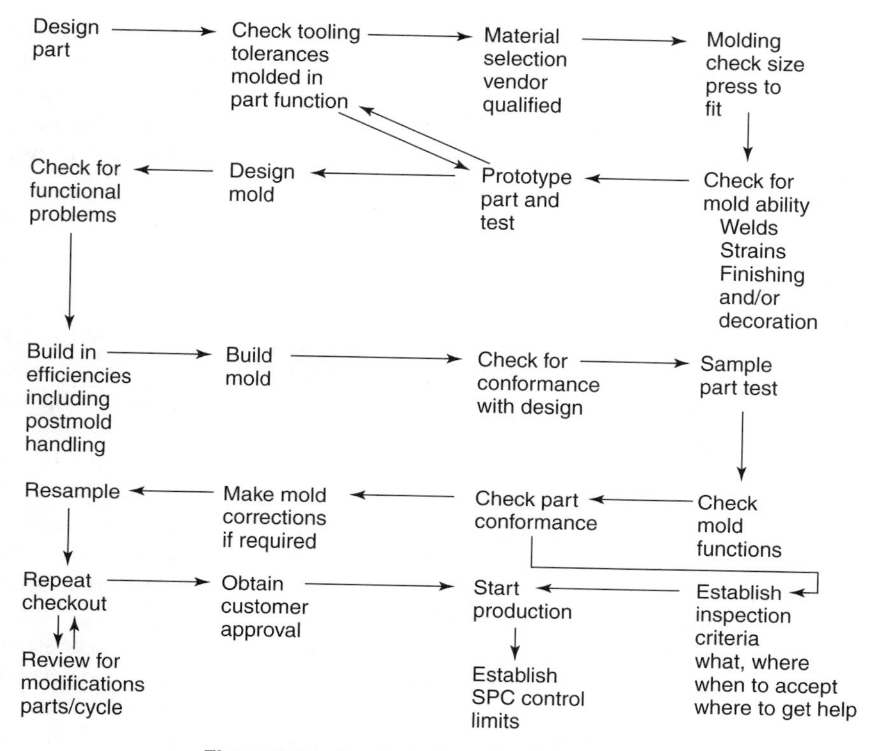

Figure 2.10 Quality in the total production cycle.

problem solving using the quality assurance methods listed in Table 2.4 to arrive at a satisfactory solution. The design, process, and quality checklist questions determine the level of customer quality requirements and manufacturing control required to meet the product specifications. Selecting the methods to measure, monitor, and always document the manufacturing process for product quality is very important, and a flowchart for these controls is shown in Figure 2.10. The team also will interact with the customer's representatives as required for maintaining part quality standards for the products throughout manufacture.

ASSEMBLY

The design team is also responsible for specifying the product assembly methods and assembling the product along with the drawings and assembly instructions. This includes purchased parts such as screws, inserts, clips, and other items packaged within the assembly. How the part is assembled may be production's primary responsibility, but the designer and the design teams are accountable for the method. Too often control over these items is lost at assembly and problems occur because tolerance control or other factors are not specified by the designer, which are critical to the success of the program. If the product design requires it, what is required should be documented so that purchasing can make the best buying decisions based on the part's specifications. This is information that the designer develops and should be conveyed to purchasing and production personnel.

DECORATING

Product components requiring decoration or information molded in or added later on the part should be fully described, detailed, and documented on a separate drawing.

If the part is colored or internally pigmented, the color system should be specified as to organic pigments, dyes, or powders with a color sample in the base resin supplied by the resin or pigment supplier. The pigment system must also be tested in the finished product to ensure that it does not degrade the resin properties. The exact mixing ratio of pigments and, when possible, their composition should be supplied by the vendor. Some organic pigments have a detrimental effect of the base resin's physical properties that is not always apparent in low-quantity use in pigments. Some plastics may require special cleaning, surface preparation, or static discharge before decoration. If special handling or product preparation is required, this should be stated on the drawing and manufacturing route card. Decoration adds value to the part but if not done correctly can cause rejects, lost profits, and customer complaints.

Decoration of a part can be done in many ways, such as in-mold decoration, painting, printing, decals, metalizing, and plating. Decoration is usually done in house but more specialized jobs such as as plating or metalizing are done at an outside vendor. If done outside, contact the vendor to determine the part surface preparation requirements for the decoration operation. This may include special cleaning, part handling to meet their requirements, or none at all. Often, special molding conditions are required to reduce molded-in stresses if the decoration medium can cause stress cracking of the parts during decoration. This is often seen when plastic parts are metalized, as they require an acid etch, which can cause the part to fail if high amounts of molded-in stresses are in the product. These specific issues should be addressed to ensure that parts are in a condition to be decorated. Also the decoration process should be discussed as to effects on the part and material, such as cleaning solutions, solvents, etching baths, paint cure oven time, and temperature. Parts from standard production lots should be decorated, as soon after molding as possible and end-use-tested to ensure that the molding process has not induced molding part stresses that may cause a problem in the decoration process or end-use properties.

PACKAGING AND SHIPPING

The packaging and shipping checklist is used to assist purchasing, manufacturing, quality assurance, and management in satisfying the customer's requirements for protection of the products sent to their warehouse. The checklist answers questions for packaging the products, storage, shipping requirements, and other important information needed to ensure that delivery of the customer's products is correct and within the terms of the contract. This is important for just-in-time manufacture for products sent directly to the customer's assembly lines for immediate assembly in their product. Suppliers are quality-rated on their ability to maintain on-time delivery of products.

The type of packaging for the product plus the materials required for there shipping must be available in the correct size, quantity, and type to meet the requirements of the program. Also, when the supplier must build ahead of product delivery, the storage of products must be protected and ready to ship to meet the customers economic order quality requirements. This cost must be included in the program cost to ensure that their program is profitable. Packaging costs may be a factor in determining the product's final cost. Packaging program costs should be added to and used in calculating the total product cost. Packaging involves two cost factors, labor and package material cost. Packaging costs are based on how you collect the products after manufacture and how they are packaged. Operators can remove the product from the mold, or it can fall free, or be collected by a robot. Parts on the runner can be separated in an automatic operation in which the products are collected on a conveyor and are transported to the next operation or packaging station.

Product packaging and collection considerations are

1. Is packaging required, or is the product part of another assembly?
2. Is intercompany transport packing required for transfer to the customer's plant for final assembly?
3. What type of packaging is required, where is it done, and by whom?
4. Can reusable dunnage or containers be used, and if so, out of what materials?
5. Does the part require protection from the environment prior to a secondary operation?

Can the products be collected and packed at the molding machine? An operator can do this; or for small parts, this can be done by a part counting and/or weigh packing operation and then dropped directly into a shipping container. Piece part counters and weigh count scales are built with automatic box filling equipment for automatic machine operation and packaging.

Some products require secondary operations, decoration, or assembly and require a method to transport them to other plant operations. These can be reusable boxes, trays, and/or dunnage rack part holders.

If cardboard box dust is a potential problem, plastic box liners should be used. This is important if parts are molded and not immediately decorated or assembled to keep them clean and dust-free. If possible, reusable containers can hold cost down for both storage and shipping parts to the customer's plant, which may then be returned for reuse. Automotive part suppliers are major users of reusable dunnage racks for parts shipped to their assembly plants.

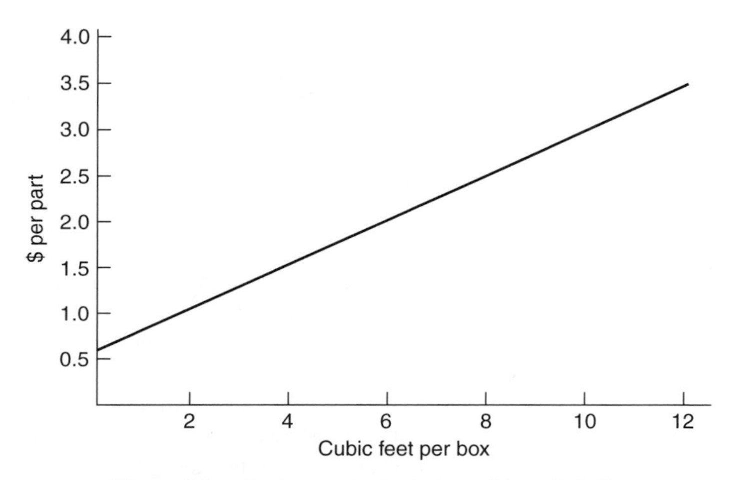

Figure 2.11 Packaging costs. (Adapted from Ref. 3).

Calculating Box Cost

Container or box costs (BC) is calculated using the following equation with data from Figure 2.11:

$$\text{Box costs} = \text{BC} = \$0.55 + \$0.24\,\frac{V}{n}$$

where BC = dollars; per part packaging cost, box, liner and taped shut
V = cubic feet of box
n = number of parts per box

Some packaging can be recycled for considerable savings. The amount of savings is based on the number of reuses. Interplant reuse is often easy to accomplish, but having your customer store and ship them back involves collection and shipping costs that must be considered. There are also shipping and storage costs to be considered. Always make sure that the figures used are up to date and as accurate as possible. Packaging should be considered in the original contract for the product manufacture when required.

For just-in-time manufacture, more sturdy containers that are recyclable for large parts should be considered if product is shipped outside the manufacturing plant. Dunnage part holders, usually made of semirigid urethane, are molded to hold and protect the products after molding and for transport to the customer plant. They are returned for reuse on returning trucks to the supplier. These dunnage racks, with parts, are often transported directly to the assembly line, thus reducing assembly plant cost and increasing product protection until installed.

WARRANTY PROBLEMS

Every product produced can eventually develop a customer-related problem. The supplier must analyze their design and manufacturing system using quality methods, FMEA, and other means to eliminate as many as possible problems before starting production. The intent with ISO/QS 9000 certification is to eliminate these potential problems. Using the concepts of preventive action versus reactive response quality methods, problems can be minimized. But when problems occur, the supplier and the design team must assist in solving the problem after it is identified. The design team is the first to assist in solving product failure problems, as they are knowledgeable in the product's development, design, and manufacture. Problems can occur any time in the part's development, manufacturing, or end-use life. Therefore a method for preventing and solving these problems should be in place to speedup an accurate solution, solve the problem, and continuously monitor the solution to ensure that it does not happen again. Problems can occur during part development, prototype, manufacture (molding), assembly, or in the worst case, field failure of the product. Often the most elusive item to determine is "What is the actual problem?" What first needs to be addressed is to collect complete and reliable data on what caused the problem and how it can be eliminated and prevented from recurring. Then solutions can be developed to analyze the failure and determine whether the entire lot is suspect or just a localized problem. After the problem has been completely identified, you have 60% of the information and the solution to problem.

To identify and document the problem, these questions need to be answered:

1. What exactly happened to cause the part to fail, and under what conditions?
2. What is the frequency of the failure: recurring only once or frequently after changes were made?
3. Does the failure occur at the same place or at multiple locations on the part?
4. What are the part's history and the lot number of the material? When was the part manufactured, and is its history traceable to known manufacturing information?
5. How exactly did failure occur, and what aere the conditions preceding or at the moment of failure?

The problem solving checklist questions must be answered very accurately, with no guessing on conditions or environmental factors. The gathering of information relies on the checklist questions designed to collect information on what occurred before and during failure.

Plastic parts fail because of poor design, material properties degraded during molding, molding processing problems, mold fill and packout, or assembly problems and even packaging and shipping problems plus excessive, unanticipated, end-use conditions. Gathering data is only as good as the paper trail (documentation of facts and records) that is available for developing information on the product's probable cause of failure. ISO/QS 9000–certified suppliers would have good records for the design and manufacture of the product to aid in solving a problem. This is because this is a prime requirement for the supplier's certification.

Failure Analysis

Failure analysis begins by developing the following information.

1. Historical—has it happened before? If so, did it occur in the same way, and with what frequency and severity?
2. Are failed parts available for analysis, and are nonfailed parts from the same manufacturing lot?
3. Visual analysis—are there visual signs that can be used?
4. Who was in charge of the failure analysis if it occurred before?
5. Have analytical, chemical, physical, and other tests been performed to resolve failure conditions?
6. Research past records of possible causes during manufacture, assembly, decorating, or even shipping and storage.

The field salesforce that learns about a part's failure from their customer usually gathers historical and mode-of-failure data. Their information gathering is essential for documenting that a problem exists and determining its severity. Information requested is on the customer corrective-action information list in Table 2.5.

Table 2.5 Customer corrective-action information checklist

Customer name, location, and contact phone number
Product number, identification, and description
Cavity number, if on part
Purchase order for part and date
Approximate time when part was in stock
Any secondary operations performed on part
Obtain and send both failed and current inventory samples
Lot number/Inspection number
Nature of defect, failure, or complaint
Seriousness or liability involved and extent
Application or use conditions
Environmental history and installation date
Unusual circumstances
Percent defective
First time or repeat problem
Possible cause
Appropriate internal and external personnel contacts
Salesperson and date

On the basis of the information collected, appropriate company personnel can be assigned to gather additional information and begin solving the problem. Failures in the marketplace are indisputable results of "something" that was not foreseen or an inappropriate, unintended, use of the product. No amount of testing can guarantee failsafe performance, as there always lurks the exception to the rule.

Examples of historical failures could be

1. Impact failures at subzero temperatures not anticipated
2. Warpage of parts exposed to high temperatures in shipping
3. Part brittleness, lack of toughness when moisture levels drop
4. Wrong type of assembly method used, sharp threads, or overtorque
5. Paint peeling, poor surface preparation, or wrong paint used
6. Sagging or collapse of parts exposed to loads not anticipated
7. Crazing and cracking of parts exposed to strong sunlight, with no UV protection
8. Chemical exposure not anticipated or changed

Visual signs on the product may often indicate that many of these conditions occurred before or were attributed to failure. Internal visual checking may reveal the probable cause, and it is very important that both failed and in-stock products be retained for analysis.

Material contamination may be a cause but is often difficult to prove. If it is visible in the part, then analytical and physical testing will be done. Contamination as foreign material or unmelted pellets causes high internal stresses, lowering the part's physical properties. Voids and porosity in parts likewise lower physical part performance. These are visible as holes, sink marks, and warpage. Flash lines on parts and lack of radii at internal corners can cause failure under lower loading conditions. Poor weld line strength due to cold melt flow or poor mold temperature control can attribute to failure when these cooled material fronts meet. Molds wear and checking gate size and part weight may lead to a solution, especially if a specific cavity numbered part is failing. Use of regrind or too much can lower a material's relative viscosity and toughness, foreign-material contamination, or use of the wrong material or pigment system or additive may have caused the part to fail. Also the appearance of the part's surface or color may indicate processing, mold, or material changes. If visual signs and documented processing record information are inconclusive, then analytical testing is recommended.

ANALYTICAL TESTS

1. Melting point to identify material, ash filler content, and specific gravity and flame (burn/smoke/odor) tests, listed in Table 2.6, is a first good analytical test.
2. Differential scanning calorimetry (DSC), thermogravimetric analysis (TGA), and infrared spectroscopy (IR) as detailed in Table 2.7 are the next tests to be considered for the analysis.

Table 2.6 Identification tests for thermoplastic materials

Resin	Burning	Odor	Melt (°F)
Acetal	Blue flame, no smoke, drip may burn	Formaldehyde	323–347
Acrylic	Blue flame, yellow top	Fruitlike	374
Acrylic rubber mod.	Yellow flame—spurts	Use control	279
ABS	Yellow flame, drips black smoke	Use control	—
Cellulose acetate	Yellow flame, sparks, drip may burn	Acetic acid	446
Cellulose acetate butyrate	Blue flame, yellow tip sparks, drip may burn	Rancid butter	365
Cellulose nitrate	White rapid flame	Sharp decomposition	—
Cellulose propionate	Blue flame, yellow tip	Fragrant	456
Cellulose triacetate	Yellow flame, drips	Acetic acid	572
Ethyl cellulose	Yellow flame, blue top, dripmay burn	Burned sugar	—
Ethylene	Blue flame, yellow top, drip may burn	Paraffin	221L
Methylstyrene	Yellow flame, black smoke, carbon in air, softens	Illuminated gas	349
Polyester film	Yellow flame, smokes and drips	Use control	482
Propylene	Blue flame, yellow top swells and drips	Sweet	334
Styrene	Yellow flame, dense smoke, carbon in the air	Illuminated gas	374
Vinyl acetate	Yellow flame, smoke	Acetic acid	140/190
Vinyl alcohol	Yellow flame, smoke	Use control soapy	446 Decomposition
Vinyl butyral	Blue flame, yellow top melts and drips may burn	Rancid butter	345
PVC-PVACET	Yellow flame, with green	HCl	261
Carbonate	Decomposes	—	430
Chlorinated ether	Sputters, bottom green, top yellow, black smoke, carbon in air	Use control	358
Nylon	Blue flame, yellow top, melts and drips, may burn	Burned wool	
6			420
6/6			490
6/10			415
11/12			351
Vinyl chloride polymers	Yellow flame, green at edges, softens and chars	HCl	302
Vinylidene chloride polymers	Yellow flame, ignites hard, green spurts	HCl	313
Fluorocarbons	Deforms	—	
FEP			554
TFE			621
Fluorochlorocarbons	Deforms, slight melt, drips	Weak acetic acid	383

Table 2.7 Material test methods

Information	Process	Data Obtainable
		Composition
Resins, polymer modifiers	DSC	Types present, plus ratio if copolymer, blend, or mixture
	GC	Polymers, oligomers, and residual monomers; amounts present
Additives	DSC	Deduce presence and amount from thermal effects (stabilizers, antioxidants, blowing agents, plasticizers, etc.)
	GC	Detect any organic additive by molecular weight
Reinforcements, fillers	TG	Amount (from weight of ash)
Moisture, volatiles	TG	Amounts (from weight loss)
	DSC	Amounts (if sufficient heat absorbed)
Regrind level	DSC	Deduce amount from melting-point shift
	GC	Deduce amount from shift in molecular weight distribution
		Processability
Melting behavior	DSC	Melt point and range (each resin if blend); melt energy required
Flow characteristics	GC	Deduce from balance of high- and low-molecular-weight polymer in molecular weight distribution
		Physical, Mechanical
Glass transition temperature	DSC	Shown by stepup in energy absorption as resin heats (marginal for some semicrystalline resins)
	TM	Detected by sample expansion
Crystallinity	DSC	Calculate percentage from heat of fusion during melting, also can find time–temperature conditions for desired percent crystallinity
Tensile, flexibility, impact	GC	Deduce from balance of high- and low-molecular-weight polymer in molecular weight distribution
	DSC	Deduce from overall composition

Source: Adapted from Ref. 6.

A good quality control technique to assist in ensuring that no factors are missed in failure analysis is the Ishikawa "fishbone" diagram as shown in Figure 2.12. Also, if multiple reasons for part failure are developed, the use of Taguchi's design of experiments (DOE) technique can be used to pinpoint the main contributor to the failure and ensure that they are corrected. It is very important when using the DOE method that all probable causes be evaluated as one may be prime but able to mask another that could be the primary reason for the part failing. A situation occurred with nylon parts stored too close to warehouse heaters in the winter that caused the parts to dry out, oxidize, and craze the part's surface. This fine surface crazing caused the parts to fail during flexing for a snap-fit assembly. The nylon parts were

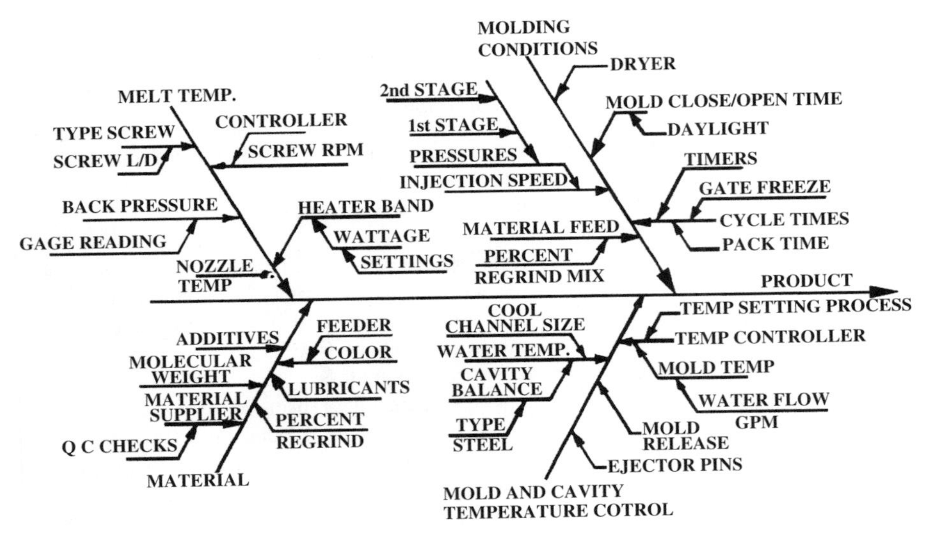

Figure 2.12 Ishakawa "fishbone" diagram. (Adapted from Ref. 7).

moisture-conditioned prior to shipment and the addition of heat created surface oxidation leading to crazing and cracking. The solution was twofold: (1) change to a modified, tougher resin grade not requiring moisture conditioning, and (2) not storing the parts near a high heat source. Design the parts to always pass the worst condition, which for moisture-sensitive plastic resins, as nylon, is dry as molded (DAM).

SPECIFICATIONS AND CERTIFICATIONS

Product Specifications

Product specifications include more than documenting the product end-use and customer requirements. They can include material, design, processing, and testing documentation required to qualify and approve product to the customer's requirements. Using these customer specifications, the design team writes the companies procedures and work instructions to produce, approve, and document the products manufacture. Specifications detailing requirements for the product decoration and assembly are often separate product drawings and/or specification sheets. These may be general or very specific for the operations required or refer to specific drawings for specific information. These drawings or specification sheets will detail specific information required for others to successfully complete their operations.

Typical part specifications will vary from the simplistic to very specific requirements. ISO/QS 9000–certified suppliers should have these requirements documented in their procedures and work instructions so that they can comply with these requirements of their customers.

In-House Specifications

These may include

1. Prime and alternate materials with regrind and percentage specified if permitted.
2. Key dimensions for the mold designer's use and referenced to one surface or point to eliminate dimensional and tolerance stackup.
3. Critical dimension tolerances, maximum of 2 to 3 with their range specified within the capability of the material, mold, and manufacturing process.
4. Noncritical dimensions specified with tolerances to match product requirements. Never use metals product title block tolerance callouts for dimensions. The product's critical dimensions will be used to size the mold cavity and used by the molder to maintain process control during manufacture.
5. Critical part sections noted to avoid stress points, weld lines, porosity, draft allowed, and mold marks for the mold designers information when designing the mold cavity for gating and fill locations.
6. Parting line acceptable location.
7. Surface finish; specify all surfaces to avoid any questions in cavity finishing. Final polishing direction for critical surfaces.
8. Allowable warpage; this may not be determined until molded prototype parts are made or the production mold finalized.
9. Flash limits; should be minimal to avoid secondary, after molding trimming requirements.
10. Performance test requirements, specific part items, agency, code, customer, and end use. Should also include purchased or assembled component tolerance requirements purchased from outside suppliers.
11. Decoration considerations, including color, printing, painting, and plating. Reference separate drawing when required to avoid questions at this critical final operation. Masks may be required to add information or block out sections during secondary decorative operations.
12. Assembly requirements, method, purchased parts if used, separate drawing when required.
13. Special product handling and surface protection when required by the operator, gloves, mold release, sprays allowed, dust protection, antistatic treatments, and dunnage type and material.
14. Packaging, type of container, count per package, where performed, labeling, and bar coding.
15. Lot control size and identification, route sheets.

The items listed are all referenced on their respective checklists.

Material Specifications

Material specifications are written with the intent of describing the part's physical performance requirements. The customer's materials or engineering department usually writes them. Material specifications can specify specific material tests and end-use requirements. The product supplier assists in determining how they are written and ensures that their materials can meet the requirement. They are also written generically for a family of materials to meet the requirements plus meeting the customer's design and performance requirements. The specifications required for a program should detail in a document what is required of the product or material to meet end-use, agency, code, and customer requirements. When the design of a product begins, the design team must review the specs to ensure that all are met or request exceptions if unable to be met in their operation.

Exceptions may not be allowed and then the supplier must seek alternate materials or modify their design to meet the requirements. If only form, fit, and function are required for simple, noncritical products, the design and manufacturing processes are easier to meet. But in no case should their quality be lowered or compromised for the product. The product requirements determine the specifications.

Data for specifications are developed using the product application and design checklists with the product meeting these requirements when tested. Specifications are used to identify requirements for the product material used and how the product's manufacture is controlled and documented, which includes the mold's design and the injection-molding process control to meet the customer's satisfaction and specifications. Specifications are also written to control the decorating, assembly, and packaging of the product, which contribute to the success of the program. The material used in the design analysis, tested, and approved must be listed exactly as described by the supplier in the material specifications. Several methods depending on product requirement's generic, specific, military, customer specifications, such as automotive specifications, can specify the material. Whichever one is used, there must be sufficient information to identify the type, grade, material properties, color, reinforcement level, and trade name of the material supplier's resin. The buyer for the product materials must know exactly what material to purchase. If the resin requires specific property or chemical analysis, specification lot data that may not be typical must be documented and requested from the supplier.

When ASTM, D4000 specifications are used (very infrequently now) list the material candidates exactly that are approved in product testing to ensure that purchasing buys the correct resin. These safeguards should be part of your procedure to ensure that the material meets product requirements. This can be a costly mistake if not followed and may lead to liability claims if the wrong product is used in manufacture and problems result. Materials should be purchased only from approved sources that have been approved by your purchasing department and meet quality standards.

Material specifications are

1. Supplier trade name and number (e.g., Dupont Zytel 101 NC-10)
2. Agency specification (e.g., FDA, CFR 177.1500 polyamide)

3. ASTM specifications (e.g., ASTM D4066 polyamide general)
4. Government/military (e.g., L-P-410/MIL M-20693 polyamide)
5. Customer specification (e.g., General Motors, GMP. PA-12.002 polyamide)
6. Generic specification (e.g., nylon, homopolymer, type 6, virgin, natural color)

The material specifications must be exact, and the buyer must know exactly what material is described. Specify or provide a copy of the supplier's Material Safety Data Sheet (OSHA standard form) or the exact specification with material alternate described if allowed, so that the correct material can be ordered. Noncritical parts can be loosely specified as item 6, and must be noted in the products bill of materials or a supplier specified for the product.

When special additives are required in the resin, for processing or part requirements, the material should be identified so that no mistake can be made when purchased as item 1. Colored resins require very precise detail in identification. Color formulations can vary from supplier to supplier for color matching the resin to the customer's color standard. Also, the color pigments and percentages can vary, which could affect the resin's physical property values. Always test each supplier's colored resin to ensure that it meets the products requirements.

Cadmium and heavy-metal pigments are not permitted today as these heavy-metal pigments are considered health hazards. Organic pigments are their replacement but lack high-temperature resistance. Organic pigments require more precise temperature control of melt temperature and less holdup time in the barrel to obtain the correct part color without burning the organic pigments out with too high a melt temperature during molding. The use of organic pigments inhibits the use of regrind. If too high a percentage of regrind is used, a color shift can occur. With each regrind pass, the pigments in the regrind will be more degraded and eventually burned out. Therefore, for exact color matched parts, if regrind is permitted, only one regrind pass should be allowed. The percentage used is kept low: 5–10% regrind to virgin resin. Then after one regrind pass, the material should be collected and segregated. This regrind can later be used for less critical colored parts as the physical properties will still be within specification. Do not try to add extra color pigments to this material, as the probability of obtaining an exact color match is almost nonexistent.

Material Certification

When supplier material or product certification is required, it should be documented exactly on the purchase order what is required from the supplier. It is important the material certification, certificate of compliance (CoC) is requested on the purchase order and exactly what information is required for certification, physical chemical properties, color, test data, and lot or historic property data. A procedure should be written to instruct incoming/receiving what material acceptance testing is required before the material or products are accepted for manufacture. Information must be supplied and available as to what specifications apply and whether any tests must be performed to accept the material. Unapproved material must be kept segregated until final acceptance is received.

The resin package type specified—bags, boxes (Gaylord's), or bulk in hopper trucks or railcars—is dependent on the quantity ordered. Larger quantities ordered have a pricing advantage. The cost of money to inventory products must be considered to ensure that not too much cash is tied up in inventory that does not turn as often as necessary to free up cash for other, possibly more critically needed materials or services. When customer-purchased materials or parts are supplied, they should be stored and segregated from other materials and accounted for in inventory and available when required for manufacture.

Purchased Parts

Specify purchased part tolerances and their suppliers. This will ensure that the correct parts are purchased and used in the product as with assembly hardware such as screws, inserts, and adhesives. This will allow the buyer to order the right materials and parts at the most economic price. Purchased parts used in the plastic products manufacturing as inserts or integral components encapsulated in the product should be specified as to tolerance, finish, edge surface finish, and cleanliness. Some metal parts may have a fine oil finish for protection against rust or never cleaned or deburred after machining. Some oils and cleaning solutions can cause stress cracking of certain plastic resins. Any molded-in or encapsulated part should be clean and free of sharp edges. This should be specified in the purchase order. Sharp corners create stress risers in the part, leading to premature failures. Also, when possible, preheat the metal inserts prior to molding to reduce molded-in material stress on cool down at the edges of the inserts. The cleaning and preheating of inserts to a specified temperature must be specified on the shop order so that the production personnel will have this done prior to molding. Since the design teams know what is required, these specific instructions must be documented in the specification and work instructions.

Tooling Specifications

The design of the tool (for injection molding, the tool is also termed the *mold*) to produce the product is a major item in the success of the product program. The design team should rely on the experience and knowledge of their mold designer, mold builder, and material supplier to develop a quality mold at a reasonable cost based on the product's requirements using the mold design checklist.

Precision parts require a high-quality mold. The materials used in its construction and the design of the runner, gating, cooling, and ejection system plus the cavity layouts are very important. The type of mold selected, two or three plates or a hot runner is important in the success of the program. The mold type selected for the product depends on many factors, such as part complexity, quantity of parts required, tolerance, and economics. Each mold type provides specific advantages in product tolerances, economics, and productivity. Hot-runner molds are much more popular than the two- and three-plate molds. The hot-runner system molds are more expensive but save on material usage as no regrind is produced. Whatever type of mold is selected for the program, the materials specified for mold construction, cavity temperature control, and ejection system are critical for consistent and quality part production.

The mold cavity temperature controller works in conjunction with the mold's cycling operation. Temperature sensors located in the mold cavity and runner system will aid in controlling and regulating coolant temperature via closed-loop coolant fluid flow in the mold's cooling circuits. Never economize on the mold's cooling system, as it is the critical control for maintaining repeatable cycle operation.

Materials used in the mold's construction—different types and grades of mold steels, other than the standard mold support plates—should be selected by the design team to ensure that the correct type of steels are specified. This is to ensure that the right mold materials are used to provide temperature, wear, and corrosion resistance and to obtain the required product volume of parts for the life of the mold. Each mold will have an estimated useful life based on the selected mold steels and number of parts to be made with the plastic resin selected for the product. Discuss the coolant circuit layout, location of temperature sensors, and individual cavity cooling circuits needed to control product dimensions. Exploring these areas early in the program will avoid surprises later during production if parts do not meet requirements.

Areas of the mold to be discussed and specified are listed on the mold checklist with specific areas of importance, such as mold types (e.g., two- and three-plate and hot runner) are listed as follows:

1. One piece mold cavity or multiple plates
2. Cavity materials, slides, and metal-to-metal contact parts
3. Cooling circuit design and layout, series, parallel or individual cooling circuits
4. Core and cavity cooling requirements, methods to control and materials
5. Temperature and coolant flow rate control
6. Draft shutoff and venting needs
7. Surface finish texture requirements and final polish direction
8. Gate type, size, and location; whether replaceable
9. Weld line considerations
10. Runner type, size, layout, and length
11. Part ejection system, pins, plate, or strippers
12. Special in-mold cam operations, core pulls, unscrewing, collapsing core
13. Insert placement, retention and special tolerance requirements
14. Cavity and part tolerance and shrinkage requirements

Manufacturing and Process Specifications

The molding operation must be very tightly controlled to ensure cycle-to-cycle repeatability to product quality products. The injection-molding machine's processing variables, controlled using a closed-loop process control feedback system, is developed during the preproduction molding trials. The process conditions such as for material, injection-molding machine variables, and auxiliary support equipment, dryers, chillers, and material handing, in conjunction with mold temperature controllers and pressure sensors plus feeders, robots, can be established using the manufacturing checklist to determine the optimum molding operation parameters to product quality products.

The injection-molding machines variable process settings are recorded on a molding record data sheet or programmed into the machines or the process control unit, with alarms set to denote process variance outside of acceptable limits. The part's, real-time quality checkpoints are recorded, as are the part's weight, critical dimensions, and other parameters so that the process can be monitored and controlled by the machine's computer or operator.

The design team will monitor and require process control charts for each shipment of parts on the molding parameters. The operator, production supervisor, and quality assurance should ensure that the molding process variables are within specified limits and should monitor them. Agreement for required data must be established before production begins to ensure that the controlling variables are documented and their times for collection and documentation are reflective of the molding operation's required control process procedures. If any product or processing deviation occurs, the data should be analyzed, adjusted, and monitored for the solution determined to keep the product's manufacturing operation in control.

Manufacturing data should have product lot number, time, date, and material lot identification with listed material package numbers. Then, if a problem occurs, information will be available to determine a solution to the problem. This is a requirement for ISO/QS 9000–certified product suppliers for total control of the product manufacture. The use of a CIM (computer integrated manufacturing) system with a closed-loop/continuous-feedback process control computer system can be used to control and document these items. This will increase the supplier's confidence in being able to certify and document process and product control in managing a successful business operation and, when required, supply these data to their customer as proof of process and product quality control.

This is very important in the medical, electrical, and automotive industries, where liability claims can be very expensive. With computer and barcoding techniques now available, accurate identification tracking and documentation is available if established prior to production. One of the greatest faults of production is to loose control of their product's manufacturing variables and documentation during production.

ASSEMBLY AND DECORATING

After molded parts are determined acceptable they are transferred to the assembly and/or decorating area. The main reason why real-time, process quality control is maintained during the molding cycle is to ensure that only acceptable parts are produced. If molding conditions go out of control, unacceptable parts are produced, but with quality control, they can immediately be segregated, the problem solved, and resolution made for the parts disposition. This will save many hours of measuring questionable parts or scrapping good parts if the problem production period cannot be accurately determined. Parts not immediately destined for secondary operations should be protected from dust, dirt, and moisture contamination. Be sure that the protection and storage used are adequate to keep the parts from being degraded before the next operation.

The product development package should have separate drawings for detail explanation on the assembly and decoration of the product. The process documents will list this information for the method used and all assembly or decorating specifications being developed from these checklists.

Assembly drawings can show an exploded view of the identified parts to be assembled and their order of assembly for the production personnel of the product. Noted in key areas should be screw type and part number, screw torque tightening limits for an automatic tool or the operator, flash limits for fusion or sonic bonded parts, and assembly speed settings for automatic snap-fit parts. No detail should be left to chance, as this could later affect the product end-use requirements and quality. Work instructions should be written to explain all operations and list the names of operators trained in the assembly process.

Inspection personnel for approving product for appearance, function, color, and fit should have samples color chips and gauges with documented instructions to verify that product meets the customer's specifications. Personnel must be adequately trained in detection of any discrepancies in these final approval areas.

Decorated parts should have their separate work instructions and support drawing if necessary for clarification of any product questions. Prior to decoration, special part handling and protection instructions may have to be written to protect the part's surface. These may be "no mold release allowed," "antistatic-treat molded parts," "box only in dust-free containers," or "operator to handle parts only with white cotton gloves." The molder must ensure that an adjacent molding machine, allowing mold release, does not collect on the molded parts manufactured in the same area. In special cases, a separate molding cell with positive airflow may have to be set up to avoid contamination. The operator may have to wear white gloves to prevent skin oils from getting on the part, or the parts must be cleaned before decoration.

Robots can easily handle these areas of product handling and protection. Dust in the plant can also contaminate surface-decorated parts. How parts are collected and stored before decorating is important. The dunnage used to hold parts must be dust-free and not contribute to a problem. Also, the airflow in the plant, generated by fans, can contribute to this problem, which is often overlooked until a problem results.

Contacting the decoration equipment and material suppliers will benefit the program, as they will know what areas need to be considered and how the parts need to be processed and handled prior to and during the decoration operation. Also, some decorating methods may require special molding conditions, to avoid part problems in manufacture such as molded-in stress that can affect the parts during decorating. Suppliers can also provide input for realistic specifications for the decorating operation. Knowing what is required and training the plant personnel to manage the production of the parts within the required specifications will keep the program successful and profitable.

Handling, Packaging, and Shipping Specifications

How products are handled and protected during manufacture before they are packaged or delivered to the customer must be considered early in the product's design program. Medical products are the most closely monitored manufactured products today. Parts are manufactured in cleanrooms with specified federal standard air quality [measured in parts per million (ppm)] of solids contaminates. The majority of other parts, excluding compact computer disks, require less stringent requirements. How the products are handled during manufacture will vary with the resin, part size, method of removal from the mold, and the plant environment, which can affect the success of a program.

When conditions require additional handling and part protection, this must be specified in the manufacturing and contract program. If a product is decorated, it must remain dust-free and protected and stored in dust-free containers until decorated. Parts should also not be made in an area where dust or resin fumes could contaminate the parts surfaced, as around grinders for sprues and runners and adjacent molding machines.

Always know what is required and take all precautions and steps to ensure that the product's quality is not lost during production through final packaging and shipment. Shipping product to market is rarely considered unless a problem develops and products are damaged. The type of packaging, and boxboard strength used can provide protection for impact-sensitive product. Temperature effects on parts shipped throughout the world should also be considered.

Temperatures in warehouses, boxcars, and trucks and bulk containers for foreign shipment can get extremely hot or very humid, such as for transoceanic shipments. If parts were molded in cold molds, these temperatures could cause molded-in stresses to be relieved, causing the part to warp or postmold shrinkage to occur, or affecting other properties of the part. In these cases the plant's shipping department must be made aware and alternate means or special shipping instructions employed to get the parts successfully to market. Humidity and heat can seriously affect the properties of some plastics, and their protection from these elements may be required. The design team should consult with their material and part suppliers to see if this is a factor in using these materials.

REFERENCES

1. *Part Cost Estimation,* E-32985, E. I. Du Pont de Nemours Corp., Wilmington, DE.
2. J. Dym, "Cost Estimating of Plastic Parts," *Plastics Design Forum* 51–52 (Nov./Dec. 1983).
3. W. Tobin, "Practical Part Costing for Custom Molders." *Plastic Machinery & Engineering* 37–39 (Dec. 1986).
4. *Modern Plastics Encyclopedia,* McGraw-Hill, New York, 1985–1986.
5. Schonberger, "Work Improvement Programs."
6. B. Miller, "Product Quality Problems? How Did You Check Your Resin?" *Plastic World* 49–55 (Aug. 1989).
7. E. Kindlarski, "Ishikawa Diagrams for Problem Solving," *Quality Progress* 26–30, (Dec. 1984).

CHAPTER 3

PRODUCT MANUFACTURING METHODS

The methods employed to manufacture a product must be carefully considered when a new or existing product is developed for the market. The design team must consider the best and most economical method to manufacture the product. Plastic products have many manufacturing options to consider that are not available with metals. These manufacturing options are injection and blow molding, rotomolding, thermal and vacuum forming, and extrusion and compression molding. The final product shape, function, and material will dictate how it shall be economically manufactured to meet sales anticipated volume and price for the product.

Some plastic materials and product type can only be manufactured in specific ways on specialized equipment. The volume of parts required also has a direct bearing on the product's raw material and method of manufacture. The cost of tooling, part complexity, and lead time all involve final product economics. Tooling cost will vary for each method of manufacture from the very low-extrusion die or rotomold cavity to the very expensive, injection mold. Compression/injection-molding tools (molds) are the most expensive tooling because of their complexity, size, and product tolerance requirements. To compensate for their high cost, only high-volume products are considered as the cost of tooling can be amortized in the product's salesprice and not a direct startup cost item paid for at the beginning of a product program.

Tooling costs are typically amortized over the life of the product or a specified time period and are a factor in the final product's piece part cost. Tool costs are increased if the product is very complex, requiring core pulls, unscrewing mechanisms, or multiple molds to meet the volume of products for sales.

SELECTING THE METHOD OF MANUFACTURE

The method of manufacture must suit the final product's end-use requirements and the tooling designed to meet the product specifications for the anticipated life of the

Table 3.1 Alternative manufacturing processes

Processes	Tooling Material	Tool Life, Number of Parts		Typical Tooling Lead Times (Weeks)	Relative Dollars	Type of Material Used
		Low	High			
Structural foam (thermo-plastic and RIM)	Steel-machined aluminum	200,000	1,000,000	18–24	100	Polyurethanes, PC, nylon, acetal, PBT, PET, PVC, ABS, PE, PS, PP
		5,000	250,000	14–20	60–80	
	Cast aluminum kirksite	500	50,000	10–18	50–70	
	Cast aluminum filled epoxy	10	500	6–10	20–30	
	Cast epoxy (100%)	2	25	4–8	5–20	
Injection molding	Steel-machined aluminum	200,000	1,000,000	18–24	100	Thermoplastics and thermosets, broad range
		5,000	250,000	14–20	60–80	
Compression molding	Steel	200,000	1,000,000	18–24	100	Thermosets, (polyesters) sheet molding compound, alkyds, ureas, phenolics, epoxies diallyl phthalates
Vacuum forming	Machined aluminum	5,000	250,000	14–20	60–80	ABS and ABS alloys, PVC, PPO, acrylic, PS, PE, PC, PP
	Cast aluminum kirksite	500	50,000	10–18	50–70	
	Cast aluminum filled epoxy	10	500	6–12	20–30	
	Cast epoxy	2	25	4–8	5–20	
Hand layup sprayed glass fiber	Machined aluminum	5,000	250,000	14–20	60–80	Thermoset polyesters
	Cast aluminum kirksite	500	50,000	10–18	50–70	
	Wooden pattern	10	1,000	4–8	3–10	

product. The materials and quality of manufacture and tooling are very important for the success of a program. To assist in this, Table 3.1 lists some of the manufacturing process available for plastics. Also, review checklists 2, 5, 7, and 9 in Appendix A to be sure that the method selected is best for your product. The majority of small and

Advantages of Process	Limitations of Process	Applications
Large detailed parts, low production costs, rigidity, complex shapes, parts consolidation; high strength: weight ratio; low density; molded-in inserts; low-pressure molding due to low viscosity; wide material selection; minimizing or eliminating sink marks; low molded-in stresses; improved chemical resistance	Surfaces usually need secondary finishing and/or painting; some sacrifice of physical properties relative to base resin; longer cycle times than injection-molded parts	Business machine housing, automotive fascias, medical and electronic cabinetry, furniture, materials handling equipment
High-volume production runs, close tolerances, molded-in color, low part cost, large material selection, parts consolidation	High tooling investment; long tooling lead times	Wide range of applications in all industries
High-strength, heat-resistant parts; high modulus; complex shapes, parts consolidation; excellent surface finish; molding large parts, polyesters full size range potential, excellent fatigue resistance	High tooling costs; deflashing needed; labor-intensive; parts generally need painting	Automotive parts, electrical connectors, business machine parts, power tool housings
Excellent for complex contours, with minimum internal details; large or small parts	Often listed by large radii, shallow depths, large draft angles, loose tolerances: exposed edges must be trimmed and buffed or milled	Signage, business machine housings, furniture, medical cabinetry, recreational (boats, campers, transportation) (interior and exterior parts) packaging, (cups/plates).
Large parts; basically shells can be molded with complex curves, excellent surface finish and high rigidity	Internal details (bosses, ribs) must be manually layed into inside wall and then overlayed with glass fiber; labor-intensive	Recreational boating; material handling, furniture, construction, transportation (truck hoods, bus seats).

complex plastic parts are injection-molded, versus the other types of plastic product manufacture. Selecting the type of manufacture depends on the product function, size, shape, volume, tolerances, and price, and the material and quality requirements. These variables must be considered in choosing the best manufacturing method to produce the product to meet program end-use and economic requirements.

SELECTING THE PRODUCT MANUFACTURER

The working relationship established with the product manufacturer, either in house or through an outside product manufacturer and supplier, will be one of the design team's biggest business partners. A good business relationship requires the team and members to communicate clearly and obtain consensus of opinions and commitment on the product's design and manufacture and delivery. The manufacturer will provide insight, knowledge, and experience to assist the design team in achieving their cost and program objectives.

Selection of one or several potential manufacturers early in the program and soliciting their input will be very beneficial. An approved list of manufacturers capable of producing the product is necessary to ensure the program's success. Most companies involved with the design and manufacture of plastic products have an approved list of key suppliers whom they can contact for their assistance and bidding on new programs.

It is equally important to ensure that your in-house manufacturing department is capable and qualified to produce your plastic product. Companies are now requiring their manufacturers to become ISO/QS 9000–certified and be held in compliance to industry quality standards. Companies with in-house manufacturing capability should require their manufacturing department to be in compliance to the same standards.

INTERFACE WITH THE DESIGN TEAM WHEN QUOTING

Often the product supplier is asked to just bid on a program, a request for quote (RFQ) without any prior input in the products design and manufacture. They are requested to bid and submit only a finished piece part price. In some situations only the low-volume or problem products are outsourced, if the company has in-house manufacturing capability. This often involves problematic or difficult-to-deliver products or products with tooling often questionable in its capability to product a quality product. The manufacturer seldom knows how the tooling was built or if it can be controlled during the manufacturing process for the product. It is difficult, therefore, for them to supply an accurate quote. This is why the manufacturer should be involved early in the product design or furnished with product specifications and tool drawings if available to provide an accurate quote.

TIE-IN MANUFACTURING WITH CONTROL

Try to select the best supplier, not always the most economical, early in the program to get the benefit of their knowledge. Also, negotiate a contract, a written description of what is expected and required with a schedule of events, with a product milestone chart. The contract and schedule may include the design of the product plus tool design, auxiliary operations, and decoration; also packing and shipping when not manufactured in house. Review the sales contract checklist plus the other relevant checklists during this process found in Appendix A. Always state what is required for the manufacture of the product and provide them with all the drawings and bills of materials for an accurate quote. The contract can take many forms but must be considered binding by all parties with real financial implications if not met. This should also be the same for your in-house manufacturing department who may be asked to quote on the program. All quotes must be binding and the manufacturer able to provide the services required by the program's design team, including in-house manufacturing.

To assist in ensuring that the product is of highest quality, the supplier may require an ISO/QS 9000–certified molding source. An ISO/QS 9002–certified supplier provides only the manufacturing expertise and support and this certification is being eliminated by the newest ISO/QS 9001-2000 requirements. An ISO/QS 9001–certified manufacturer offers the total service and support from design to manufacturing and service and problem-solving support.

Areas for selecting a qualified manufacturer are:

1. Manufacturing capabilities to provide the service and support required for the program in
   a. Business practices, sound financially, and dependable
   b. Engineering support, experience, and knowledge
   c. Real-time quality process control, ISO/QS 9000–certified desired
   d. Secondary operations capability for assembly/decoration desirable
   e. Tool maintenance, in-house capability and experience, personnel, and machine shop
   f. Problem-solving capabilities from an experienced staff
   g. Manufacturing machines in the size and capability range to produce the product
2. Other areas to be considered and to be documented along with establishing rapport with key contacts are
   a. Document statistical (real-time) process control (SPC) requirements or quality standards early in the program, material, lot, and part tracking, and real-time documentation
   b. Establish delivery schedules in writing negotiate deadlines. If altered schedules, compensation in time or money specified.
   c. Determine tool maintenance responsibility, who owns the tool, costs related to repair and maintain
   d. Identify potential manufacturing problems using quality assurance techniques as manufacturing control plans, failure mode and effects analysis (FMEA) early in the program and methods that will eliminate future problems as six sigma
   e. Assign responsibility to key individuals and their alternates in each company
   f. Determine who furnishes the tool and raw material.

Unless prior arrangements have been established with a manufacturer, the bidding process begins with a request for quote (RFQ). A good RFQ will specify in detail all requirements of the supplier from part volume, delivery schedule, who provides the tool, material, production, schedule, secondary operation provider, and equipment for assembly, decorating, and packaging. A list of approved part suppliers and the required quality level and part specifications with procedures and documentation to verify the product quality.

With a well-designed RFQ the customer can ensure that all quotes are as equal in detail as possible and a sound business decision can be made to select the best quote. Price alone is not always the deciding factor. Support services, past track record, manufacturing, capability, and response to changes are all important when making the final selection. Obtaining customer references for a new manufacturer is a good method to determine who is the best candidate for your program.

Finally, when all factors, service, and prices are obtained, a solid business decision can be made and submitted to upper management by the design team for the products manufacture, either in house or using an outside manufacturing company.

INJECTION MOLDING

Injection molding is the most widely used method of manufacturing a diverse variety of plastic products. Products can range from less than a gram to over 50 lb per part. There are over 25,000 injection-molding product and part manufacturing facilities in North America and many thousands more worldwide. These injection-molding plants range from a single injection-molding machine to multiple banks of injection-molding machines. A major safety razor company has over 50 machines of varying clamp tonnage producing disposable blade holders in one plant. Many dedicated self-contained molding cells are used to manufacture specific products from a specific grade of plastic material. A wide range of plastic materials are processed daily with mold changes required to meet their customers' mix of products. The majority of injection-molding machines in these plants are in the 150–500-clamp tonnage range with a barrel melt capacity of 15–30 oz of material.

Injection-Molding Machine Basics

Injection-molding machine size (see Fig. 3.1), is classified in ounces of maximum melt capacity of polystyrene resin melt that can be processed (melted) and stored in the barrel during one cycle of operation. Each generic resin capacity per machine will vary according to the plastic materials molecular weight and specific gravity and also if it is modified with fillers or reinforcements. The molding machine's barrel for melt capacity can vary from fractions of an ounce of material in minijet ram machines up to 3520 oz of molten melt in the 7250-clamp tonnage range machines. The molding machine's mold clamping force to hold the mold closed during the injection and mold cavity packout stage ranges from 1.5 to 7250 tons of clamp force. This clamping force is either of hydraulic or toggle clamp exerted on the machines plattens that support the mold. The design team should select, with input from the manufacturing and tool design team members, a molding machine size within the melt capacity and clamp tonnage range for the manufacture of their specific product.

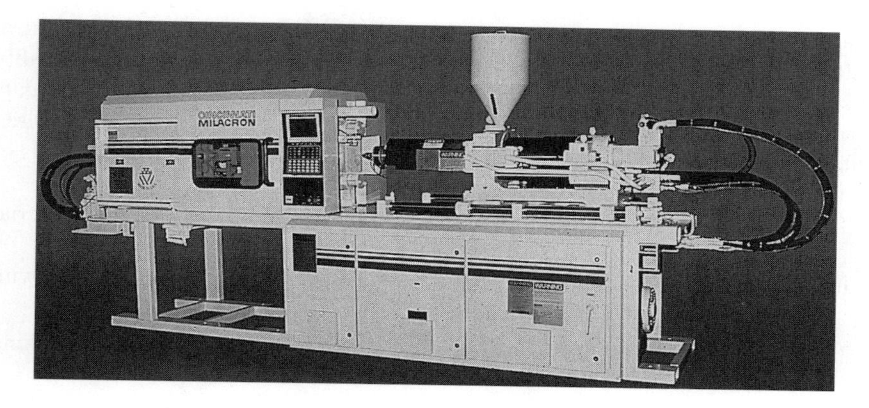

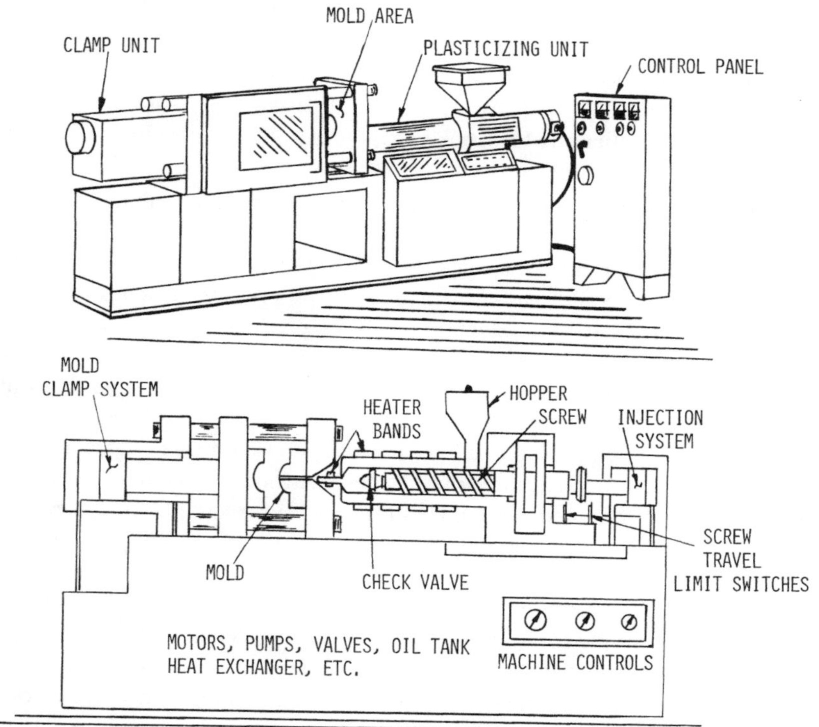

Figure 3.1 Injection-molding machines. (Courtesy of Cincinnati Milacron.)

It is also advantageous to have a continuous process control feedback and control of processing variables for the machine cycle to analyze the molding cycle for the manufacture of the product. The molding machine's control and feedback system is used to maintain control of the molding process variables by analyzing each cycle's process variables. The computer control system is capable of analyzing the last cycle data, and if off specification, adjust control variables for the next cycle. These control systems make minor variable adjustments for the next cycle, which control the quality variability of the material, molding machine fluctuations, and mold variables. The data of adjustments are documented in memory, and if too many adjustments are required within a number of cycles, the operator is alerted to the problem with a diagnosis of the recurring process control problem.

The selection of the injection-molding machine is based on the following criteria:

1. Mold must fit between machines tie bars and not extend beyond platens with sufficient opening stroke to remove the part.
2. Mold clamping capacity to keep the mold closed under injection and packing pressure
3. Melt generating capacity equals 2–3 times product and runner system weight of resin in the barrel to fill the mold.
4. Molding machines screw design to meet the plastic resins' melting and processing parameters
5. Closed-loop/continuous-feedback process control for controlling the quality of the machine's molding parameters.
6. Controlled molding machine and mold cooling system to maintain ±2°F.

The manufacturing equipment must always be in excellent operating condition. There are over 25 injection-molding machine, material, and process variables that must be tightly controlled for quality production of plastic products. It is management's responsibility to ensure that manufacturing has the assets, knowledge, support equipment and trained personnel to always produce a quality product to their customer's quality requirements.

Injection molding requires a reliable and quality conscious material supplier for a source of certifiable resins. The mold must be designed and built with quality materials to produce the number of parts required to the tolerances specified, cycle to cycle. The mold cavity temperature control must be designed to control the cavity temperature to within ±2°F.

The auxiliary support equipment necessary to produce a quality part—including material handling, drying, mold temperature control systems, chillers, temperature control of the injection-molding machine's barrel and hydraulic oil temperature system, mold, sprue and part grinders, part handling and separating, and a process control system, computer-integrated manufacturing (CIM), linking all systems together to act in a closed-loop processing control unit—must be in good working order. The plant's support equipment, clean dry air filter system, and temperature-controlled cooling water systems must also be in operation. These plant systems must be maintained in good operating conditions to support the molding operations.

Material and machine suppliers are constantly working on new materials and manufacturing methods to improve product economics. New methods of manufacture are developed to keep plastics competitive with other materials and methods of construction.

Gas-Assisted Injection Molding

Gas-assisted injection molding (GAIM) is one of the newer methods used for the manufacture of injection-molded plastic products even though it was being developed in the early 1970s. Gas injection can increase part strength while reducing product weight. This is accomplished by producing controlled gas-blown hollow sections during the injection-molded process of a product. GAIM is a relatively low-injection-pressure process versus standard injection molding high pressure. GAIM can reduce parts surface blemishes, sink marks at thick sections, and produce a part with less molded in stress and lower postmold warpage. The reduced injection pressure and gas blowing produces products with lower mold-in-stresses by not overpacking the material in the mold cavity, which can produce molded-in stresses. Gas-assisted injection molding allows the design of thinner sections that require less cooling time, resulting in faster molding cycles. Designers using GAIM principles for their products can create controlled uniform blown hollow sections in the product to increase part rigidity, reducing section thickness and ultimate product cost. The design team must learn which materials are the best candidates for GAIM. GAIM was developed in Europe, is patented, and requires licensing approval to use. It was developed to reduce product weight, better utilize materials, and increase a product's section modulus and strength. GAIM employs the use of very dry air or an inert gas, usually nitrogen, as the blowing medium. The gas is injected in a controlled manner into the center of the melt stream in the parts section using a hollow metal pin extending into the center of the melt stream, while filling the mold cavity. The gas is under a controlled minimal pressure to create an internal gas bubble in the melt stream as the cavity is being filled with plastic. The gas injected into the melt stream at predetermined points is timed after the head of the melt front has passed the air injection pin. Then a pressurized gas is injected into the melt stream for a preset time or predetermined volume of gas that expands in the center of the moving resin flow, forming a bubble. The pressure inside the plastic presses the resin against walls of the mold cavity, where it continues to solidify. In correlation with the gas blowing pressure, the resin's melt temperature and injection pressure and rate of filling the mold cavity, a controlled wall thickness (of a hollow section) is formed within the products section in the mold cavity.

Gas-Assisted Injection-Molding Precautions

The gas is injected through a hollow pin that protrudes into the melt stream of the mold cavity. The hollow pin must be of sufficient strength to resist the injection pressure and resin melt flow so that it does not bend or fracture. It must also be located in the direction of part ejection to permit easy removal of the part from the mold cavity. The gas bubble can be controlled to form hollow sections in specified areas of the product. The bubble formed follows the molten resin as the cavity is filled by maintaining a positive pressure. This results in a fairly uniform wall thickness as the melt stream fills the mold cavity.

The amorphous resins are the most widely used today. Because of their slower solidification or setup time, they permit better control of the part's wall thickness as the gas is injected, creating hollow sections. Crystalline resins have a faster setup time, and control of melt and mold temperature is more critical for obtaining uniform wall and section thickness. Always ask your material supplier for examples of their recommend resins for your design and their processing parameters. Products in each family of resins have been manufactured successfully. Examples of recommended wall sections are shown in Figure 3.2.

Both large and small parts for machine housings to hammer handles have been successfully manufactured using the GAIM method. Multiple injection points are specified for large parts to control the location and size of the hollow sections. GAIM requires precise timing and pressure control for injection of the gas. Obstructions in the mold cavity can be used to divide the melt flow and gas injected in multiple areas to form an internal ribbing system in the product's section.

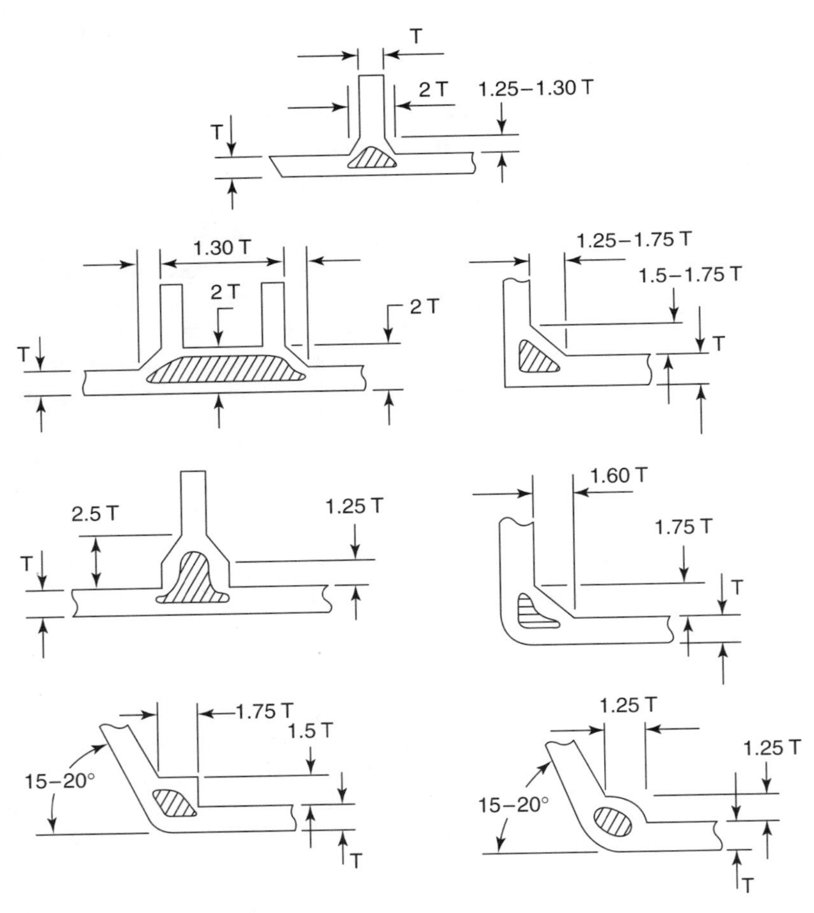

Figure 3.2 Gas injection design guide. (Adapted from Ref. 2.)

The internal gas pressure forcing the melt under pressure against the mold cavity surface during injection can enhance surface finish. Control of material shrinkage and dimensions can also be increased as the material will be denser and on cooling, in a thin section, experience less shrinkage. When part weight is used to monitor mold part quality, it can vary slightly as a result of resin and molding machine variability. This will require sectioning parts to ensure that the wall thickness results are sufficiently thick, and then a part weight range for acceptance can be determined for acceptable part weight. The use of ultrasonic thickness measurement technology can be a tool to measure wall thickness without destructive analysis. Reinforced resins can also be manufactured using GAIM. The fiber orientation is in the primary flow direction and at nonblown sections fiber orientation is more random adding strength, distributed unidirectionally at these thicker sections.

An advantage with gas injection molding is that the reinforcements are more evenly distributed throughout the part. Weld lines should be avoided if possible, as there is no reinforcement crossing the boundary or weld line. This can be reduced by having melt stream divider plates located on one side of the mold cavity set at a lower height to avoid blocking the resin's flow on one surface. This allows the melt to flow around and over the divider to form an internal rib. This will eliminate weld lines and increase product section strength. If this new technique is of interest, contact your resin supplier to find out who has the capability, knowledge, and license to do this type of molding and assist you in designing your part.

Types of Injection Molds

The product design team should also be knowledgeable in mold design for the different types available, two/three-plate, hot-runner and other mold types for manufacture of their products. They can learn this from their mold designers, mold component suppliers, and their own or custom molding product suppliers plus material suppliers who are always on the forefront of new manufacturing technology. Examples of mold types are shown in Figure 3.3.

EXTRUSION

Extrusion (Fig. 3.4) is used to produce a shaped product in a continuous operation. Products can be tubular (tube or pipe), profiles (vinyl siding and window frame parts), or sheeting and film products. The product shape is controlled by the contour of the die used to form the product as material is forced through the die, forming the shape of the product as it exits the die. Extrusion is used to manufacture pipe, tubing, fibers, sheet, blown and extruded films, custom profiles, wire and cable insulated, and jacketed products. With multiple head die coextrusions, multilayered products are made with multiple resins for multiple-ply films and wire and cable and jacketed products in single-pass operations.

Extrusion grades of thermoplastics and thermoelastomers are formulated as high-molecular-weight polymers with matching viscosity. The mix of resin

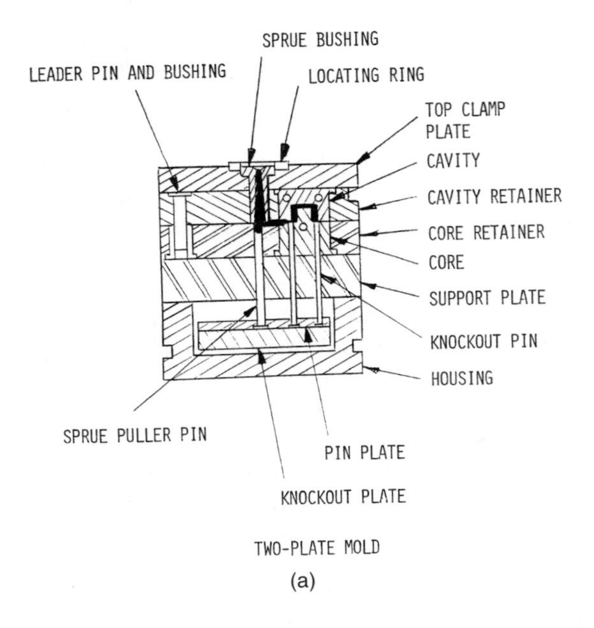

(a)

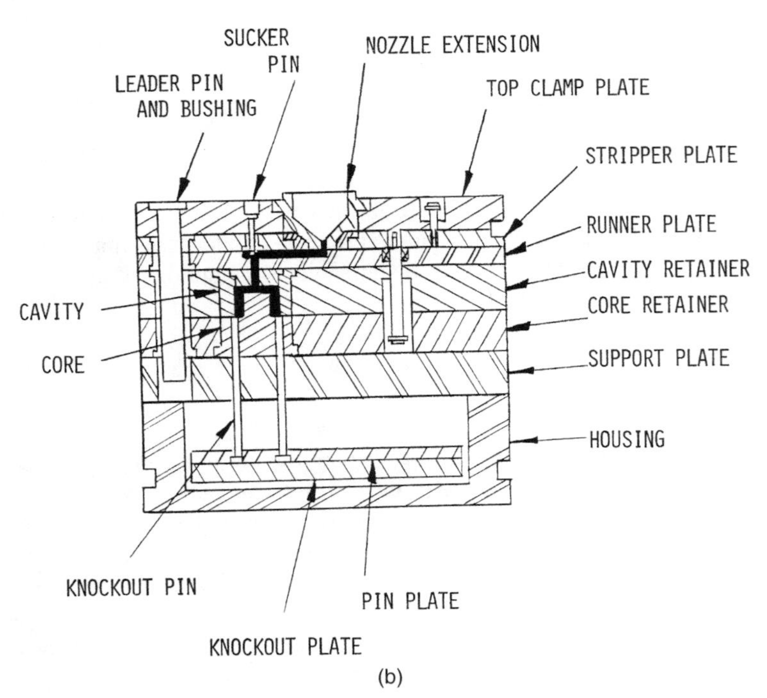

(b)

Figure 3.3 Mold types: (a) two-plate mold; (b) three-plate mold.

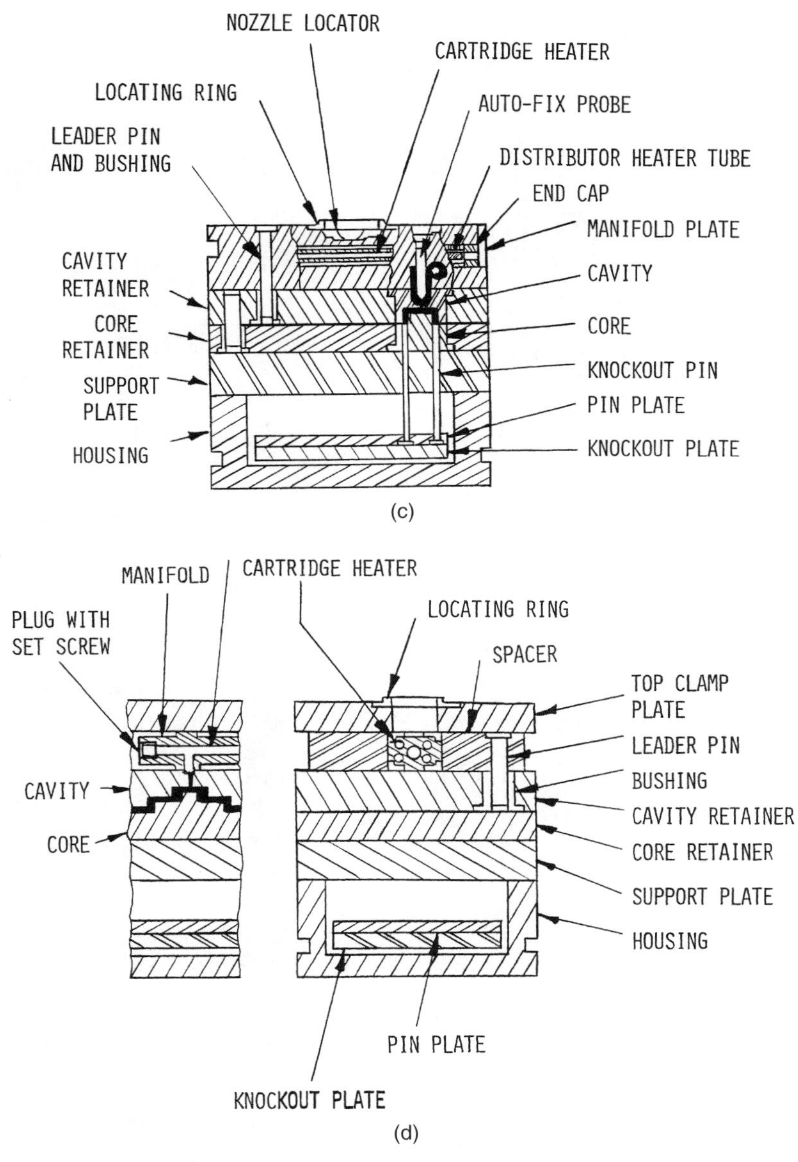

Figure 3.3 (*Continued*) Mold types: (c) hot-runner mold; (d) hot-manifold mold.

grades is very large with polyvinyl chloride (PVC) resin at millions of pounds consumed annually for pipe, conduit, tubing, wire insulation jacketing, sheeting, and house siding. Polyethylene is used for packaging films and ABS sheet resins for thermoformed products as refrigerator and freezer appliance shells.

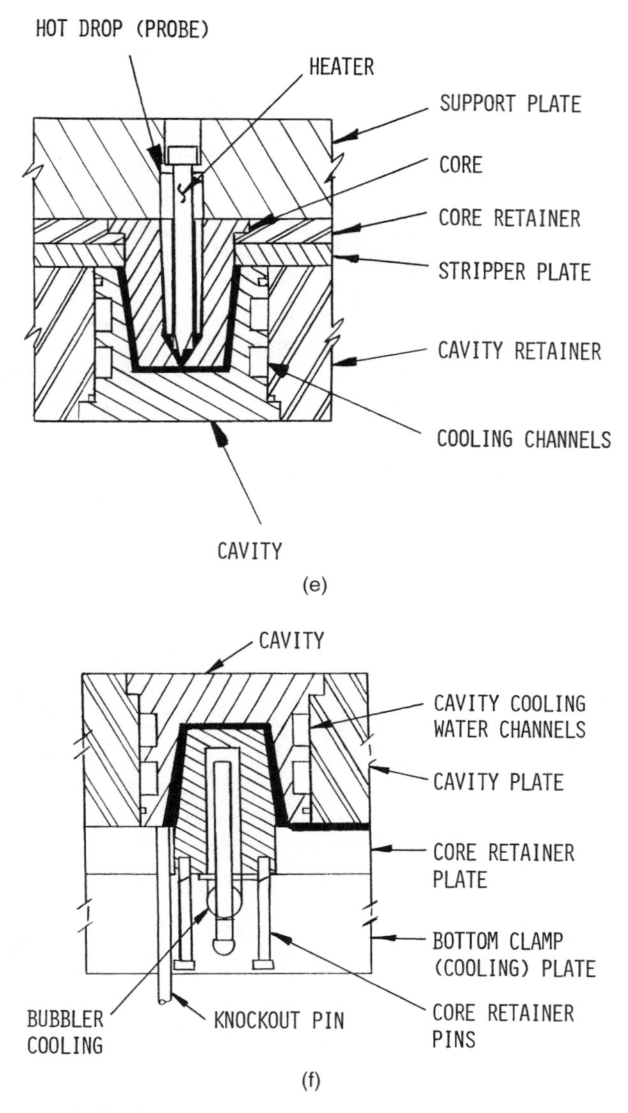

Figure 3.3 (*Continued*) Mold types: (e) inside center-grated modified hot-runner stripper ejection mold; (f) edge-gated conventional ejection mold.

Extrusion is a specialty manufacturing operation with 9 out of 10 extruders custom-built for specific product manufacture. The machines have special screw designs and closed-loop/continuous-feedback process quality controls to maintain consistent machine operations to product hour after hour of quality products. Extrusion is a very economical method of producing a finished product. The design team should explore the possibility of this method of manufacture if the product so lends itself. Tooling

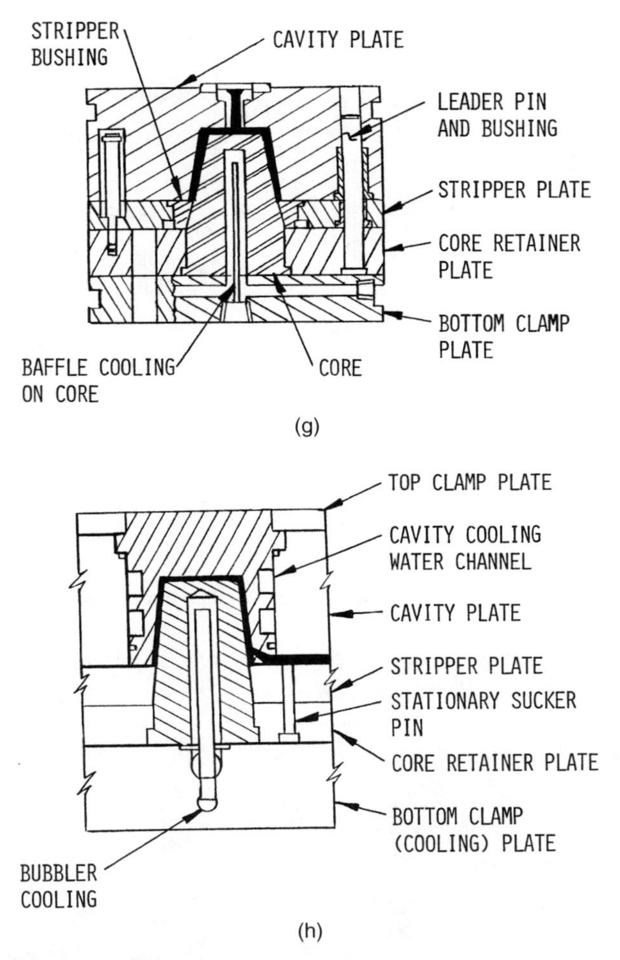

Figure 3.3 (*Continued*) Mold types: (g) stripper mold; (h) submarine-gated stripper ejection mold.

is less expensive than injection molds with very economical long production runs. There are many excellent product design and manufacturing books on this subject in case this is an option.

BLOW MOLDING

Blow molding is a manufacturing method used to rapidly form blown bottles and containers of various shapes and sizes of an ounce to many gallon containers for a multitude of bottled products. It has replaced glass and metal containers in a multitude of applications in the bottling and packaging industries.

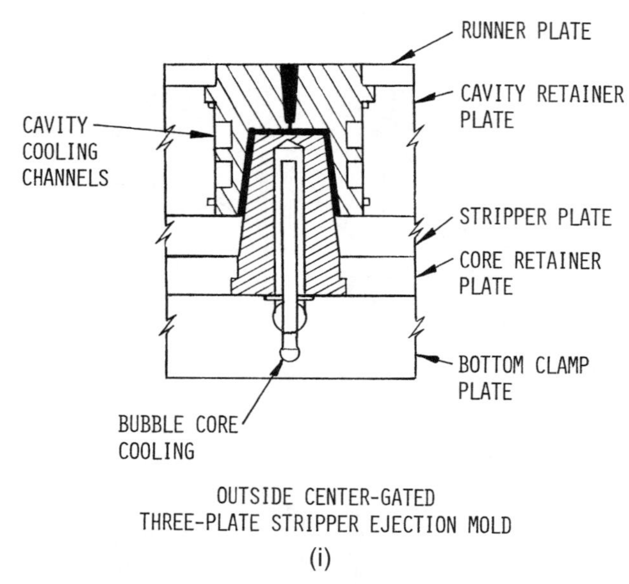

Figure 3.3 (*Continued*) Mold types: (i) outside center-gated three-plate stripper ejection mold.

Figure 3.4 Photograph of assembly line showing the extrusion process. (Courtesy of Welex Incorporated, Ref. 4.)

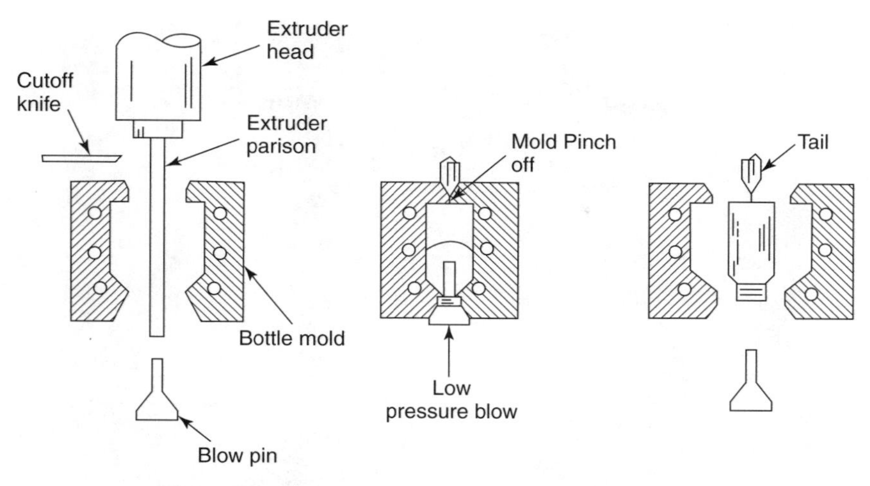

Figure 3.5 Blow-molding process. (Adapted from Ref. 4.)

The developments in resin blends and blow-molding equipment technology since 1991 has greatly increased the applications for blow-molded products. Automotive applications are leading this move for large blown products with developments in multilayered fuel tanks, spoilers, and fascias; the latter require a Class A, mirror finish surface.

The beverage, small-container, and milkbottle consumer product industry are the prime users of blow molding. Blow-molded bottles are used to package all types of products for food, medical and consumer products.

An example of blow molding is shown in Figure 3.5, the blow-molding process.

Conventional Blow Molding

Blow molding is an automatic semicontinuous operation using a multistation mold indexing setup. Plastic bottles and containers of less than an ounce to several gallons are blow-molded in continuous automatic operations. Blow molding involves a tube of hot thermoplastic being extruded continuously downward from the die as shown in Figure 3.6. When the extruded parison is of the correct length, the mold is closed, pinching off the top and bottom ends and a cutoff knife cuts the parison at the die face. The mold is then indexed to the blowing station. Air or an inert gas is injected into the hot parison through the blow pin and then expands, replicating the mold shape. The mold is then indexed to a cooling station, where it solidifies and is then ejected. The "tail" formed on the bottom of the part at pinchoff is then trimmed automatically (or by an operator for large parts). Using a multiple mold arrangement, the molds are indexed continuously in a Ferris wheel or carousel arrangement. Mold indexing systems can produce thousands of parts very rapidly.

Reinforced glass fiber resins can also be blown using this technique to produce high-strength/temperature products for use in high-temperature and extreme

Figure 3.6 Blow molding. Hard–soft–hard PP sequential parison just before mold closing. (Courtesy of D.M.E. Co., Ref. 3.)

environment applications. Products for automotive radiator overflow and coolant reservoirs use this manufacturing method. Extruders with process control can be programmed to extrude a controlled variable-wall-thickness parison using the controlled movement of the torpedo centered in the die. The torpedo, a tapered moving die pin, controls the formed parrison wall thickness during the initial extrusion process. The wall thickness of the parison is initially thicker than the container's final wall thickness as the parison is expanded by air or an inert gas by two to four diameters during the blowing operation. Wall thickness can be controlled in almost all sections of the container. This is based on the container size and shape with the sidewalls typically a uniform thickness in most applications. Where the container must expand and stretch more, the extruded parison is programmed to be extruded thicker. The programmable parison extrusion program for thickness utilizes just the correct amount of material for the container. Parison thickness control is achieved by close control of the resin's melt properties and the extruder's screw acting as the speed control pushing the melt through the die face forming the parison or the torpedo's movement back and forth in the die body controlling wall thickness. The resin's melt strength and viscosity must be carefully monitored and controlled to keep them within process specifications.

Blow-molded parts are made with undercuts and are easily removed from the mold, eliminating the need for expensive in-mold cams and slides. Blow molds are also less expensive since they are not subjected to the high injection and packing pressures of injection moldings. Therefore, cast aluminum, zinc, and steel alloys plus beryllium copper molds are used, as they are subjected only to a few pounds per square inch (psi) of air pressure required to expand the hot resin to the mold cavity.

The most frequently used materials are the amorphous and semicrystalline resins as PVC, PET, PE, and PP. For applications requiring higher end-use heat deflection temperature, as under the hood automotive use, temperatures approach $\geq$350°F, specially formulated engineering resins such as acetal, nylons, and other crystalline materials have been developed. This is for engine air-intake manifolds, ductwork, and pressure and vacuum vessels under the hood.

Extrusion Blow Molding—Multishaped

Complex corrugated and flexible tubing and ducts with expansion joints formed using corrugated sections are extrusion blow molded in a continuous process.

The process begins with a horizontally extruded tube exiting the forming die with a positive air pressure maintained inside the tube. The tube then enters the forming mold with the pressure expanding the tube out against the mold walls, forming a predetermined wall section thickness, against a continuously moving forming mold in the shape of the product. This type of mold is called a "caterpillar continuous tracked mold." It is made in two halves that meet at the horizontal centerline and is centered on the hot plastic extruded tube. The caterpillar mold moves at a slightly faster rate than the extruded tube to pull and stretch the parison from the sizing die. The product is then wound on reels or cut to various lengths by an automatic cutting operation as the formed product is continuously extruded.

Injection Blow Molding

Injection blow molding (IBM) has grown producing products for the replacement of metal cans and other similarly shaped containers. IBM is a very rapid automated method of manufacturing to produce small size containers, of 8–12-oz sizes.

Injection blow-molding manufacture involves an injection-molding machine instead of an extruder with a three-station autoindexing mold table. The first stage of the operation involves filling the mold cavity under low pressure around a mandrel, forming a preform. The mold table then indexes and at stage 2 in the operation, the parison is blown. The table indexes again and the part cools and is ejected from the mold. These methods eliminate all postmold finishing operations plus pinchoff and weld lines, trapped gas, and mold erosion. This operation reduces cycle time by 10–15%.

IBM is capable of forming many complex parts with formed sealing surfaces. Very complex finished parts are blown, not requiring a secondary trimming operation as with conventional blow molding. Parts are made with self-contained handles, pour spouts, sealing surfaces, threaded seals, and other components.

ROTATIONAL MOLDING

Rotational molding produces hollow, seamless products of very complex shapes from single- to multiwalled, colored products. Product range from small hollow toys to industrial gasmasks, through 22,000-gal containers, totes bin playground equipment, military vehicle fuel tanks, and boat hulls. Rotation molding is on the increase as a result of the development of new resin. It has applications for commercial, industrial, and structural product applications and has proliferated in children's toy markets, large industrial containers, military fuel tanks, and outdoor sports equipment such as kayaks. Rotational molding requires a lengthy manufacturing time, from 15 min to an hour or more, to produce a hollow product. Products of all sizes, complexities and geometries can be made in multiplane opening molds to reproduce the most complex shapes. Rotational molding is also known as *rotomolding, slush molding,* and less frequently, *casting*. Rotomolding manufacture of products is increasing at 15–20% annually.

Products are made from resin ground to a 12–35 or smaller mesh size and from liquid plastisols. Amorphous, semicrystalline and engineered resins are the base resins used, including vinyls (PVC compounds), crosslinkable polyolfins, and nylons depending on the requirements of the end-use part.

Molds are inexpensive and very cost-effectively made from cast or welded aluminum for small to medium-size products. Aluminum molds have good heat transfer characteristics and for large parts, sheetmetal molds are used. Fine detail reproducible products are possible using electrically and vapor formed molds.

Rotomolding requires the resins or plastisols be placed inside the mold and the mold closed and sealed. The amount of material used is determined by calculating the desired wall thickness over the surface area of the mold for the part. The mold is then placed into a heated oven while the mold is slowly rotated on a multiaxis spindle. This causes the material in the mold to soften, flow, and fuse to the heated inside surface of the mold. During heating a low internal pressure is maintained in the mold to keep the air inside from oxidizing the resin, which can cause color problems and lower the quality of the material's physical properties. Nitrogen gas or dry ice is used to maintain this inert atmosphere inside the mold. The product density and wall thickness are controlled by the speed of rotation of the mold, heating time, amount of resin used, and temperature of the oven.

After the product is formed, the mold is removed from the oven and slowly cooled by a fine cooling water spray while still turning. As the material cools, it shrinks, effecting its release from the mold's surfaces. Typically rotational molded parts have lower tolerance requirements than other manufactured plastic products. The molds used for rotomolding require a mold release agent applied prior to heating for easy release of the product. A fluoropolymer coating on the mold surface can be used to eliminate the mold's product release surface preparation.

Rotational molding involves a four-stage manufacturing operation consisting of the following stations as shown in Figure 3.7:

1. *Loading Station.* The mold surface is sprayed with mold release, and the mold is filled with a predetermined amount of fine-mesh resin or liquid plastisols.

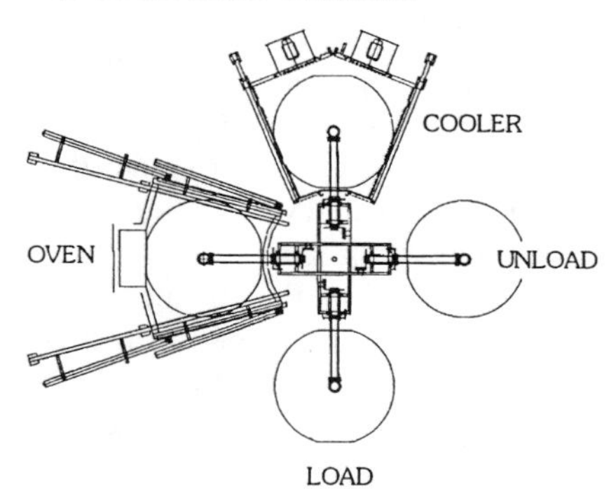

Figure 3.7 Four-arm rotational molding operation with double-load stations. (Copyright © 1999 by Association of Rotational Molders, Ref. 5.)

2. *Fusion or Heating Station.* The mold, mounted on an articulating arm, is indexed or rotated into a temperature-controlled oven and rotated biaxially at a predetermined rate. The resin melts, flows, and fuses on the hot mold's inner surfaces while a positive inert air pressure is maintained inside the mold cavity.
3. *Cooling Station.* After all the resin has fused, the mold is indexed to a controlled air and/or water spray cooling station. While rotating, the mold is slowly cooled down to avoid thermal shocking the product.
4. *Unloading Station.* Once cooled the mold is indexed to the unloading station, the product is removed, and the cycle repeats. The loading and unloading stations are often the same, as sufficient time is available because oven time controls the overall cycle.

The most common rotational molding machine used is the "carousel" style with three to six operating arms for mounting molds and providing rotation.

Another option is the latest-model "clamshell," which is a self-contained unit with a built-in oven and cooling station. It has only one articulating arm, and all operations are at the same location. A third is the double shuttle with a common oven. The mold frames are mounted on a biaxial rotating arm bed that shuttles alternately from opposite ends into the oven. After fusion the mold moves on the track to a cooling station, and after the product is cooled, demolding occurs and the mold is recharged.

A microprocessor controls the manufacturing process for heating and cooling time and rotation speed. Mold costs are low, based on part size, finish, and complexity. Multicavity and different-size molds can be used on the same rotational arm. Scrap is essentially eliminated, and secondary finishing is minimal. Rotational molding is adaptable for using metal inserts for mounting the completed product after manufacture. Very smooth contours are possible plus stress-free parts with a minimum of cross-sectional deformation and warpage.

THERMOFORMING

Thermoforming transforms a two-dimensional sheet of plastic into a three-dimensional product. It is a highly automated form of manufacturing for complex large and small products. Products range from highly automated multicavity blister packaging to the manufacture of single-piece refrigerator/freezer shells and door liners. The tooling is inexpensive and manufacturing costs are low compared to those for injection molding. Many thermoformers now extrude their own thermoplastic stock sheet, which allows them to control the thickness quality of the sheet and utilize an in-line manufacturing operation. Thermoforming is competing for some small injection-molded products and some traditional blow-molded containers.

Markets now using thermoforming are the appliance, automotive, electronic, industrial, lighting, and medical fields. With twin-sheet thermoforming equipment, the large blow-molded and rotomolded techniques are looking more attractive for parts with reduced cost and higher manufacturing rates.

The engineering thermoplastic resins and alloys are also replacing fiber-reinforced thermoset resins in the recreational vehicle markets. If a thermoplastic can be extruded into sheet form, it is a viable candidate for thermoforming.

High-pressure thermoforming is now competing for business and computer machine housings. Lead-time and tooling costs can be about 50% less than injection molding. Thermoforming can be done in many ways, including airslip, drape, free, matched-die, plug-assist, plug and ring, straight vacuum, and vacuum snapback forming with examples shown in Figure 3.8. Thermoforming consists of a preformed sheet of thermoplastic heated to soften it and then using air pressure, mechanical drawing, or vacuum, or combinations thereof, with the sheet forming the product by being contoured over a core block or in a mold cavity. Thermoforming (using a multiple indexing station setup similar to that in rotational molding) involves the following operations.

Thermoforming begins with the sheet clamped into a holding frame. The sheet is indexed or depending on the type of thermoforming setup, and a heater is positioned around (normally above and below) the sheet, to uniformly heat it. Convection, conduction, or radiant heating and newer methods can accomplish this.

A computer using oven infrared sensors on setpoint temperature control controls the heating of the sheet. This guarantees a uniform sheet temperature for optimum part quality, cycle to cycle. Controlled heating of specific sheet sections is also done when required for deeper draws of specific sheet sections. Computer control of the forming equipment also guarantees precise machine movements and timing for exact and repeatable operations. When the sheet has reached the required temperature and time of heating, the heaters are withdrawn and the softened sheet indexed to the forming station. To ensure the required stretching, the entire or specific sections of the sheet are uniformly heated by gradually and evenly raising the temperature of the sheet. Often before the product is formed, the heated sheet is prestretched to the desired wall thickness. This is done by heated air blowing a bubble in the sheet or the mold core moving into the softened sheet preforming, or locally stretching the sheet. The sheet is then clamped in place over the mold and the sheet formed in

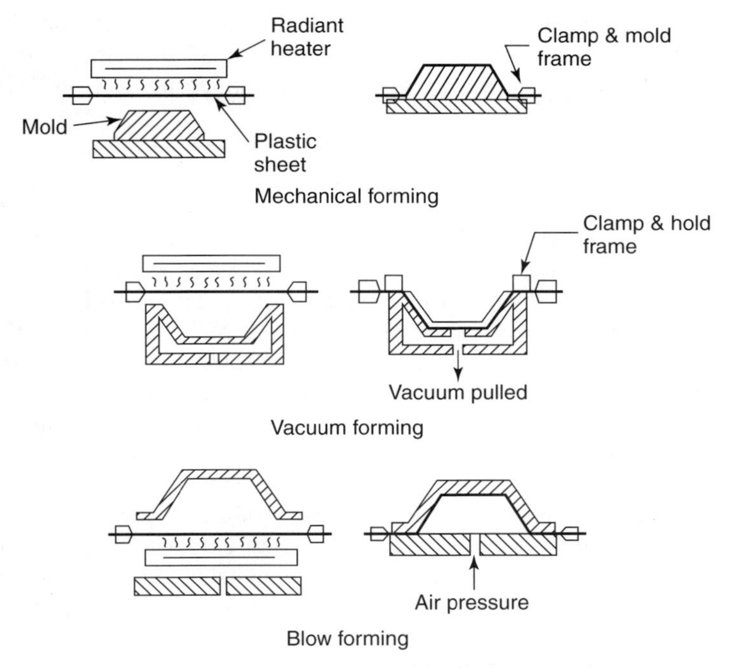

Figure 3.8 Typical thermoforming systems.

the tool. A vacuum is then pulled in the mold, and the sheet conforms and replicates the cavity contour. After a short cooling period the mold and formed sheet are indexed to a cooling station. This usually requires only a few minutes depending on the sheet material and temperature of the mold's surface. The product is then removed from the mold frame and trimmed to final size by hand or by a stamping press. Thermoforming can be used for prototyping some products. The cost is low, tooling is simple, and lead time is short.

Thermoform molds are made from any material that can support the forming forces and temperatures. Molds can be made from plastic and wood for small runs, to cast aluminum and metal molds for long production runs that provide better thermal conductivity and temperature control. Molds also have cooling lines to assist in cooling and cycle control.

COMPRESSION MOLDING

Thermoset products are manufactured using both compression and injection molding. The method of manufacture depends on the thermoset compound. Thermosets are usually classified as wet materials. They are unsaturated polyesters with reinforcing mediums such as glass fibers and classified as bulk molding compounds (BMCs), sheet-molding compounds (SMCs), and (TMC) thick molding compounds that allow fast processing into finished products. Thermoset materials used are phenolic, urea, melamine, DAP, epoxy, or polyesters in preblended composites as BMC, SMC, and TMC.

BMC uses different resins filled and mixed with wood fibers, flour, minerals, and cellulose. The matrix resins are basically polyesters with resol phenolic, hybrid resin systems and epoxies. BMC can replace various metal products when high strength and low weight are required as automotive underbody panels. Compression molding is the oldest manufacturing operation for plastics and process advances have kept it as a major method to rapidly produce plastic products and it is growing.

Compression molding involves placing a preheated BMC (preform) or SMC (sheet) into a heated, matched die mold. Under high pressure (5000 psi), a finished product is produced. Some of the newer SMC compounds require only 150–200 psi to form the product. The entire molding cycle varies from under a minute up to 20 min or more depending on the product size, material, and cure time. Finished parts usually require deflashing, with small parts, tumbled with nonabrasive materials such as glass beads to break and smooth the brittle flash from the edges of the product.

Products can be produced nearly stress-free with elimination of gate and flow marks. Products requiring a Class A automotive paintable finish are possible with excellent reproducibility cycle to cycle. Compression molding is ideally suited for close tolerance and short production run of parts. Compression molding is widely used in automotive applications for long production runs of show surface and under the hood body panels.

Bulk Molding Compound Molding

Compression molding using BMC begins with a prepared mixed and blended charge. The charge is composed of thermoset resins, fillers, and/or reinforcements premixed before molding. The prepared charge is placed in a heated mold (300–400°F). The mold is closed, compressing the charge at about 500 psi, causing the material to flow in the mold and form the product. Products made from BMC include trays, equipment housings, and high-performance electrical components. BMC can also be injection-molded with the blended charge loaded into a ram machine. The charge is then injected into a mold and the fart formed similar to injection-molded parts.

Sheet Molding Compound Molding

The SMC method is similar, using preimpregnated resin fillers, catalysts, and reinforcements in sheet or preform charges that cure and harden in heated molds of <1000 psi up to 2000 psi molding pressure. Parts made from SMC are automotive body panels, bathtubs, outdoor electrical parts, and septic tanks.

New molding and technology such as drawing a vacuum on the oven during preheating of the preform or sheet and mold has almost eliminated defects in products caused by trapped air, residual catalyst contaminants, and moisture remaining in the charge after preheating. This has increased product quality and advances in materials, equipment, and automation (robotics) handling charges and finished products. Tight manufacturing process control continues to keep compression molding cost-effective and efficient.

These are the major methods used to manufacture plastic products. The essential decision that the design team must make during the design of a product is to determine which manufacturing method is best for their product. Factors to be considered are product complexity, material, reproducibility, quality, volume, and cost. Each manufacturing method has their own benefits, requirements, and economic limitations. The final decision on which method to use may be obvious or could involve two or more manufacturing methods to complete the whole product. Once the product factors are known and all inputs analyzed with the design team using the design and manufacture checklists, the final decision for the method of manufacture will become obvious. Contacting equipment and material suppliers may help answer these questions to your satisfaction.

REFERENCES

1. Courtesy of Cincinnati Milacron (slide of injection-molding machine).
2. E. Galli, "Design Tips for Gas-Assisted Injection Molding," *Plastic Design Forum,* 35–42 (July/Aug. 1990).
3. Courtesy of D-M-E Co.
4. Courtesy of Welex Incorporated, Blue Bell, PA.
5. Courtesy of Association of Rotational Molders.

CHAPTER 4

VERSATILITY OF DESIGN AND ASSEMBLY WITH PLASTICS

Plastics offer the industrial product designer many design latitudes in the use of a material to express and incorporate their ideas into a single product or an assembly. Multiple part functions can be incorporated into a single plastic product. This was virtually impossible prior to the introduction of plastics when using metals and other available materials. This problem of design was overcome by using plastics for the simplest of designs. An example is a segmented gear where the designer incorporated multifunctions into a single part. The design also included natural material properties of the base resin. The design consisted of a molded-in bearing surface, natural lubricity, shock absorption resulting from an abrupt or planned motion stoppage, color coding, and a cam to operate other components in the system as shown in Figure 4.1. Plastic products also incorporate ease of product assembly and repair with reduced cost using the material's multipurpose properties in the design such as molded-in bosses and threads, snap and press fits, molded-in hinges, and other assembly methods, including fusion and sonic bonding. The versatility offered using a plastic material during the product's design phase must be considered for economic reasons and product enhancement and combining of functions into a single part. Plastics use the mechanical, thermal, and sonic assembly techniques as well as adhesives and solvent methods. The material's properties and the designer's ability to incorporate these design ideas into new and innovative products for today's markets are virtually endless when designing with plastics. Plastic products can also provide chemical, electrical, environmental, and safety protection by selecting a plastic resin grade compatible with and/or resistant to its operating environment. Plastics are natural insulators to heat and cold, plus electricity. Plastics can be molded in colors and the parts finished for a show Class A surface or textured to add beauty and appearance appeal. Information can be molded on the part's surface or later added through the use of inks, paints, transfer decals, and foils or can be painted to match adjacent metal or plastic parts.

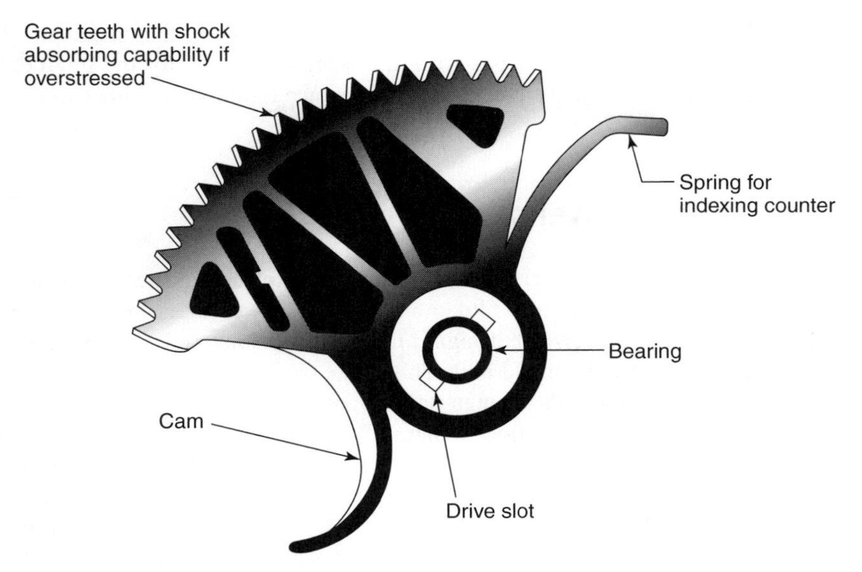

Figure 4.1 Multiple part function in one molded part.

Some plastics can be metalized or plated as headlamp reflectors to replicate metal parts and to provide EMI/EMP shielding for sensitive internal electronic components.

Specific grades of plastics are approved for food contact, potable water, and medical applications. These resins are manufactured with special care and quality to requirements for specific applications meeting agency specifications, (FDA, NSF, etc.) for products requiring their certification for use. The testing and certification of these special resins are controlled by the resin supplier and agency specifications where they are used. When these resins are required for a product the designer must specify this requirement in their design and material specifications. This will ensure that purchasing buy only the approved material and request the required certification information from their raw-material supplier. Once the material is received, it is the product manufacturer's responsibility to ensure that this material is identified, stored, and properly handled so that the material and product will maintain the specified certification.

PRODUCT DESIGN ANALYSIS

The analysis of a product begins with defining the product's end-use design requirements based on the product development and design checklists. Questions to be answered are on these checklists 2 and 7 in Appendix A. The product development checklist does not differentiate between a pencil holder and a complicated housing for a power tool where part mounting, assembly, electrical insulating properties, and support strength, dimensional stability, plus toughness are required. Questions also address product, appearance, and aesthetic value, customer appeal, and information for the product use when molded into the product and visible to the customer. The selection of the material is developed from the design checklist for the product to meet the end-use customer requirements and acceptance for the marketing success of the product.

METAL PRODUCT CONVERSION TO PLASTIC

Plastics are daily replacing metal products in the marketplace. Plastics are typically less costly, weigh less and in most situations perform as well and often better than metals. The most common error a designer makes when converting a metal part to plastic is to copy exactly the metal part. This is the easiest for the designer but the most ineffective way to use a plastic material. Before beginning to convert an existing metal part to plastic, ask the question "What if?" Do not copy the metal product, but use the plastic resin to add saving, safety, long life and physical strength, mounting and support with fewer parts, and aesthetic beauty to the new redesigned product. An example of how this is accomplished was the redesign using a nylon resin for a motorcycle metal fuel tank assembly. The tank was redesigned to add new features the metal fuel tank was lacking. The design was to incorporate impact resistance, color, and shape to fit the existing space with additional fuel capacity and reduced cost. The planned design was to mold the fuel tank in two halves with mounting brackets and fittings molded into each half. Assembly of the two tank halves was accomplished using vibration and angular welding. Testing produced a weld joint that passed all pressure, burst, and crash survival tests of the assembled tank. The molded-in tank mounting brackets and fittings reduced assembly time, and the tank was produced in the company's color. This product satisfied all the "What if?"s for the company product requirements and exceeded the product's prior safety requirements.

Even more severe applications were being explored and found acceptable for reinforced thermoplastics as glass-reinforced nylon and other engineering materials. Reinforced engineering plastics are now being used for small gasoline engine motor crankcases. Engine crankcases have always been die-cast metal until someone asked the question "What if?" The plastic part can withstand the temperature and stress created by the engine and is insulated from the engines block's heat transfer to the plastic by gaskets that allow it to perform as the metal counterpart.

Almost every industry has converted some or their entire product to plastic that was originally a metal or other material first. Only in very rare or extreme cases where high heat, severe wear, exact precision, and continuous high loading are required will a plastic not yet fit. It has been proved in the plastic industry that as new resin families are developed, it is only a matter of time before these plastic resins will replace metals in even these severe applications.

Consider the "What if?"s, as plastic is now the leader for automotive components. Consider the radiator system and brake fluid reservoir, one subjected to high water temperature and pressure for cooling the engine and the other newer system, redesigned to avoid the high pressure that the older brake system experienced. The redesign of the car engine and brake system has lead to downsizing and less weight to the car, allowing these redesigns in plastic to reduce the cost of the cooling and brake system with no loss in safety and cooling efficiency.

The "What if?"s were answered in these two specific cases, leading to the correct redesign of these parts with weight and cost savings previously believed impossible. A plastic product designed correctly will add many advantages to existing metal redesigned components and systems.

PROCESSING TO MAINTAIN PHYSICAL PROPERTIES

Physical property retention for each plastic resin's properties must be maintained during processing. During manufacture each resin loses a very small percentage of its physical properties as a result of processing temperatures that slightly degrade the material while it is being processed in the manufacturing equipment. The design team should consider this possible variance when designing the product. It is also the manufacturer's responsibility to control the plastic materials processing variables to avoid reducing the material's physical properties by overheating the material. Plastic resins rely on their thermal properties for processing to produce the product. This means that they must be converted from a solid pellet to a viscous hot liquid, which is then forced, pushed, or injected into a die or mold under high pressure. The plastic then solidifies by a loss or transfer of heat through the tool steel, causing the material to shrink. When heat transfer from the resin through the tools walls cools the part, it is ejected or exits from the tool. The manufacturing cycle is controlled by the plastic resin's setup time or solidification rate and is a major factor in the product's cost. This includes the resin's raw-material cost per pound used to estimate the product's manufacturing piece part price.

Engineering resins, crystalline or semicrystalline structured materials, have higher physical properties and faster setup time because of their crystalline structure. They are also more expensive and have a higher specific gravity, weight per volume value. The amorphous resins have lower physical properties and slightly longer manufacturing cycles with a lower cost per pound and specific gravity. Engineering resins, when compared with amorphous resins in similar applications, will have a thinner section thickness due to their higher physical properties. This results in a faster manufacturing cycles and a possible equivalent or even lower part cost. This can result in lower product weight, faster cycles, and possibly a lower total product cost. The engineering resins are more expensive and weigh more, but their advantages must not be overlooked during the analysis stage of the product design. They may also be able to more readily combine multiproduct functions, thus eliminating other parts in the assembly such as metals, bushings, springs, and screws.

Specific plastic resins provide excellent optical properties for light transference, reflectance, or visible viewing as a lens or reflector. Special grades of acrylic, nylon, PC (polycarbonate), PE (polyethylene), PP (polypropylene), PVC (polyvinyl chloride), PS (polystyrene) and other resins and alloys of various resins can provide strength, optical characteristics, toughness, and beauty.

Table 4.1 Typical fillers, reinforcing fibers, and modifiers

Fillers	Reinforcing Fibers	Modifiers
Glass spheres	Glass fibers	UV stabilizers
Carbon black	Carbon fibers	Plasticizers
Metal powders	Aramid fibers	Lubricants
Silica sand	Jute fibers	Colorants
Wood flour	Nylon fibers	Flame retardants
Ceramic powders	Polyester fibers	Antioxidants
Mica flakes		Antistatics
Molybdenum disulfide		Preservatives
		Processing aids
		Fungicides
		Smoke suppressants
		Foaming agents
		Viscosity modifiers
		Impact modifiers

PLASTIC MATERIAL MODIFIERS

All plastic resins can be modified, to some extent, to change or enhance the base resin to meet more stringent product requirements. The additives listed in Table 4.1 can increase resin strength, toughness, flammability, weathering, and other properties, usually at minimum cost. This is primarily why there is such a proliferation of resin grades in today's markets to meet specific product end-use requirements. Many base resins serve many markets and through modifiers and property enhancers become more valuable and multimarket servers. To list them all would fill a large volume, but the basic materials can be located and described in the *Modern Plastics Encyclopedia,* published annually by *Modern Plastics Magazine.* The best source of information on materials versatility is the suppliers of these resins. Material suppliers have case histories on the use of their resins in many markets. They also have the design, physical property, and processing and test data for their resins.

Most suppliers have available a computer software database on their resins properties and it is available to their customers in computer disk format. The information they furnish the designer is invaluable, as the designer must use the material supplier's resin property data generated under typical end-use product conditions when designing their products and selecting a plastic material. Table 4.2 lists the force/design material property relationship.

Table 4.2 Stress and property identification

Type of Force	Design Property
Bending	Flexural modulus
Compression	Compressive modulus
Tension	Tensile modulus
Torsion	Shear modulus

An excellent beginning reference source for reviewing a material's typical, dry-as-molded (DAM), properties is the yearly edition of McGraw-Hill's *Modern Plastic Encyclopedia.* The property data are presented per The American Society for Testing and Materials (ASTM) test methods, and the designer can initially compare similar material property values of different resins generated under identical conditions, usually 73°F and DAM conditions. (Designers, please note that these are only typical product reference data and not to be used in the design of a product. Use only the "specific property data" for the resin selected for the conditions the product experiences in its end-use application. Your material suppliers can furnish specific resin property data for the product's resin selection with recommendations and lists of products now being manufactured using these resins.)

RECYCLING

A new area to consider is the ability and ease of recycling the used product. Products identified by their recycling codemark with the material of manufacture noted within the mark can be easily segregated for recycling. Many communities are now recycling blow-molded plastic bottles. The material identification used for identifying recyclable plastics can now be incorporated into the product's marking system. This will aid in their disassembly and make recycling easier when implemented.

To conserve our natural resources, recycling is gaining in use. The ability and ease of recycling a used product is now a factor to consider in the design of a product. How easily identified are the recyclable plastics and other components, and can they be easily separated? The material separation techniques can be incorporated into the assembly, making service, repair, and replacement for recycling of parts easier and possible when designed into the product. Plastic products are suitable for recycling when they can be identified and separated from other plastic materials. The designer of plastic products is now more aware of the reuse of materials with the energy savings, costs, and value that can result with recycling plastic materials. Special recycling codes are now used on some materials to assist in identifying specific generic resins for recycling. The symbols shown in Figure 4.2 are used mainly for the commodity, blow-molded, resins for containers such as PE, PP, PET, and PVC, but will soon be on all resins, aiding recycling companies in correctly identifying and segregating materials for recycling.

The automotive industry now uses approximately 150–200 lb of plastics per vehicle. They anticipate the future use to increase to 300 lb per vehicle. They and their suppliers are developing current and new applications to recycle and use old auto parts, bumpers, instrument panels, door panels, and other parts for other remanufactured auto parts from these materials. Thermoplastics will be separated, reground, repelletized, and possibly reprocessed back to their "original basic" elements for reprocessing into pellets for new parts. Other recyclable thermoplastic products will be ground up, reextruded, pelletized, and blended back with virgin resins at varying percentages to obtain properties approaching the virgin material, such as vinyl/foam moldings recycled and fed at 15% into virgin resin yielding 90% virgin material strength. Nylons, acetal, and thermoplastic polyesters can be depolymerized to their basic building blocks, and it is economically viable based on current technology and market demands.

Figure 4.2 Recycling symbol for plastics.

Recycling is being done more today in business machine toner packaging and other internal parts. The units are returned for recycling, and their replacement cost is substantially lower than that of throwaway replacement. For some products that are more difficult to recycle, suppliers have, added during manufacture, biodegradable additives, such as flour, which, when exposed to the elements will degrade more readily in landfills and not contaminate the environment. When recycled resins are used in a product, either reclaimed or reground and fed back into the system, their new physical property values must be reestablished. As long as their values are known and end-use testing is conducted to establish their suitability in an application, their reuse can be profitable.

Using recycled resins as the base material or as a percentage mixture of the products material, as regrind blended into the base virgin resin, can substantially help reduce cost. It is important to know the source of the resin and initial properties to ensure that the new product will have good quality and processing consistency. If the reclaimed resins properties or processing qualities vary too much, they may prove to be more of a problem than a solution. The percent of use must be controlled to ensure no loss of properties or processing capability. The resins used must fit the application, processing, and cost structure to be successfully used in any program.

DESIGN, DEVELOPMENT, AND ASSEMBLY CONSIDERATIONS FOR PRODUCTS

In multicomponent assemblies, as a paper copier, the following guidelines can be used:

1. Repairable parts should not use permanent assembly methods, such as adhesives, or sonic or heat staking.

1a. Use snap/press fits, interlocking designs, screws, or similar for frequently disassembled parts. Use threaded metal inserts in mating parts to avoid screw reassembly problems, mismatched threads, stripping, and other irregularities.
2. Color-code replaceable and repairable items. Color match need not be exact if hidden. Color-code similar materials for assembly and repair and to identify for ease of recycling.
3. Mold in instructions or directions for part removal and repair or use adhesive labels.
4. Standardize purchased fasteners and minimize their use for key areas. For frequently disassembled parts use threaded inserts in mating parts to avoid screw reassembly problems.
5. Use permanent assembly methods, sonic/heat welding, or staking for one-time-use, throwaway parts.
6. Make diagnostic points easily accessible for service or identifying the problem instantly without disassembly of other components.
7. Ensure that mating components are clearly marked in high visible areas and keyed to fit together for easy assembly. Use matched markings on mating parts.
8. The design team must constantly look to reduce product costs for manufacture, warrantee, and service life.

Use the industry recycling identification markings for the resins. The design team must constantly try to reduce part costs both during manufacture, warrantee, and the parts service life.

ANALYSIS OF RESIN PROPERTIES

What analysis or test results are needed to qualify a recycled resin or compound is seldom known by the design team. They need to know what quality control tests will specifically identify and predict performance and process ability of the material. These resins may contain varying amounts of impurities, plus a degree of uncertain composition and uniformity. One technique that can answer their questions is thermal analysis (TA). TA can tell how a material will behave in processing and in part performance.

Thermal analysis monitors a small sample of material for a fundamental characteristic. The sample is cycled through a precise program, and a graph is developed that can be compared to the virgin resin graph to determine specific property or additives. It is not a continuous monitoring process but in some quality control operations can be established as such to monitor material or finished parts, such as analyzing a runner sample or part from the molding cycle. Tests for characterizing resins are tested listed in Table 4.3 and may take less than 10 min to one hour to complete. One limitation besides only specific samples is that it cannot identify small percentages of amorphous thermoplastics, such as PVC, which could affect material processing and performance. Specific tests could be run on a production line if dedicated to one or two specific test parameters. Output of analysis results is shown in Figures 4.3–4.7.

Table 4.3 Material test methods

Information	Process	Data Obtainable
		Composition
Resin, polymer modifiers	DSC	Types present, plus ratio if copolymer, blend, or mixture
	GC	Polymers, oligomers, and residual monomers; amounts present
Additives	DSC	Deduce presence and amount from thermal effects (stabilizers, anatioxidants, blowing agents, plasticizers, etc.)
	GC	Detect any organic additive by molecular weight (MW)
Reinforcements, fillers	TG	Amounts (from weight of ash)
Moisture, volatiles	TG	Amount (from weight loss)
	DSC	Amounts (if sufficient heat absorbed)
Regrind level	DSC	Deduce amount from melting-point shift
	GC	Deduce amount from shift in MW distribution
		Processability
Melting behavior	DSC	Melt point and range (each resin if blend); melt energy required
Flow characteristics	GC	Deduce from balance of high- and low-MW polymer in MW distribution
		Physicomechanical
Glass transition temperature	DSC	Shown by stepup in energy absorption as resin heats (marginal for some semicrystalline resins).
	TM	Detected by sample's expansion.
Crystallinity	DSC	Calculate percentage from heat of fusion during melting; also can find time-temperature conditions for desired percent crystallinity
Tensile, flexibility, impact	GC	Deduce from balance of high- and low-MW polymer in MW distribution
	DSC	Deduce from overall composition

Source: Adapted from Ref. 1.

THERMAL ANALYSIS TECHNIQUES

Several methods are used for thermal analysis:

Differential scanning calorimetry (DSC)—used to measure melting, crystallization, curing, melt profile for semicrystalline, thermoplastics and the amount and effectiveness of antioxidants by heat flow to or from a sample as it is heated or cooled. It measures specific heat and energy of thermal events.

Figure 4.3 1020 Series TGA7 thermal analysis system. (Courtesy of Perkin-Elmer Corp., Norwalk, CT.)

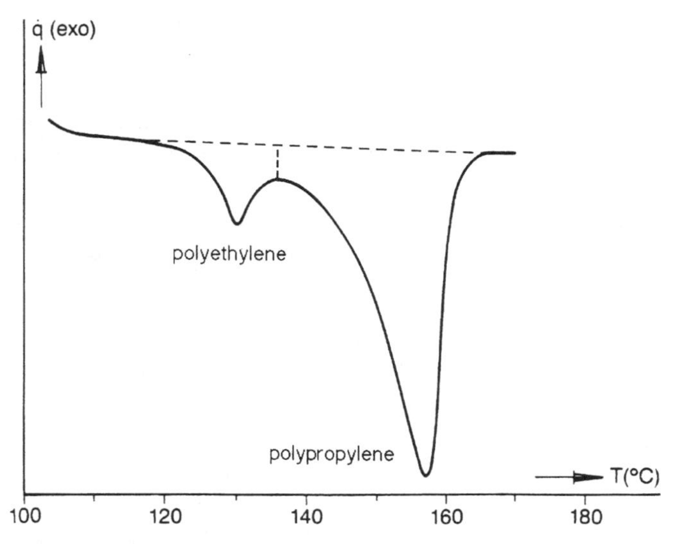

Figure 4.4 Polyethylene/polypropylene copolymer curve. Polyethylene melts at a lower temperature than polypropylene, and consequently the relative peak areas can be used to determine the percentage of polyethylene in the blend. All blends of thermoplastics having different melting points can be analyzed this way. It is a very attractive feature of the DSC technique. (Courtesy of Mettler Instrument Corp., Highstown, NJ.)

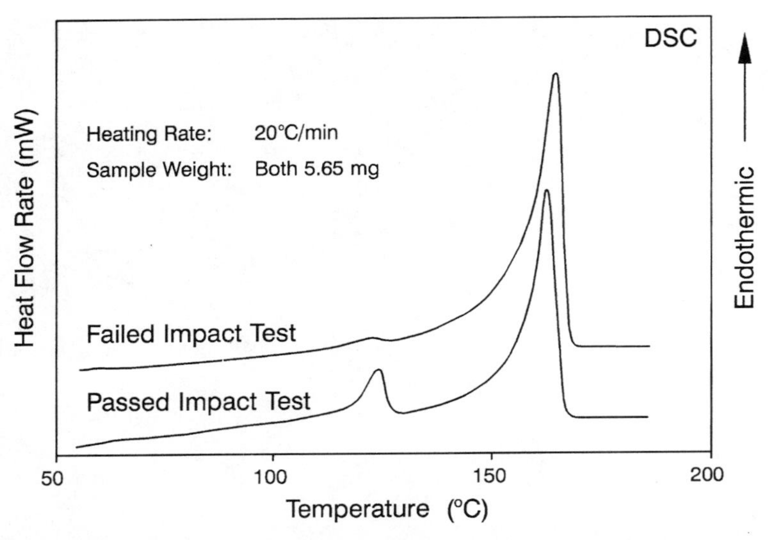

Figure 4.5 DSC analysis of polyethylene/polypropylene copolymer. (Courtesy of Perkin-Elmer Corp., Norwalk, CT.)

Thermogravimetric analysis (TGA)—determines the weight loss of a sample as it volatilizes or decomposes at a specific temperature as it is heated. It separates components in a polymer or elastomers by relative thermal stability and can quantify the resins thermal stability. It quantitatively determines carbon black, filler levels, moisture, oil extenders, plasticizer, and polymer types.

Thermomechanical analysis (TMA)—measures changes in mechanical properties of a sample as it is heated under a constant force or with no applied force. Determines the coefficients of thermal expansion and softening points plus a modified heat deflection test.

Dynamic mechanical analysis (DMA)—determines mechanical property (tensile, flexural, compressive), changes in a sample as it is subjected to a variable force. It can operate in different modes to deform the sample in an oscillatory of force

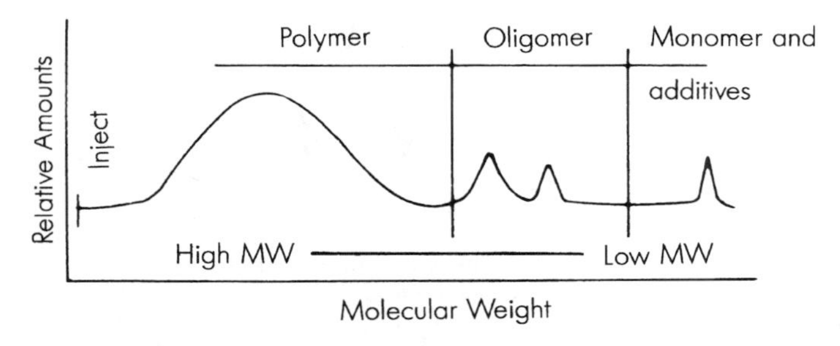

Figure 4.6 An idealized chromatogram shows the elution order of various components. The largest molecules elute first, followed by the oligomers, and finally by the unreacted monomer and additives, (Courtesy of Millipore Corp. All rights reserved.)

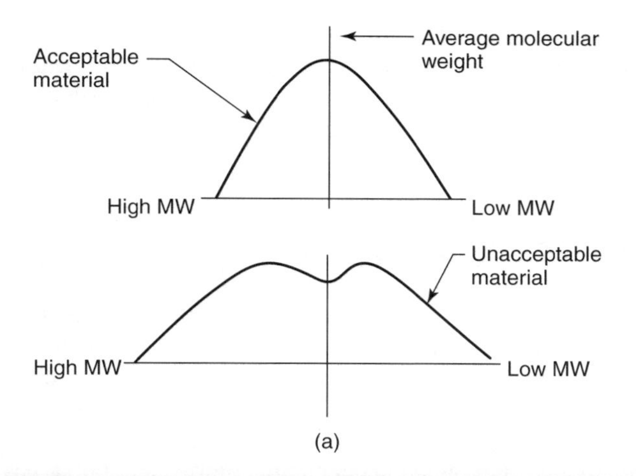

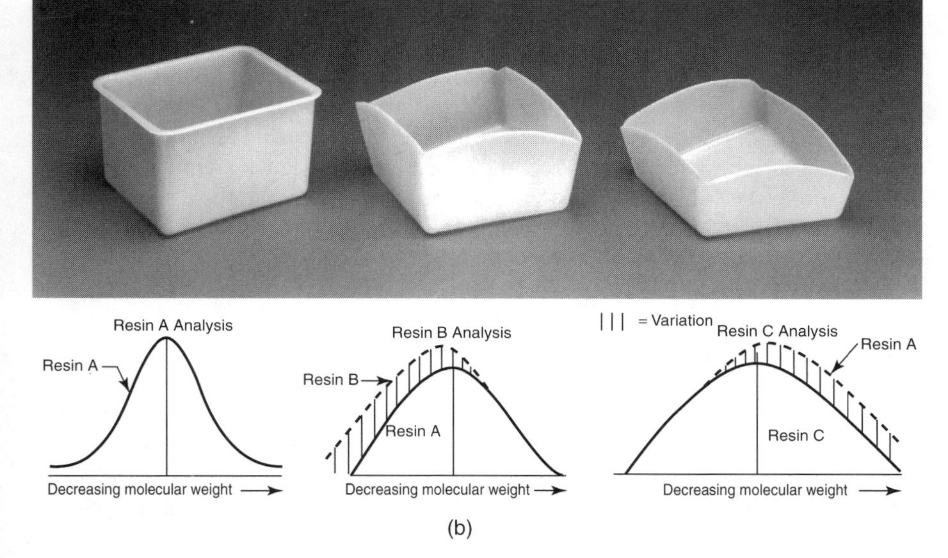

Figure 4.7 The effects of molecular shifts on part quality: (a) molecular weight distribution; (b) GPC comparision of high-density polyethylene resin variances after meeting incoming melt index specifications. Slight shifts can cause part quality problems. (Courtesy of Millipore Corp. All right reserved.)

ramping style while varying a single parameter such as time, temperature, stress, or frequency while holding the other variables constant.

Differential thermal analysis (DTA)—similar to DSC but with higher temperature operation ranges, above 730°C, but with less accuracy.

Thermal analysis is not inexpensive; typical equipment costs $25,000.00. The material supplier should supply the material analysis for their resins if they want to be your supplier. The use of recycled resins is volume-cost-driven. If the savings can be realized and a reliable source obtained, then the use of recycled materials may prove beneficial to your program. Before making this decision, be sure that all questions are answered and volume and material quality is attainable at a price within the program objectives.

MANUFACTURING CONSIDERATIONS

The method of manufacture—injection-molding-specific in this text—should be decided early in the design program. There are many methods of manufacturing plastic parts, including injection molding, extrusion, blow molding, thermoforming, rotomolding, and compression molding and casting, which the designer can select to manufacture their products. The majority of plastic parts lend themselves to injection molding except bottles, hollow products, tubing, profile, and film types of products. Therefore, each product must be evaluated independently as the other manufacturing methods may be the best or only method of producing the product economically. The product's complexity and volume are all considerations, as well as the material candidate selected to perform the product's end-use requirements. Some parts in an assembly may use one or several of these different manufacturing methods to produce the finished assembly. Each manufacturing method has its own strengths and benefits that add to the versatility of using plastic in product design. Also, each material may not be adaptable for all manufacturing methods on the basis of its chemical and physical structure and thermal properties.

MANUFACTURING METHODS

The criteria for selecting a method of manufacture are based on the manufacturing and product development checklists in (checklists 5 and 7 in Appendix A). Part complexity plus cost play a major factor in the selection of a manufacturing method, including material and part volume. Blow molding is the leader in mass volume production of single items such as cola, milk, and home and personal-use containers. Next is extrusion of food packaging and consumer film products followed by injection molding with a more widely diverse mix of plastic resins and types of products.

Products manufactured by injection molding require the highest quality for the design and building of the mold. The mold, often referred to as the "tool," should be the entire design team's responsibility. The mold designer is a member of the design team and will be aware of the critical areas of the mold that must be met. Early in the design program the design can specify tooling requirements to ensure that the best mold design for the cost, product, and manufacturing requirements are attained. The correct mold design must be specified for the product, which includes, the number of cavities, mold steels, gating, draft allowance, type and location, cooling, layout, and control and tolerance control requirements for the product. Failure to consider these items can produce out-of-tolerance product or other more serious manufacturing problems, such as inability to control processing tolerances, and end-use performance problems. Use of the mold design checklist can avoid these problems (checklist 17 in Appendix A).

Also, if the wrong plastic resin was chosen for the product and the mold built for its shrinkage rate and then changed, the mold cavity and other items may have to be reworked because of differences in the resin's molding requirements. This is why prototype end-use product testing in the selected resin is so important before the production mold is finalized and built.

In multicomponent assemblies, where purchased parts are involved, such as a power tool, if bearing, shaft, motor, or other internal parts are changed after the mold is built, it may have to be reworked. This has happened and resulted in weakening and early internal mold core failure. It can also affect other sections in the mold, such as the rerouting of cooling lines, which has resulted in a disruption of the temperature balance in the mold cavity.

Each portion of a product design must be carefully evaluated, finalized, and approved by the design team before the production mold and the products mold cavities are built. The industry rule is that the product can only be as good as the mold, process equipment, and quality standards according to which it is manufactured.

TOLERANCE CONSIDERATIONS FOR MANUFACTURING

How a plastic product is manufactured, specifically, the tooling and manufacturing equipment, will determine the tolerances that the part is capable of attaining and the manufacturing cost and quality factors. Each method of manufacture and each material has its own manufacturing tolerance limit. The tooling, processing equipment, plant support equipment and services, plus the operation personnel all affect the product's manufacturing tolerances. How well the manufacturing process is controlled, from the beginning of the design program impacts on the parts finished quality. The move to world-class ISO/QS 9000 manufacturing methods will assist in ensuring products are maintained within control and specification limits when performed in real-time process control manufacturing.

The design team must also consider the thermal expansion and contraction of the plastic resin both during manufacture and in service. Each plastic resin and product will have its own manufacturing tolerance limits based on the product's end-use requirements. Thermoplastics exhibit higher rates of thermal expansion than do metals and thermoset resins. During manufacture they experience their own specific material shrinkage on solidifying in the tool. This occurs as the plastic goes from a molten liquid, when forced into a tool and when cooled returns to a solid-phase shrinking in volume an amount based on the material's generic molecular makeup. Even after manufacture, additional plastic resin product shrinkage will occur for hours and even days. Also, products exposed to end-use temperatures above their manufacturing tool temperature will experience additional postmold shrinkage.

In a reverse manner hygroscopic materials, such as nylons, will grow dimensionally as they pick up moisture after manufacture. This is a time–temperature–thickness–moisture percentage situation in which only end-use testing can verify the amount of dimensional variation for each specific resin. In the case of nylon, a hygroscopic material, a product will experience shrinkage after manufacture and if exposed to high humidity, will actually grow. In a tight tolerance product, end-use testing is required to determine final product end-use dimensions for the application due to the variable's prior to assembly.

ASSEMBLY CONSIDERATIONS

Plastics are ideally suited to be joined with other materials. They are manufactured to precise tolerances and have some flexibility to adjust in their assembly by bending or yielding, whereas a rigid metal material may not. It is essential during design that mating parts made in different manufacturing methods, materials or tools be referenced or keyed to the same product surface for dimensional reference. Always reference each mating part from the same starting or reference point or product surface.

The design team needs to maximize to the product's advantage the specific properties and capabilities of each plastic resin used in their product's design. Each plastic material, either unreinforced or reinforced, has its own unique assembly features. Low-modulus resins to reinforced engineering materials can be assembled using the following assembly techniques:

1. Press and snap fit
2. Insert screws and molded threaded fittings
3. Sonic, spin, or angular welding
4. Fusion, adhesive, and solvent bonding

Product assemblies can be made repairable or tamperproof. Selecting the correct assembly method can reduce cost, number of pieces, and assembly time. The product assembly can also be automated within process quality checks designed to detect product component problems before final assembly is completed.

Incorporating assembly features in the plastic product can give added value and serviceability to the product. Like and dissimilar plastic materials can be assembled when the designer utilizes the "total design concept" for their assembly. Plastics can provide assembly features right out of the mold that other materials cannot provide when considered during the design phase of a product program.

With the correct analysis and utilization of assembly techniques, dimension tolerance stackup in mating parts can be minimized. The use of slotted holes or elongated bosses in the plastic part using screws can allow for minor variation in mating a plastic part to a metal component. Never relax the dimensional tolerance requirements for the plastic products mold cavity dimensions. Do not force manufacturing to adjust the plastic product to fit mating parts by adjusting the manufacturing cycle, machine, and mold variables to meet the new requirements. A simple mold change, such as relocation of a core pin, may be more economical and practical. Product assembly versatility can be attained by using sonic heading of a molded-in stud and sonically

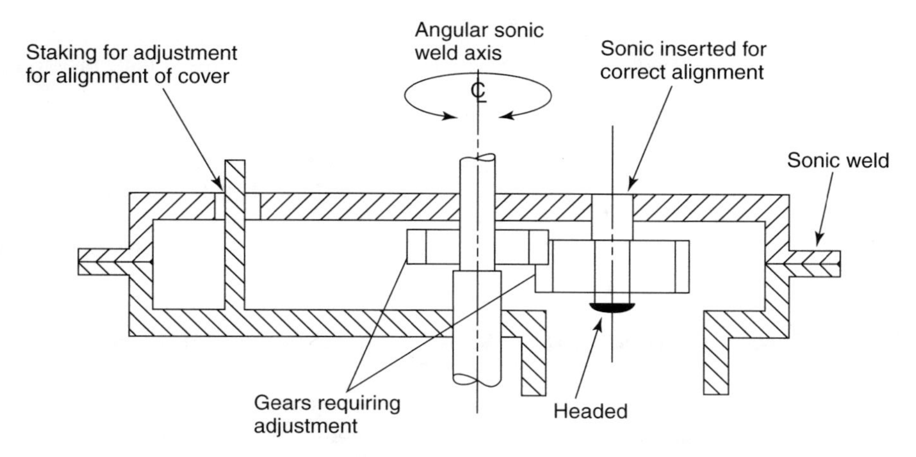

Figure 4.8 Sonic assembly for tolerance adjustments.

inserting a shaft in the support plate to provide adjustment to a mating gear's tolerance requirements as shown in Figure 4.8. The speed of assembling the product is important if a snap or press fit must be considered. Plastics can absorb high amounts of energy during assembly without failing, even with highly loaded or reinforced plastics. The area under their stress–strain curve, the maximum DAM elongation, and limits on the amount of energy absorbed are important factors. This curve, as well as elongation data, function as an indicator when comparing assembly methods and different materials for assembly force absorption, bending, and impact. When the assembly forces are too high and outside the material's stress–strain curve's allowable limits, or forces are applied too rapidly, not allowing the material to absorb the assembly energy, the product may fail by stress cracking during assembly. Therefore, with highly loaded, filled, or reinforced plastic resins, with high assembly stress, the assembly time of the product, which allows more time for the material to deform or bend, may have to be extended. This additional assembly time can allow the plastic to absorb this energy without failure, by deflecting slightly to reduce the stress during assembly. This is a critical factor in snap-fit, press-fit and thread-forming screw assemblies.

Screws

Screws are the most common fasteners used for assembling plastic parts, used for both permanent assemblies and when access is required for repair of the product. Selection of the wrong type of screw often leads to assembly, service, and quality problems. Screws are secured in a boss, molded on the part's mating assembly. Screw type determines how threads are formed and the driving and holding force. Mechanical screw fasteners are either thread-forming or thread-cutting. Screws are economical and join and secure the parts with strong, tight, and problem-free assembly over the life of the part.

Table 4.4 Guidelines for selecting screw types

Resin Modulus of Elasticity	Screw Type	Thread Configuration
≤200,000	Thread forming	AB–B–BP–U
200,000–400,000	Thread forming	Trilobe—high–low
	Thread cutting	L–BF–BT
400,000–1,000,000	Thread cutting	High–low B–F–G–BF–BT
≥1,000,000	Thread cutting	Type T

Source: Adapted from Ref. 2.

Thread-Forming/Thread-Cutting Screws. Thread-forming screws deform the plastic into which they are driven and form threads in the part. Thread-cutting screws physically remove material, such as a machine tap, forming the screw thread when inserted. To select the correct screw, the designer considers the resin's modulus of elasticity, which is a direct function of the material's elongation. Table 4.4 provides guidelines for selecting the correct screw. As noted, only the lower-modulus materials are selected for thread-forming screws, as they impart considerable hoop stress into the boss area.

The different types of screw thread designs are shown in Figure 4.9. Each type and thread provides a unique assembly or holding benefit. The supplier must also consistently maintain the quality of the screws used. The plastic part must not have to absorb variances in screw quality. The selection of thread configuration depends on the material and ultimate part function. The U-type screw is for permanent assembly. Other types—AB, B, and BP—are removable, but thread damage may occur. Thread-cutting screws are not recommended for removal. If on reinsertion the threads are not accurately lined up, they may recut the threads and ruin the joint assembly. When the type T screw is used and removed, the threads will always be destroyed. Reassembly is possible only by using the next-larger screw size. If screw removal is anticipated, the boss diameter must be initially designed larger to handle the next-larger size. If repeated assembly and disassembly are contemplated, always use a molded-in metal threaded insert to permit repeated removal.

The guidelines for self-threading screws, as shown in Figure 4.10, are

1. Thread engagement length should be 2.5 times the screw diameter.
2. Hole diameter should be based on 50–70% thread engagements. This can vary with type of fastener and resin.
3. Holes should be counterbored (preferred) or chamfered to aid alignment during insertion and reduce cracking.
4. Boss diameter should be 2.5 times the screw's diameter for best performance. If disassembly is anticipated, use 3 times the screw's diameter. Allow space in the bottom of the boss for thread-cutting screws to deposit debris.
5. Strip : drive torque should be at least 3 : 1 for hand assembly. With power tool assembly, 5 : 1 is preferred to reduce stripping the threads if torque cutoff settings go out of calibration.

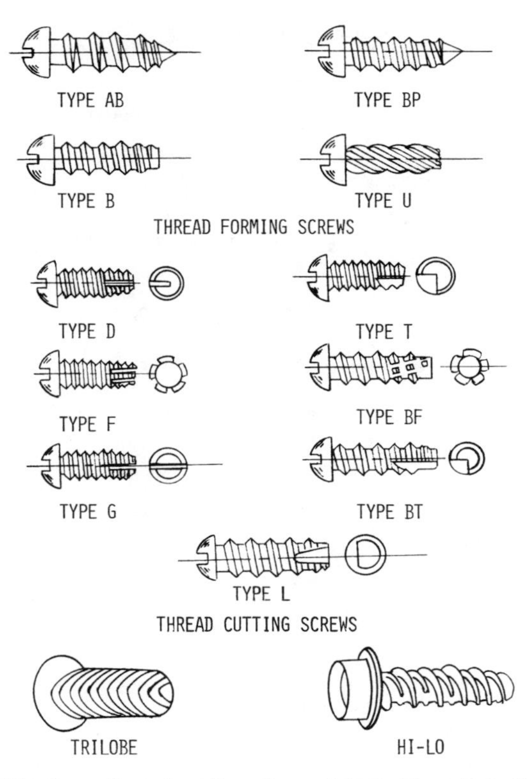

Figure 4.9 Screw thread configurations. (Adapted from Refs. 2 and 3.)

6. Special screws for plastics, such as Trilobe and Hi-Lo Plus, cost more but give improved holding strength and performance.
7. Screw manufacturers and material suppliers can make recommendations for a particular application.

The strip:drive torque calculation is important. During assembly—either by hand or using a power tool—the screw is tightened to the correct torque without stripping.

When a self-tapping screw is tightened, it produces a torque–engagement curve (see Figure 4.11. As the screw penetrates the plastic up to point *A*, the driving torque slowly increases. At point *A*, the head of the screw seats. Any further tightening to point *B* is used to torque the threads into the plastic for a strong attachment. If torquing continues to point *C*, the plastic yields and the threads begin to shear. From points *C* to *D*, the threads strip and the fastener fails. It is important to reach, but not exceed, point *B*. Workers need to be properly instructed not to exceed this torque level and power tools must be calibrated to cut out when the drive-torque setting is reached. This can be calculated and verified by running prototype tests on bosses or flat plaques molded in the plastic selected for the product.

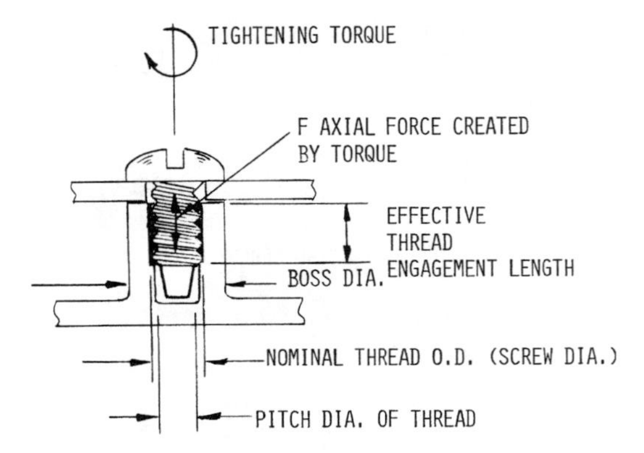

Figure 4.10 Self-threading screw. (Adapted from Ref. 4.)

Since a screw's strip torque is so important, it can be calculated at point *C* by using the following equations:

$$T = \frac{Frp + 2\text{fr}}{2r + fp}$$

where T = torque needed to develop pull out force
F = pullout force
r = pitch radius of screw
p = reciprocal of threads per unit length
f = coefficient of friction

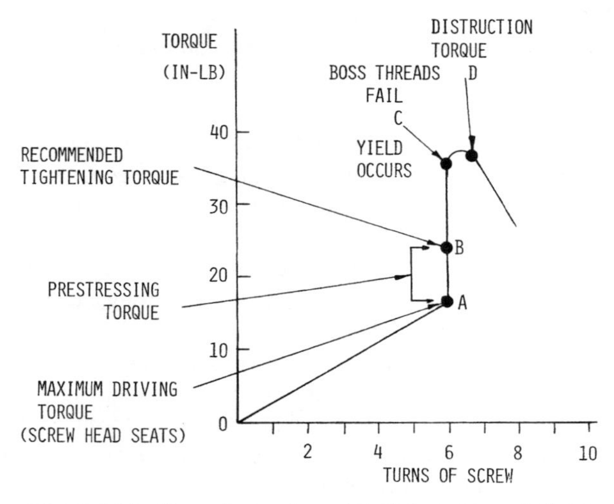

Figure 4.11 Screw torque-turn plot. (Adapted from Ref. 2.)

The pullout force is

$$F = S_s A = S_s(3.14)D_p L$$

where F = pullout force
S_s = shear stress ($S_t/1.732$)
S_t = tensile yield stress of resin
A = shear area = $(3.14) \times D_p \times L$
D_p = pitch diameter
L = axial length of full thread engagement

As with other assembly techniques, elevated temperatures can cause creep, and testing will verify whether the threaded fastener will provide the required holding strength. If testing proves that the self-threading screws are not adequate, should expansion, molded-in, or ultrasonic threaded inserts be used? Threaded inserts should be used if frequent disassembly of the part is anticipated. If the wrong type of screw is used, stress cracking of the boss or thread shearing can result at or after assembly. Each supplier recommends a specific type of screw for their material's assembly.

PRODUCT SURFACE CONSIDERATIONS FOR DECORATION AND ASSEMBLY

The product surface must also be considered when surface-applied adhesive bonding or decoration is used. If the product surface lacks sufficient adhesive attraction between the adhesive pad or surface-applied decoration medium, such as paint, labels, or foils, and it adheres poorly, such as flaking, or bubbling up, the program can fail as a result of product returns or liability reasons. The product's surface energy and applied medium for adhesion must be considered. If not, adhesion problems can occur, a days or even month after the product is in the consumer's hands. To obtain good adhesion, the surface energy of the decorated surface must be lower than the medium applied. This surface energy information is available from both the material and the medium supplier.

Even in cases where the surface energy of each component is very close, this can be handled. The surface of the part to be decorated can be treated, cleaned, or even chemically etched. Nylon products require surface preparation for plated metalized products. In a worst-case situation, selecting a different decoration medium, changing additives in the material, or using a primer or product surface treatment has proved successful. Some material additives, used for processing aids, can migrate or form on the surface of a manufactured product as a plateout after manufacture. This can cause a change in the surface energy of the product's surface. In some cases this has required changing to a different adhesive or decorating system. Therefore, always consult with your material, adhesive, paint, and decorating medium supplier before finalizing the selection of the product's decorating medium. Material suppliers have product design recommendations and processing guidelines established for their resins. I recommend that the design team obtain their material suppliers respective materials design and processing handbooks, which are readily available.

Table 4.5 Adhesive systems

Adhesive Key No.	Type	Physical Properties: Tensile Strength (psi)[a]	Physical Properties: Peel Strength (psi)	User Factors: Cure Time to Handle	User Factors: Price per Liter[b]
1	Acrylic, methyl methacrylate	3000–5000	20	5 min	E
2	Acrylic, other monomers	3000–5000	25	30 s	E
3	Epoxies, RT[c] cure	3000–5000	3	5 min	L
4	Epoxies, heat cure	3000–5000	3–25	5 min–72 h	M
5	Cyanoacrylates	3000–5000	2–3	5–10 s	VE
6	Urethanes, with solvent	1500–2500	25–40	4–24 h	M
7	Urethanes, 100% solids	1500–2500	25–40	5 min–24 h	M
8	Solvent	1500–4500	5–20	2–5 min	L

[a] Pounds per linear inch squared.
[b] Room temperature.
[c] *Cost factor:* VE = very expensive, E = expensive, M = moderate, L = low cost.

Adhesive selection guide

Substrate, Plastic	Recommended Adhesives[a]	Substrate, Nonplastic	Recommended Adhesives
Acrylic	1, 6, 8, 5	Ferrous metals	2, 1, 7, 6, 4, 3, 5
Glass-reinforced TPs	1, 2, 4, 6, 7, 3, 1, 5	Nonferrous metals (except copper)	2, 4, 7, 6, 1, 3, 5
Nylons	1, 6, 7, 5	Copper and alloys	4, 7, 6, 2, 3, 5
Polycarbonates	2, 1, 5	Glass	2, 4, 3
Polyesters, T/S	1, 4, 6, 7, 5	Wood	1, 2, 6, 7, 3, 5
Polyolefins	6, 1, 5		
Polyurethanes	6, 7, 5		
Styrenics, including ABS	1, 6, 4, 8, 5		
Vinyls, flexible	1, 8, 5		
Vinyls, rigid	1, 6, 8, 5		

[a] Listed in order of preference, toughness being the most preferred property. To pick adhesives for dissimilar substrates, select only from the matching key numbers listed for both materials. (For styrenics to steel, for example, select 1, 4, or 5.)

Source: Adapted from Refs. 6 and 7.

Adhesives

Adhesives can be used for joining plastic materials as listed in Table 4.5 with different joint designs such as those shown in Figure 4.12. To achieve maximum bonding strength the adhesives must have a lower surface energy than the surface energy of the plastic to be bonded. If not the joint's bond strength may fail over time when exposed

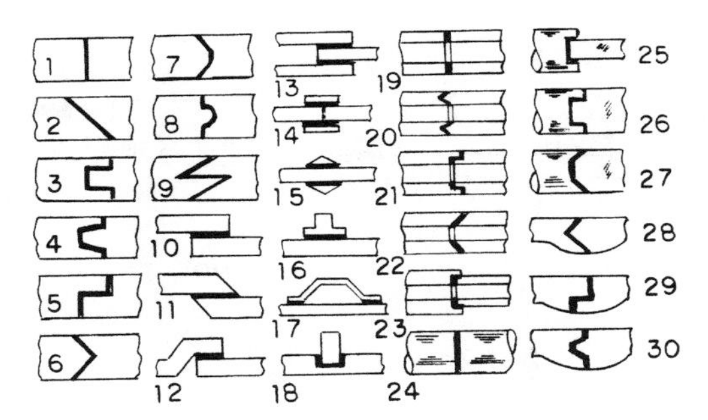

1. A BUTT JOINT.
2. A SCARF JOINT
3. A SQUARE TONGUE AND GROOVE JOINT.
4. AN ANGLED TONGUE AND GROOVE JOINT.
5. A HALF-LAP JOINT.
6. A "V" JOINT.
7. A "V" TYPE JOINT WITH A FLAT.
8. A ROUND TONGUE AND GROOVE JOINT.
9. A DOUBLE-SCARF LAP JOINT.
10. A SIMPLE LAP JOINT.
11. A TAPERED SIMPLE LAP JOINT.
12. AN OFF-SET LAP JOINT.
13. A DOUBLE-LAP JOINT.
14. A DOUBLE-STRAP JOINT.
15. A BEVELED DOUBLE-STRAP JOINT.
16. A "T" SECTION JOINT.
17. A HAT SECTION JOINT.
18. A RECESSED RIGHT ANGLE JOINT.
19. A TUBULAR BUTT JOINT.
20. A TUBULAR "V" JOINT.
21. A TUBULAR HALF-LAP JOINT.
22. AN ANGLED, TUBULAR, HALF-LAP JOINT.
23. A TUBULAR LAP JOINT.
24. A ROD-BUTT JOINT.
25. A TONGUE AND GROOVE JOINT IN A SOLID ROD.
26. A LANDED TONGUE AND GROOVE JOINT IN A SOLID ROD.
27. A SCARF TONGUE AND GROOVE JOINT IN A SOLID ROD.
28. A "V" TYPE JOINT WITH INCREASED BONDING AREA FOR ADDITIONAL STRENGTH.
29. A HALF-LAP JOINT WITH INCREASED BONDING AREA FOR ADDITIONAL STRENGTH.
30. AN ANGLED TONGUE AND GROOVE JOINT WITH INCREASED BONDING AREA FOR ADDITIONAL STRENGTH.

Figure 4.12 Adhesive joint designs. (Adapted from Ref. 3.)

to high temperature and/or severe chemicals in use. When the surface energies of the plastic and the adhesive are not compatible, the surface of the plastic substrate must be modified for good adhesion. Corona or plasma surface preparation or treatment of the plastic part's surface will usually provide a surface that will yield a very good and strong adhesive joint. Sometimes only a light sanding or roughening up of the plastic parts surface will be sufficient. Always test and try the easiest method first.

The methods used to test joint strength are shown in Figure 4.13, for the type of forces applied on the joint, tension, shear, and cleavage or peel stress. Fusion joints are also subjected to these similar stress factors, and all joints must be tested to ensure that the bonding method is satisfactory for the product's requirements.

Thermal Bonding

Thermal bonding is the fusion and joining of two or more materials by an external heat source. Thermal bonding of plastic materials requires that the thermal, melting,

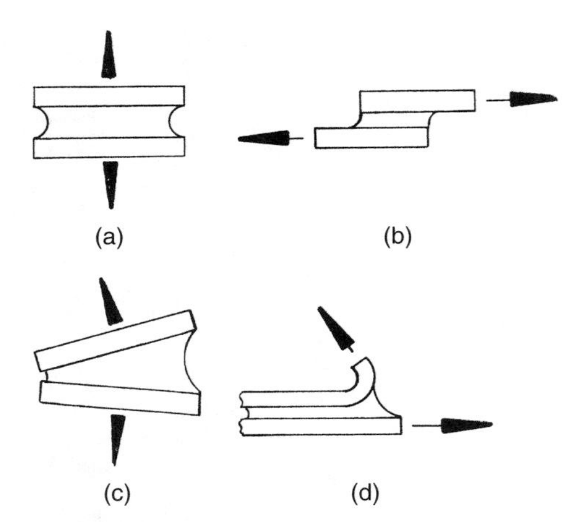

Figure 4.13 Stresses on Adhesive joints: (a) tensile stress; (b) shear stress; (c) cleavage stress; (d) peel stress. (Adapted from Ref. 7.)

or softening properties of each resin be known. For the assembly of amorphous resins by fusion bonding, the difference in fusion temperature for each part can be as much as 40°F. For crystalline resins, with a very sharp or well-defined melting temperature, only a 5°F difference in melting temperature is allowed. If the temperature of fusion is greater than these temperature limits, bonding may not occur, or the weld and joint strength will be very weak. Plastic resins and techniques of joining are compared in Tables 4.6 and 4.7. Always check with your material supplier for their input to the best method of thermal bonding for your product's design using their materials.

Flexible Hinges and Sections

The unique features of plastics allow molding in flexible sections, such as hinges or forming them after molding. Flexible hinges can be designed to be molded into the parts or formed by coining the hinged area by the machine operator directly out of the mold while still hot. The hinge can also be formed later in a secondary coining operation. Coining involves localized compression or the thinning of a section of the part to form a flexible hinge. For a molded-in hinge, the material in the mold at the hinge is locally thinned down from the adjacent section thickness. Then, immediately after the part is molded and still very hot, the operator flexes the hinge section a minimum of 90°, 3 or 4 times, orienting the material at the joint and forms the hinge. This dramatically increases the hinge's flex and tear strength. Coining the hinged joint after the part is cooled is similar to using a heated coining die, or if the material has high elongation, it is done at room temperature. Each form of coining causes a clouding of the material at the coined area, more visible in colored parts, which orients the material in the direction of flexure. Crystalline and semicrystalline materials, such as polyethylene, and polypropylene, nylons, acetal, and polyester, nonreinforced resins

Table 4.6 Plastic joining weldability and compatibility[a]

	1	2	3	4	5	6	7	8	9	10	11	12	13	14	15	16	17	18	19	20	21	22	23	24
1 ABS	0	0	0		0	0	X	X												X				
2 ABS/PO		0			X	X	X		X				0											
3 ABS/PVC			0		X	X	X		X										X					
4 Acetal				0																				
5 Acrylic					0		X						0							X	X			
6 Acrylic multipolymer						0	X		X											X				
7 Acrylic/PVC							0		X															
8 ASA									X											X				
9 Cellulosics (CA) CAB, CAP									0															
10 PRS										0														
11 PPO											0						0							
12 Nylons (H)												0												
13 Polycarbonate (H)													0	X										
14 PC/Polyester (H)														0							X			
15 Polyethylene															0									
16 Polypropylene																0								
17 Polystyrene (GP)																	0			X				
18 Polysulfone (H)																		0						
19 PVC (Rigid)																			0					
20 SAN/NAS																				0				
21 Polyester																					0			
22 Structural Foams																						0		
23 Polymide																							0	
24 Polyimide-Imide (H)																								0

[a] Weldability—read across; compatibility—read across and up. *Key:* 0 = good compatibility; X = compatible at times based on material blend; H = hygroscopic, must be kept dry before welding.

Source: Adapted from Ref. 5.

Table 4.7 Ultrasonic weldability charts

Material	Resin Type	Welding	Inserting	Staking	Swaging	Degating	Spot Welding	Near Field	Far Field
ABS	A	G	G	G	G	G	G	E	G
ABS/PO	A	G	G	G	G	G	G	E	G
ABS/PVC	A	G	G	G	G	G	G	F	P
Acetal	C	G	G	F	F	F	F	E	P
Acrylic	A	G	G	G	G	G	G	F	F
Acrylic Multipolymer	A	G	G	G	G	G	G	G	F
Acrylic/PVC	A	G	G	G	G	G	G	P	P
ASA	A	G	G	G	G	G	G	E	P
Cellulosics (CA, CAB, CAP)	A	F	G	G	G	F	G	F	P
PRS	C	G	G	G	G	G	G	G	F
PPO	A	G	G	G	G	G	G	G	G
Nylon	C	G	G	G	G	F	G	E	P
Polycarbonate	A	G	G	G	G	G	G	G	G
PC/polyester	C	G	G	G	G	G	G	G	F
Polyethylene	C	G	G	G	G	P	G	F	P
Polypropylene	C	G	G	G	G	P	G	F	P
Polystyrene (GP)	A	G	G	G	G	G	G	E	F
Polysulfone	A	G	G	G	G	G	G	G	P
PVC (rigid)	A	G	G	G	G	F	G	G	P
SAN/NAS	A	G	G	G	G	G	G	G	P
Polyester	C	G	G	G	G	F	G	G	P
Structural foams	A	G	G	F	F	P	G	F	P
Polymide	C	G	G	G	G	F	G	G	P
Polymide-imide	C	G	G	G	G	F	G	G	P

[a] *Key:* A = amorpous resins; C = crystalline resins. Weld characteristics: E = excellent; G = good; F = fair; P = poor.

Source: Adapted from Ref. 5.

are ideal candidates for hinged products. The designer's imagination and the selection of the correct plastic limit only these design and assembly methods. Using the inherent properties each plastic has can greatly reduce the number of parts in an assembly, bringing added value and function to the part and reducing manufacturing costs and increasing supplier profitability.

Decorating and Material Modifiers

Plastics are naturally adaptive to all decorating processes. Products can be molded in colors and user information molded on the product surfaces. Plastics can be internally colored or externally painted, plated, or printed on and have decorative films and foils attached both in and out of the mold. Information molded either in or on a plastic part can eliminate a secondary and expensive information-added transfer operation.

Pigmented and Colored Products

Internally coloring plastic resins is done in several ways. Color pigments can be compounded in by the supplier or added by the molder at or before the molding operation using color concentrates, dyes, powders, liquids, or pigments. The design team must be aware that some internal pigment and colorant systems, even in very low concentrations, may drastically reduce a plastic resin's physical property or alter the resin end-use operating characteristics. Therefore, the color and pigment system selected must always be end-use-tested to ensure that it does not affect the product requirements. The color pigments and formulation must be selected and specified to ensure that it is never changed in any manner, including type or percentage of pigment, during manufacture. This is done to avoid any future litigation should a change unknowingly be made by the material or colorant supplier and created a product liability problem for the product. Today, most heavy-metal and cadmium-based pigments are not used because of health and environmental reasons. This requires natural or organic, nonmetallic, pigments to be used that have lower heat processing stability during product manufacture as shown in Table 4.8. The resin material suppliers and pigment manufacturers may be aware of the pigment's lower heat stability, but they may not always convey this information to the design team and molder.

When exact product color matching is required, the product's surface finish can affect the color when viewed on the part after manufacture. This is caused by the manner in which light is reflected from the product surface. A slight variance in surface finish will affect how the color is perceived to the viewer's eye. Part section thickness varying in the part, as at ribs, bosses, or other internal sections of varying thickness, or in some plastic resins will affect the shade of the products color to the viewer. These section thickness variations must be consistent for the product using design and compensated for other means as texturing the part area or using uniform thickness throughout the part where possible.

To eliminate color variance between similar material mating parts and when identical parts are made in multiple molds, always state in the mold finishing or polishing

Table 4.8 Organic pigment processing temperature limits

Color[a]	Pigment	$/lb	150	200	250	300	350
Red	Perylene	50			280		
	Quinacridone	30				300+	
Yellow/orange	Diarylide	5		230			
	Disazo	20+			280		
	Pyrazolone	20+				300	
	Islondoline	20+				300	
	Anthrones	20+				300	
	Hansas	20+				300	
	Iron oxide	1		230	280		
Green	Phthalic	10				300	
	Chrome oxide	2					380+
Blue	Indanthrone	35			280		
	Phthalo	10				300	
	Iron	2		230			
	Ultramarine	2				300	
Violet	Carbozole	35		220			
	Quinacridone	30				300+	
	Ultramarine	5				300	
Brown	Iron oxide	1			280		
	Ferrite	1				300+	
Black	Carbon black	0.5				300+	
	Iron oxide	1		230			
White	Titanium dioxide	1				300	
	Barium sulfate	0.5				300	

Resin	Process	Processing Temperature Range 150	200	250	300	350
Polyethylene	Molding		210	290		
Polypropylene	Molding		230	290		
Polystyrene	Most			260		
PVC	Most		230			
Engineering resins	Most				300	350

[a] Adequate colors cannot be achieved for high-temperature polymers, particularly for yellows, oranges, and reds due to pigment burnout during processing.

Source: Adapted from Ref. 8.

instructions, all final polishing must be in the same direction to achieve the required cavity finish.

Product color matching for mating parts in different base resins or a matching painted metal surface can be a problem. Use the color-match request form shown in Table 4.9 for all color matches to ensure that the necessary data are available for correct color matching. Each base resin may require different pigments and color systems to obtain exact color matches. Therefore, before approving any color match, be sure that all parts were recently molded and assembled in the product for final approval. Then view the assembly under the anticipated lighting conditions to be sure that they meet the customer's color-matching requirements. If this is not done prior to production, the products can look like a collage of the same basic color but in varying shades of the same color. Varying surface finish on a product and texturing will also reflect the same color in varying shades and result in uneven color matching.

Other factors can affect a plastic parts color over time. Exposure to high humidity and UV (ultraviolet) exposure can change a part's color. To ensure minimal color problems, mating parts should be processed from the same lot of compounded color, within the same timeframe and preferably by the same molder. This will ensure more uniform color control of products. Processing temperature will also affect the product's final color. For organic pigment processing, maximum holdup time in the machine barrel without degradation is approximately 15 min without degradation of the organic pigment system. Any longer holdup time at processing temperatures will cause the organic pigments to burn out, decompose, and cause a shift in color. If a problem occurs during molding, turn down the barrel temperature, or better yet, slowly purge the barrel during the problem solution by taking air shots. This will keep new material coming through and prevent degradation of the resin. The design team's selection of the plastic resin, pigment system, and any modifiers affecting the product's manufacture must be considered and manufacturing instructions written if a problem occurs during introduction to not degrade the resin in the machine's barrel. Products decorated after molding should not use mold release. If mold release is required to assist in releasing the part during the ejection cycle, the part has another problem that must be solved with the mold design, not solved with the use of mold release. But if product release is a problem, there are mold releases especially formulated to avoid decorating problems. Only in special situations must parts be cleaned or degreased before decorating. Any decorating should be done directly after molding to avoid part contamination from dust or foreign particles in the plant's air system. This is especially true for metalized products as head, and taillamp reflectors. Also, question your material supplier to be sure that there are no internal processing aids in the resin that will exude to the surface, causing decoration problems.

The design team is responsible for specifying the decorating medium for the product. Drawings and specifications for decoration should have specific and separate drawings and specifications sheets to instruct and show where the decoration is on the product. Never leave these important decoration items to chance.

Table 4.9 Color-match request form

Customer:__________________ Contact:__________________
Address:__________________ Position/Phone/email__________________
____
Application:__________________ (__)________-________
Product:__________________ ________@ ________
Material:__________________ Material source:__________________
Material replacement: Y/N, What:__________ Prior material:__________
Part thickness:________ Show surface: Y/N Class of finish required:__________
Estimated product weight:______ pounds/kg
Color match: Exact: Y/N, color coding only: Y/N,
Color target standard enclosed or part to match: Y/N
Color sample required: polished, textured, type:________, ribbed, what:________
Date desired:__________ Send to whom?:__________________
Molding sample for evaluation required: Y/N, Number of pounds required:__________
Must be compounded: Y/N, salt and pepper acceptable: Y/N, other/what:__________
Color match requirements:__________________

An exact match requires a flat, glossy target standard. For critical colors, use a master target standard. If plastic masters are not available, will supplier provide them? Actual parts available supply sample, adjacent parts or an assembled product and a pellet sample of 1–10 lb of current material.

Color tolerances: Match Exact ( ) Close, or ( ) Wide
Parts have to match paint: Y/N Source available: Y/N Who:__________
Matching parts materials:__________,__________,__________
Are parts adjacent: Y/N, Separated: Y/N, How far away:__________

Color-match light sources customer will use for color approval: Mark 1 for primary, 2 for secondary light source.
( ) Macbeth Daylight ( ) Macbeth Horizon ( ) Tungsten
( ) Cool White Fluorescent ( ) Other__________________

Physical property requirements: Test results must accompany sample: Y/N
Mechanical requirements:__________________
Electrical/chemical/other:__________________
Pigment restrictions if any:__________________
Agency requirements:__________________
UV required: Y/N, percent of carbon black allowed:__________
Color specifications:__________________
Other comments:__________________
____
____

Approval release:
Design team leader:__________________ Date:__________
Production manager:__________________ Date:__________
Purchasing contact:__________________ Date:__________

REFERENCES

1. B. Miller, "Product Quality Problems? How Did You Check Your Resin?" *Plastic World,* 499–555 (Aug. 1989).
2. *General Design Principles—Module 1*, 201742B, E. I. Du Pont de Nemours Corp., Wilmington, DE. (Sept. 1992).
3. R. D. Beck, "Plastic Product Design," Van Nostrand Reinhold, New York, 1970.
4. *Designing with Plastics: The Fundamentals,* Engineering Plastics Division, Hoechst Celanese 46-93 15M/490 (1992).
5. *Ultrasonic Weldability—Compatibility Chart for Thermoplastics,* Dukane Corp., Form 10348-K-84.
6. A. J. Klein, "Update on Adhesives." *Plastics Design Forum,* 59–65 (May/June 1989).
7. B. Miller, "Adhesives Toughen Up, But Stay User-Friendly," *Plastics World,* 39–43 (May 1987).
8. C. Kirkland, Sr., "Shop Wisely for Heavy-Metal Free Colorants," *Plastics World,* 49–54 (Oct. 1990)

CHAPTER 5

MATERIAL PROPERTY CONSIDERATIONS

The design team must be knowledgeable in what benefits plastics bring to their product. This begins with understanding what advantages physical, electrical, chemical, economical and environmental plastics offer over conventional materials and applying them to their program. Basically a designer can chose from three distinct families of plastics. These are the thermoplastics, thermosets and thermo-elastomers that are available in pellets, granules, or powders. An excellent reference is the McGraw-Hill *Modern Plastics Encyclopedia* or the Internet for specific resin suppliers and their properties.

THERMOPLASTICS

Thermoplastics are described as being able to be repeatedly heated and cooled. This involves their becoming softened or liquefied, and on cooling returning to a solid. However, during each controlled heating and cooling cycle the plastic material suffers a small loss in physical properties as the processing heat breaks down the material's molecular bonds. This property loss is shown in a 100% regrind study (where sprue, runner, and test sample parts are repeatedly reground and reprocessed into new test specimens for property retention) (Fig. 5.1).

THERMOSETS

Thermosets, unlike thermoplastics, have essentially only one heating cycle before solidifying in a mold. When thermosets are heated, under high pressure and temperature, in an ejection-molding machine or in a ram compression-molding machine, they

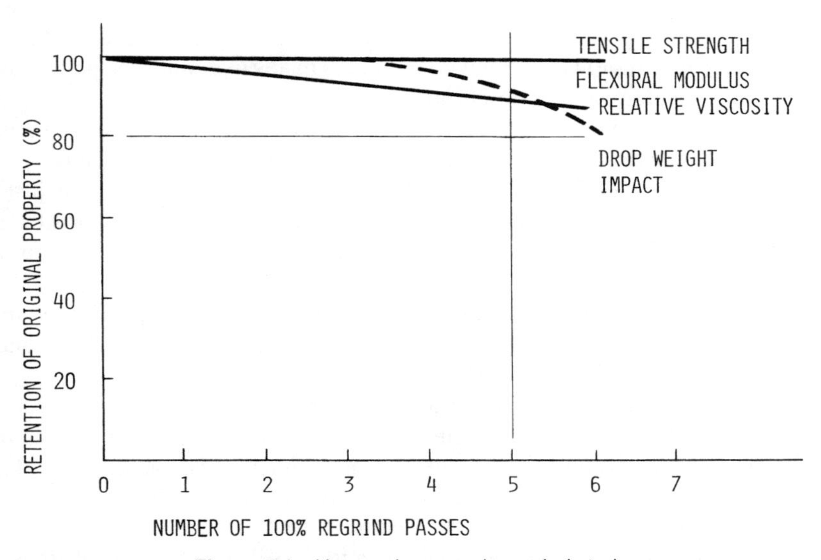

Figure 5.1 Homopolymer resin regrind study.

soften, flow, and fuse in a heat-induced crosslinking chemical reaction in the mold. Thermoset compounds are available in pellets, powders, liquid, and/or gel and are composed of two or more component systems.

Each material system can be enhanced and modified with varying amounts of process additives and material enhancers as fillers and reinforcements. Thermoset compounds are usually in liquid or gel form and, when mixed in specified proportions, cure with an exothermic reaction. The reaction of the system (chemically creating heat) that causes the crosslinking chemical process and the materials solidify. Once thermosets have reacted and cooled, they cannot be remelted or reformed. Thermosets are a one-time reaction polymer system. Thermosets are compounded in many raw material forms pellets, powders, and liquid systems. They can be processed by injection, blow, and compression molding and cast, extruded, and rotomolded depending on the product application and raw-material system.

THERMOELASTOMERS

Elastomers are a combination of thermoplastics and thermosets. Depending on their base resin, they offer toughness and higher elasticity for a wide variety of product applications. Elastomers can be made with high-energy absorption properties and others very flexible with high cut and tear resistance. They are like rubber that is formed and cured by high heat and pressure in a thermoset reaction. Their hardness property is measured in Shore hardness durometer readings. Their tensile properties are lower than thermoplastics and thermosets but have excellent toughness, high elongation, and sound-deadening properties plus excellent chemical resistance. Thermoplastic elastomers (TPEs) are a hybrid material with modifiers that can produce the performance properties of conventional thermoset rubbers.

TPEs consist of two or more blended polymer resin systems, one a hard thermoplastic and the other a soft elastomeric phase system. Each has their own phase and softening temperature (T_S). Their mechanism of useful temperature range is just below the hard thermoplastic phase's T_S. The hard phase anchors and supports the soft thermoelastomer phase that provides the materials ability to be formed into usable products.

There are six classes of TPEs: block copolymers, styrenes, copolyesters, polyurethanes, and polyamide or elastomeric/thermoplastic compositions; thermoplastic elastomeric olefin; and thermoplastic vulcanizates.

TPE materials have replaced numerous under-the-hood automotive thermoset rubber components. Used for air dams, bumpers, and grommets plus higher demanding applications as airducts and grease-filled convoluted boots. TPEs have also replaced thermoset rubbers in appliance and home and construction products in business machines, appliances and tools and in the electrical field as insulation and jacketing materials. TPEs can be recycled and used again for less demanding applications.

Thermosets, thermoplastics, and thermoelastomers are available in many grades with specific chemical, electrical, and physical properties and price variations. Within each family of resins are hundreds of compounds for general and specific applications.

END-USE MATERIAL SELECTION

The selection of the type of plastic used for a product depends on the end-use requirements for the product. At the beginning of a design program the design team must select a family of plastics to meet these requirements. In general, material suppliers try to produce resin formulations that will meet a wide variety of product applications. As product applications become more demanding, proliferation of a material supplier's product lines increases to meet more demanding product applications. This provides a wider selections for the design team to consider but makes the specific resin selection more difficult.

Material suppliers facilitate material selection with their literature listing resin types and grades for products now using their materials. They also provide online computer access to their property databases that assist the design team in making their material selection. Also consumer, government, local, and national, agencies, such as UL, FDA, USF, CSA, and VDE, and major OEM companies list plastic materials approved for specific applications that meet their material requirements. This assists the design team's search for resins that are already approved for applications.

The general classification of plastics is shown in Table 5.1, which lists some typical resin physical property values. Property values vary from grade to grade and with additive modifier. Property values also change with temperature and other environmental conditions that may affect the specific plastic resin.

Table 5.1 Generic and property classes of plastics representative properties[a]

Materials	Specific Gravity	Flexural Strength at 10^3 psi	Notched Izod Impact Strength (ftlb/in.)	Deflection Temperature at (264 psi)°F
Thermosets				
Alkyd polyester	1.2	12	0.3	80
Epoxy, general-purpose	1.25	16	0.6	350
Phenolic, general purpose	1.5	10	0.3	370
Urea formaldehyde, black	1.5	10	0.3	270
Thermoplastics				
Commodity				
Low-density polyethylene	0.92	1.6	No break	<100
Polypropylene homopolymer	0.92	7	0.8	130
Crystal polystyrene	1.05	12	0.4	180
Rigid polyvinyl chloride	1.3	13	1.0	155
Intermediate				
Polymethyl methacrylate	1.18	14	0.4	180
ABS, high heat	1.07	11	4	230
Cellulose-acetate-butyrate	1.18	4	3	150
Thermoplastic olefin elastomer	0.93	5	No break	na
Engineering				
Acetal copolymer	1.41	13	1.4	230
Nylon 6/6, 30% glass fiber	1.3	30	3	480
Polycarbonate	1.2	13.5	5	270
PBT, 30% glass filled	1.54	28	1.7	405
High-performance				
PPS, 40% glass filled	1.64	35	1.6	500
LCP, 30% glass filled	1.61	37	2.8	450
Polytetrafluoroethylene	2.17	1.7	3	100
Polyetheretherketone	1.31	16	1.6	320
Polyethersulfone	1.41	18	1.4	395

[a] Most properties have ranges, often wide, that depend on grade and modifiers. Typical values are shown.

Thermosets are divided into four main family classifications: the alkyd polyesters, epoxies, phenolic, and urea formaldehyde resins. Thermoplastics are arranged in performance categories of commodity, intermediate, engineering and high performance. As the resin performance properties increase, the materials price per pound also increases. The design team takes this into consideration when calculating the product's cost based on the resins cents per cubic inch material volume cost. The design team determines which material is best for their application by the following methods:

1. Knowledge of past experience and material knowledge
2. Use of the competitor material after identification

3. Use of material supplier and consultant recommendations
4. Use of agency-, code-, company-, or other specified and approved materials
5. Selecting a material by using supplier software and material selection information

Sources for Material Selection Data

During the initial design phase the team develops a list of potential material candidates. The key required physical properties for the product are listed in matrix form, including cost per pound of each resin. Also, any special or specific product requirements such as electrical, chemical, or impact, flexure requirements are listed. The material candidates are then evaluated and ranked against the required product requirements. This requires a good knowledge of each material as all materials are ranked against the anticipated end-use requirements of the product using only their key 73°F physical properties. When necessary, contact the material suppliers to obtain specific data, seldom available on material property data sheets or in their general material computer data files. Once the ranking is completed, consider combining of functions in the assembly, decoration, and processing data for the final material selection. Selecting the correct material takes time and should not be rushed; each candidate should be evaluated to select the best performing material meeting all performance, processing, and cost objectives. If information is lacking or inconclusive for a particular resin, obtain the necessary information from the material supplier before making the final selection.

An excellent initial source for information is McGraw-Hill's *Modern Plastics Encyclopedia,* (MPE). It is a good reference for the early selection of a material to determine which family of plastics can meet their requirements. Listing by requirement a materials mechanical, processing, thermal, physical, chemical, and electrical properties and ranking each against the other to meet the requirements of the product. Other tables in the MPE list materials that are approved to automotive, ASTM, federal, and military specifications and other agency requirements.

A series of performance property charts in Figure 5.2–5.6 list generic resin families, ranked between two variables to select likely material candidates. These broad property charts can assist the design team in learning how each generic resin family fits, in comparison to others, to specific material property values.

Computer database property information is available from consulting companies with built-in software support for selecting and evaluating the data spreadsheets, graphics, and tables. This information is updated on a set schedule and is available, for a fee, from these companies for the 5000-plus engineering thermoplastics or the 11,000-plus grades of thermoplastic, thermoplastic elastomers, and thermosets available from material suppliers.

Many large material suppliers offer their own material selection software programs for personal computers. One program is BASF's jointly developed program "CAMPUS," a trademark of CWFG, with the Bayer, Hoechst & Huels, databank, which stands for computer-aided material selection by uniform standards. Material characteristics are based on DIN (Deutscher Normenausschuss, 4-7

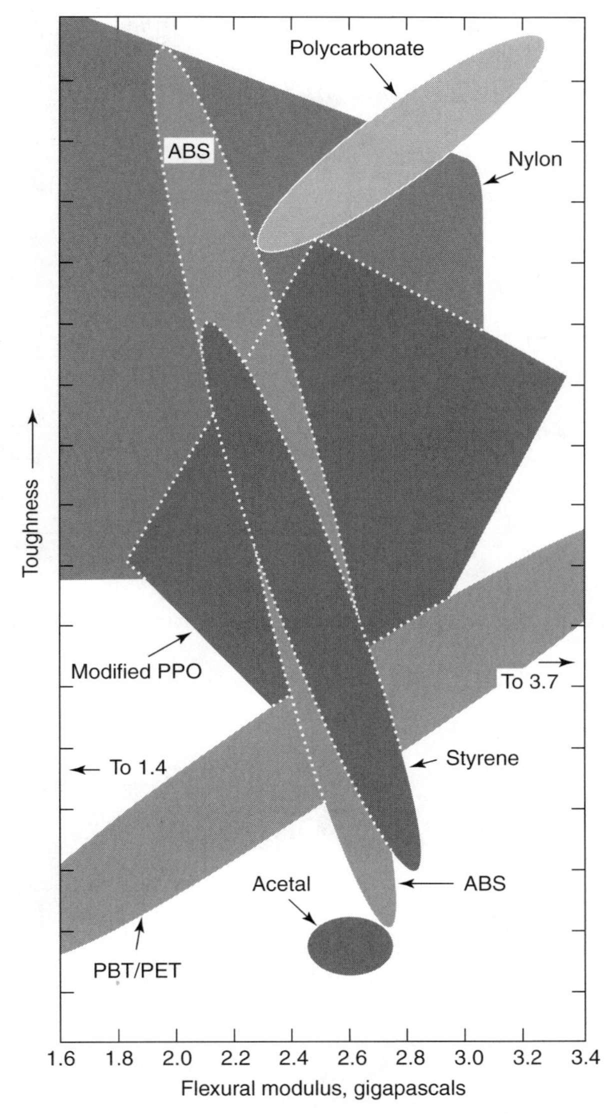

Figure 5.2 Impact strength and flexural modulus of engineering thermoplastics. (Adapted from Ref. 1.)

Burggartenstrasse, 1 Berlin 30, Germany) and conform to International Standards Organization (ISO 9000) for uniformity. Information on their resins is stored in database form and can be used to print out stress–strain curves for the designer to select the modulus of elasticity or scant modulus of a material in the nonlinear portion of a materials curve. Also for materials under long-term loading, creep is a

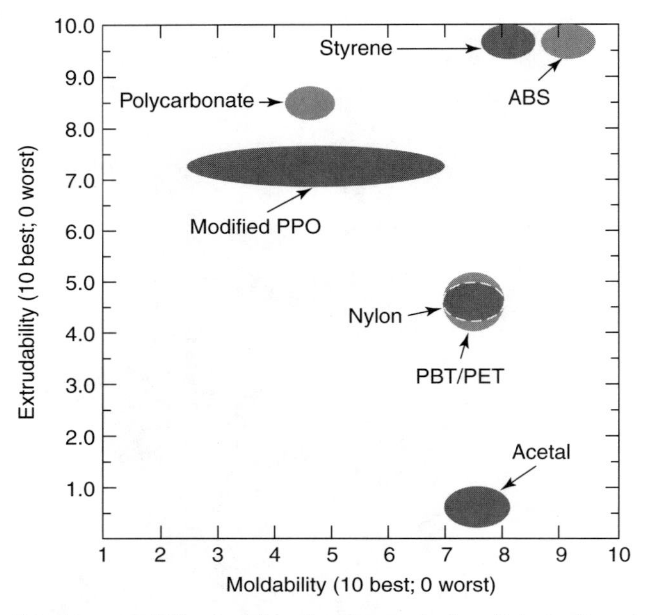

Figure 5.3 Processability of engineering thermoplastics. (Adapted from Ref. 1.)

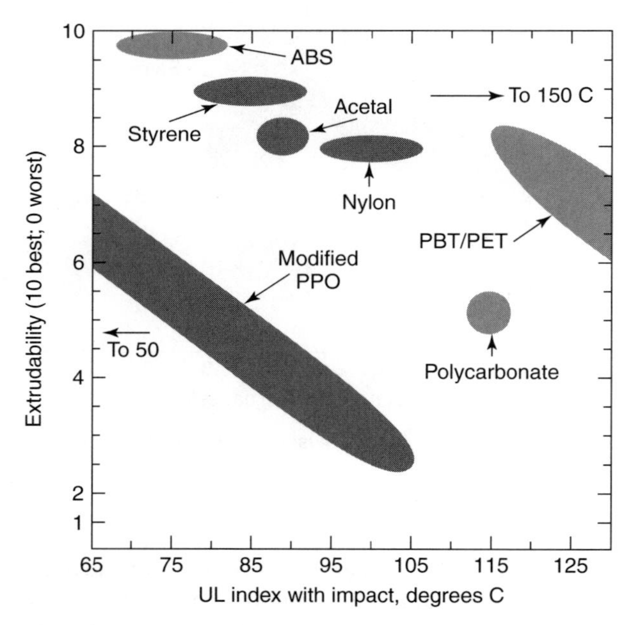

Figure 5.4 UL index with impact. (Adapted from Ref. 1.)

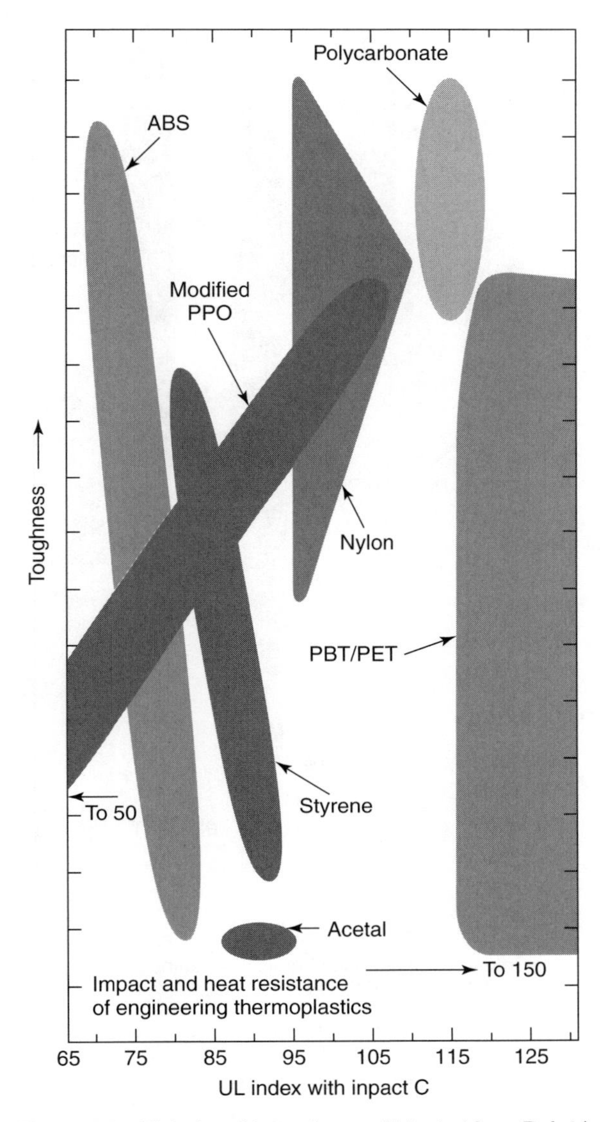

Figure 5.5 UL index with toughness. (Adapted from Ref. 1.)

factor and the modulus of creep replaces the modulus of elasticity for temperature ranges from −40 to +200°C depending on the material. There is also information available for simulating the plastic's melt flow during injection molding with the aid of the mold material flow and fill program.

If the material you are considering does not have a material software databank, the supplier should have the necessary property data for the end-use conditions you are

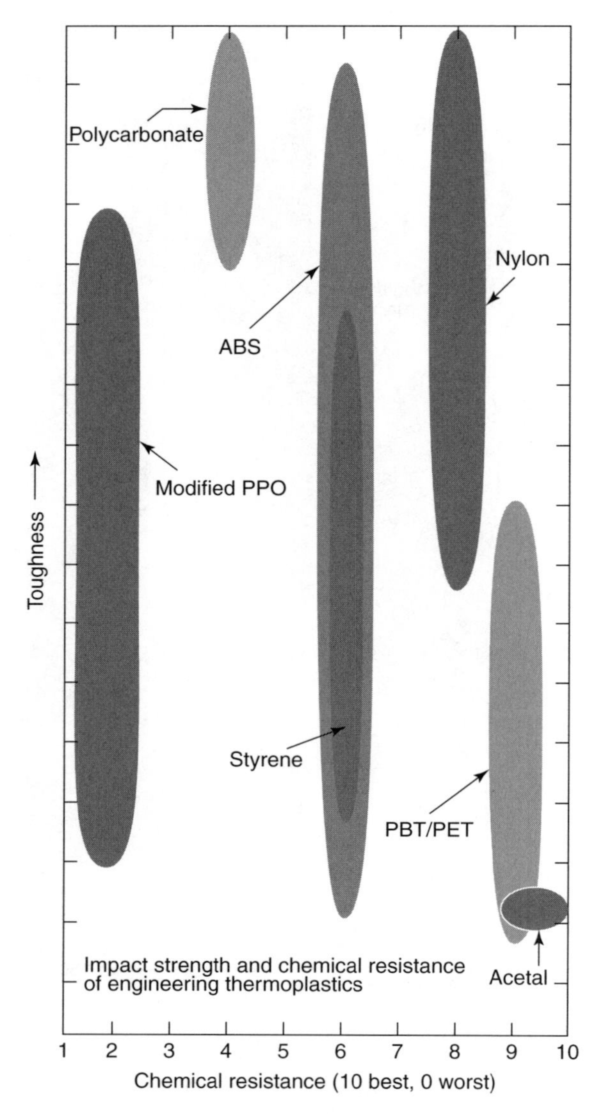

Figure 5.6 Chemical resistance and toughness. (Adapted from Ref. 1.)

considering for your design program. If not, contact their competitor to see if they have the required information. Many product manufactures use the same materials over and over as they have found that they meet their product requirements. This is both good and bad—*good,* since it allows purchasing to maximize quantity pricing and avoid the possibility of cross contamination of resins in their plants, plus processing problems, but *bad,* when product requirements change, as the current material may not fit the application or be as economical. The need to consider other or new resins is the design team's responsibility weighed against all factors, such as product design requirements, processing, material cost, product and company image, and long-term liability.

UNDERSTANDING THERMOPLASTICS

Thermoplastics listed in Table 5.2 can be further classified as amorphous, semicrystalline, and crystalline polymers. A material's characteristic property and processing will vary according to this classification. The comparisons of their dry-as-molded (DAM) properties are shown in Table 5.3. The DAM listing of material properties is used to equate them to a common base property line. The properties of these materials will vary greatly once exposed to the environment and temperature.

Amorphous Plastics

Amorphous plastics when subjected to high heat and pressure, will soften and flow. They do not have a distinct glass transition melting temperature (T_S), melting like an icecube from the outside in to the center. Amorphous plastics have a good low end-use temperature range (0–75°F), when compared with their price per pound cost. Their physical properties do decrease fairly rapid as temperature is increased. Most amorphous resins have high impact strength and some are transparent in their unmodified resin state. Examples are ABS (acrylonitrile butadyne styrene), acrylics, PVC (polyvinyl chloride), polycarbonate, polystyrene, and polysulfone. Amorphous resins experience longer molding cycles than crystalline resins because they do not have a distinct T_s melting temperature. On heating a gradual softening of the material when, going from a solid, initially a pellet, to a molten or softened state. Amorphous resins have a longer time period from softening to molten state temperature range, and it takes longer, after injecting the molten material into a mold cavity, for the heat to be removed from the material before it is rigid enough to be ejected from the mold cavity without damaging the part. Mold temperature control is very important for amorphous resins to minimize processing cycle time. Amorphous resins can fulfill most product applications in the consumer field for appliances, handtools, and communication equipment and less critical automotive and building products. They are dimensionally stable with good impact resistance over a wide temperature range extending from 32 to 120°F. They can be easily colored and made flame-retardant with many hundreds of product grades available for demanding product applications.

Their transparency and optical qualities are used in automotive headlamps and taillamp products, plus lenses and reflectors for many decorative and illuminating applications. A drawback is that they often lack good chemical resistance and care must be exercised to protect them from harsh chemical environments. But one amorphous and clear, transparent, resin—polysulfone—can withstand some severe chemical agents and operate in harsh environments up to 325°F in air without degradation and with 75% of its flexural modulus retained up to 320°F. Each material must be evaluated for its strength and weaknesses for each application. A very versatile amorphous resin

Table 5.2 Basic thermoplastics property dry as molded (DAM) (range)

Polymer	Degrees (°C)	Specific Gravity	Shrinkage	%	(M PSI)	(PSI)	Required	Range
ABS	A 110–125	1.01–1.08	0.004–0.009	1.5–125	300–1800	2500–16,000	Yes	0.10–0.15
Acetal copolymer	C 160–175	1.41	0.020 average	40–75	370–450	7200–7800	Yes	0.10
Homopolymer	C 175–181	1.42	0.180–0.025	25–75	380–430	8800–9700	Yes	
Acrylic	A 85–105	1.09–1.20	0.001–0.008	2–70	200–500	7000–11,000	Yes	0.02–0.10
Cellulosic acetate	C 230	1.22–1.34	0.0030–0.0010	6–70	1200–4000	1900–9000		0.40 (maximum)
Acetate butyrate	C 140	1.15–1.22	0.003–0.009	40–80	90–300	2600–6900	Yes	
Polyamid								
Nylon 6	C 210–220	1.12–1.14	0.003–0.015	30–100	390–2000	6000–24,000	Yes	0.05–0.20
Nylon 6/6	C 255–265	1.13–1.15	0.007–0.018	15–80	410–2000	13,700	Yes	0.05–0.20
Nylon 11	C 191–194	1.03–1.05	0.012	300	150	8000	Yes	
Polycarbonate	A 140–150	1.02	0.005–0.007	110	340	7000–20,000	Yes	0.02 (maximum)
Polyester	C 220–267	1.30–1.38	0.009–0.022	50–300	330–400	8200–27,500	Yes	0.02–0.20
Polyethylene	C 122–124	0.918–0.94	0.020–0.022	100–965	40–105	1900–4000	Yes	0.02 (maximum)
Polypropylene	C 160–175	0.90–0.91	0.010–0.025	100–600	175–250	4500–6000	Yes	0.10
Polystyrene	A 74–105	1.04–1.05	0.004–0.007	1.2–2.5	380–490	5200–7500	Yes	0.02
PVC	A 75–105	1.16–1.58	0.003–0.050	40–450	300–500	1500–7500	Yes	
Polyimid	A 310–365	1.36–1.65	—	3–10	450–500	10,500	Yes	0.02

Table 5.3 General polymer properties

Property	Crystalline	Crystalline Reinforced	Amorphous	Crystalline Filled	Liquid Crystal
Specific gravity	Higher	Higher	Lower	Higher	Higher
Tensile strength	Higher	Higher	Lower	Highest	Highest
Tensile modulus	Higher	Higher	Lower	Highest	Highest
Ductability/elongation	Lowest	Lower	Higher	Lowest	Lowest
Creep resistance	Higher	Higher	Lower	Highest	High
Maximum usage temperature	Higher	Higher	Lower	High	High
Shrinkage and warpage	Lowest	Higher	Lower	Low	Lowest
Flow	Higher	Higher	Lower	Higher	Highest
Chemical resistance	Higher	Higher	Lower	Higher	Highest

family is the PVC grades. PVC, polyvinylchloride, resins with specific modifiers can be made clear, flexible, and rigid depending on the additives used in their formulations. PVC is used in all industry products from automotive to consumer, including building, food, and medical products. Amorphous resins are being used successfully for all except the most demanding high temperature applications, and their use is growing, with new PVC compounds being developed. Their raw-material and processing costs are very economical for most product lines. But for the more strenuous and high-temperature applications, subjecting the product to heavy loads for long time periods, the crystalline plastics are the prime candidates.

Crystalline and Semicrystalline Materials

Crystalline plastics, including semicrystalline, have high-molecular-chain bond strength that result in higher physical property strength, with good chemical and temperature resistance. They are usually opaque except for the semicrystalline polymers, which can be optically clear but have lower physical properties. Special grades can be made clear in thin (0.10-in.) thickness. They can also be modified for improved impact resistance with minimal loss of tensile properties. Crystalline resins experience a minimum reduction in physical properties as temperature is increased and are rigid (no melting) until they reach their glass transition melting point temperature T_S. The plastic families, crystalline and thermoset, retain their structural integrity as temperature increases, but their physical properties will decrease, with the part creeping under continuous loading. Crystalline resins have a low melt viscosity and high flow in very thin (0.010-in.) section products. Special crystalline resin grades are formulated for higher viscosity, permitting them to be extruded into profile shapes, cable jacketing, and tubing. Nylon is used for abrasion-resistant coatings over PVC primary insulated conductor wire. Comparing a crystalline resin strength:stiffness ratio to product section thickness of amorphous and semicrystalline resins, crystalline resins are superior in thinner sections.

Crystalline resins process faster than amorphous resins in hot molds, 140–350°F. Heat increases the resin's rate of crystallization, and phase change from a liquid to a solid, which increases the material's setup or solidification time. Crystalline and semicrystalline resins are usually molded in a heated mold that increases the material's flow in the mold. This makes them ideal candidates for very thin (0.005–0.010-in.) section thickness products as printed circuit board (PCB) connectors. High mold temperatures also produce a more stable end-use product for the customer. Hot molds stabilize the material by limiting their maximum material shrinkage to the hot mold cavity and not allowing it in high-temperature service.

Crystalline resins can be reinforced to increase physical properties, such as tensile, flexural, and modulus strength, and to reduce creep to meet a variety of product application requirements. Their material costs versus specific end-use property values is shown in Figure 5.7. Crystalline resins are used for high-performance gears, housing, bearings, automotive parts, power tool housings, electrical switch components, connectors, and industrial products.

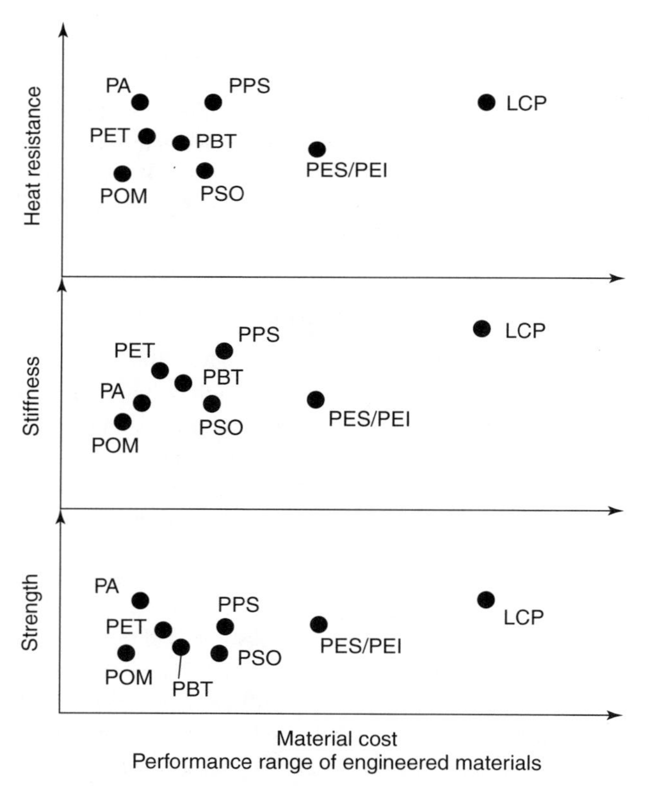

Figure 5.7 Performance range of engineered materials. *Legend:* LCP, liquid crystal polymer; PA, polymide (nylon); PBT, polybutylene terephthalate; PEI, polyetherimide; PES, polyarylsulfone; PET, polyethylene therephthalate; POM, polyacetal; PPS, polyphenylene sulfide; PSO, polysulfone.

Liquid Crystal Polymers

Liquid crystal polymers (LCPs) bring a new class of injection-moldable engineering resins to the marketplace. Their high unreinforced physical properties parallel those of the glass-reinforced engineering resins and can be reinforced with 30% glass fibers in many resin grades. Their resistance to almost all chemical compounds and solutions at elevated temperatures make them excellent candidates for severe environments. They are nonburning and form a char, with no ignition of the sample, at ignition temperatures. They have a high heat deflection temperature (HDT) of 446°F at 264 psi, with specialty grades that can attain 645°F. They are processed on conventional injection molding equipment using very fast injection times (0.1–1 s). LCPs flow is increased with high shear rates. This allows lower injection pressure, resulting in lower barrel temperatures for the same flow length in very thin sections (0.008' in.) Mold cavity temperatures range from 130 to 350°F and can be lowered to 130°F with excellent ejection of hot parts. The higher mold temperatures require mold oil heaters with reinforced hoses to handle the elevated temperatures at pump pressures.

LCP crystal structure is rodlike and orients in the flow direction in the melt and solid state. Since LCPs have an order orientation in the mold and there is no crystallization, cooling time in the mold is minimized. Their mold shrinkage for 30% glass-reinforced grades is in the range of 0.001 in./in. for the flow direction versus 0.005 in./in. in the transverse direction, across the flow direction. Nonreinforced LCPs have 0.001–0.006-in./in. typical mold shrinkage.

LCPs are used for electrical products requiring high heat resistance at elevated temperatures under load. Their end-use electrical properties are very high, as the resins do not absorb moisture, nonhygroscopic, after molding. Underwriters Laboratories, Inc. rates them, UL94, V0 to 500°F for continuous electrical use with intermittent temperature spikes to > 600°F. LCPs oxygen index rating (percent of oxygen required to support combustion) is 42, which is extremely low for smoke generation. These resins also exceed the current FAA standards for aircraft interior parts. LCP's exhibit resistance to alcohol base fuels, making them candidates for fuel systems, fuel rail, and so on in the automotive industry.

Their room temperature properties are high, not affected by the environment or long-term effects of UV radiation, as are those of many engineering plastics. They retain a high flexural modulus at 575°F of 600,000 psi and even higher at subzero temperatures. Their high flow, even with fillers and reinforcements, make them ideal candidates for both thick- and thin-wall products. Molding cycles are very fast, with 5–30-s molding cycles and 30 seconds maximum for most products.

The LCPs are *anisotropic,* meaning that their properties are not the same in all directions, due to the alignment of the rod-shaped molecules. The typical orientation in the mold is 60% lined up in the flow direction with 40% oriented in the cross-flow direction, like glass-reinforced resins. LCPs exhibit very low elongation, 1.3–4.5%, for nonreinforced grades and 1–2.2% for reinforced grades. LCPs are not recommended for impact product applications; therefore, the elimination of weld lines in the product is very critical when designing the mold cavity layout and gate locations.

FILLED, REINFORCED, AND MODIFIED RESINS

There are distinct differences between filled, reinforced, and modified resins; different additives are listed in Table 5.4. All plastics can have foreign materials added to the matrix resin, such as minerals, glass beads, fibers, talc, cellulose, wood flower, and rags, to increase the resin's physical properties and/or aid in lowering material costs. Each additive is selected for a specific purpose and produces different results for each resin depending on the desired end-use requirement for a product.

Fillers

Fillers are inexpensive and used to reduce the amount of resin while stiffening the material, which results in lower cost per pound, and usually reduced molding cycle time. Fillers are not chemically or physically coupled or bonded to the resin as illustrated in Figure 5.8. The scanning electron micrograph shows an impact-fractured surface of (a) an untreated glass-fiber-reinforced (35% by weight) reaction injection-molded polyurethane composite. There is no bonding between the fibers and the base resin. The material is filled only, which imparts no appreciable increase in tensile strength, with only minimal increase in flexural property stiffness in the base resin. View (b) shows a treated glass fiber that is physically bonded to the base resin that increases the physical properties of the base resin.

Reinforcements

Reinforced materials have higher physical properties and greater creep resistance at elevated temperature and loads. Their Izod impact strength, DAM, is typically

Table 5.4 Typical fillers, reinforcing fibers, and modifiers

Fillers	Reinforcing Fibers	Modifiers
Glass spheres	Glass fibers	UV stabilizers
Carbon black	Carbon fibers	Plasticizers
Metal powders	Aramid fibers	Lubricants
Silica sand	Jute fibers	Colorants
Wood flour	Nylon fibers	Flame retardants
Ceramic powders	Polyester fibers	Antioxidants
Mica flakes		Antistatics
Molybdenum disulfide		Preservatives
		Processing aids
		Fungicides
		Smoke suppressants
		Foaming agents
		Viscosity modifiers
		Impact modifiers

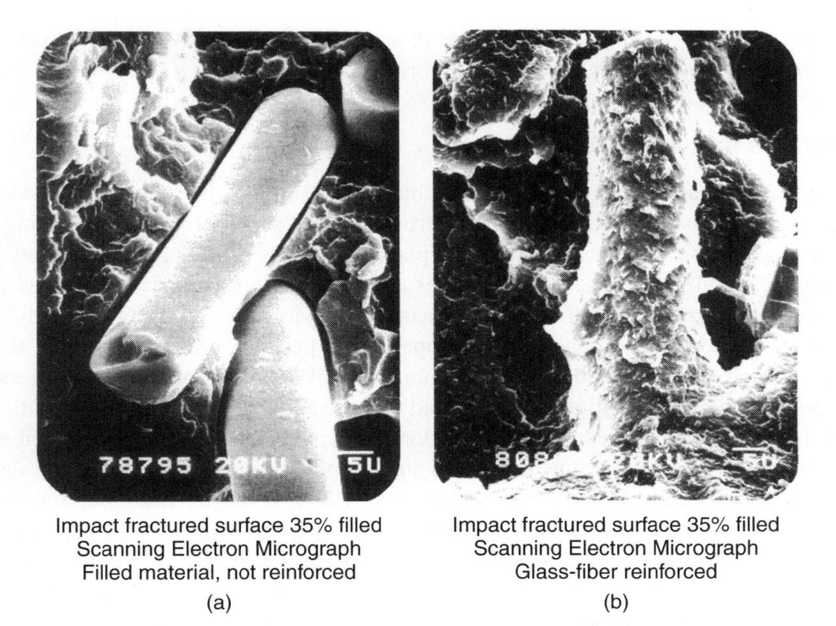

Impact fractured surface 35% filled
Scanning Electron Micrograph
Filled material, not reinforced
(a)

Impact fractured surface 35% filled
Scanning Electron Micrograph
Glass-fiber reinforced
(b)

Figure 5.8 Scanning electron micrograph of impact fractured surface: (a) 35% filled material, not reinforced; (b) 35% filled; glass-fiber-reinforced. (Ref. 2.)

equivalent to the base resin tested unreinforced. The reinforced resin impact value has obtained results from the strength of the reinforcement medium and the bond of the reinforcement to the base resin. It is not a result of the base resin's elongation property, which absorbs the energy of the blow. During the impact test the reinforced material absorbs the energy of the impact. The reinforcement holds the base resin together and assists the material in absorbing the impact energy on the test specimen. This implies that the material is tougher, but stronger and able to absorb more energy only without failure and able to resist the impact due to the reinforcement. This is evident by comparing the higher tensile strength of reinforced versus filled materials, which have very low Izod impact properties. Filled materials, of the same resin, will exhibit higher creep under load than will the reinforced resin. Reinforced resins have a higher flexural and tensile modulus with greater resistance to creep as the reinforcement is bonded to the base resin. Filled resins exhibit a more uniform shrinkage rate in all directions than do reinforced resins since the fillers are more random in the matrix, which results in less orientation of fillers in the flow direction.

Reinforced resins use a combination of reinforcing materials, long and short glass fibers, mica, carbon fibers, and minerals that are chemically bonded to the base resin. These reinforcement materials increase the physical properties of the base resin 2 to 3 times; restrict creep, raise the base resins heat deflection temperature, and increase flexural modulus.

The material's impact strength as discussed earlier with Izod impact strength, may be equal to the base resin or increased 1–4 times. Long-glass- versus short-glass-length reinforced resins has even higher values in some base resins. Always investigate all possible choices before selecting the final resin grade.

Modifiers and Additives

Modifiers for plastic resin are used to enhance and modify the material's physical, chemical processing, and aesthetic properties. This is accomplished by blending or compounding the base resin with other plastics such as ABS, or specific material enhancing additives. These additives alter the resin to improve or assist processing or meet requirements not inherent in the base resin. Most plastic resins are modified, filled, or reinforced to enhance their properties to perform multipurpose functions and aid in processing. With the right combination of resin and property modifiers, a material can perform multiple functions and end-use applications. Resins are modified with lubricates and processing aids; UV inhibitors; glass, mineral, and metal fibers; carbon; and Teflon fibers, plus flame and smoke inhibitors, which allow the product to meet special end-use requirements. Always check with your material suppliers to be updated on new resins and molding compounds that may fit your current and future product requirements. Additives consist of enhancers for improving impact, flow, lubricity, wear, flame retardant and smoke generation, weather (UV) ultraviolet resistance, EMI/EMP protection, and color for the application.

Processing aids are used for increasing flow in the melt state, protection of the base resin from oxidizing during processing; adding colorants, lubricants, and mold release agents; and increasing the number of nucleation sites for faster molding cycle time.

PCV resins require processing aids because without them, the base resin is unusable. PVC resins are produced in selected molecular-weight ranges and with modifiers, plasticizer, antioxidants, fillers, lubricants, and other agents, are blended into clear, flexible, rigid compounds for a multitude of automotive, consumer, industrial, and medical products. PVC requires antioxidative additives during processing so that the base resin will not turn color. Modifiers make PVC one of the most widely used materials.

Other types of modifiers provide protection to the product when exposed to elevated temperatures as heat stabilizers for nylon. A misunderstood modifier is the heat stabilizer. This modifier does not raise the material's end-use temperature rating. Heat stabilizers protect the surface of the product (the chemical bonds) from prematurely breaking when exposed to elevated temperatures.

Lubricity of a material can be increased with additives that are compounded into the base resin in the form of dry powders (molybdenum disulfide), Teflon fibers and flakes, and oil-based modifiers. These additives increase the wear life by reducing friction of a product against mating parts. Impact resin modifiers use rubber, thermoplastic elastomers, and other energy-absorbing additives or plasticizers that modify the base resin's elongation to increase the resins energy absorbing characteristics and ability to absorb impact loads. These modifiers decrease the base materials tensile and flexural strength and increase fracture and tear strength. Plastic material additives and modifiers as shown in Table 5.4 use fibers, minerals, pigments, flame and smoke inhibitors, UV absorbers, and more specific metal fibers or flakes for EMI/EMP protection of enclosed electronic components.

THERMOSETS RESINS

Thermoset resins were the most widely used materials before the use of thermoplastics for automotive, consumer, construction, and industrial products. Uses included handles on ovens and irons that have now been replaced with high temperature thermoplastics. The current major use is electrical power handling items as switches, circuit breakers, power meter box support frames, relays, and automotive battery cases. It is still the prime material for high-energy electrical products for its dimensional, current carrying, and price per pound use by the power and industrial products companies.

Thermoset resin products are the most thermally stable plastic materials available. Thermosets have very low elongation, 1–2%, with excellent creep resistance at elevated temperatures and under high loads. They are not used for products subjected to impact forces, as their energy absorption properties are low, especially in thin sections. Thermosets have excellent electrical properties, as they are mainly nonhygroscopic after being molded. Thermosets have very good to excellent chemical resistance because they are chemically crosslinked when processed, which produces very strong material molecular bonds. They are used in automotive engine fuel delivery systems with pure methanol or methanol–gasoline blends. They do not support combustion and char only under severe flame conditions. They can be modified to reduce smoke generation in flammability tests. Thermosets can be filled and reinforced with a wide variety of low cost materials such as minerals, cotton, cellulose, wood flower, rags, glass, and graphite fibers.

Thermosets are available in many grades and used in many applications from liquid cast systems for room temperature cure, to bulk and sheet molding (BCM/SMC) compounds.

BULK MOLDING COMPOUNDS

Bulk molding compounds (BMCs) are typically used for transfer compression molded products as fender skirts in automotive applications. They are unsaturated polyesters mixed with varying amounts of reinforcement mainly long- and short-glass fibers and calcium carbonate extenders to control shrinkage. The percentage of glass can vary from 65 to 75% with this higher loading compression molded in single-cavity molds. BMCs with lower glass percentages can be injection-molded from preformed compounds (logs) that have extruded forms. These forms are placed in the injection-molding machine barrel and under high pressure, injected into multicavity molds. BMC has a limited flow length; therefore, the part size and number of mold cavities must be sized for the anticipated flow.

SHEET MOLDING COMPOUNDS

Sheet molding compounds (SMCs) are prepared in blanket matting form, 36–48 in. wide. The glass matting is mixed with resin, allowed to stabilize for 24–72 hours, and then cut into shapes. These cut shapes are then placed in a compression mold and under high heat and pressure, form the product. Pressure vessels and large tanks (used for gasoline storage, etc.) use reinforce fiber gel coated windings of two or more components over a collapsible inner core form. This procedure produces lightweight, tough, strong, and longlasting storage and pressure vessels for industrial use. BMC and SMC are used for automotive body panels and grills requiring low weight, high strength, and toughness. Used in early iron handles and stove knobs in the past, thermosets are used today in many aerospace programs. In fact, thermosets are finding new applications daily for the more vigorous applications that thermoplastics cannot meet. Their pricing structure also makes them very attractive for many applications.

MATERIAL PROPERTY VALUES

The design team must always remember the listed property values of a resin on their physical property data sheet are only typical values. The data were developed from test specimens (molded ASTM test bars) DAM, (dry as molded), and tested at 73°F usually 24 h after molding. These bars are tested under industry-standard test conditions in accordance with ASTM, DIN, or other agency-standard test methods. This data is used to rank and ensure lot-to-lot uniformity of the resin and/or compounds with others under standard testing conditions. This data are also used for comparison of properties with similar materials. For ASTM detailed test information see the ASTM Standards, Plastics Methods of Testing.

The designer should never use the resins typical physical property values from a data sheet such as modulus, tensile strength, Izod, etc. as the information is only one averaged point on the materials property value data curve. A materials physical property will vary considerably over temperature, physical loading spectrum, chemical exposure and environment versus these single laboratory developed data points. Material suppliers can provide for their major resin environmental stress/strain/creep graphs tested under various end use conditions. These data must be used for the product design to ensure the product design will meet the end-use requirements of the program. Plastic material physical properties versus metals will now be developed.

Physical Properties Defined for Plastics

This section is presented to explain the difference between metal and plastic physical property definitions and how the plastic material property value is selected for design.

A plastic resin's physical properties, including modulus, tensile and compressive strength, Izod impact, and elongation, are developed from standard molded test bars as specified by ASTM and DIN standards. The test bars are molded individually or in a test bar family mold under optimum, with typical molding conditions for each resin. Each lot of raw material produced will have samples molded for testing under a specified manufacturing quality sampling plan. Testing is conducted by the material supplier and becomes part of each lot's quality control data file. The test bar manufacturing quality control data, chemical analysis, and process control records also becomes a documented part of each lot of material.

Since many base resins are alloyed with other resins or modified with additives, the final lot data become the quality and physical control record for each lot of material. The size of a lot of material can vary from a few hundred pounds up to 100,000 lb or more. Lot size is determined by the material supplier's method of manufacture. When multiple small lots are made, they can be blended together to average out the material's property values. This is frequently done to ensure that the property values are always within the manufacturing limits, especially when some lots have higher or lower property values. The final lot data are then developed from the blended materials test data.

Each material supplier is responsible for complying with and maintaining their quality control laboratory to meet industry standards. If lots of materials do not meet required standard values, they are not sold as standard-grade material. These lots are sold as off-specification material at lower prices or blended to meet the standard with other lots of the same material. Companies in compliance with ISO 9000 certification are requiring their suppliers to certify each order of material. They also certify material to special end-use customer requirements when requested by their customer. Certificate of compliance documentation with or without property values provides the customer with documentation that the material purchased complies with the regulations and specifications that the product must meet for their industry requirements. This is a requirement for the automotive, federal government, UL, FDA, or other agency requirements involved in consumer health and safety products.

Physical Property Data

A material's "complete" physical property data are generated over a range of specified loading conditions by the material supplier. Data are collected using varying temperature, time, chemical, and environmental conditions. Physical property test data are usually presented in stress–strain (tensile or compression) and creep curves developed using variable loading conditions after the test specimens have been conditioned in various chemical solutions and environmental condition states. The designer uses these data in the appropriate design equation to select the material and section thickness for the design of the product. The design analysis evaluates the material's physical properties to the product's design requirements and end-use conditions. Typically a design safety factor is used during the design evaluation to ensure that the material will meet the product end-use requirements for the anticipated life of the product.

MATERIAL SELECTION

Determining Product Material Requirements

The end-use requirements for a product are the main items that the design team must consider when designing and selecting a material for an existing product. Each product's end-use requirement may not always be known, or the product may be exposed to unanticipated forces. How, then, should the design team proceed?

The proven method is to establish product design limits based on function, liability, and agency and code criteria testing and cost. Typically, these items are known or can be established using the design checklist. Because of product liability concerns, all efforts must be made to ensure that the product is safe and conforms to agency and/or code requirements. For example, electric power handtools were not originally designed to withstand a 6-ft drop onto concrete. Operator safety requirements now require this test anytime after the product is assembled, if the power tool is to receive UL recognition seal of approval for safety. The UL test criteria is that, if the product is dropped, no component or wiring will be exposed that could cause an electrical shock or death to the user. This is a UL consumer safety requirement as the tool may still be energized. The test does not differentiate between being able to operate tools that are only electrically safe. The test is typically conducted in the manufacturer's plant with the products housing surviving the drop, without shattering, but cracking is allowed. No electrical component can be exposed to cause electric shock or more serious injury. Therefore, the material for the plastic housing must be extremely tough and strong in the DAM (dry-as-molded) state. This requires first evaluating a materials elongation and notched Izod toughness test in the DAM state for the housing. A product designer using DAM physical properties will have a higher probability of surviving this drop test as toughness is lowest in the DAM state. Prime materials selected for this application were nylon, with a lower value in the DAM condition, and polycarbonate that has a high DAM toughness right out of the mold. These prime material candidates are also often reinforced with glass fibers for additional strength requirements. ABS was another material considered for the housing as it has very good toughness because it is alloyed with elastomers and can survive drops on hard surfaces.

When existing metal products are converted to plastics, their modes of failure should be known. This information may be available from your quality assurance department or engineering files documenting customer returns and their reasons. Using this information, the new plastic product can be improved, made lighter and tougher at lower cost and with more safety features if necessary. A new product's design can benefit from these data when their requirements are established. The product application and design checklist can assist in establishing answers for the product's end-use requirements. The checklist considers the following items: forces, impacts, use temperature, and creep for the product, plus manufacture, assembly, and repair. The design team must know these answers to establish the product's requirements. These requirements can then be compared and with the material checklist to verify whether all requirements are met for the product.

Product Life Considerations

A products lifespan may be very short or last for many years. The automotive and consumer industries have plastics performing for many years in mild to severe environments. The anticipated performance life of a part will have a major impact on selecting the product's material. The conditions potentially leading to a product's failure must also be considered, and it should be determined whether the product is repairable or a one-time use, throwaway item. Many copier and computer printer machines have components that are recycled and repairable. This assists the consumer in keeping repair time and cost lower when it runs out of toner or a moving part fails. This allows the consumer to repair the unit and quickly return it to service.

As a result, plastic parts have gained a positive reputation in the marketplace for performing longer while keeping turnaround time for repair of units low by sending a recycled unit to them in less time and effort if they bring the unit in for repair and often at no charge. However conditions seldom never anticipated do occur, and a product first believed to solve a problem may suffer a new problem. It is difficult to anticipate all potential problems, but solutions are always to be found.

Consider the following example. Glass bus windows were subjected to stones, and other hard objects thrown by vandals were replaced with unbreakable extruded polycarbonate sheet. The polycarbonate (PC) windows worked very well against thrown objects until sprayed with aerosol paints. The vandals found out that polycarbonate sheet sprayed with common spray paints used for graphic art were very sensitive to the solvents in the paint. When the PC bus windows were sprayed, the plastic windows shattered catastrophically. This was caused by the paint solvent attacking the highly stressed, extruded sheet that had high manufacturing extrusion stresses in it. A solution was found using a protective clear coating, on the window surface, DuPont's Absite, which was resistant to these chemicals. The Absite coating was first developed by DuPont to provide a protective coating for acrylic parts that have very poor scratch and chemical resistance. This also gave the PC windows better scratch resistance, over glass windows, during washing the buses with rotating brushes for improved long-term clarity.

The design team must use the materials property data that represent the severest conditions that the product will be exposed to in service or must meet for agency approval. Never use the material's product data sheet physical property values. These values are averaged single-point properties for the material under DAM and controlled test conditions usually at ambient, 73°F temperature and 50% RH (relative humidity).

Never design a product for a single environment unless it is so specified. The product's environment may be controlled, as in an office, or variable with climatic conditions, as for automotive in all parts of the world. Plastics exposed to the elements for months on end will condition to their surroundings with the material properties reflecting their surroundings, which affects the product life.

Plastics exposed to sunlight, specifically UV (ultraviolet) radiation, for long periods of time can cause rapid degradation of a material's physical properties. Effects are recognized as surface crazing or chalking. The surface of the material may fade, chalk, craze, or roughen, due to surface breakdown of the polymer. Surface degradation can cause serious property problems leading to early failure. Using UV absorbers in the material can reduce surface problems. Carbon black is an UV absorber, and when blended into a plastic resin at low levels, 2% by weight, provides good UV protection for the material.

The product can also be externally coated with UV-absorbing paint to protect exposed surfaces. The paint's solvent system must produce a good bond with the plastic surface. Always verify whether any exterior coating is not susceptible to cracking and has a coefficient of thermal expansion as close to the plastics as possible. If not, the paint may crack and introduce a stress crack into the surface of the plastic material. Always test a paint's solvent system with the plastic substrate with which it is to be used, or use solvent-based paint systems, which bond chemically to the plastic materials surfaces, resulting in good adhesion and product protection. Also, ensure that the solvent does not attack the plastic over time. Some paint solvents can create problems in your plant involving OSHA guidelines for fumes or emissions. To eliminate this problem, consider using water-based paint systems, although they are not as tough and have lower bonding strength with the plastic surface.

Products exposed to elevated temperatures should use resins formulated for high temperature end-use exposure. Resins, such as nylon, can have a heat stabilizer additive added that inhibits oxygen from attacking the product's surface at elevated temperatures. The heat stabilizer does not extend higher-temperature use; it provides extended use time at temperatures for the material only within its operating temperature range. Be sure to ask your material supplier the intended results for each material additive as with heat stabilizers for nylon resins. Only use additive packages that add value, not cost, to the product.

Product Part Function Consolidations

During the product's initial design phase the design team should consider the possibility of combining any product function or at least the assembly into a single multifunctional product. Product, part, or function consolidation can aid in reducing product cost by reducing the number of parts and assembly time and also provide extra benefits and value to the application. This may require reevaluating the initial product material, based on combining new and multipart functions into a single plastic part.

Part consolidation is often possible when converting metal products to plastic. Mold cost may be higher but is often offset with cost saving from functions combined in the part, resulting in fewer parts for the product. Often several product functions and features as snap- and press-fit assembly, part support sections and bosses, and molding in springs and ratchets can be combined as one part by using a multipurpose engineering resin that can be used to mold these parts into the single product, replacing many other costly parts and assembly items. Discuss part consolidation options with your material suppliers to see if their materials are able to reduce the number of parts. Part consolidation may be accomplished when the design team performs an analysis on the product. This is done by breaking down the product's functions into what it must actually do and how best to incorporate these operations into one plastic part.

Explore the product design by asking "What if?" questions before the design is too far along. What functions can be combined, what parts eliminated or combined, requirements for assembly, mounting components, information molded into the part, and so on into one part. Consider assembly options using snap/press fits to eliminate screws. Also evaluate the part's layout to include bearing surfaces, springs, and other internal components and items. Lay out all components and evaluate combining them into the part. Suppliers can assist in recommending and offering the design team new ideas on how to use their resins. A well-informed material supplier can assist in improving a product at the conceptual stage. Part consolidation of functions can save considerable money by reducing inventory, number of molds, and production and assembly time.

FORM, FIT, AND FUNCTION

Many plastic products require only the customer's form, fit, and function requirements. This requirement can be demanding according to what the product must perform. As a result, replacing a metal part with plastic is not always successful since end-use requirements must be duplicated or value-enhanced to be successful. Many metal product properties are not well defined since the metal part is overdesigned and little design data may be available. Also, the customer's specifications may be based on metal requirements that could eliminate the use of any plastic until the actual forces are developed and evaluated for a plastic.

METAL REPLACEMENT

Plastics have replaced metals in many applications and now have more than 15% of the existing metal's product markets. Automotive applications have spearheaded the way with engine components; valve covers, brakes and radiator parts, air cleaners, fans, and other components, plus interior and exterior hardware and body panels to lower car weight and cost, while maintaining reliability and meeting safety standards.

New resin compounds and grades are developed yearly that become the prime materials for more demanding applications. To replace metals, the design team begins by evaluating the design criteria that met the original product end-use requirements. This includes how the product is currently manufactured and is driven by cost. Product improvements should be considered as part consolidation plus enhanced performance and potential lower liability risks. To be economical, a metal part cost reduction in plastic should be a minimum of 20% in finished product savings. This means that the design team must find improvements and savings in part design, performance, manufacturing, assembly, and decorating cost. The price of the base plastic material in dollars per cubic inch is the first evaluated cost. Plastic raw materials are more expensive in price per pound but when evaluated in dollars per cubic inch may yield lower finished product costs. This comparison is shown in Table 5.5.

Base material cost should not be the prime consideration when metal-to-plastic conversion is considered. All areas of the product's design, manufacture, and assembly

Table 5.5 Raw-material cost—plastics versus metals[a]

	Price		
Material	\$/lb	lb/in.^{3}	\$/in.^{3}
Plastics			
Acetal (copolymer)	1.79	0.051	0.091
LCP (30% glass)	8.55	0.058	0.496
Nylon 66 (30% glass)	2.42	0.050	0.120
PBT (40% glass/mineral)	1.56	0.058	0.091
PPS (40% glass)	3.31	0.059	0.196
Metals			
Aluminum	0.74	0.098	0.073
Bronze	1.75	0.315	0.551
Cast iron	0.30	0.280	0.084
Magnesium	1.40	0.066	0.092

[a] Based on 1999 prices with volume pricing up to 500,000 lb annually.

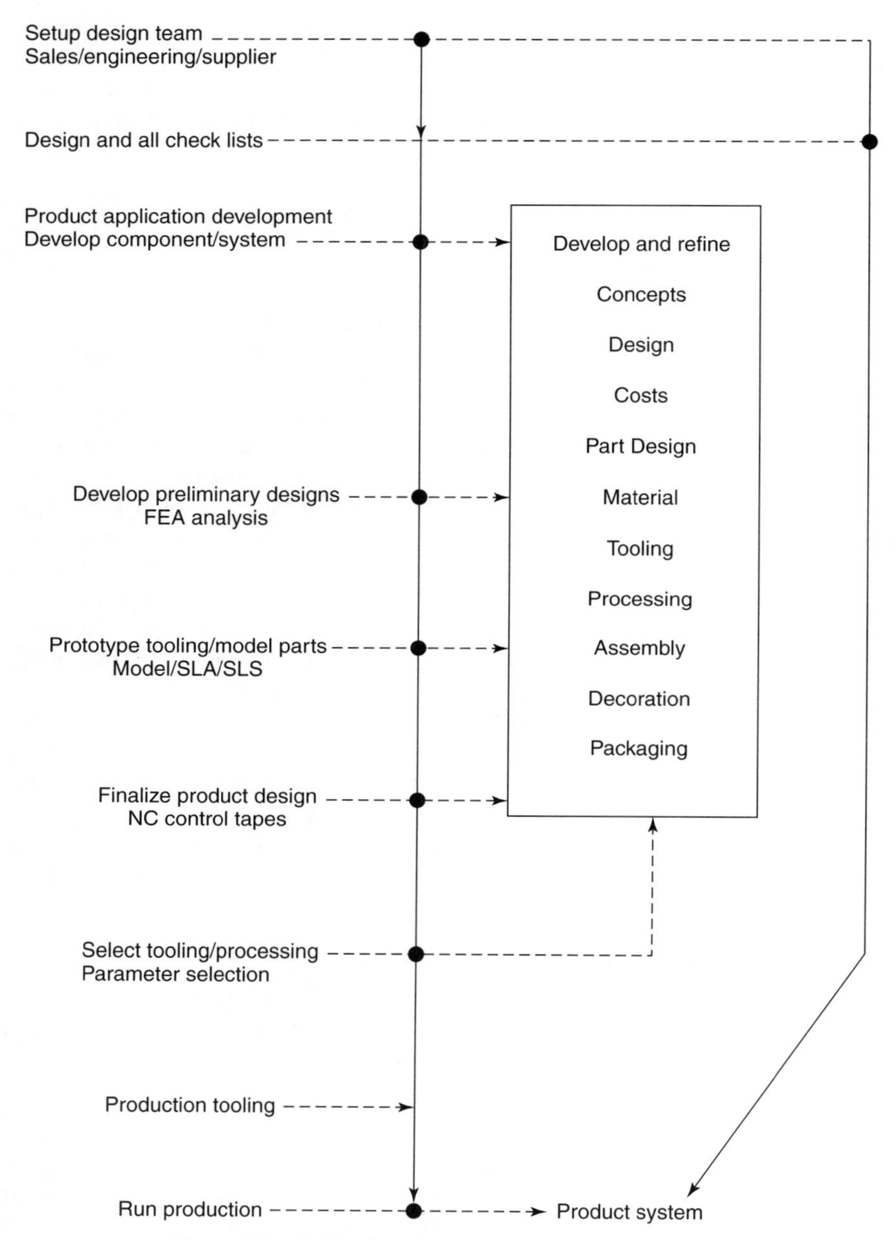

Figure 5.9 Design flow conversion from metal to plastic.

plus shipping factors must be considered during the conceptual design analysis to determine whether plastic will meet and beat the products cost and performance requirements in metal.

Using these criteria, decisions can be made involving product design, material, manufacture, tooling, assembly, and decoration. Prototype products can be made, tested, and evaluated in prime candidate resin systems to see if they will perform, lower cost, and prove equal to or better than metal in the application. The typical path for converting metals to plastics is shown in Figure 5.9, and follows typical part design criteria.

The key to developing a product that will meet customer satisfaction and end-use performance is teamwork and "total Involvement" of the design team. This cooperation will assist in ensuring that all pertinent design and manufacturing areas are explored and questions answered. It will also speed production with the knowledge that the part's tooling, manufacturing process, assembly, and decoration are all within the program guidelines to produce quality parts.

DESIGN FOR ASSEMBLY AND SERVICE

Consolidation of multifunctional parts, required for a metal products design, into one plastic part should enhance the product's end-use value. Using a material with more design capable of using function consolidation can result in elimination and/or reduction of individual parts for a metal product. Consolidation of multiple functions into one molded part can reduce production and assembly time. These improvements can result in lower warranty problems and improve serviceability to increase the product's value to their customer.

This analysis is known as, *designing for manufacture, assembly, and service* (DMAS). DAMAS encompasses the review critical operations during the products early design stage and, when used, anticipates, eliminates, and solves production assembly and service problems. DMAS brings to the design team, the field- service repair technician, who typically is never in the design loop. Their input can assist the design team to make the product user-friendly and reduce service warranty costs. Easily replaced, used-up, or worn-out parts can be stocked by the customer and replaced as required, yielding a user-friendly product. The quality team can perform a failure mode and effects analysis (FMEA), prior to finalizing the design to further analyze the manufacturing operations for process control and problem area solutions. The objective of DMAS is to increase reliability and quality while reducing the number of components, simplifying assembly and repair, plus reducing design time, service, product liability, and part costs. After part consolidation has been considered, the assembly of the parts into the unit is explored for both assembly and customer service and repair requirements if applicable.

MATERIAL IDENTIFICATION AND SPECIFICATION

It is very important to correctly identify the specific plastic resin for a product. This is recommended in ISO 9000 certification in purchasing data [i.e., ISO 9000-1987

(E) 4.6.3]. Material used for a product must identify the type, class, style, grade, or other characteristics to ensure that purchasing buys the correct materials. Material identified by generic name is insufficient, as it is not specifically identified. This is especially true for modified materials, colored resins as additive, and/or processing aids, which can vary with each supplier. Therefore, each specific resin or material grade tested in and approved for a product must be accurately identified. Alternate materials tested and approved must also be identified. Material substitutions should never be allowed without design team approval and after end-use testing.

Each material supplier's resins are identified with a specific alphanumeric or similar identification product code as DuPont, Zytel 157HSL BK010. This is a 6/12 nylon, heat-stabilized, lubricated, and colored black. The numbers (157) identify it as a 6/12 nylon from DuPont. The (HS) is a heat stabilizer system used to protect the nylon's surface from crazing at high service temperatures. The letter (L) indicates the material is internally lubricated to provide improved machine feed and provide a degree of part release in the mold. The color (BK010) uses a specific percentage of carbon black, approximately 2% by weight, to obtain UV resistance. Another supplier providing a similar 6/12 nylon would use a totally different product identity code with likely different additive and loading weight percentages for their additives.

Other acceptable methods are used to ensure that the resin specified will meet the product end-use requirements. The designer may specify a range of material property values known to meet the product requirements using ASTM D4000-82 material identification and specification callout. The ASTM D4000 callout is a material property line code callout system used for all plastic resins. It can be very general (with basic requirements) or an in-depth listing (cell and suffix requirements) to specify specific resin grade property values. An example of the callout for a 6/6 nylon, heat-stabilized, with 33% glass fiber reinforcement with cell and suffix specified, is ASTM D4000 PA120G33A53380GA140.

The design team specifies and the buyer decodes this material callout using the ASTM D4000 reference booklets. This callout method for material can be confusing, and good communication within the company is necessary to approve and use the material when received. The problem with ASTM D4000 is to identify the supplier and ensure that their materials that meet the specification. Military or government agencies designing a product for open-bid manufacture mainly use this method of material identification. The ASTM D4000 specification is available for a fee by writing to the ASTM (American Society for Testing and Materials, 1916 Race Street, Philadelphia, PA 19103).

When special material or additive packages are specified, the selection of approved material suppliers becomes more difficult. Therefore, many OEM product suppliers develop their own internal material specifications and approve one or more material supplier resins to these requirements. The automotive and consumer product companies do this extensively to ensure that only the approved material will be used in their products.

Plastics fill specialty needs, and without them our lives would be dramatically changed. The proliferation of uses for plastic continues to grow daily. The hardest task of a designer of products is to be kept current in existing and new materials that they can specify for their products. Selection and identifying the best and most economic material for their use is not easy. As a result, material suppliers and design and plastic magazines list guides for selecting materials based on these many and often specific requirements.

The designer can use the information as shown in Figures 5.2–5.6 to rapidly screen generic plastic materials for meeting their product requirements if the applications meet the figure's reference material. After selecting a material candidate it is advised to consult with a material supplier of the resin to obtain additional and specific physical property product data to see if they meet their specific product's end-use requirements.

When ISO 9000 documentation for the certification of a material is required to meet agency or specific customer specifications, the customer must identify and verify that the material received at incoming inspection meets the specification by testing and/or document verification. The material supplier will usually, if required, issue a "certificate of conformance" CoC for the resin shipped, stating that it meets the customer's specifications and material callout. The CoC becomes a product quality document, and the material grade is identified on the manufacturing work order to ensure that the correct material is used in manufacture.

When alternate resins are specified, the design team must make sure that they are also evaluated, tested, and approved to be sure that it meets the product end-use requirements. This will ensure that a quality product is not jeopardized and a backup certifiable resin is always available. Often special product resins are required that must meet specific requirements or have special properties. If ever in doubt, request the alternate supplier to furnish certifiable test data that their material meets the competitors.

REFERENCES

1. *Plastics Engineering* (Jan. 1985).
2. Union Carbide, RIM Polyurethane Organofunctional Silane A-1100 Treatment.

CHAPTER 6

PROPERTY CONSIDERATIONS WHEN DESIGNING PRODUCTS IN PLASTIC

The design team, when beginning the design for a new or redesigned product, must remember that the engineering design equations were developed assuming linear and isotropic material behavior. This means that the equations were developed and based on the response of metals and how their properties responded under conditions of force. Metal properties, for the majority of their basic materials and alloys, are uniform in all directions, below their yield point. This is not true for plastic materials.

The plastics designer is cautioned to remember that plastic materials are nonlinear or anisotropic in behavior as represented by their stress–strain curves. This means that a plastic material will not always return to its original dimensions after the release of a force when the strain exceeds 0.05%. Some creep or material elongation will have occurred. Plastic and metal property behaviors differ in this area. Metals exhibit essentially linear dimensional behavior as demonstrated with their stress–strain curves up to the metal's yield point, where the metal begins to elongate and creep. Temperatures up to about 180°F have marginal effect on a metals stress–strain curve and the metals yield point. Whereas a small temperature change has a significant effect on plastic properties as shown for modulus of elasticity in Figure 6.1. Therefore, a plastic under a continuous load will experience some degree of strain and with a modest temperature increase of only 5–10 degrees, can experience a higher amount of creep. The amount of creep is proportional to the temperature and the force on the part.

END-USE EFFECTS ON A PLASTIC'S PHYSICAL PROPERTIES

Designing a plastic product requires the design team to consider their products exposure to the entire range of end-use conditions. Temperature extremes; type, location,

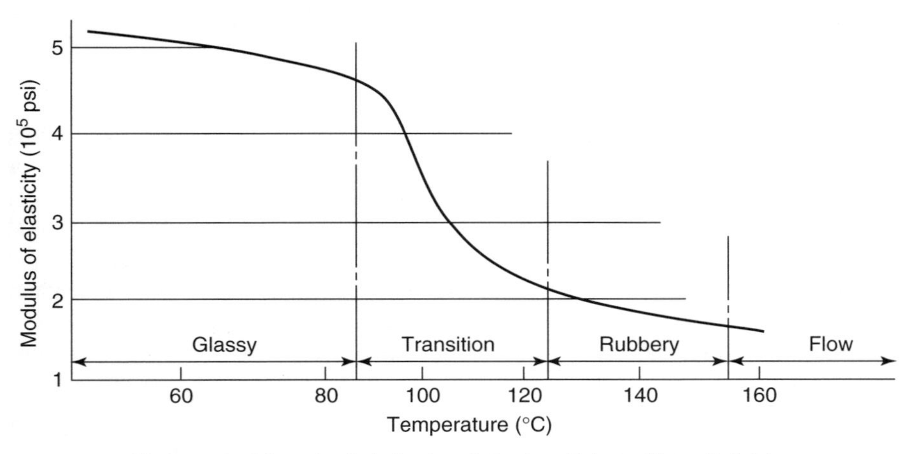

Figure 6.1 Viscoelastic behavior of plastics. (Adapted from Ref. 1.)

and forces on the product, whether one-time, continuous, or alternating; impact loads; frequency; time of application; and severity must be known and the product designed to handle these items while in service.

The designer seldom has the specific end-use material data on the product's end-use conditions. It is hoped that a material supplier will develop the material properties required by the design team for the material candidates at end-use conditions. If this information is not available or is time-dependent, the design team can approximate (worst condition) or use prototype parts and test the product in different materials to determine the most successful material. This is expensive and time-consuming, but necessary for a successful product. All material suppliers recommend, even when their end-use material data are used to end-use-test the product, verifying whether the design and materials meet the requirements.

The design team must consider whether the product's material could crack, craze, discolor, and/or suffer a loss of its physical properties in other ways. Thermal aging (oxidation) can occur with plastics over varying periods of time at elevated temperatures. This is often visible as color changes, chalking, and crazing of the product surface. With some plastics and colors, this may not always be visible and recognized only when the part fails, often catastrophically.

Some plastic materials can be modified initially to reduce aging affects but not always if conditions are to extreme. Thermal aging, with exposure to elevated temperatures for varying lengths of time, can cause the surface of a part to craze, with fine hairline cracks developing on the surface. Ductile plastic resins can turn brittle on the surface, and under a heavy load the material may crack and propagate through the part, resulting in brittle failure. Ultra-violet radiation can also cause crazing. Suppliers age their resins in Arizona and Florida to develop long-time property information for product exposed to direct sunlight with associated weather conditions. Always determine the material's aging effects with the material supplier when elevated temperature and sunlight exposure are involved.

PRODUCT FINISHING AND ASSEMBLY OPERATIONS

Product Preparation for Decorating

Finishing operations for decorating must be considered if the process could affect the plastic product properties. These operations include cleaning the product for decoration and the decoration process. Depending on the method of decorating, strong cleaners, chemicals, and the decoration medium could affect the plastic, which could result in property loss in the material and product. Some paint solvents and etching solutions, to prepare the surface, could attack highly stressed products, and the paint cure temperature and time can reduce molded-in stresses, causing parts to warp. The decoration process must be analyzed and controlled to ensure that the product retains its properties and appearance.

Product Assembly Forces

The design teams product responsibility continues through all phases of the product development to the customer. The method of assembly must be considered to ensure that the assembly forces do not exceed the material's properties and the assembly method is taken into account when the part is designed. Some assembly methods require minimum requirements of the part to be assembled with their method or equipment. As an example, for vibration welding the welded surfaces must be stiff enough to not deflect when the welding is occurring. The force and magnitude of movement of the top-welding surface against the base must not cause the walls of the base to deflect. If they do, a poor weld will result. This is a major requirement for vibration welding. Mechanical assembly forces for screws, snap and press fits, sonic welding fixture forces, and other factors, can overstress the product's material and must be considered during design. Materials selected for press fit assembly must have sufficient elongation to withstand the hoop stress forces of interference fit on the boss. Snap fits have cantilever sections deflecting and must not exceed the elongation of the material.

If the material's elongation is too low for the press interference fit or snap-fit deflection is too much, a very rapid automatic or manual assembly can produce very high local stress levels in the boss or cantilever lug. Too rapid assembly forces the material to elongate very quickly, and the material may not have enough mechanical strength to overcome the higher induced stresses and can fail. Therefore, the amount of interference and deflection must be considered and assembly instructions written and equipment built to limit and/or eliminate these high stress levels. For example, electronic printed circuit board (PCB) connectors are often used in reinforced materials with low elongation, with thin-walled sections having electrical contacts, metal pins, pressed into molded sections. The insertion should be uniform and at a speed such as to avoid creating high insertion stresses in the plastic connector. The design team must also consider that the cores forming the contact holes will wear, getting smaller over the life of the mold. The mold cavity metal core pins forming the holes for the metal press-fit contact pins will decrease in size by resin wear on their surfaces during filling of the mold cavity, causing loss of tolerance control. As the tooling wears, the press fit becomes tighter, requiring more force to press the metal pins into their cavity, and if the metal pins have sharp corners, this will create even higher local stress levels in the plastic pin sockets. Production, mold maintenance, and quality assurance must monitor the core pin dimensions to ensure that wear does not cause an assembly problem. The mold dimensions must always be checked when it is removed from service for key dimensions and wear areas as the gate.

THERMAL EXPANSION AND CONTRACTION ATTACHMENT CONSIDERATIONS

Thermal expansion and contraction between dissimilar materials can create product problems. Plastic parts exposed to extreme temperature variations, when attached to metal plastic parts, must have attachment expansion/contraction slots or be designed to withstand the forces, or they may buckle, shear, and fail. Plastics have a higher rate of thermal expansion than do metals, and any growth or contraction must be considered during product design if the product is firmly anchored and unable to move. Expansion slots without a firm anchor can allow parts to move independently to other parts in the assembly. If expansion slots are not possible, then the plastic product must be insulated or made strong and rigid enough to withstand the expansion forces with mating part.

SHIPPING CONSIDERATIONS

Product quality problems may occur during shipping and warehousing. Products can be subjected to shock handling loads, being dropped or mishandled if not correctly packaged. Once the product is off your shipping dock, you have very little control of the product, just the packaging and method of shipping.

Temperature extremes can occur in storage and transit. Temperature and handling stress must be considered once the product leaves the plant. Problems can be eliminated during the design phase by using the checklists for handling and packing of the product to ensure that the customer receives it in good condition.

CHEMICAL EFFECTS ON PLASTICS

Chemicals, in various concentrations, can affect some plastic product property. Plastics have good chemical resistance to most chemicals except for strong acid and base solutions. The effects of chemical compounds, fuels, oils, and cleaning solutions are complicated, and knowing exactly what solutions a product will be exposed to may be very difficult. An increase in temperature will accelerate a chemical effect on a material. Also, the same chemical solution may react differently with modified resin compounds of the same generic family depending on the additives and fillers in the resin.

Similarly, a resin's chemical resistance can vary in different supplier's chemical solutions and products that are considered the same in their industry. But they may use different additive packages, or concentrations of additives such as, motor fuels and oils, or lubricating and cleaning fluids that may react differently with the same plastic material. Plastics are carbon-based molecules, and the majority of chemicals are absorbed, which can cause physical property and dimensional changes.

Chemical product suppliers are continually modifying the formulation and ingredients of their products to be more competitive in their markets. This information is difficult to obtain and not always communicated to their customers as long as the end uses of the chemical product remain relatively similar. Therefore, the design team, if their product is exposed to this chemical product or solution, must test their product for suitability. This is to ensure that the chemical product's active ingredients have not been altered or percentage changed, which could cause a problem with their product. If they know that their product is used in the chemical environment, they must contact the supplier to find out what active ingredients are used which could affect the product. They can also request, if the formulation is altered in any form, a sample to be sent to them to verify that the product is still satisfactory in the chemical environment.

The chemical resistance data of a generic plastic, as shown in Table 6.1, are developed in different ways. These data are expensive and time-consuming to develop and not always available from the supplier. The results can vary on each plastic compound using the same base resin depending on their additives. Chemical resistance data are developed using standard flex bars $4 \times \frac{1}{2} \times \frac{1}{10}$ in. These bars are exposed to concentrations of pure or diluted chemical agents at different temperatures that can accelerate the effects in both stressed and nonstressed conditions. Both conditions are tested since some unstressed plastics may not exhibit noticeable effects when exposed to a chemical. But when stressed, the effects are accelerated, and the bar may fail catastrophically. This is known as *stress cracking*. Testing may require exposures for days to determine their effect, while others may show the chemical's effects within minutes. More physical ASTM test bars are immersed in varying chemical concentrations plus common fuels and lubricants from various suppliers. From these bars, physical properties, tensile, elongation, flexural modulus, and dimensional growth of the resin are developed. Chemical effects on plastics are a time, temperature, and concentration occurrence. The more active the solution, the sooner the effects on the material will occur. Testing is usually obtained only on the supplier's standard base resins, for both unreinforced and reinforced grades. After the test bars have been exposed for a predetermined period of time, dimensional measurements and weight are taken. These bars are then tested using ASTM test methods for the chemical effects on the resin's physical properties. On the basis of initial test results, the exposure of the test bars may be for long or short time periods. Some results are known immediately, as with acrylic and polycarbonate as even weak and low concentrations of solvents cause immediate stress cracking of these resins. Others, such as nylon, when exposed to moisture and most fuels and lubricants, take weeks to see the effects. Three main tests are used to determine a material's chemical compatibility: immersion unstressed, immersion stressed, and the gas bomb, for accelerated testing, used mainly for fuels.

Table 6.1 Chemical resistance of various plastics by chemical class

Chemicals	1	2	3	4	5	6	7	8	9	10	11	12	13	14	15	16	17
						Acids and Bases											
Acids, weak (dilute mineral)	1	2	3	1	1	1	1	1	1	1	2	1	1	1	1	1	3
Acids, strong (concentrated mineral)	3	3	3	2	—	3	2	1	—	3	3	—	1	1	2	3	3
Base, weak (dilute sodium hydroxide)	1	3	1	2	2	1	2	1	—	3	1	1	1	1	1	2	3
Base, strong (concentrated sodium hydroxide)	1	3	3	—	—	2	3	1	—	3	1	—	1	1	2	2	3
Acids, organic (weak, acetic vinegar)	1	2	3	1	1	1	1	1	1	1	2	1	1	1	1	3	3
Bases, organic (strong, trichloroacetic)	3	3	3	2	—	3	2	1	—	3	3	1	1	1	2	3	3
Automotive																	
Auto, fuel	1	1	1	1	1	1	1	1	3	3	1	3	3	1	1	1	1
Auto, lubricants	1	1	1	1	1	1	1	1	3	3	1	1	1	1	1	2	1
Auto, hydraulic	1	1	—	1	1	—	1	1	3	3	3	1	1	1	—	—	—
Aliphatic hydrocarbons																	
Heptane, hexane	1	1	1	1	1	1	1	1	1	1	1	2	3	1	1	1	1
Ethylene chloride	1	2	3	2	2	1	1	1	3	3	3	—	—	—	2	2	2
Halogenated chloroform	1	1	2	1	1	1	1	1	1	1	1	1	1	1	1	1	1

Table 6.1 (*Continued*)

Chemicals	1	2	3	4	5	6	7	8	9	10	11	12	13	14	15	16	17
Alcohols: Ethanol, cyclohexanol	1	1	1	1	2	2	1	1	—	3	2	—	1	—	1	2	1
Aldehydes: Acetaldehyde, formaldehyde	—	—	—	—	—	—	3	2	—	3	2	—	1	—	1	2	2
Amines: Aniline, triethanolamine	1	2	1	1	2	2	1	1	3	3	3	3	3	3	1	1	1
Aromatic hydrocarbons Toluene, xylene, naphtha	—	—	—	—	—	3	—	1	3	3	3	—	—	—	1	1	1
Aromatic Halogenated chlorobenzene	3	3	3	3	—	3	1	1	—	3	3	—	1	—	2	3	1
Aromatic hydroxy, phenol esters: Ethyl acetate	2	2	1	2	2	2	1	1	3	3	3	—	3	—	2	2	2
Ethers: Butyl ether	2	—	1	1	—	—	—	1	—	1	2	—	3	—	1	1	1
Ketones: Ketone, acetone	2	2	1	2	2	2	1	1	3	3	3	—	2	3	1	1	1

	1	2	3	4	5	6	7	8	9	10	11	12	13	14	15	16	17
Miscellaneous																	
Detergents: laundry, dish soap	1	—	1	—	2	—	—	1	1	1	—	2	1	—	1	1	2
Inorganic salts: zinc chloride, cupric sulfate	2	2	2	—	1	—	—	1	—	1	—	—	1	1	2	2	2
Oxidizing agents (strong, 30% hydrogen peroxide, bromine)	3	3	3	—	3	—	2	2	—	3	—	—	1	—	3	3	3
Oxidizing agents (weak, sodium hypochlorite solution)	3	3	3	1	—	1	1	1	—	1	—	1	1	1	2	3	1
Water, ambient	1	1	2	1	1	1	1	1	1	1	1	1	1	—	1	3	2
Water, hot	2	3	2	3	3	2	1	1	—	3	—	1	3	—	1	3	2
Water, steam	3	3	3	3	3	3	2	1	—	3	—	—	3	—	1	3	—

Key: 1—acetal, copolymer; 2—acetal, homopolymer; 3—nylon 6/6; 4—PBT, polyester; 5—PET, polyester; 6—elastomer, polyester; 7—LCP; 8—polyphenylene sulfide; 9—polyarylate; 10—polycarbonate; 11—polysulfone; 12—modified polyphenylene ether; 13—polypropylene; 14—ABS; 15—316 stainless steel; 16—carbon steel; 17—aluminum.

Source: Adapted from Ref. 4.

Immersion

Immersion testing of a material involves using standard molded test bars placed in chemical solution for a specified time and temperature. The bars are later removed, cleaned, measured, and tested for physical property changes and dimensional and weight changes. These data are too general for design purposes, and the designer must obtain specific test data and graphs from the material supplier.

Stressed Immersion. Immersion of a test bar under a load in a chemical speeds up the reaction on the material. It is a more severe test than the product may experience while in service but is very representative of the final effects. Stressed immersion bars provide information that the design team can use in material selection. This test can detect if a plastic is susceptible to a chemical, causing cracking. Test bars are immersed in the chemical solution usually in flexure, at ambient temperature, and under a fixed stress or strain along the length of the bar. After exposure, the samples are examined for dimensional and visual changes. The effects of change may be crazing, cracking, stress whitening (cold flow elongation), or total failure during the test. Unless the concentration, temperature, and time under stress and/or strain are similar on the plastic test bars, the results can vary in actual service. Always test the material or compound, natural or pigmented, to be used for the product to ensure that the color system or any additives will not cause the product to fail. Changes in pigments and additives, even in very small amounts, can have a very adverse effect on the product. Always test the final material to ensure that the chemicals do not affect the product's/resin's properties.

Gas Bomb

The gas bomb test uses a pressure vessel that accelerates the testing time of plastic materials' compatibility in fuels. Test bars are exposed under pressure and elevated temperature to the test solution or fuel. They can be stressed or unstressed, and after removal, dimensional and weight changes are recorded and their physical properties tested.

These tests show chemical compatibility with a plastic but not the performance properties required for design purposes. Only creep rupture tests conducted under actual chemical, time, and temperature conditions provide the appropriate information for design. These tests are expensive and difficult to run and are seldom available to the designer. Therefore, end-use testing is always recommended.

Mechanisms of Chemical Attack

The mechanism of chemical attack on plastics occurs in the following ways. Chemical interactions occur in three distinct ways: reaction, solvation, and plasticization.

1. *Reaction* is a chemical attack on the polymer chain. This produces a progressive reduction by breaking the molecular chain lengths in the polymer and lowers the polymer's molecular weight. This chain breaking causes a reduction of the

polymer's physical properties, which is irreversible. Molecular chain length weight is directly proportional to the material's physical properties. Interaction causes chemical bonds to fracture and go into solution with the chemical agent.

2. *Solution* (solvation) causes the polymer to dissolve. In high-molecular-weight polymers it occurs slowly, but the end effects are the same, a lowering of the material properties exposure over time.
3. *Plasticization* occurs when a chemical is miscible with a plastic, or absorbed. Absorption causes swelling, dimensional growth, and physical changes in properties, typically a drop; for instance, in tensile and modulus strength, but impact resistance can increase if the chemical acts as a plasticizer. Also warpage of the part may occur as a result of relaxation of molded-in stresses, a byproduct of plasticization.

Plastics that are hygroscopic (absorb moisture) will swell and grow with dimensional changes to a product. When exposure is eliminated, the moisture will slowly evaporate, returning the material to its original physical properties, shape, and state.

Chemical tests provide valuable information on plastics performance. When chemical exposure is anticipated, the designer should use the creep rupture test curve data. The data should be gathered from molded samples that were exposed to the same temperature and chemical environment that the end-use product will experience. End-use product testing is always recommended to ensure that the product will survive.

Reinforced grades of some resins tend to appear superior to the unreinforced base resin. This is due to the coupling effect of the reinforcement to the resin. Fillers and reinforcements reduce the chemical effects according to the amount of additives in the resin. This leaves less resin to be affected, and dimensional and weight changes are also less. Degradation of physical properties can vary between the type of reinforcements and coupling agents for the glass and mineral. Both long and short glass fibers tend to retain their property strength better than do small (mineral) particle reinforcements.

Most material suppliers have chemical resistance test data in tables developed from exposing test bars in chemicals at varying concentration temperatures and lengths of time. The physical property data generated are from test bars at DAM conditions, which allows comparison to the original property data. From these data they record weight and dimensional changes plus any color or surface effect changes.

Plastics can be used in many mild chemical environments. Because of the environmental and hazardous concerns of some chemicals, the more severe chemicals have been removed or replaced by less active ingredients in many chemical compounds and solutions. But if temperature and concentration data are different from the test data, end-use testing on molded parts must be performed.

ENVIRONMENTAL EFFECTS ON PLASTICS

Metals are not affected by sunlight or to the effects of radiation, microbes, and burning. Plastics are affected to some extent by the environment. Plastics exposed to weather extremes, radiation, bacteria and soil fungi, nonflammable gases, and flames react differently, and each must be evaluated independently.

Weathering Effects

The effects of weather are usually associated with the plastic withstanding "sunlight," namely, UV) predominately with exposure to direct sunlight. Most plastics experience some degree of property degradation from the effects of UV. This is usually visible on a product by chalking, crazing, and fading, or through physical tests with a loss in toughness usually caused by surface affects. Plastics suppliers offer UV-resistant grades of their popular resins. The most common UV absorber for plastics is carbon black. It is used in most resins at 1–2 wt% compounded into the base resin at time of manufacture. This causes a slight decrease in the resin's physical properties since carbon black is a foreign material used as a filler/colorant and not bonded to the base resin. Carbon black is hardly noticeable with reinforcement grades. Other UV-blocking additives are available for plastic resins, and this information is available from their material suppliers.

A resin's UV test data are obtained with direct outdoor exposure in Arizona and Florida, or in a "weather-o-meter," which is a special test cabinet with xenon or carbon arc lamps capable of generating high levels of UV exposure in a short period of time. The weather-o-meter produces alternating, wet/dry cycles, within the chamber on timed intervals to replicate the environmental exposure quickly. The correlation of weather-o-meter test data to actual exposure is not precise. It is an expedient way to develop data but must be evaluated with actual outdoor exposure. Most resins suffer loss in impact resistance in the first 6–12 months of UV exposure. Actual testing is the only way to ensure that requirements can be met.

Moisture Effects

All resins must be dried to low moisture levels before processing; otherwise the effects of moisture during processing causes the molecular chains to break with a drastic lowering of the material's physical properties. Hygroscopic resins, specifically nylons, will pick up moisture before and after molding. Moisture absorption in molded nylon parts reduces tensile properties and increases dimensional growth and toughness. When a product is used in a wet environment, the design stress and property values must be adjusted for the conditions that the product will experience. Molded products will stabilize over time within their environment and can be exposed to extremes of wet and dry conditions that can cause the product's properties to vary accordingly in each wet or dry situation.

Products exposed to boiling water or reusable medical products requiring frequent sterilization must be evaluated independent of just moisture absorption as repeated sterilizations can cause detrimental surface effects on some plastics. These can be rough surface finish, crazing, chalking, and color changes. Certain specific plastics can meet these requirements. Suppliers of these resins have data that define their properties over a wide range of conditions.

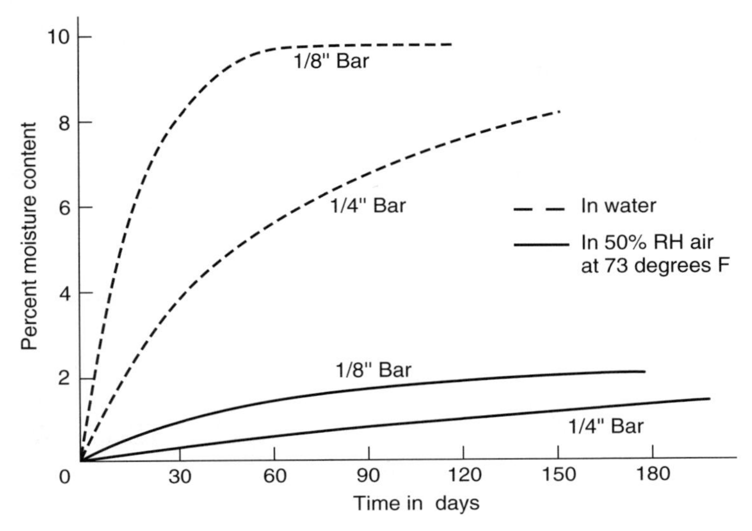

Figure 6.2 Moisture absorption rate, type 6 nylon (unfilled). (Adapted from Ref. 7.)

Moisture Consideration

Humidity (moisture) must be taken into account for hygroscopic plastic products. The design team is responsible for knowing the effects of moisture and water on the product and obtaining the physical property data to design for these effects from the material supplier. Any product used in a wet environment, for pump housings, impellers, propellers, containers for food products, or other applications, must be designed for these wet conditions, with respect to the level of exposure, moisture concentration, temperature, time, and end-use requirements. The resin selected must be evaluated for moisture effects on its physical properties. A plastic resin is either hygroscopic, absorbing moisture, or resistant to its effects. Nylon resins physical properties are affected by moisture, and the design team must know the product's operating conditions to correctly design the product. Different families of nylon, such as 6, 6/6, 6/10, 4/6, 11, and 12, all behave differently to moisture exposure. Each family of nylon absorbs moisture in varying amounts when exposed to wet conditions. This varies for the saturated state from 0.75% to 10% moisture absorption for each family of nylon. Moisture causes a lowering of tensile and flexural strength but increases impact resistance by plasticizing the resin. This is shown in Figures 6.2 and 6.3.

Nylon also expands as it absorbs moisture, and this effect must be considered in product design. Nylon, which is hygroscopic, absorbs moisture with a time/temperature rate occurrence. It takes about 6 months for a 0.125-in.-thick nylon 6/6 test bar to reach 50% moisture level in a controlled atmosphere held at 50% relative humidity and 73°F.

Other plastic resins, such as PC, PE, PP, PVC, Acetal, and PET, are not affected by moisture in the molded product state. Plastics are used in plumbing applications, with

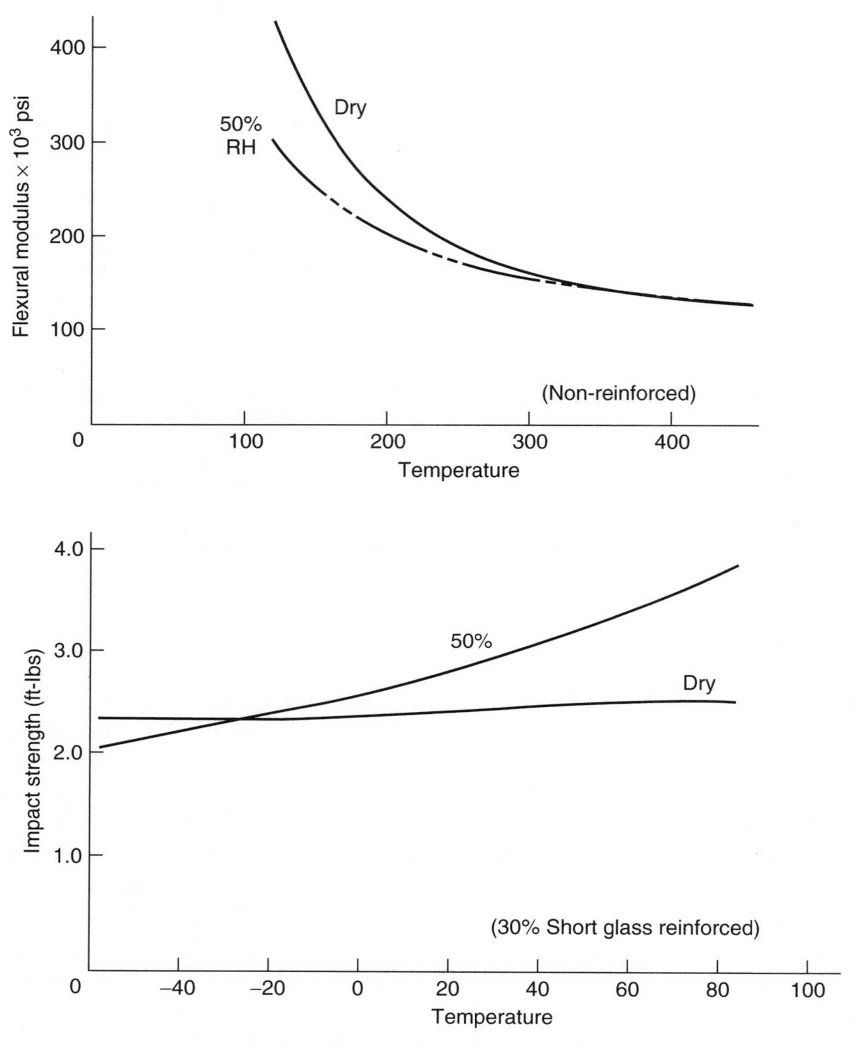

Figure 6.3 Moisture effects on nylon type 6. (Adapted from Ref. 5.)

the largest usage PVC for flexible and rigid potable water applications and drainage for both hot and cold applications. Special grades are available for each application, which meet rigid building and agency code requirements.

When designing a product for a wet environment, the designer has several options. Select a material not affected after molding by moisture or a hygroscopic material designed for all the anticipated effects the product will experience in service.

Hygroscopic materials absorb moisture, and in dry environments this moisture evaporates from the product over a period of time. It takes approximately 6 months in an environmentally controlled atmosphere at 73°F and 50% relative humidity (RH) for an $\frac{1}{8}$-in. nylon (type 6) bar to reach equilibrium at 50% RH. Submerging the bar in hot water reduces time or if exposed to high humidity levels at elevated temperatures. Moisture conditioning molded nylon products is done in special controlled areas to rapidly condition them and increases the impact resistance of the product from its DAM state. One company specified their molded nylon power tool housing to be moisture-conditioned by passing a specific drop test that the design team did not consider when designing the part. Moisture conditioning increases the product's drop impact strength. Moisture conditioning implies that the original design was poor and/or the material selected was marginal. Moisture conditioning can create new problems with dimensions and internal part alignment. Designing hygroscopic materials for impact always requires DAM data impact values. Then, when a product picks up moisture in service, it can only become tougher. But you should also consider dimensional changes when alignment of the motor's armature shaft and plastic housing bushing retainers are critical. Always test the final product in the worst end-use environment

Humidity Effects

Moisture can affect all plastic resins during processing, assembly, and decorating operations. All material suppliers recommend their plastic resins must be dried before processing, with some as low as 0.02% moisture as PET resins. Wet resins degrade in the high heat of the injection-molding machine's barrel, lowering the material's physical properties. Injection molders consider processing quality their responsibility and dry resins before processing to the manufacturers moisture content. See Table 6.2 for recommended drying moisture requirements for the generic resins. During the parts manufacture, assembly, and decorating it is the responsibility of the design team and production personnel, to ensure that the part's quality is not sacrificed (by moisture) during these operations.

ASSEMBLY CONSIDERATIONS

The design team must implement procedures to protect molded parts before assembly, including how the products should be handled, scheduled for follow-on operations, and protected to ensure that the finished product meets the requirements of the customer. All plastic products can be affected by moisture through absorption or surface contamination when the relative humidity in the plant is high. Moisture can affect assembly operations when thermal methods such as welding or fusion bonding, by spin, sonic, and vibration welding, are used to assemble parts. Parts must be kept dry-as-molded, or moisture at the joint area will turn to steam on heating, creating a weak and porous bond. Try to schedule fusion assembly operations immediately after molding when parts are at their maximum DAM condition.

However, if assembly is delayed, store parts in moistureproof bags with desiccants to keep moisture from collecting on the part surface. Paying attention to details early

Table 6.2 Recommendations for resin drying time and processing moisture level

Resin	Drying Temperature (°F)	Time (h)	Initial Moisture (% Virgin)	Maximum Residual Moisture (%)
ABS	180–200	2–3	0.45	0.02–0.10
Acetal	185	3	0.25	0.05–0.20
Acrylic	160–180	3–4	0.4	0.10
Cellulose acetate	160	2–3	0.7	0.10
Cellulose acetate butyrate	170	2–3	0.7	0.10
Cellulose propionate	170	2–3	1.0	0.10
Ionomer	150	8	0.32	0.05
LCP	300	8–24	NA	0.01
Nylon 6	180	4–5	0.50	0.10–0.20
Nylon 6/6	180	4–5	0.50	0.10–0.20
Nylon 6/10	185	4–5	0.50	0.15–0.20
Nylon 4/6	250	2–4	0.40	0.05
Nylon 11	170–200	4–5	0.70	0.02
Nylon 12	150–230	5	0.70	0.10
Polyaryletherketone	300	3	NA	0.10
PBT	284	3	0.25	0.02
PC	250	3–4	0.16	0.02
PET	250	2–3	0.25	0.02
PETG	160	4–6	NA	0.08
Polyester elastomers	175–220	2–4	NA	0.03
Polyethersulfone	300	3–6	0.43	0.02
Polyethylene	195	2	0.20	0.05
Polyethylene (40% black)	195	3	NA	0.05
Polyethylene (30% glass)	185	2.5–3.5	NA	0.05
Polyimide	360	5–10	0.32	0.05
Polypropylene	195	2	0.20	0.02–0.20
Polystyrene	180	2	0.10	0.02
Polysulfone	250	2–4	0.10	0.02
Polyurethane	180	3	0.90	0.01
PPE	190–230	2–4	NA	0.02
PPS	300	3–4	0.20	0.04
PVC	160	1	0.40	0.08
SAN	180	2	0.30	0.10
Styrene butadiene	140	2	0.60	0.02

Note: Drying time and processing moisture levels for hygroscopic materials obtained from resin supplier data sheets. For specific materials, filled, reinforced or color modified, refer to the manufacturers recommend drying and moisture levels including processing temperatures.

Source: Adapted from Ref. 6.

in the design and manufacturing process will save time and money and prevent future problems.

DECORATION CONSIDERATIONS

Decorating a plastic product can involve many different mediums. Molded-in colors with pigments and dyes, foils and labels, printing and silk screening, metalizing and painting are the main medias used to decorate and apply information on products.

Moisture in the molded part and on the surface can affect some of these processes. Hygroscopic materials, such as nylons, with molded-in colors can experience a change in the shade of the color with moisture absorption over time. The manufacturing schedule of exact color-matched molded parts should never allow an interval of more than a few days between treatments to ensure that the exact colors match. This problem occurred when a company did not follow a schedule for manufacture for mating parts that were the same color except in varying shades, which was very noticeable after assembly.

BACTERIA AND SOIL FUNGI

Many plastics are used in plumbing, including ABS, PVC, CPVC, and Acetal, both above and below ground, where they must retain their integrity and not be affected by their environment. Materials that carry potable water require NSF (Natural Sanitation Foundation) ratings when used in these applications. NSF lists these products under various use applications. Material suppliers have developed resins and compounds chemically resistant to attack or deterioration from bacteria, fungi and belowground insects, such as termites.

These materials are often under pressure ($\geq$30 psi typical home line pressure) and are exposed to hot water temperatures. They may also be subjected to water hammer, with air trapped in the lines, which magnifies line pressure several times when this problem occurs. Plastics used for pressure vessels or control valves must resist repeated shock loading while under constant pressure. They often have to be resistant to various gases. The permeation rate of different gases is usually performed on films of resins deemed suitable for these applications. Some of these data are available in the McGraw-Hill *Modern Plastics Encyclopedia,* but the majority of permeation data are from the material suppliers. The typical data available are on water vapor, oxygen, carbon dioxide, nitrogen, and helium.

FOOD PRODUCTS

Materials that come in contact with food products require FDA certification. Food containers or serving dishes used in microwave heating ovens must also meet FDA requirements. Additives must not leach out during normal storage, or when heated, they could contaminate the food products. The manufacturer of the product must be told that the FDA must approve any external mold release.

STERILIZABLE PLASTICS

Plastics are used extensively in medical and food applications, with over 60% designed for one-time use, as disposable items. Resins used for medical products must be sterilizable and free of additives or compounds that could leach out, during manufacture and come to the surface while in use. Specific formulated FDA-approved plastic resins are compounded to meet agency requirements for their use in food and medical products.

Food- and packaging-grade resins are often chemically washed to eliminate any internal manufacturing additives and ingredients that could contaminate a packaged product. After packaging, medical products require sterilization to ensure that they are free of all forms of contamination and bacteria and are safe to use. The methods used for sterilization of some packaged food products are also used extensively for medical applications.

Types of Sterilization

The three sterilization methods used for comparison in Table 6.3 are

1. *Heat*—dry and steam [100°C and (270°F)] (minimal use)
2. *Gas*—ethylene oxide (ETO) (decreasing usage)
3. *Radiation*—gamma or electron beam (gaining usage)

Heat Sterilization. Heat or steam sterilization is used at the point of use (hospital or clinics) to reuse products. It is not used by major medical suppliers to sterilize their products before shipping. Steam sterilization is used selectively since the sterilization temperatures are too high for most resins to be used as syringes. These high-use products are now single-use items. Also, steam heat can cause surface effect problems for reusable products such as surface crazing, roughening of the surface, and making some materials brittle, eliminating them for reuse. Only selected plastic materials are deemed heat/steam-sterilizable. Therefore, the majority of medical parts are one-time use for health, liability, and cost reasons. The sterilization of these parts employ either ETO or radiation sterilization methods.

ETO Sterilization. ETO is a colorless toxic gas. It is mixed in a pure form with carbon dioxide, an inert gas, to sterilize products. Products are packaged in a porous membrane to allow the ETO gas to penetrate the package and kill any bacteria or foreign organism on the product. Products must be moisture-conditioned at 30–60% relative humidity to assist the ETO gas in permeating the product surface and subsurface in a pressurized chamber. The time required for ETO to sterilize a product

Table 6.3 Comparison between ethylene oxide, gamma radiation, and heat/steam sterilization

Basic Feature	Ethylene Oxide	Gamma Radiation	Heat/Steam
Physical parameters	Time Temperature, RH Pressure, ETO Concentration	Time	Point-of-use time & temperature
Microbiological control	Essential	Not required	Not required
Product mix	Single product	Mixed product	Mixed product
Material of construction	Most satisfactory	Most satisfactory	Selected
Packaging	Porous essential	Hermetic seal possible	None
Radiation	Batch only	Batch or continuous	—
Cost	Good	Very good	Low
Reliability	Good	Very good	Good

depends on the part's density, material, and thickness. After sterilization the product must remain in the chamber until the ETO gas has been aerated from the package. This is a major drawback and the reason why ETO sterilization is rapidly decreasing in use.

Radiation Sterilization. Medical products use a wide range of plastic resins, and the type of radiation used for sterilization must be known to eliminate any product problems when exposed to the sterilization process. Plastics exposed to a heterogeneous radiation flux of atomic pile experience increased strength but when exposed for longer periods of time became brittle with a loss in toughness. Gamma radiation is used predominately for preserving packaged food and sterilization of medical parts. Studies have been conducted on films that show the effects on a plastic's chemical structure and resultant physical properties. Should your part be subjected to radiation in its end-use environment, your material suppliers should be consulted to ensure that the resin radiation type and exposure time are within the materials capability.

Election-beam (E-beam) and gamma radiation are the two basic types of radiation used for sterilization. Packaged products are easily automated for this method of sterilization, and the package is an airtight container. The radiation easily penetrates the package and quickly sterilizes any plastic inside. The advantage with radiation sterilization is that the product can be shipped immediately after exposure.

Gamma-Radiation Sterilization. Gamma-radiation sterilization is the least expensive and has fewer process variables. Sterilization involves hermetically sealing the product prior to sterilization. Gamma-radiation sterilization takes several hours, with the parts exposed to cobalt 60 or cesium 137. Gamma rays easily penetrate thick and thin products of varying density, traveling in all directions through the part. Product orientation to the radiation is not a factor during sterilization. Sterilization time is determined by the time necessary for the product to absorb the required megarad dosage to kill all foreign organisms and bacteria. Gamma sterilization is a low-temperature operation, under 100°F and suitable for all plastic materials.

E-Beam Radiation Sterilization. E-beam radiation is faster than gamma radiation and uses a single focused beam of electrons projected onto the product, which causes a 10–20°C temperature increase in the product's material. It is very fast, taking only minutes to sterilize the product. E-beam radiation is highly suited for automated production line sterilization. It is directionally aimed on the part and controlled so that the time of exposure and effects on materials of thin part sections are minimized. It is easier to control than ETO. Table 6.4 shows the effects of radiation on some plastics used for medical applications.

Each plastic may exhibit a side effect from either ETO, gamma-, or e-beam sterilization. The material supplier for the product should be contacted to obtain their guidance for the correct resin best suited for a product sterilization requirement. Major material suppliers have developed specific resin grades for food and medical applications.

TRANSPARENT PLASTICS

Clear or transparent plastics have found new applications to replace glass in products requiring clarity and increased toughness. New transparent resin grades are tougher, lighter, and more chemical-, heat-, and abrasion-resistant than glass. Transparent plastics are amorphous in structure and are modified for impact strength, high heat distortion, and high-temperature use with better chemical resistance, which these resins lacked when compared with opaque crystalline resins. Polycarbonate, acrylic, polysulfone, and polystyrene are the major transparent materials. New grades and compounds are being developed that provide the designer with more product options. All of these resins have their good and weak points, and designers should consult with their material suppliers to find out which is best for their application. A list of the major resins available is shown in Table 6.5 with comments on their notable characteristics and suppliers for specific product information.

Several resin grades are colorless and have light-transmitting properties close to, or better than, those of glass, especially in UV transmission. These materials are acrylics, polystyrene, SMMA, polycarbonate, and polymethylpentene. A material's optical properties are expressed in terms of its refractive index (RI), light or luminous transmission, and haze. RI is the ratio of the velocity of light in free space to the velocity of light in the material. Light transmission is the ratio of light exiting from an optical material to the light entering the material. The degree of light transmission is affected by the amount of color inherent in the resin and the part's thickness. In some resins, the thicker the product, lower the clarity. The material supplier closely controls this material's property during the manufacturing process. Areas to

Table 6.4 Effects of radiation on selected plastics

	Radiation Dose to Produce Significant Number[a]		
Plastic Material	Radiation Stability	Damage (Mrad)	Potential Sterilizations
ABS	Good	100	M
Acetals	Poor	1–2	0
Acrylics		5	
PMMA	Fair	5	1
Others	Fair	10	1–2
Amides			
Aliphatic	Fair	50–100	1–2
Aromatic	Excellent	1000	M
Cellulosic	Fair	20	1–2
Fluoroplastics			
PTFE	Poor	1	0
PCTFE	Fair	10–20	1–2
FEP	Fair	20	M
PVF, PVDF, ETFE, ECTFE	Good	100	M
Polycarbonate	Good	100+	M
Polyesters (PET, PBT)	Good	100	M
Polyolephins			
Polyethylene	Good	100	M
Polypropylene	Fair	10	1
Polymethypentene	Good	30–50	M
Copolymers	Good	50	M
Polystyrene	Excellent	1000	M
Copolymers	Good	100–500	M
Polysulfones	Excellent	1000	M
Polyvinyls			
PVC	Good	50–100	M
Copolymers	Fair	10–40	1–2
Polyphenylene sulfide	Excellent	1000	M
Styrene/acrylonitrile	Very good	100+	M
Silicone	Good	20–100	1–2
Polyetherimide	Excellent	1000+	M
Epoxies	Excellent	100–10,000	M
Phenol (or urea) formaldehyde	Good	500	M
Polyesters (thermoset)	Good	1000	M
Polyimides	Excellent	100–10,000	M
Polyurethanes	Excellent	1000+	M

[a] Sterilization potential of polymers divided into four categories: 0 = nonsterilizable; 1 = one procedure; 1–2 = somewhat more than one; M = multiple sterilizations possible.

Table 6.5 Transparent polymers[a]

Generic Family	Trade Name	Notable Characteristics
ABS, transparent	Magnum (Dow)	Good impact and processability
Acrylic (PMMA)	Plexiglas (Rohm & Hass) OP (Continental) Lucite Acrylite (Cyro)	Excellent UV, crystal-clear
Allyl diglycol carbonate (thermoset)	CR39 (PPG)	Good abrasion and chemical resistance
Cellulosics (acetate, butyrate, propionate)	Tenite (Eastman)	Good impact, limited heat and chemical resistance
Nylon 6, 6/6, 6/10/11/12, ect.	Zytel, Dupont Grilamid (Emser) Trogamid T, ect.	Excellent toughness and abrasion/wear
PET, PETG, PBT	Tenite (Eastman) Petra, Honeywell Selar (Dupont), ect.	Good barrier, clarity, toughness,
Polyarylate	Durel (Hoechst) Arylon (Dupont) Ardel (Amoco)	Good UV, high heat distortion
Polycarbonate	Calibre (Dow) Lexan (GE) Makrolon (Mobay)	Excellent toughness, thermal and flame resistance
Polyetherimide	Ultem (GE)	Good chemical/solvent, thermal and flame properties, inherent high color
Polyester (polyphthalate) carbonate	Lexan PPC (GE)	Good thermal autoclavable
Polyethersulfone	Victres (ICI America)	Excellent thermal and resists creep
Polymethylpentene	TPX (Mitsui) Crystalor (Phillips 66)	High Crystalline melt point, lowest density of all TP, UV, and moisture-sensitive
Polyphenylsulfone	Radel (Amoco)	Excellent thermal stability resists creep
Polystyrene	Styron (Dow) Polystyrol (BASF) Hostyren (Horechst) Chevron, Mobil, ect.	Brittle, poor UV, excellent clarity
Polysulfone	Udel (Amoco)	Excellent thermal and hydrolytic stability, poor weather and impact
PVC	Geon, (BF Goodrich) Multiplesuppliers	Excellent chemical and electrical properties, weather resistance, toughness
Styrene acrylonitrile	Tyril (Dow) Lustran (Monsanto) Luran (BASF), ect.	Good stress crack and craze resistance, clarity, brittle
Styrene butadiene	K-Resin (Phillips 66)	Good stress crack, craze resistance, brittle
Urethane, rigid	Isoplast 30(Dow)	Excellent chemical and solvent resistance, good toughness

[a] For an updated listing of transparent plastics, see the *Modern Plastics Magazine,* New York, NY.

consider when selecting a clear plastic other than the materials optical and physical properties are

1. Chemical resistance to the environment
2. Abrasion resistance (coatings are available but may not be economical as it is a secondary operation)
3. Part thickness (some resins have a critical thickness that changes as thickness increases from being ductile to brittle)
4. Process ability, heat stability in the machine to not turn color and flow length in the mold color and property retention over the life of the part (UV exposure can seriously reduce some resins' physical properties and light-transmission properties)

PRODUCT DESIGN CONSIDERATIONS

These considerations are as follows:

1. Mold in all holes or opening, as secondary machining operations will show in the part.
2. Abrupt changes in part thickness can cause voids to form in these sections.
3. Gate marks will show on the part, so locate the part away from visible surfaces.
4. Flow or weld lines may be visible in the part, resembling cracks. The mold must have excellent temperature control to eliminate or reduce these problems.
5. Assembly operations should avoid.
   a. Self-tapping screws (mold- in threads if required).
   b. Use inserts, molded-in, for secure attachments.
   c. Any heat staking or sonic welding of parts will be visible as cloudy areas when the material melts and resolidifies
   d. Solvent bonding may be possible, check with your supplier.

The choice of selecting the correct clear plastic for an application is based on the total program requirements, clarity, performance, process ability, assembly, and finally cost.

AGENCY AND CODE REGULATIONS

There are regulatory considerations for some product applications. It is the designer's responsibility to work with the customer to ensure that these regulatory requirements are achieved. This may involve various government and private agencies that require the base plastic material to meet certain chemical and/or physical property and end-use test requirements or specifications. Each material supplier is responsible for submitting their resins for approval in these specific product applications. When this

Figure 6.4 International Safety Agency logos.

is a requirement for a product, only approved resins and suppliers will be considered. The testing and approval times are both lengthy and often very expensive to obtain approval for a new formulation. Also, if products are sold abroad, the designer must be aware of foreign government and agencies where material approval or certification is required.

There are over 22 agencies that have responsibility for material specifications for plastic resins and compounds; the major international safety agencies are listed with their logos in Figure 6.4. There are even more when specific company requirements are considered for their product requirements. The major agencies—federal, foreign

Table 6.6 Regulating agencies

Specification	Definition
ASTM	American Society for Testing and Materials, 1916 Race Street, Philadelphia, PA 19103. ASTM defines standard test methods for determining the essential properties of plastic resins. The data published by material suppliers conform to ASTM test methods unless noted otherwise. The *Standard Guide for Identification of Plastic Materials* is D4000-82. The last two digits are the year of issuance or latest revision. This guide provides a classification system for tabulating the properties of unfilled, filled, and reinforced plastic resins that are suitable for parts manufacture. Each generic resin family will have its own ASTM designation for identifying specific grades of material.
DIN	Deutscher Normenausschuss, 4-7 Burggalenslrasse, 1 Berlin 30, Germany. This is the European counterpart to ASTM, which is the German Institute of Standards.
ISO	International Organization for Standardization, Central Secretarial, 1, Rue de Varembe, 1211 Geneva 20, Switzerland. This is the current universal, world-class, ISO 9000 company quality certification standard for the design, production, and inspection of products now being adapted.
QS9000	Automotive harmonization of DaimlerChrysler AG Supplier Quality Assurance Manual, Ford Q-101 Quality System Standard, and General Motors NAO Targets for Excellence, with input from the truck manufacturers.
QS-TE	New tooling manufacturer's requirement released in January 1999 for suppliers of machines and tooling involved in the manufacture and molding of automotive products. This will be a supplement to QS 9000.
FDA	U.S. Department of Health, Education, and Welfare, Food and Drug Administration, Washington DC, 20204. Involves "individual additives," such as any substance that could migrate from the plastic into food contacting the surface. FDA regulates drugs and medical devices, not materials. Therefore, suppliers and designers must be aware that the use of plastics in some applications must be in full compliance with the Medical Service Amendments of 1976.
USDA	U.S. Department of Agriculture, Washington DC, 20250. The WSDA controls the use of materials in federally inspected meat and poultry processing facilities and packaging materials used for these products.
UL	Underwriters Laboratories, 1285 Walt Whitman Road, Melville, NY 11746. UL is an independent, nonprofit testing laboratory primarily involved in evaluating equipment safety for general sales. Many state and local governments require UL recognition of products, especially electrical, before they can be sold and/or installed in their jurisdictions.
CSA	The Canadian Consumer Safety Administration equivalent of UL. Their requirements are independent of UL tests but may be similar in scope. Products must meet CSA requirements to be sold in Canada.

Table 6.6 (*Continued*)

Specification	Definition
NSF	National Sanitation Foundation, PO Box 1468, NSF Bldg. G., Ann Arbor, MI 48106. Plastic materials are listed after testing for acceptable taste, odor, and toxicity ratings. NSF listing is necessary for food processing equipment and pipe and fittings for potable water.
3-A	3-A Sanitation Standard Committee comprised of the International Association of Milk, Food and Environmental Sanitarians, the United States Public Health Service and the Dairy Industry Committee. Controls the use of plastics for multiple-use product contact requirements and the clean ability requirements established for these industries.
MIL, FED, AEC, GSA, ANSI (NAS), AS, CS/PS, OR/OS	These government specifications for plastic material approval can best be researched for acceptability in applications by contacting the material supplier directly.
VDE, VDI	See DIN.
NEMA	National Electrical Manufacturers Association, 2101 L Street NW, Washington DC 20037. NEMA controls the use of plastic materials that carry electrical power require their certification and recognition.
AUTO	Each automotive supplier certifies and lists supplier materials to their individual material specifications. Suppliers will have a list of their specific resin grades approved to automotive specification.
Others	Each firm requiring material certification to meet their specifications must be contacted to determine what supplier resin grades are their parts.

government, and private—are listed in tabular form in Table 6.6 for easy reference. The others are listed in the McGraw-Hill *Modern Plastic Encyclopedia,* mid-December 1992 issue in their "Specification/Materials Chart" (pages 224–225) for suppliers submitting information. The designer must always have the latest revision for the regulations they must meet. They are often being revised with new data, with materials added and deleted, and possibly rewritten to suit changing requirements.

Your material supplier is your first source of information for material and design requirements and can supply information on the regulations and approved materials.

REFERENCES

Designing with Plastics, Design Handbook, Hoechst Celanese (TDM-1) HCER-92-313/10M692: 4-4, Chatham, NJ.

L. English, "Some Common Problems with Thermoplastics," *Materials Eng.* 47–50 (Aug. 1989).

A. Klem, Sr., "Plastics that Withstand Sterilization," *Plastics Design Forum* 45–58 (Nov./Dec. 1987).

Designing with Plastics, Design Handbook, Hoechst Celanese (TDM-1) HCER-92-313/10M692:8-7, Chatham, NJ.

General Design Principles—Module II, H-44600, E. I. Du Pont de Nemours Corp., Wilmington, DE. (Sept. 1992).

"Purchasing, Resin Dryers," in *Injection Molding Magazine, Almanac,* Abby communications, Inc., Chatham, NJ, May 1999, pp. 22–24.

Designing with Plastics, Design Handbook, Hoechst Celanese (TDM-1) HCER-92-313/10M692:5-1, Chatham, NJ.

CHAPTER 7

TEMPERATURE AND ELECTRICAL PROPERTY EFFECTS ON PLASTICS

The design team must understand how temperature rise and fall affects the properties of a plastic product. Plastics are affected more by both high and low temperature than are metals. The majority of metals are not seriously affected by a change in temperature, as are plastics. Metals at elevated temperature lose only a small percentage of their physical properties, unlike plastics. Metals have a very minimal loss of property strength, in contrast to plastics, where the properties of each are comparable to the same change in temperature. Plastics are more sensitive to temperature changes, and when exposed to elevated temperatures, the amorphous resins soften and quickly lose their room temperature (73°) physical property values as the temperature increases above 150°F. The engineering resins, semicrystalline, crystalline, and LCPs, have higher room temperature physical properties and experience a lower decrease in physical properties as the temperature is increased above 150°F. The LCP polymers experience an even lower loss of properties as temperature is increased. Therefore, the product end-use temperature range must be known when designing the product. The same is true for all conditions that the product must perform in the end-use environment.

TEMPERATURE AND THERMAL EFFECTS ON PLASTICS

The end-use temperature that a product experiences has a dynamic effect on how it is designed. Some plastics are sensitive to minor temperature changes of only ±20°F, unlike metals. Therefore, the product's end-use temperature range must always be considered during design. A plastic material's physical properties, such as creep, modulus, tensile strength, and toughness, are all affected by increasing temperature

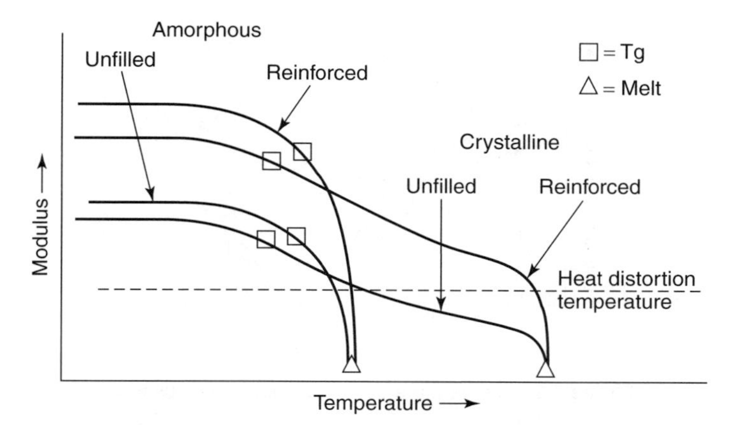

Figure 7.1 Temperature effects on plastic resins. (Adapted from Ref. 1.)

as shown in Figure 7.1. Amorphous thermoplastic resins' physical property values decline more rapidly than do those of crystalline resins because of their weaker molecular structure and bonds. They soften and lose their physical properties more rapidly when subjected to increasing temperature. Crystalline, semicrystalline, and LCP resins are less affected by increasing temperature because of their crystalline molecular structure, which has strong molecular bonds. As a result, these materials show a lower decrease in physical properties with increasing temperature. They also retain their physical shape, when not under load up to their melting point (glass transition temperature, T_g), at which time they begin to melt like an icecube. These materials will bend and creep as temperature reduces their physical properties, but not soften and flow like the amorphous resins.

A common question asked by designers is "What is the material's maximum or end-use temperature capability?" To better understand the characteristics of plastics, the following thermal terms are defined.

Heat Deflection Temperature

Heat deflection temperature (HDT) is not the material's upper-limit use temperature. HDT is the temperature at which an ASTM standard test bar, supported at each end and loaded in the center, under a constant stress of 66 or 264 psi, has deflected 0.010 in. as the test specimen is submerged in an increasing temperature-controlled oil bath. A material's HDT value is used by the designer to access the capability of a material's property at elevated temperatures under a specific load. The test is performed with a specified hold time at each temperature to equalize the test bar's temperature to obtain repetitive and standardized test results. Submersion eliminates the effect of oxidation on the sample's surface, and oil is a very uniform heat transfer medium. Some resins oxidize rapidly at elevated temperatures, which may cause surface crazing and a rapid decrease of their physical properties.

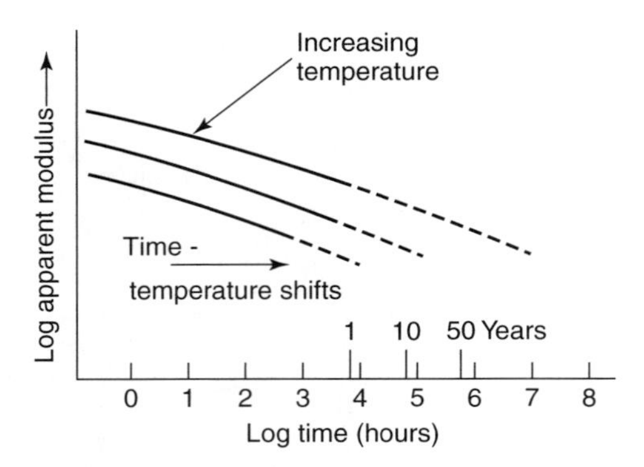

Figure 7.2 Temperature effect on modulus versus time.

Thermal Aging and Oxidation Effects

Effects of thermal aging and oxidation on plastics are determined by oven-aging test bars. Nonstressed test bars are placed in an oven at various temperatures and aged for extended lengths of time. Test samples are removed at set intervals and their physical properties tested in tension and flexure. Data are plotted on graphs as specific physical property values versus aging temperature and time. The values are determined at room temperature. These data are used to measure the thermal stability of materials and are also used during design as shown in Figures 7.2 and 7.3 for estimating the flexural modulus of a material over temperature and time. Oxidation may cause surface crazing, which is visible as fine cracking of the outer skin surface caused by oxygen breaking the molecular bonds of the material at elevated temperatures.

Chalking is a longer-term oxidation effect that is not as serious as and is more aesthetic than crazing, but the surface discolors. Nylon parts used as boat hardware and exposed to direct sunlight will chalk over time and can be returned to as-molded condition by cleaning with a mild abrasive cleaner. Oxidation surface effects can be reduced by blending antioxidant additives in the resin during processing and seldom affects the product's physical properties when added in small amounts. Carbon black is the most commonly used and effective antioxidant additive, blended into the resin by the manufacturer.

Heat Resistance

There are essentially two types of heat resistance additive agents. The first is to protect the material from surface oxidation when exposed to elevated temperatures. Heat stabilizers are blended into resins to reduce oxidative effects on the product surface when exposed to elevated temperatures. Heat stabilizers are used for all plastic materials used in high-temperature applications. The heat stabilizer in nylon 6/6, from the DuPont Company, gives the resin a slight greenish hue that is in contrast to the material's natural milky white and opaque color. The second type of heat stabilizer is used to protect the resin during processing. Heat-sensitive resins need protection

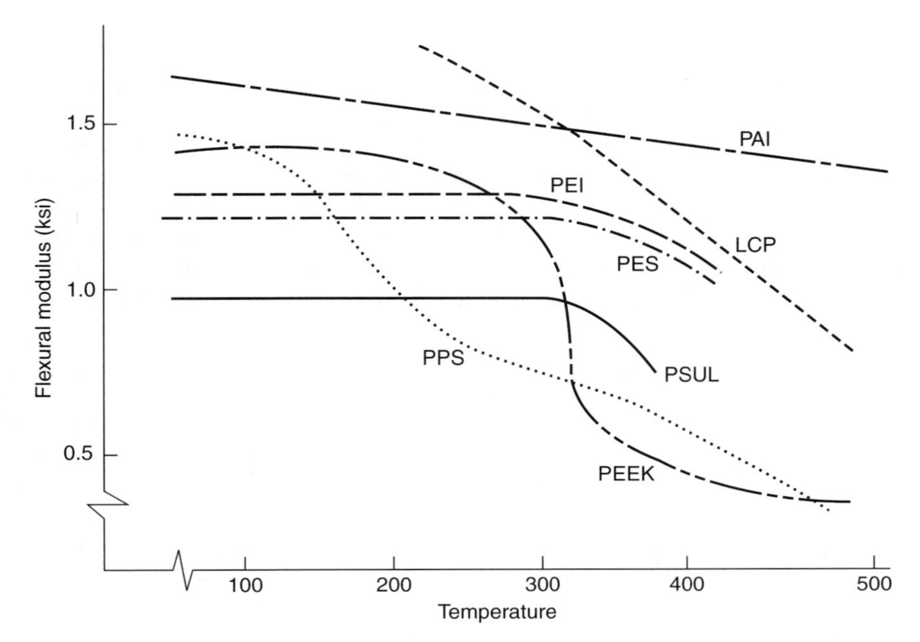

Figure 7.3 Material property ranges for 30% glass (short)-reinforced materials. *Legend:* LCP, liquid crystal polymer; PAI, polyamide imide; PEI, polyetherimide; PES, polyarylsulfone; PPS, polyphenylene sulfide; PSU, polysulfone; PEEK, polyetheretherketone. (Adapted from Ref. 2.)

during processing to keep the material from discoloring and degrading due to the high heat and shear temperatures in the injection machine's barrel used to melt the resin.

PVC resin requires this protection as it will rapidly oxidize and discolor during processing. The resins requiring process protection are supplied with the additives in the resin when purchased. However, some processors blend and make up their own compounds and are told that the best processing aids to use are available from the base resin supplier, based on their experience. The heat stabilizer processing additive inhibits resin oxidation but does not provide any protection for the product at elevated end-use temperatures.

End-Use Temperatures

A plastic resin's end-use temperature limit is based on the material's molecular structure, reinforcement, and forces on the product, environment, product additives, and product life. New plastic resins, such as LPCs, have been developed that can survive 400–500°F in specific applications. Each resin will have a threshold use temperature limit that should not be exceeded. Be sure to discuss with your resin supplier the product requirements when selecting the material type to meet the product's end-use applications.

Temperature Index

Temperature index is another method used to rank a material for an end-use application. Satisfactory continuous end-use temperatures for electrical and mechanical applications are listed on the resin's property data sheets and assumed acceptable for the resin tested. Temperatures are listed separately for electrical and mechanical applications, with and without impact. Data are listed in the resin's DAM property sheet in tabular form for evaluating a material for specific end use applications and available from UL and material suppliers. Data are also available on some resins for brittleness temperature use in cold applications and found on the resin's property data sheets.

Coefficient of Linear Thermal Expansion

All materials when heated or cooled experience a linear and volume change in dimensions. Plastics experience greater thermal contractions and expansion than do metals. When they are attached to or mate with metal or other dissimilar materials, the design

Table 7.1 Typical coefficient of linear thermal expansion values[a]

Material	(in.) (in.)/°F $\times 10^{-5}$	(cm) (cm)/°C $\times 10^{-5}$
ABS	4.0	7.2
Acetal	4.8	8.5
Acrylic	3.8	6.8
Nylon	4.5	8.1
Polycarbonate	3.6	6.5
Polyethylene	7.2	13.0
Polypropylene	4.8	8.6
Polypropylene sulfide	2.0	3.6
TP polyester	6.9	12.4
ABS (GR) (glass-reinforced)	1.7	3.1
Acetal (GR)	2.2	4.0
Liquid crystal (GR)	0.3	0.6
Nylon (GR)	1.3	2.3
Polycarbonate (GR)	1.2	2.2
Polypropylene (GR)	1.8	3.2
TP polyester	1.4	2.5
Epoxy	3.0	5.4
Epoxy (GR)	2.0	3.6
Aluminum	1.2	2.2
Brass	1.0	1.8
Bronze	1.0	1.8
Copper	0.9	1.6
Steel	0.6	1.1
Glass	0.4	0.7

[a]Plastic materials unfilled unless noted by (GR) (TP = thermoplastic). Typical values for generic resins not accounting for grade difference. Values can vary, request information for specific grade from supplier.

team must consider their expansion and contraction rates. Table 7.1 lists typical values for various material coefficients of linear thermal expansion. The designer needs to be aware that many plastics, especially fiber-reinforced and liquid crystalline resins, will have different coefficients of expansion in the flow and cross-flow directions. This must be considered when the product is composed of various materials.

Thermal Stresses

A product subjected to a force has physical material stresses in its structure. When a product is anchored in place and restricted in its movement, and then heated or cooled, thermal stresses are created in the product. When materials contract or expand, they create forces within the product known as *internal thermal stresses*. The induced stress can magnify the forces acting on the product and affect the product and attached components if they are not free to move to release the stresses in the part. These stresses, if not considered and compensated for during design, can cause poor part performance and possible product failure.

When joining dissimilar materials, as shown in Figure 7.4, each material should be designed to move independent of the other components whenever possible. This is accomplished by providing slotted attachment holes or flexible adhesive systems or selecting mating materials with similar coefficients of thermal expansion. If this is not possible, the design of the plastic product must be reinforced and stiffened so that it will not buckle when anchored. This condition creates stresses in the anchored part that are transmitted to the adjoining parts and must be considered in the product's design, or else one of the other adjoining parts may fail.

The mathematical relationship for thermal expansion of a product is shown in Figure 7.5, and the following equations apply:

$$\Delta L = \alpha L \Delta T$$

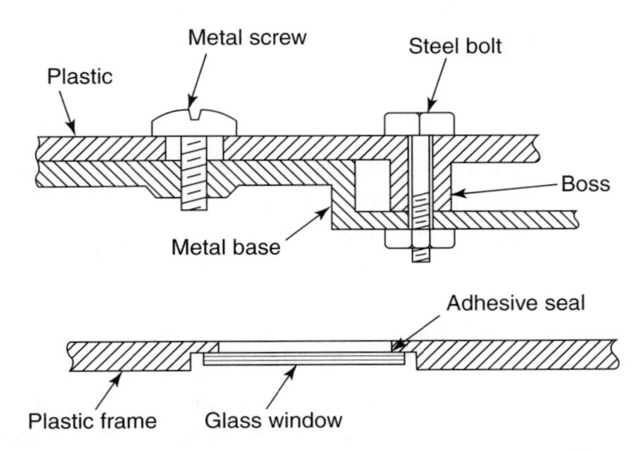

Figure 7.4 Assemblies with possible thermal stress problems.

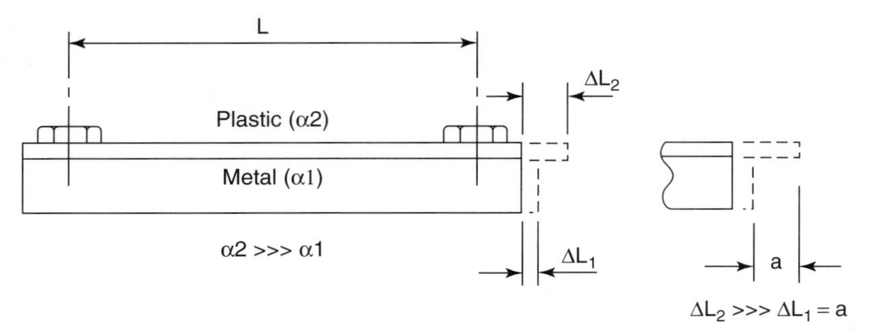

Figure 7.5 Design to eliminate expansion stress (to prevent buckling).

where ΔL = change in part length
α = coefficient of linear thermal expansion (see Table 7.1)
L = length of part (include hole diameters)
ΔT = temperature change

The strain induced with fixed parts, with no expansion or contraction, by the temperature change is:

$$\epsilon T = \frac{\Delta L}{L} = \alpha \Delta T$$

Stress is then calculated by multiplying the strain by the tensile modulus of the resin at the specified temperature.

$$\sigma = \epsilon T E_{\mathrm{p}}$$

where E_{p} is the flexural modulus of plastics at design temperature

When a plastic part is firmly anchored to a metal part, which expands less, the approximate thermal stress in the plastic is

$$\sigma T = (\alpha_{\mathrm{m}} - \alpha_{\mathrm{p}}) E_{\mathrm{p}} \Delta T$$

where α_{m} is the coefficient of thermal expansion of metal or mating material and α_{p} is the coefficient of thermal expansion of plastics.

It is possible to reduce expansion stresses by controlling the product's section thickness and allowing controlled buckling of the plastic part.

Controlling Differential Thermal Expansion

Thermal expansion effects in a plastic part firmly mounted to a metal surface in a hot environment can be accommodated in the product design. Plastics are used for engine valve covers, which must not buckle or warp, and require a tight seal against the gasket-sealing surface. The use of a reinforced plastic with a low coefficient of thermal expansion will limit the amount of differential growth of the plastic and seal surface. The designer now has two methods to ensure that the seal is tight. Design the valve cover walls and sealing surface to resist the thermal buckling forces with ribs to strengthen the cover's sidewalls and a channel section with internal ribs to add strength to the gasket's sealing surface.

Plastic fuel rails mounted on the engine deliver gasoline to the engine's fuel injectors. Fuel injectors require precision alignment, and any expansion of the fuel rail must be compensated for in their attachment to the engine. During engine operation, the fuel cools the fuel rail, limiting the amount of thermal expansion. The thermal expansion problem occurs only when the hot engine is turned off and heat soak increases the temperature of the fuel rail and it expands in length. The design of the fuel rail and anchor points limits the amount of thermal expansion between the plastic rail and each fuel injector. Fuel rail anchor sites were calculated using a glass-reinforced plastic resin rail to limit and control linear growth of the fuel rail. The fuel injectors used compression seals to absorb the minimal amount of expansion, which solved this differential expansion problem. The current materials of choice are filled thermoset and polyphenylene sulfide, which are resistant to alcohol-based fuels. Another plastic engineering resin used is 33% glass-reinforced, heat-stabilized 6/6 nylon.

In the fuel rail example, the rail has a solid anchor at each of its ends for dimensional control. The plastic part, therefore, must be designed to resist expansion stresses that can cause the part to buckle or bend. Referring to Figure 7.6, the problem is developed accordingly. The plastic rail must be designed to resist the stress of expansion (S_{exp}):

$$S_{\mathrm{exp}} = \frac{a}{LE} \qquad \text{(expansion differential)}$$

where

$$\begin{aligned} a &= \Delta L_2 - \Delta L_1 \\ \text{Expansion } \Delta Ln &= \alpha n \Delta T L \\ \Delta T &= T_{\mathrm{max}} - T_{73^\circ\mathrm{F}} \\ \alpha n &= \text{coefficient of linear expansion of material at } 73^\circ\mathrm{F} \end{aligned}$$

Critical expansion stress (S_{cr}) is defined as:

$$S_{\mathrm{cr}} = \frac{P_1}{\mathrm{area}_p} \quad \text{(cross sectional area of the plastic part)}$$

where $P = 4\,\pi^2 EI/L^2$
E = flexural modulus plastic at elevated engine soak temperature
I = moment of inertia of plastic parts cross section.

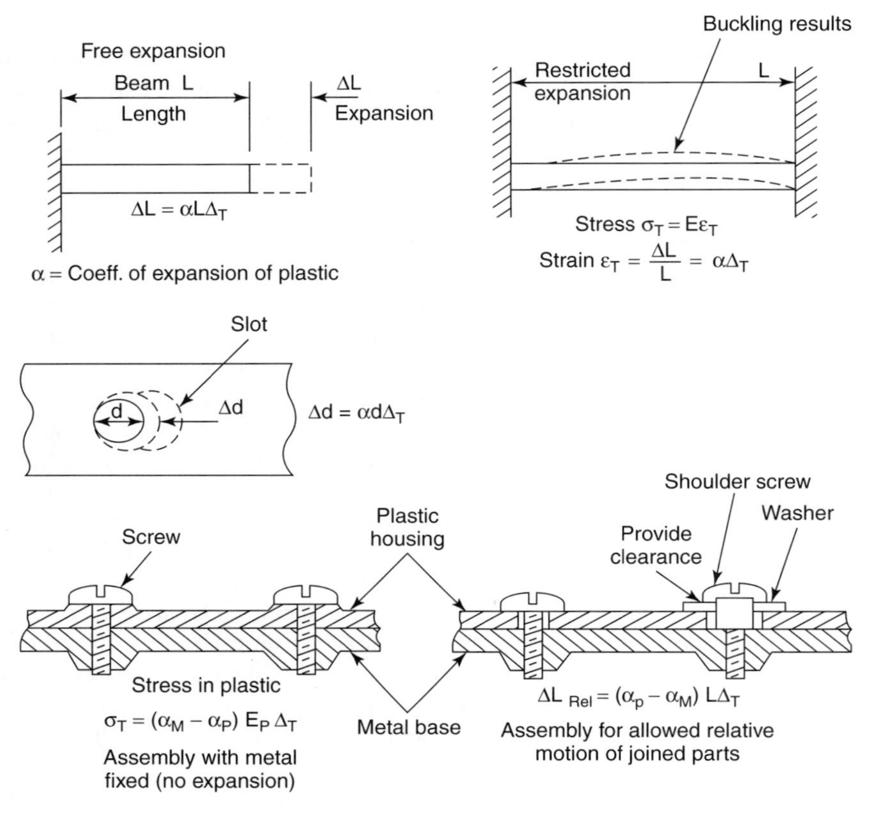

Figure 7.6 Thermal expansion considerations.

The design criterion states that the critical stress (S_{cr}) must be greater than the stress of expansion (S_{exp}), so the part will not deflect or buckle. Thus S_{cr} is greater than S_{exp}.

The flexural modulus selected for rigidly attached plastic parts must be selected for the maximum expected temperature. The plastic will be put in compression as the rail expansion (L) is restricted and buckling forces maximize. Conversely, as temperature drops, the plastic part shrinks more than the metal does, and the stress situation changes to tensile stress in the part. Therefore, each condition of operation must be evaluated. Also, additional anchor points on the plastic part may be necessary to control segment expansion and contraction values between fuel injectors for dimension control and stress reduction.

ELECTRICAL PROPERTY ANALYSIS

Thermoplastics and thermosets are excellent insulators of electrical energy. They are used extensively in the electrical and electronic industries to provide insulation for current-carrying applications for electrical and electronic equipment. PVC compounds provide the primary jacket over the copper conductor, with insulation for commercial and industrial electrical wiring. Thermosets and thermoplastic have proved their values in all except the highest-energy transmission line applications. Their excellent mechanical properties combined with their electrical insulating properties make them an excellent material for switches, plugs, connectors, coil forms, and other current-carrying devices. Important electrical properties of plastics used for electrical application are described in the following sections.

Volume Resistivity

How well a material conducts an electrical current (conductivity) is a measure of the electrical resistance of the material when a direct potential is applied to it. This measurement, known as *volume resistivity,* is the resistance measured in ohms times the area of the smaller electrode divided by the thickness of the specimen, reported as Ω·cm as shown in Figure 7.7a. Test results can vary depending on sample test conditions such as temperature, relative humidity, test voltage, and moisture content of the sample. Plastics are considered good insulators with values above 10^{14} Ω·cm, while

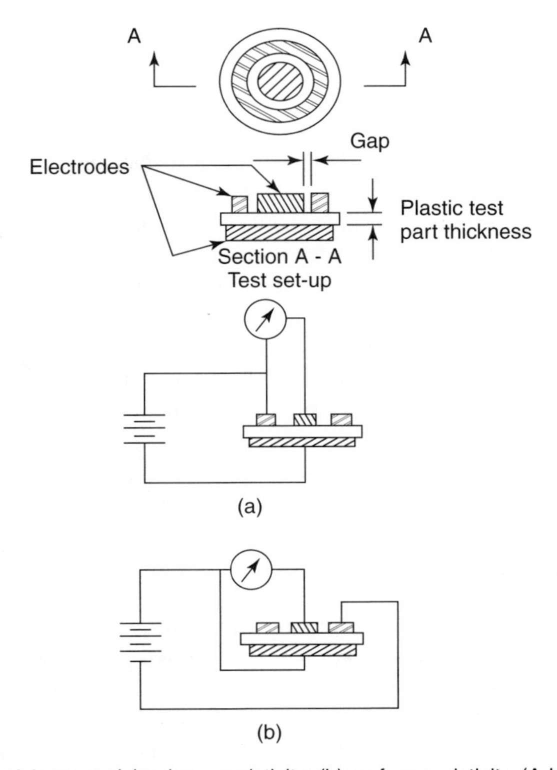

Figure 7.7 Resistivity tests: (a) volume resistivity; (b) surface resistivity. (Adapted from Ref. 3.)

Table 7.2 Typical volume resistivity for thermoplastics

Material	Volume Resistivity Ω·cm
ABS	10^{16}
Acetal	10^{14}–10^{16}
Acrylic	10^{16}–10^{18}
Nylon	10^{12}–10^{16}
Polycarbonate	10^{15}–10^{17}
TP polyester	10^{14}–10^{17}
Polypropylene	10^{14}–10^{17}
Polysulfone	10^{15}–10^{17}
Modified PPO/PPE (Noryl[a])	10^{15}–10^{17}
Polyphenylene sulfide	10^{16}
Polyarylate	10^{16}–10^{17}
Liquid crystal polymer	10^{15}
Phenolic	10^{17}

[a]General Electric trade name.

materials with values between 10^3 and 10^{12} Ω·cm are considered semiconductors. Values for typical plastics are shown in Table 7.2.

Surface Resistivity

The measurement of current to flow across the surface of a material due to surface contamination, especially moisture, is a measure of surface resistivity (see Fig. 7.7b). Applications in which surface leakage (conductivity) may be a problem, such as distance in length of the plastic insulator between pins on a connector, use these data, which are subject to large errors. Therefore, a safety factor in the form of path distance, a rib or valley molded in between pins to increase the pathlength between pins, should be considered in the design analysis of parts carrying high current or voltage. Critical current-carrying applications should specify only virgin resin is to be used in molded parts. Regrind should be used only in less demanding applications as it may contain very small carbonized resin particles, which are excellent conductors. Always specify on the drawing and in the contract whether regrind is or is not allowed. Also, mold releases, external or internally, should not be permitted unless nonconducting, as this could provide a conductive surface path.

Dielectric Strength

Dielectric strength measures the insulating capability of a plastic subjected to increasing voltage just prior to breakdown. Breakdown occurs when current flows through a specified thickness of material expressed in volts per mil when tested as in Figure 7.8. Variables that can affect the results are temperature, sample thickness, moisture concentration, and rate of applied voltage and test time. Any contamination on a sample's surface or porosity in the molded product will lower a resin's dielectric

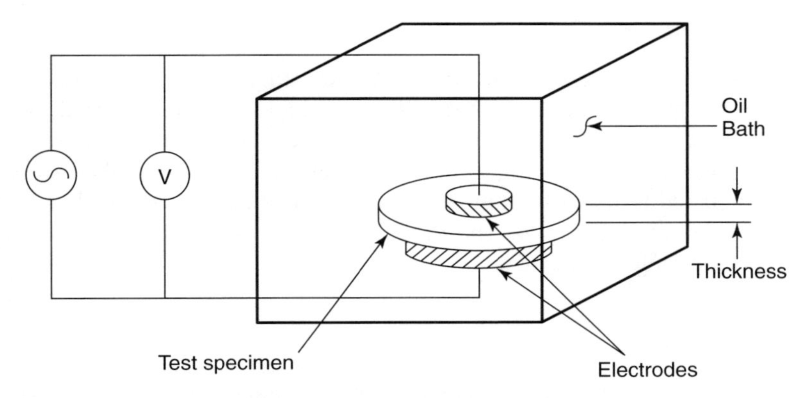

Figure 7.8 Dielectric strength.

strength. These can cause arcing or a conductive path through the sample, resulting in premature failure.

Dielectric Constant (Permittivity)

The dielectric constant is important for insulating materials used in high-frequency applications as radar and microwave equipment. This property is the susceptibility of a material to become polarized. The presence of molecules in a material, when an electric field is applied around the material or to line it up in the direction of current flow, is called *polarization*. When the field is reversed, as with alternating current, the molecular polarization reverses. A dimensionless constant called *permittivity* measures the ease of this reverse polarity. Permittivity varies with changes in temperature, part thickness, current reversal frequency, and moisture level of the sample. Table 7.3 shows values for various insulating materials.

Table 7.3 Dielectric constant and dissipation factor for various thermoplastics at room temperature

Material Description	Dielectric Constant	Dissipation Factor
Acetal	3.7–3.9	0.001–0.007
Acrylic	2.1–3.9	0.001–0.060
ABS	2.9–3.4	0.006–0.021
Nylon 6/6	3.1–8.3	0.006–0.190
Polycarbonate	2.9–3.8	0.0006–0.026
TP polyester	3.0–4.5	0.0012–0.022
Polypropylene	2.3–2.9	0.003–0.014
Polysulfone	2.7–3.8	0.0008–0.009
Modified PPE (PPO)	2.4–3.1	0.0002–0.005
Polyphenylene sulfide	2.9–4.5	0.001–0.002
Polyarylate	2.6–3.1	0.001–0.022
Liquid crystal polymer	3.7–10	0.010–0.600

Dissipation Factor

When polarizing reversals occur rapidly with alternating current, 60 Hz (cycles per second), heat is produced in the test sample as the molecules are activated by rapid changes in their polarization. The measure of the heat dissipation in the sample is reported as the dissipation factor (DF). DF is expressed as the ratio of the energy lost as heat, compared to heat transmitted from the sample and is measured at 1 MHz (10^6 cycles per second) or other specified frequencies. A low dissipation factor is desired for electronic part designs.

Arc Resistance

Arc resistance is important where the possibility of arcing across a surface exists in applications involving insulation of a switch, circuit breaker, or automotive ignition components and in high-voltage applications. Thermosets were the first plastic to utilize this property in product applications. Arc resistance is a measure of a conductive path formed on the surface of a material by electrical energy or heat decomposing the material on the surface of a part. The test uses electrodes that are spaced apart at a set distance on the surface of the test sample to create localized heating as shown in Figure 7.9. The test measures the time in seconds that a conductive path can develop across the surface of an insulating material. Materials with high time to breakdown values have good arc resistance.

Comparative Tracking Index (CTI). Underwriters Laboratories (UL) has developed a more realistic test to determine arc resistance on a contaminated surface. In the UL test, a solution of ammonium chloride wets the electrode's contact surface. The

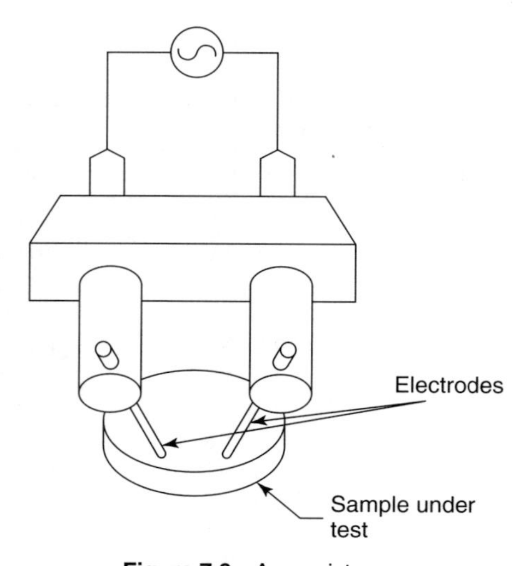

Figure 7.9 Arc resistance.

CTI numerical value is the voltage required causing a conductive path to be created between the electrodes at a predetermined current.

Other UL tests for electrical and electronic equipment are more specific. These are in the area of both material and end-use application. Before a consumer or industry can certify a product with the UL approval seal for use, the entire product must be tested to UL requirements. The designer is responsible for knowing the requirements for UL certification if the UL seal is necessary for sale of the product. Other product testing agencies, such as Canadian' Consumer Safety Administration (CSA) (similar to UL in the USA), and other foreign government testing agencies have their specific and often similar requirements for electrical products sold in their countries.

Electrical Properties of Plastics

Plastic resins are excellent insulators. They exhibit good mechanical, chemical, flame/smoke, and heat resistance characteristics to meet the demands of commercial and industrial electrical applications. Thermosets and some thermoplastic resins are recognized as suitable materials for all but the most severe electrical applications.

Thermosets have been used for commercial electrical meter frames, housings, circuit breakers, and power distribution blocks and switches for decades. Thermosets are least affected by moisture, dimensionally stable within a wide temperature range, and nonburning. Engineers and designers often find it difficult to select the right plastic material for electrical applications. This should not be a detriment since material suppliers offer assistance in these areas. The designer needs only recognize the electrical properties required, select a material to meet them, and continue the product design with good design practices. Electrical terminology and material properties will be discussed in detail later.

Material suppliers have developed electrical property data on their resins used for the major product areas. The available data address the electrical properties of each material for the environment and use impacts such as humidity, temperature, chemical, and UV effects on their resins. All of these items can affect a part's end-use mechanical and electrical performance. Product liability is a major issue for consumer and industry electrical products, including how well the product performs, meets and exceeds requirements, and if it should fail, does not cause injury to the user. Therefore, quality of the design, material, and product manufacture should be of the highest quality. To ensure that liability of a product is not compromised, there are government and private agencies to prove, verify, and certify that a product and its material of construction meet the customers' and agencies' safety and reliability standards. Many companies such as the automotive electrical industry test material to specific requirements and certify them for specific end-use applications. National testing agencies such as Underwriters International, UL, CSA, the FDA, NSF, EIA, (Electrical Industry of America), and FAA (Federal Aviation Administration), test products for both consumer and/or industrial safety. They can approve materials and/or products to requirements or specifications to assist the designer in making knowledgeable material selections for their companies' products.

FLAMMABILITY AND SMOKE GENERATION OF PLASTICS

One area of major importance for plastic products is a material's flammability and smoke generation potentials. Plastics support electrical heating elements and are used in the housings for many appliances and tools used in the home and workplace. Consumer safety and fire codes are more demanding in a material's ability to support a flame. A plastic flammability is rated for the percentage of oxygen termed, oxygen index (OI), required to support combustion at sea level. Air contains 20% oxygen. The higher the plastic OI value, the more oxygen that is required for the material to support combustion. The lower the OI, the easier for a material to support combustion. Some plastic resins can be made or are inherently flame-retardant. Flame-retardant additives are used in these resins to keep oxygen away from the material's surface when exposed to flame or ignition sources.

The elimination of the possibility of a plastic igniting, burning, and generating smoke is also very important. Smoke occurs from nonburning gases given off during combustion. On the basis of the material's chemical composition, gases are given off during combustion, even during smoldering or nonburning situations as with thermosets and flame-retardant plastics. This is of prime concern for plastics used in aircraft interiors, such as seating, housings, and seat trays. Now nonburning and low-level smoke-generating materials are specified for use in these products for passenger safety.

One family of plastics inherently flame-retardant is PVC, but during high heat situations, a smoke byproduct released is chlorine gas, which is very toxic. Many plastics can be rendered flame-retardant to some degree, but others may not, such as acetal and cellulose resins. Tests are conducted on standard grades to determine how readily they support and sustain combustion, smoke generation, and ignition temperatures. Underwriters Laboratories (UL) has established a plastic material evaluation and rating system. Their UL94 flammability classes range from allowing a very small amount of combustion when the flame is applied and then extinguishing (as V0 and V5) to others as V1, V2, and HB (horizontal burn), which allow a higher amount of combustion, based on the requirements of the test. Each UL94 test selectively rates each material's ability to support and/or maintain combustion after the ignition flame is removed. This is shown in Figure 7.10, illustrating the UL flammability test method.

UL94 HB: Horizontal Burn

Plastics that burn slowly and do not self-extinguish. This classification is UL's lowest rating, usually awarded to plastics that fail vertical V0, V1, or V2 requirements. This is the provisional or often the final rating given to a plastic if the aging and testing required by UL is not completed before the plastic is introduced in the market or it fails the other ratings.

UL94 V0, V1, V2, and V5: Vertical Burn

The Underwriters Laboratories Flammability, classification system is broken down as follows:

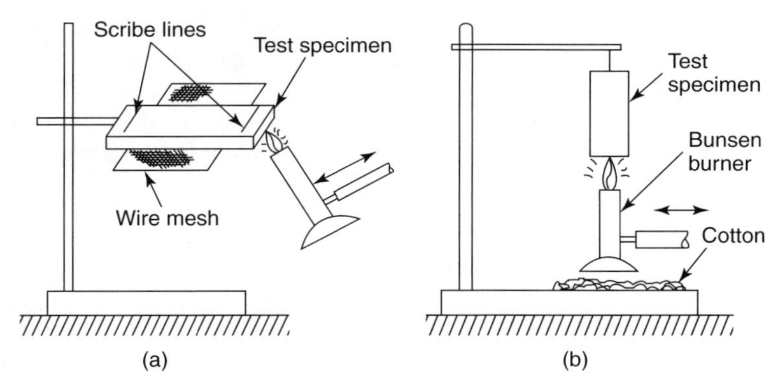

Figure 7.10 UL94 flammability tests: (a) UL94 HB—ASTM flammability and test for slow-burning plastic; (b) UL94 V0–V1–V2—self-extinguishing burn tests.

1. UL94 V0 and V5 are the "premium" classification for plastics that quickly self-extinguish and for a specified time period do not drip flaming particles (i.e., do not start new fires by allowing flaming particles to fall onto a cotton pad, one foot below the test specimen) after removal of the flame source. The flame is $\frac{3}{4}$ in. long and applied twice for each flammability test for UL94 V0, V1, and V2. (See UL94 V5 for this more stringent test method.)
2. UL94 V1 rating is similar to V0 and takes longer to extinguish. This test allows particles to drip on the cotton, but ignition does not occur.
3. UL94 V2 is the same as V1, but allows burning particles to ignite the cotton one foot below the flame application point.
4. UL94 V5 is a brutal test that relates to real-life performance. This test uses a 5-in. flame applied 5 times to a test plaque. No dripping of particles is allowed, no significant distortion of the test plaque is permitted, and plaques cannot have any burn holes.

Table 7.4 lists the various UL flammability tests and the requirements of each on the basis of a specific material thickness.

Table 7.4 United laboratories UL94 requirements

	Criterion for			
Condition of Test	V0	V1	V2	HB
Burning time after first ignition	<10 s	<30 s	<30 s	Sample burns up to clamp
Burning time after second ignition	<10 s	<30 s	<30 s	
Total burning time 5 samples per 10 ignitions	<50 s	<250 s	<250 s	
Time for glowing combustion—second ignition	<30 s	<60 s	<60 s	
Ignites cotton	No	No	Yes	

If a plastic is not generically flame-retardant as PVC, it can often be made flame-retardant. Flame-retarding a plastic often results in a change in physical properties. The modified material is seldom as tough for impact applications, and physical properties may be lower. Therefore, the designer needs the specific design data for the material when designing the product. Contact your material supplier for this information.

REFERENCES

1. *Designing with Plastics, Design Handbook,* Hoechst Celanese (TDM-1) HCER-92-313/10M692: 4–2, Chatham, NJ.
2. *Designing with Plastics, Design Handbook,* Hoechst Celanese (TDM-1) HCER-92-313/10M692: 5–2, Chatham, NJ.
3. *Designing with Plastics, Design Handbook,* Hoechst Celanese (TDM-1) HCER-92-313/10M692: 5–3, Chatham, NJ.

CHAPTER 8

DESIGN ANALYSIS OF MATERIAL PROPERTIES

Designers have a variety of material options and many property decisions to make when designing a product in plastic. Plastic provides the designer with more design options than can be realized with metal and other materials. The majority of metal parts are overdesigned solely on the basis of the available thickness of standard materials of manufacture for their products. Sheetmetal and die-cast materials place limits on the product design for the optimal utilization of the material's properties. Metal product designers using the standard isotropic behavior engineering equations for stress and strain calculations with metal room temperature physical property values, typically overdesigned products even when these products were subjected to end-use temperatures above 100°F.

The design for a plastic product uses these same equations, except the plastic material's physical properties are anisotropic. Therefore, the plastic material's behavior and design stress values will vary considerably depending on the product's end-use temperature and environment. For these reasons, the typical published plastic material physical property values listed on material supplier data sheets should not be used for the design of a product. The physical property values listed are only typical single data points developed using DAM (dry-as-molded) test samples that do not accurately represent the material's properties at end-use conditions. DAM physical testing of material samples is more accurate at the plant and test laboratories to generate physical property data on their plastic resins. These data reference point values should be used only for the initial evaluation of potential material candidates suitable for the product.

Material suppliers have test data on thousands of material samples used to obtain these single-data-point physical property values for their resins. They then use and report these values as typical reference properties for their resins. They use these physical property values as the standard room temperature property value and material specification for the manufacture of all future similar products.

The physical property and design terminology for plastics is similar to that for metals, but some terms need to be redefined specifically for plastic materials. A plastic material's mechanical engineering properties are described as

1. Tensile strength
2. Elongation at break
3. Yield strength
4. Elongation at yield
5. Flexural modulus
6. Shear strength
7. Deformation under load
8. Compressive stress
9. Poisson's ratio
10. Izod impact

A plastics thermal property require further definition for

1. Melting point
2. Heat defection temperature
3. Brittleness temperature
4. Specific heat
5. Thermal conductivity
6. Thermal expansion

Material suppliers test each lot of material manufactured. This is not necessarily done for all physical properties of the material, but for the critical, lot-specific, identification of physical properties the material must be performed in an end-use product. The property values are generated using the standard ASTM test methods. These tests use DAM test bars conditioned and tested at 73°F under laboratory-controlled testing conditions. Not all of the material supplier's resin grades will have these data available for all temperature ranges and test conditions where the materials may be used. Typically, mechanical physical property values are tested at both high and low temperatures, from −40 to 250°F, depending on the specific resin grade. These tests include tensile strength, elongation, flexural modulus, and often Izod impact. These values are typically generated for each individual lot of material manufactured in their production plants for reasons of quality control. These single-data-point values will not describe how the resin properties behave over the product range of end-use requirements but are only single-point values that certify the material as meeting the product's physical values for sale. Because of the effects of their environment such as temperature variations, moisture effects ultraviolet radiation, and/or chemical exposure, the product values for these standard tests will vary from the standard test values. These nonstandard effects on the resin's property values are independently tested. Test data relate only to the specific property and conditions under test, and this is why end-use testing of a plastic product (subjected to the same or other conditions) must be completed in the product's end-use environment.

PLASTIC PROPERTY TERMINOLOGY

Homogeneity

Homogeneous means uniform throughout a body. An unfilled thermoplastic or thermoset has a reasonable homogeneous physical property distribution. This means that physical properties such as tensile strength and elongating are the same in any small section of the body as in the whole body. But property values vary in a plastic product. As part thickness increases, section density of the molded part will vary. This is caused by the material's rate of solidification as it fills the mold. The materials flowing against the mold cavity walls are the first to freeze and are usually in the densest section of the product. The center of the section is still fluid and filled with molten resin, under pressure, still being forced into the product cavity to fill the mold. As the plastic cools in the mold, it shrinks, and if more fluid resin is not forced into the mold cavity, the center section of the part becomes less dense. Therefore, the cavity's gate must remain open to uniformly pack out the product sections to obtain a uniform product section density. A uniform product section thickness is recommended to ensure uniform packing pressure throughout the product cavity as it is filled to reduce the formation of voids and porosity on cooling in the product.

The gate for a molded product must always be located at the thickest product section. This will ensure that the gate will stay open as molten plastic material flows into the product cavity, and this section will be the last material area to solidify. Good product and mold design will ensure a uniform product section density after molding. The locations of the cavity gate controls the material's flow or fill path into the product cavity. Gate location has an effect on the product's physical property values in the flow and transverse flow directions. Physical property values are typically higher in the flow direction, especially for glass-reinforced and LCP resins. The reverse is true for shrinkage values: less in the flow and more in the transverse direction.

Heterogeneity

Heterogeneous means nonuniform throughout with different material physical property values in one section versus other sections and in opposing directions in the same section. Unreinforced plastics experience this behavior to some degree depending on how the material flows when filling the mold cavity. But for design purposes, the material's physical properties and shrinkage are assumed equal in all directions.

Glass fiber and very fine mineral-reinforced plastics are heterogeneous. The fiber reinforcements line up in the flow direction, and for $\leq$0.125-in.-thick sections the physical property values in the flow direction are higher, compared to 90° to the flow direction. In thicker parts fiber orientation is mixed or more random, yielding more uniform property values throughout the part. Symmetric reinforcements and fillers such as, glass beads, talc, and very finely ground minerals have more homogeneous property values and similar shrinkage in all directions. Reinforced resins have greater transverse shrinkage than in the flow direction. Typically heterogeneous resins are treated as homogeneous. The reason is that physical property values are normally tested in only one direction from end-gated molded physical property test bars. Therefore, physical properties transverse to the flow are approximately 30% lower. To compensate for this, a safety factor is used in the design to compensate for an actual lower property value in the cross section's transverse direction versus the flow direction. When designing with reinforced materials, end-use testing is always required to ensure that the product will meet the requirements.

Isotropy

Isotropic materials have identical physical property values in all directions throughout the part. Unreinforced thin wall plastic parts up to 0.100 in. thick are considered isotropic for design purposes. Thicker parts exhibit more anisotropic property values due to varying product density and surface flow orientation effects. The crystalline polymers in all thickness are not isotropic as LCP resins with tubular crystal structure that line up in the flow direction. Metal castings and unreinforced plastics are close to being isotropic materials when test samples are cut and tested in any direction.

Anisotropy

Anisotropic materials have different physical property values in all directions. Rolled and forged metal products develop crystal orientation in the worked or rolled direction that increases their strength. This results in higher physical properties in these worked directions versus their transverse direction. The same is true for extruded film and sheet in the takeoff direction versus side to side. These products are considered oriented axially or anisotropic in property values.

To solve anisotropic problems in some film products, biaxial film was developed to yield isotropic properties. This property is produced in the product as it is continuously extruded. While the film is still hot, it is clamped on its edges and stretched or oriented in the transverse direction. This orients the film, resulting in higher property and tear strength values in all directions. Products can be designed to have stronger properties in one direction by controlling the location of the gate. This causes the material to flow in a primary direction and for reinforced resins directs the fiber orientation or material strength properties where required. SMC composite materials can direct the fiber orientation in one direction as the ply of materials is laid up in the mold as preformed and fiber-oriented sheets of material. The ply can then be placed in the mold to maximize fiber orientation and reinforcement. This allows the manufacturer to maximize reinforcement in a controlled and highly oriented, fixed property direction in the product.

Wood products are considered anisotropic with three distinct property directions as shown in Figure 8.1 and a typical log cut pattern shown in Figure 8.2. In the growth or longitudinal direction, wood beams are loaded in compression only to hold up

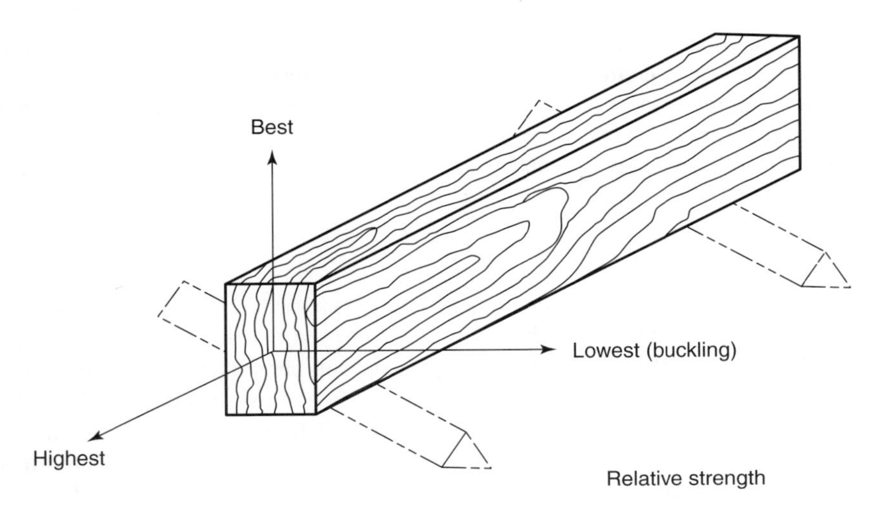

Figure 8.1 Anisotropy of materials (wood).

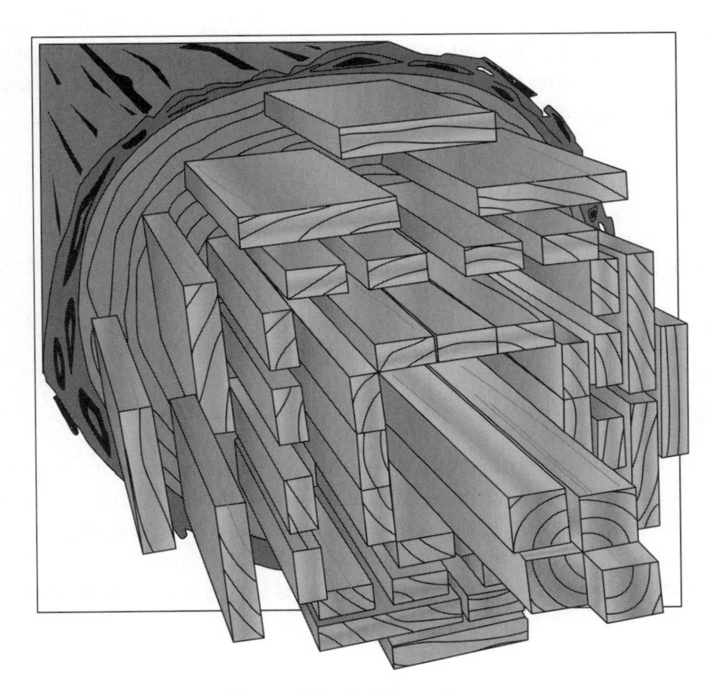

Figure 8.2 Cut log patterns.

bearing surfaces such as ceiling, floor, and roof rafters. This is similar to the flow or reinforcement-oriented direction for a thin-walled product. In the transverse direction, with the grain, the better properties are developed for horizontal load bearing, floor joists, and roof rafters. At 90° to the grain orientation the lowest properties are obtained with the beam easily bending, and when a compressive load is applied, the beam buckles.

Density

Density is the mass per unit volume of a material expressed in pounds per cubic inch or grams per cubic centimeter. This information is used to calculate a part's weight and piece part material cost using the resin price in cents per pound. Density translated to part weight is also used during molding, to verify and use as a quality check for the weight of each molded part, or at an established frequency, to ensure that the product is packed out uniformly and weighs the same, cycle to cycle. It is important that each cycle product parts weigh the same. Part weight is used as a quality and process control checkpoint during the molding operation.

Specific Gravity

Specific gravity is a unitless value (a ratio) used to compare the mass of a given volume of material to an equal mass and volume of water at 73°F, specifically, the density of a specific plastic to the density of water. Density and specific gravity are used to calculate part weight and material cost:

$$\text{Specific gravity (Sg)} = \frac{\text{density of water at } 73^\circ\text{F}}{\text{density of material at } 73^\circ\text{F}}$$

To estimate a part's weight, for comparative purposes, different resins are used:

$$\text{Part weight} = 0.0361 \text{ lb/in.}^3 \times \text{material's Sg} \times \text{part volume (in.}^3\text{)}$$

Elasticity

Elasticity is the material property that describes a material's ability to return to its "original" shape and size after being deformed by a force. Plastics exhibit a degree of elasticity at lower strain levels, typically 1% or less. Their degree of elasticity is dependent on their base resin and additive level and types. Rubber and thermoplastic elastomers have excellent elastic properties over a wide temperature range of essentially 50–180°F.

Plasticity

A material that does not return to its original shape but instead flows or creeps under a force before failure has plasticity. Thermoplastics exhibit plasticity by elongating when subjected to a force that exceeds the material's properties with the resulting material change denoted as *strain*. Reinforced and filled resins exhibit low plasticity, often failing in brittle fracture at low strain levels. Metals do not flow until stressed to their yield point, where they flow and exhibit plasticity, which in this context is termed *ductility*. As temperature increases, a thermoplastic will exhibit greater plasticity. At low temperatures plastics and other materials become brittle with less plasticity. Reinforcement lowers elongation as it is coupled chemically and physically to the base resin. In high-percentage (>20%) reinforced plastic materials, for the material to yield or flow, the fibers must fail. At lower levels of reinforcement the base resin and fibers can flow slightly more because of lower reinforcement levels. Not all areas of the material have fibers that restrict material flow or creep. This occurs at elevated temperatures and/or under high forces that may ultimately result in a fracture and failure of the product. The percentage and type of reinforcement have a direct relationship to elongation of reinforced materials. Elongation is a good measurement of a material's plasticity. Filled resins usually have low plasticity, again dependent on percentage and type of filler, and in the majority of situations fail by brittle fracture. The thermoset materials, phenolic materials in particular, have very poor plasticity. They have very low elongation and fail by brittle fracture in almost all cases. Thermosets are used in high-temperature and compressive stress applications because of their low plasticity. A filled or reinforced thermoset is more resistant to flow, exhibiting low to zero plasticity even at high temperature and under high-load-bearing applications.

Coining

Coining relies on the plasticity of a material to allow it to flow under a localized high force. Coining involves compressing a small area of a material's section beyond its yield point, causing the material to flow at the point of contact. Coining uses plasticity, a material's ability to elongate without failure to form repeatable operating flexible joints in a molded product. Coining orients the material's molecules under the forming tool, increasing flexural and tear strength at the coined area. The term "living hinge" is often used to describe this coining operation as shown in Figure 8.3. The semicrystalline and crystalline resins are often coined to form hinges and flexing sections for parts. Plastic materials such as ABS, PVC, and other amorphous plastics can also be coined, but their flexural and tear strength are usually lower than that of the engineering resins. Flex life tests using coined sections versus noncoined hinges show a 100% increase in flexure strength and also have higher tear strength values at the coined section. Coined hinges can also be formed in the mold by creating a narrowed thin section in the mold cavity at the desired hinge point. The resin is then forced to flow through the restricted area, which orients the material before joining it on the other side of the hinge section in the mold cavity.

On removal of the product from the mold, this section must be immediately flexed 2–4 times, back and forth through 60–90° to further orient the material at the hinge section. This flexing action greatly increases the hinge's flex life and tear strength.

Stress Whitening

Stress whitening is caused and visible on a plastic product where it has been locally overstressed by a force. It could be caused by flexing beyond the material's yield

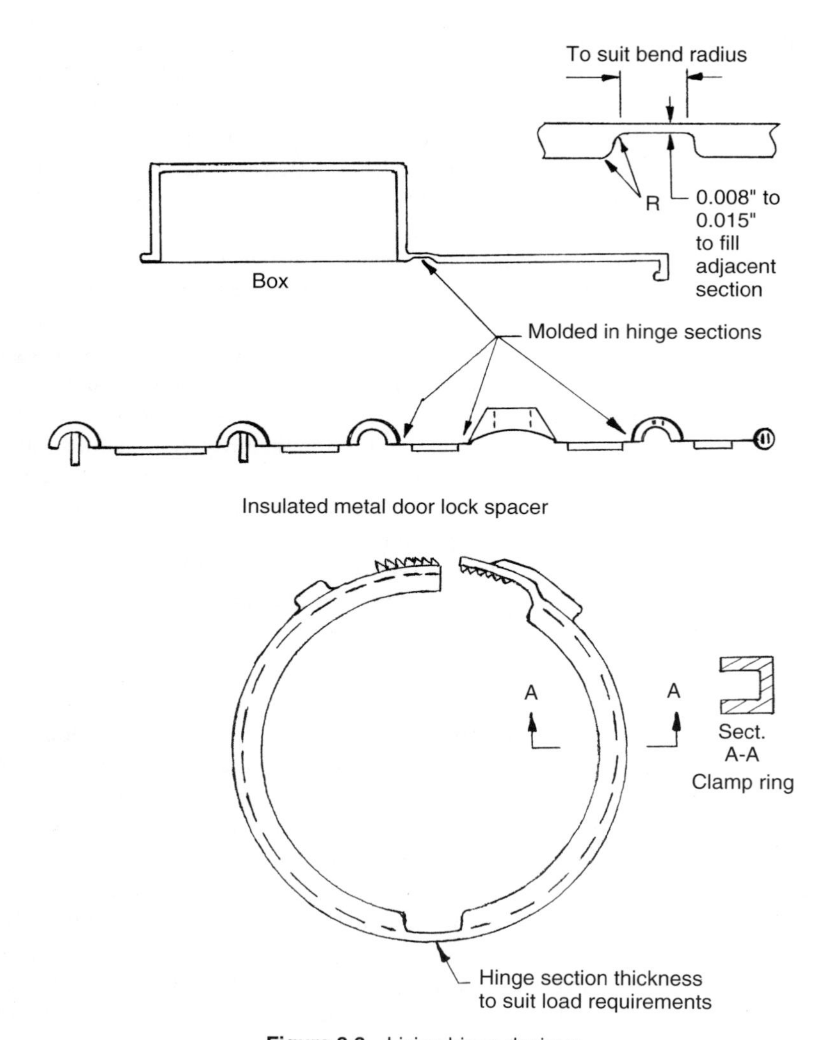

Figure 8.3 Living hinge designs.

point without causing failure, or by other methods that do not initially lead to failure. At failure points, there are often stress-whitened areas showing that the material experienced high forces and yielded prior to failure. Severe impacts and bending overstress the material, causing local high elongation, often without part failure. This is visible at coined areas where the material was deliberately overstressed to form a coined hinge. These high stress points cause a discoloration, visible on the part's surface as a whitening or optical change in the appearance of the surface. Stress whitening can be used to analyze a product that has or is about to fail, as it appears where the material has been overstressed, exceeding the material's physical properties.

Ductility

Ductility of a material allows the material to be pulled, rolled, or stretched into another shape without destroying the material's physical property integrity. Thermoplastics exhibit this property when heat, mechanical, or sonic energy is used to reform or modify their shape. Thermoforming relies on the heated material's ductility for the forming operation. Ductility of a material is a function of the material's elongation and the rate of applied force to alter the material's shape, usually with the application of heat. Molded and extruded products use the material's ductility to assemble it to other parts or to modify the product while still very hot for value-added features. For example, extruded rigid and highly filled PVC pipe, after exiting the forming die, has one end mechanically expanded to form a bell-coupling joint. The material is highly filled but has good high-temperature elongation capability to be formed before it cools. PVC drainage pipe is a very rigid and a strong product able to resist compaction pressures when buried in the ground. The coining operation eliminates a separate coupling to join the pipes in the field using an elastic interference O-ring to seal with the straight end of the adjoining pipe section.

Toughness

Toughness is the ability of a material to absorb physical energy without failure. Energy is absorbed by either elastic or plastic deformation. A material's ability to absorb energy is measured in several ways, described in the following paragraphs.

The area generated under the curve of a material's stress–strain graph is the visual representation of the material to absorb energy and is illustrated in Figure 8.4a–d.

In Figure 8.4a, the curve generated is for a high elongation material as polyethylene, polypropylene, unfilled PVC, and nylon. The force is applied, represented as stress, using an Instron tensile testing machine; the material yields (represented by strain), and at the material's yield point, the material's flow is often indicated by a drop in the stress until the force causes the material to continue flowing until rupture. The test bar necks down, reducing the stress but flowing and work hardening, resulting in an increase of stress again until the material's elongation is reached and then fails by ductile failure. Failure by ductility results in a high amount of energy being absorbed in the sample. In Figure 8.4b, the material initially exhibits a high tensile strength with low strain until the material's yield point is reached. Then the material quickly fails with a ductile break similar to that seen in Figure 8.4a. Materials exhibiting this stress–strain curve are Acetal, nylon, and polyester, all unreinforced, engineering resins. In Figure 8.4c, the material has low strength and elongation, resulting in early failure under lower internal stress. This is typical of highly filled, unreinforced resins, such as rigid PVC, styrene, acrylic, and some compounds of ABS–filled blends, including most thermosets generating this type of stress–strain curve.

Failure is usually a brittle fracture since the material cannot absorb a large amount of energy before failing. Figure 8.4d illustrates a reinforced material with high strength and low elongation. Reinforced nylon, polyester, PC, and other reinforced resins generate this type of stress–strain curve. Increasing test temperature will have a major

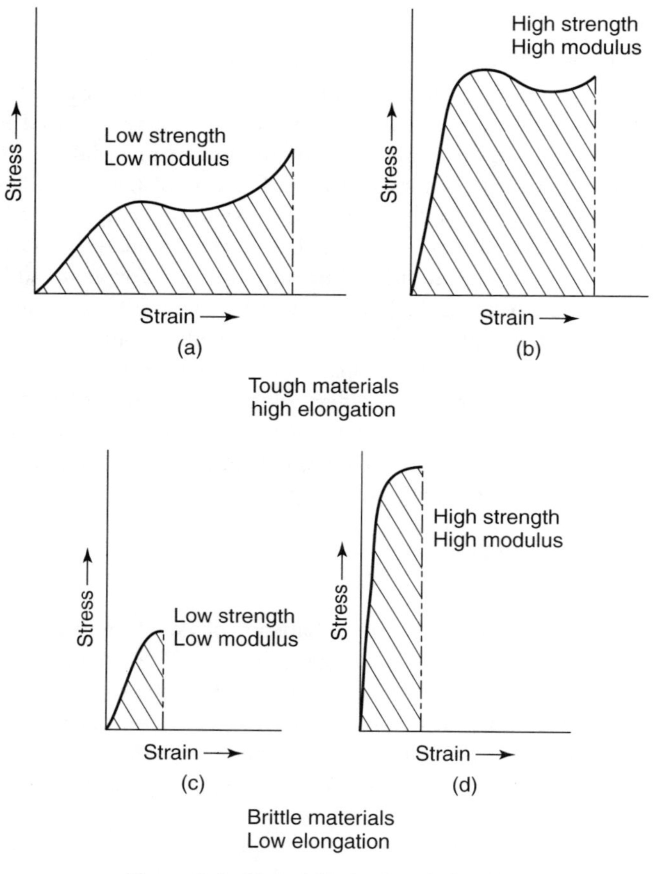

Figure 8.4 Material behavior of plastics.

effect on materials generating the graphs shown in Figure 8.4a,b. Temperature can increase a material's ductility, which, under a low stress, allows the material to absorb energy more slowly and elongate more before failure. This is less so for materials used to generate the graphs in Figure 8.4c,d.

Gardner "Drop Weight" Impact

The falling dart impact test is similar to the "Gardner drop weight" test reporting impact strength in feet or inch-pounds of energy absorption by the material as shown in Figure 8.5. This is a very severe and instant impact test performed on a molded plaque or disk of specified thickness. It is rarely run for all materials but is one of the best tests for evaluating a material's energy-absorbing capability, reported as toughness.

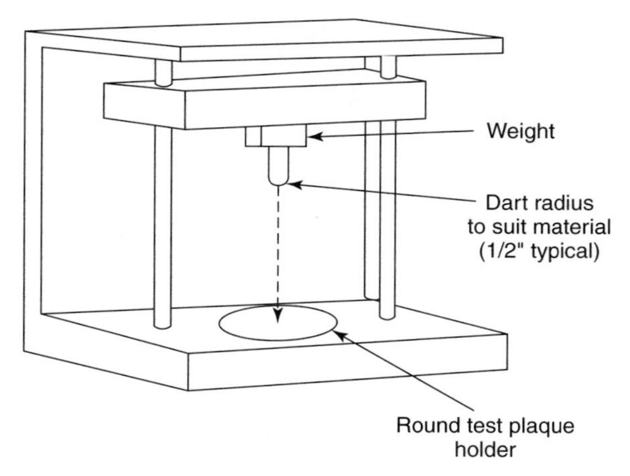

Figure 8.5 Laboratory dart impact material tester.

Izod and Charpy Impact

Izod and Charpy impact tests measure a material's ability to absorb impact energy at a molded-in or machined notch in a test sample. Impact can be a single, multiple impact and on notched or unnotched test samples. These two tests are similar and use a pendulum that swings from a set release height and impacts the test specimen as shown in Figure 8.6a. The difference between the two tests is that the Charpy test impact point is a knife edge that impacts the test specimen. The Izod test uses the flatside of the impact pendulum surface against the test bar. The Izod test method is preferred in the United States and the Charpy, in Europe.

The test has a pendulum arm falling through an arc with the head impacting the specimen with the energy absorbed by the material on impact recorded on the dial scale. The reading is recorded as foot-pounds per inch of the test specimen thickness at the impact point [ft·lb/in. or J/m (joules per meter)]. If the specimen does not fail, it is recorded as "no break." The test specimens can be notched, machined or molded in, or unnotched, as shown in Figures 8.6c–e with the test bar notch in the direction of the impact or reversed.

The values of force shown on the dial indicator scale after each impact are reported as unnotched, notched, or reversed-notch impact values. Both falling dart and swinging impact tests are used to measure the notch sensitivity of a material and the resistance to crack propagation through a material.

Tensile Impact

Tensile impact is used to measure the toughness of a plastic material in sudden tension impacts with the test apparatus similar to the Izod impact test. Tensile impact examines the material's impact tear strength by testing a rectangle or round, dumbbell-shaped

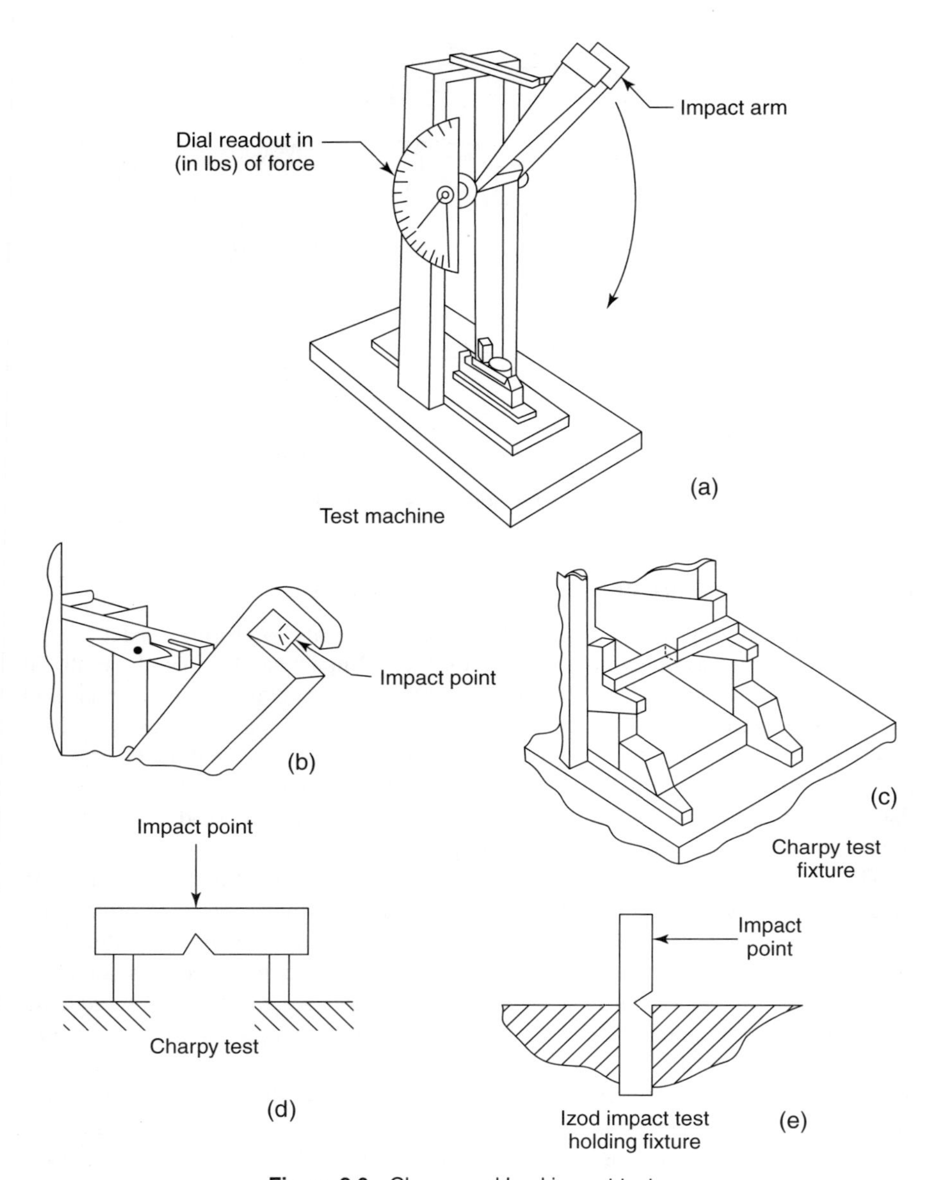

Figure 8.6 Charpy and Izod impact tests.

tensile bar as shown in Figure 8.7. Energy absorption is measured and reported at the thinned section of the test bar in pounds per inch squared lb/in.^{2} or kJ/m^2. The energy absorbed in the test sample is reported as axial toughness when the test sample is instantly subjected to an axial impact by a stop attached to the test bar. Many engineers believe that tensile impact is more representative of a material's actual

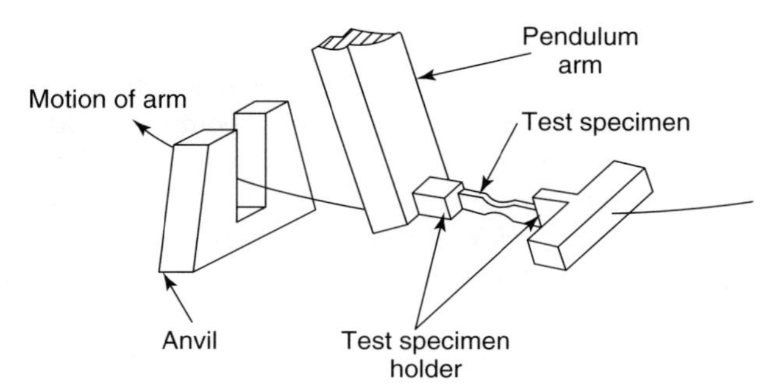

Figure 8.7 Tensile impact similar to Figure 8.6 for pendulum.

toughness versus the notched Izod and Charpy tests. It is as severe as the falling-dart test in the axial plane of the test sample versus vertical bending impact on the sample in a 360° strike zone as the dart strikes the plaque. Plastic resin suppliers do not always report tensile impact values on their data sheets.

Thermoplastic materials with high molecular weight or modified with impact additives (plasticizer and elastomer modifiers) have reported good to excellent tensile toughness. Reinforced resins also report higher than anticipated impact strength, which is attributed to the fiber reinforcement holding the test samples together on impact, which assists in absorbing the impact energy. Thermoplastics are inherently tougher than most thermosets, but impact-toughened grades of thermosets have been developed for under-the-hood applications as fuel rails and drive pulleys that perform satisfactorily in many intermittent, unanticipated, impact situations as a dropped tool on the part during a repair situation. Thermosets have excellent high-temperature dimensional stability with chemical resistance to engine fuels, especially the alcohol blends now being used worldwide.

Brittleness

Brittleness is a lack of toughness or material ductility usually associated with resins with low elongation properties. Thermosets, especially phenolic, exhibit brittleness if not modified by energy-absorbing additives and fillers. Many filled and reinforced thermoplastic resins that exhibit good to high tensile strength but have low elongation, and fail suddenly under high forces, are described as brittle materials represented by their stress–strain curve, (Fig. 8.4c,d). Factors affecting brittleness are the material's molecular weight, and modifiers, such as plasticizer, carbon black, fillers, rubbers, and reinforcement materials. The ability of a material to absorb energy is an indication of its toughness or lack of brittleness. Many base resins are inherently tough and not brittle, such as PE, PP, PET, nylon, Acetal, and PC. Always end-use-test the product to ensure that the correct material selection meets the product's requirements.

Notch Sensitivity

Notch sensitivity is the term used to describe the ease with which a crack can propagate through a material. Cracks are created at high stress points from force impacts, by shrinkage, at internal or surface imperfections and sharp internal corners. They usually occur at points of high impact or stress, called *stress cracks,* which may propagate throughout the section. Stress cracks appear at sharp, 90-degree corners, no radius, on a part. High-elongation resins exhibit better notch resistance. Notch sensitivity is reported on data sheets as *notched Izod impact.*

Lubricity

Thermoplastics have an inherent lubricity that is the load-bearing characteristic of a material under relative motion. Plastics with good lubricity have a low coefficient of friction value measured by dynamic and static tests. Lubricity imparts low wear characteristics to a material, usually requiring minimum external lubrication and long wear life depending on the PV (pressure velocity) factor against a mating surface. Thermoplastics can be modified to increase their lubricity with additives and fillers. The amount required is small, a percent or two by weight, which will not appreciably affect the material's physical properties. Internal modifiers used in plastic compounds and blends are oils, Teflon fibers and flakes, carbon black, molybdenum disulfide, and other lubricant additives.

Abrasion, Wear, and Friction

When there is motion between parts, gears, bearings, pulleys, contact surfaces, and other components, careful selection of materials is required to reduce abrasion and wear. The forces on materials with continuous or intermittent contact generate a PV value that is used to select materials with coefficients of friction suitable for the application. Material suppliers develop friction and wear information for their resins against different and commonly used mating materials and surface finishes.

To reduce contact wear between parts in motion, always use dissimilar materials. Like materials rubbing together produce higher wear rates usually by galling and abrasion of the contact or wear surface than do dissimilar materials.

Abrasion and Wear. A material's rate of abrasion is obtained using several test methods. Abrasion tests are run on plastic materials using the Taber and ball mill abrasion tests with results recorded as weight loss of material or wear. Material suppliers provide their data ranked against generic plastics, thermosets, and metals. Typically, reinforced materials have shown greater wear in these tests than have nonreinforced resins. The wear rate of a material is a function of the plastic material's pressure and velocity on the contact surface, and the environment. Abrasion tests must also consider the surface finish of mating materials and surface hardness. There are too many variables affecting abrasion and wear rate, and each situation must be independently evaluated.

Table 8.1 Product considerations for wear evaluation

Material	Processing	Environment	End Use Conditions
Tensile strength	Surface finish of mold	Temperature	Pressure/force
Flexural modulus	Packout	Moisture	Velocity/speed
Toughness	Mold temperature	Chemicals	Lubrication
Hardness	Process temperature	UV	Surface finish
Coefficient of friction	Roundness	Dust/dirt	Surface hardness
Wear rate	Dimensions	Life of product	Type of mating surface contact
Additives and fillers			Lubricity
Crystal structure			Type of contact, impact continuous or intermittent

Plastics are resilient and can deform under a load, which can attribute to a lower wear rate. Plastics also have a natural lubricity inherent in their carbon molecule structure and composition. An example illustrates this for a nylon worm gear outlasting a metal gear, 3:1, for one application. The nylon with natural lubricity was able to deform under the load and not wear, as did the metal worm gear. Factors affecting wear of plastics are summarized in Table 8.1 for selecting a suitable material for an application involving wear.

When all answers to the items noted are obtained, a material can be selected that should have a high rate of success. Other questions should be raised if the environment has a more severe requirement.

Friction. Plastics do not behave according to the classic laws of friction. Plastics are not rigid bodies, including reinforced resin grades as do metals, on which the laws of friction were developed. Adhesion and deformation characterize frictional forces between a metal/plastic interface. This results in forces on contact surfaces not proportional to the load but proportional to the speed in the classic pressure time's velocity (PV) equation. This relationship between rigid or deforming surfaces with PV produces an unstable contact surface between the rubbing surfaces. Each contact situation of a material will determine that material's individual maximum PV limit. This maximum PV limit is developed by testing the materials with and without lubrication in various types of contact and velocity situations.

Adhesive Wear. The primary mechanism for wear is adhesive wear, characterized by the removal of fine surface particles of material from the contact or wear surface. This type of wear is anticipated, and the fine particles indicate that the rubbing surfaces are wearing correctly. If other than these fine particles appear, and one surface melts or larger particles are removed, this indicates that the material's PV limit was exceeded. When an increase in operating temperature occurs at the contact surface, the material's PV limit value is reduced. As the contact surface finish is abraded, the wear rate increases and the dynamic materials coefficient of friction is changed.

Coefficient of Friction. The coefficient of friction of a material is a function of the relative motion of two surfaces in contact with each other. The coefficient of friction for a plastic is a function of wear on the contacting surfaces as a result of the PV conditions. When similar plastic materials rub against each other, their wear rate is high. This leads to galling, the excessive breakdown of the contact surface of each material by the removal of surface particles. Unless the operating temperature and pressure are low, premature failure of the parts results. Avoid like material contact surfaces except in very-low-PV applications. For plastic contact applications with high PV values, use of crystalline resins from different generic families is recommended. When amorphous or identical resins are used in high-PV applications, a 15–20% loading of PTFE (polytetrafluoroethylene) as an internal lubricant in the resin is recommended if not externally lubricated. Other internal lubricants can be used, such as silicone fluids, graphite powders, and molybdenum disulfide (MoS) when testing validates that their performance is acceptable for the application.

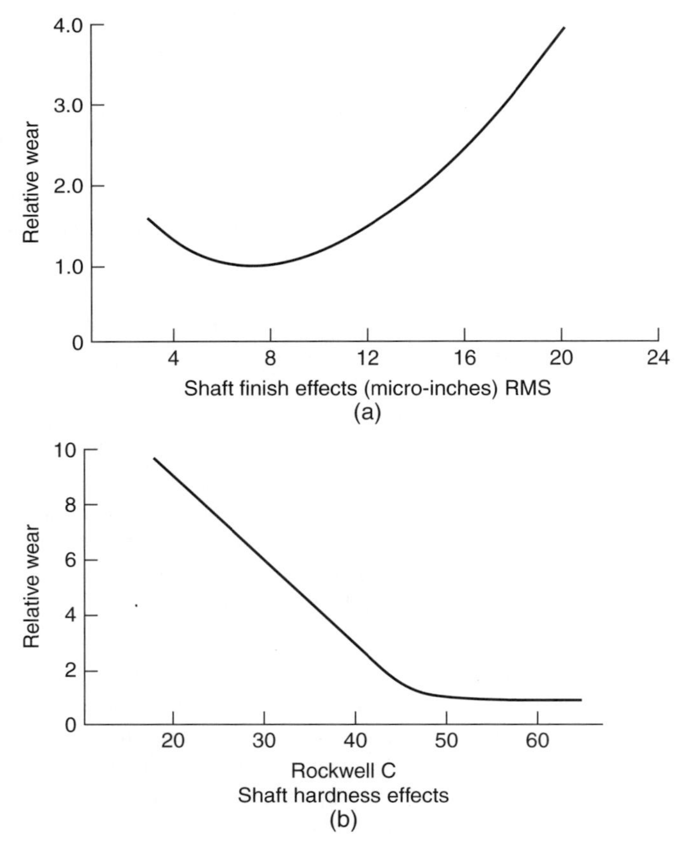

Figure 8.8 Wear effects on materials lubricated with Teflon (PTFE): (a) shaft hardness effects; (b) shaft finish effects. (Adapted from Ref. 1.)

Internally compounded lubricants will lower the plastic's physical properties, specifically flexural modulus, and must be considered in the parts design analysis. In most applications an external lubricant on the contact surfaces can be specified to reduce friction and wear. Always verify the chemical compatibility of the lubricant with the plastic material at the operation temperature.

Reinforced resins are used for low-PV-rate applications and experience low wear rates. A reinforced resin, once its resin-rich surface is abraded, can quickly wear an unreinforced material. However, when run against another reinforced resin, it may obtain a lower wear rate, as with glass fibers against softer carbon fibers. The carbon fibers act as a lubricant, and the mating surface experiences less wear. Again, this is a function of the applications PV values and the materials rubbing against each other.

When plastics run against metals, the metals' hardness and surface finish affect the wear behavior of the plastic. Rough metal surfaces rapidly abrade the plastic surface by preventing a lubricating surface film of plastic particles from forming on the metal. Similarly, a very smooth metal surface inhibits the lubricating plastic film from developing and reducing the force of friction between the mating surfaces. There is a fine line between metal contact surface finishes with a plastic surface to obtain minimum wear. Shaft finish and hardness effects of a metal shaft against a plastic are shown in Figure 8.8a,b. A study of plastic versus metal wear of various resins, reinforced and lubricated, is presented in Table 8.2. For surfaces other than steel, similar testing must be performed, and results may vary. Material suppliers should be able to furnish recommended PV values for their materials in contact applications. Comparison wear tests on generic base resins using the ball mill test method are presented in Table 8.3, for comparative purposes.

Before selecting any material in a wear application, determine all the factors on the application in its end-use environment. Remember to use dissimilar materials in wear applications to reduce friction, galling, wear, and fretting erosion.

Shrinkages

Material Shrinkage. When thermoplastics are heated, they become fluid and expand. When cooled, they solidify and shrink from their original molten volume state. This change in material volume and density going from a liquid to a solid is known as *material* and/or *mold shrinkage*. The average material shrinkage value of each plastic resin is listed on the materials data sheet as mold shrinkage. A plastic materials shrinkage is reported in inches per inch (in./in.) of dimensional shrinkage, or change, from the original cavity dimensions. A materials mold shrinkage value is obtained using a standard ASTM molding sample test bar mold. The supplier generates the values under optimum material molding conditions. Values are averages and will vary in flow and transverse directions according to the molding conditions.

Amorphous resins, because of their molecular structure, experience less mold shrinkage than do the crystalline and engineering resins. The mold designer compensates for mold shrinkage by adjusting the cavity dimensions to suit the gating and part layout in the mold cavity. The material suppliers provide typical plastic resin shrinkage values, typically in only the flow direction. The Society of the Plastics Industry,

Table 8.2 Wear[a] and coefficients of friction of graphite-, bronze-, and MoS[b] -lubricated thermoplastics against steel

Material	Type of Reinforcements and Fillers	Wt%	MoS Wt%	Graphite Wt%	Wear Factor ×10 (min ft.lb^{-1} h^{-1})	Coefficient of Friction: Static (40 psi)	Coefficient of Friction: Dynamic (40 psi, 50 ft/min)
SAN	Glass beads	30		10	105	0.17	0.21
Polysulfone	Mineral	20		15	530	0.18	0.18
Nylon 11	Glass beads	25		10	30	0.18	0.22
Nylon 11	Bronze	85			82	0.15	0.15
Nylon 11	Bronze	83	3		65	0.12	0.12
Nylon 6/12			<5		145	0.33	0.33
Acetal				10	60	0.16	0.22
Nylon 6				5	60	0.16	0.19
Nylon 6			<5		160	0.28	0.30
Nylon 6	Glass fiber	30	<5		80	0.26	0.32
Nylon 6	Glass fiber	40	<5		75	0.28	0.34
Nylon 6/10			<5		145	0.30	0.31
Nylon 6/6				6	55	0.15	0.20
Nylon 6/6			<5		150	0.28	0.30
Nylon 6/6	Glass fiber	30	<5		75	0.24	0.31
Nylon 6/6	Glass fiber	40	<5		70	0.26	0.33
ETFE			<5		145	0.11	0.18
ETFE				10	165	0.13	0.23
Nylon, high-impact			<5		60	0.20	0.33
Modified PPO	Glass fiber	15		10	11	0.09	0.11
Modified PPO	Glass beads	30		10	145	0.29	0.22
Modified PPO	Mineral	20		15	850	0.27	0.21

[a] All wear factors are times (×10).
[b] Molybdenum disulfide.

Source: Adapted from Ref. 1.

Table 8.3 Comparative abrasion weight loss tests[a]

Material	Test Method	
	Taber	Ball Mill
ABS	9	10–20
Acetal (homopolymer)	2–5	4–6
Cellulose acetate butyrate	9–15	10–20
Nylon 6/6	1	1
Melamine formaldehyde (molded)	—	15–20
Phenol formaldehyde (molded)	4–12	—
Polystyrene (several grades)	9–26	15–20
PVC	9–12	—
Die-cast aluminum	—	11
Hard rubber	—	10
Leather	22	—
Mild steel	—	15–20

[a]The weight lost factor in the same time period for each material when compared to nylon 6/6 as the standard, based on its weight loss as unity.

Source: Adapted from Ref. 2.

Inc., publish a *Standards and Practices of Plastic Molders* book listing guidelines and resin tolerances to consider when sizing a product's mold cavity based on the class of molding tolerance required for the product: commercial or fine. An example is shown in Figure 8.9 for a plastic ABS resin.

Material shrinkage, typically called "mold shrinkage," refers to a plastic's dimensional change in the mold during the molding operation. Mold shrinkage is the volumetric and dimensional change that a molten thermoplastic experiences going from a liquid to a solid in the mold cavity during the molding operation.

All thermoplastic resins experience shrinkage when going from a hot molten melt in a molding machine's barrel and then being injected under high pressure into a cooler mold cavity and on cooling, experiencing reduced its volume, and shrinking during solidifying.

Amorphous resins with random molecular structure and liquid crystal polymers (LCPs) exhibit lower shrinkage in the flow direction, typically of 0.001–0.010 in./in. Shrinkage in the transverse direction, 90° to flow, will be slightly higher. The semicrystalline and crystalline resins have a high molecular order or crystal packing and on cooling experience a higher amount of shrinkage, 0.003–0.030 in./in. in the flow direction depending on the material's family and grade. Shrinkage transverse to the flow direction will be even higher, especially for LCP resins whose crystal structure, cylinders, line up in the flow direction.

Variable effects on material shrinkage during and after manufacture are as follows. Mold temperature for semicrystalline and crystalline resins affects the amount of shrinkage on cooling. Higher mold cavity temperature creates a more ordered material

Standards & Practices of Plastics Molders

Material
Acrylonitrile Butadiene Styrene (ABS)

Note: The *Commercial* values shown below represent common production tolerances at the most economical level. The *Fine* values represent closer tolerances that can be held but at a greater cost. Any addition of fillers will compromise physical properties and alter dimensional stability. Please consult the manufacturer.

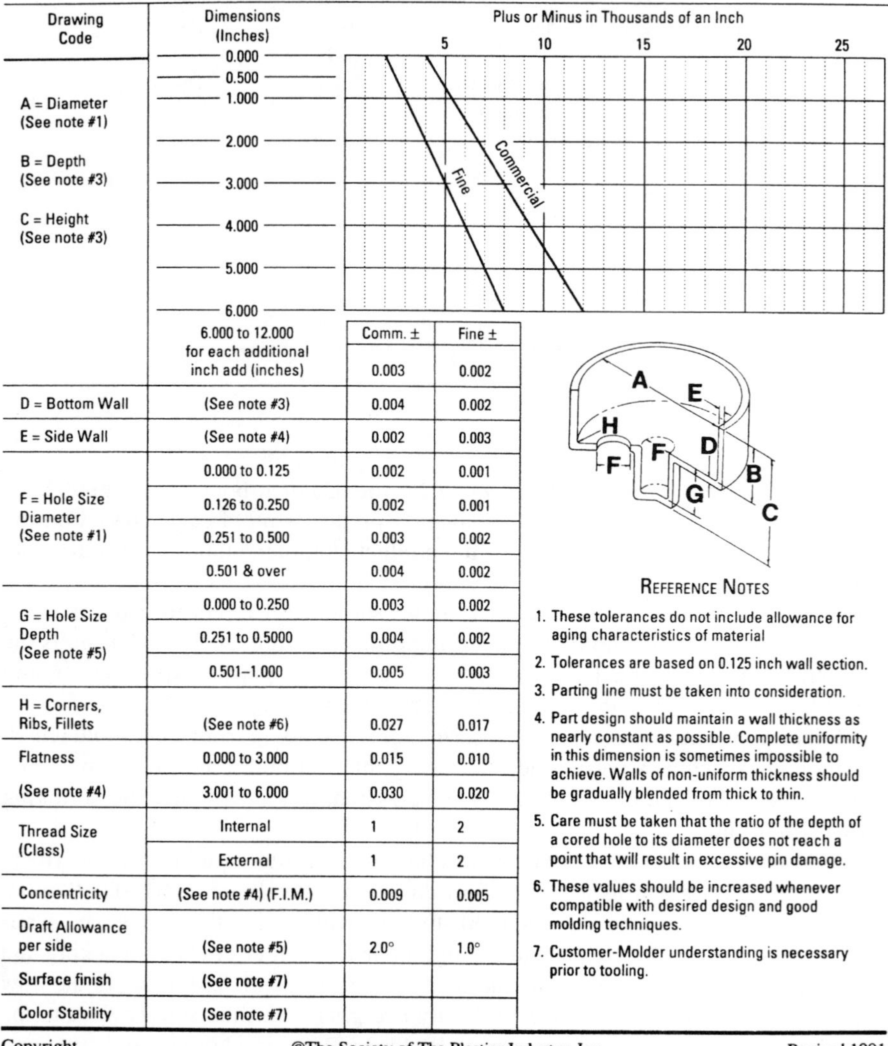

Drawing Code	Dimensions (Inches)	Comm. ±	Fine ±
A = Diameter (See note #1) B = Depth (See note #3) C = Height (See note #3)	0.000 / 0.500 / 1.000 / 2.000 / 3.000 / 4.000 / 5.000 / 6.000 (see chart: Plus or Minus in Thousands of an Inch)		
	6.000 to 12.000 for each additional inch add (inches)	0.003	0.002
D = Bottom Wall	(See note #3)	0.004	0.002
E = Side Wall	(See note #4)	0.002	0.003
F = Hole Size Diameter (See note #1)	0.000 to 0.125	0.002	0.001
	0.126 to 0.250	0.002	0.001
	0.251 to 0.500	0.003	0.002
	0.501 & over	0.004	0.002
G = Hole Size Depth (See note #5)	0.000 to 0.250	0.003	0.002
	0.251 to 0.5000	0.004	0.002
	0.501–1.000	0.005	0.003
H = Corners, Ribs, Fillets	(See note #6)	0.027	0.017
Flatness (See note #4)	0.000 to 3.000	0.015	0.010
	3.001 to 6.000	0.030	0.020
Thread Size (Class)	Internal	1	2
	External	1	2
Concentricity	(See note #4) (F.I.M.)	0.009	0.005
Draft Allowance per side	(See note #5)	2.0°	1.0°
Surface finish	(See note #7)		
Color Stability	(See note #7)		

REFERENCE NOTES

1. These tolerances do not include allowance for aging characteristics of material
2. Tolerances are based on 0.125 inch wall section.
3. Parting line must be taken into consideration.
4. Part design should maintain a wall thickness as nearly constant as possible. Complete uniformity in this dimension is sometimes impossible to achieve. Walls of non-uniform thickness should be gradually blended from thick to thin.
5. Care must be taken that the ratio of the depth of a cored hole to its diameter does not reach a point that will result in excessive pin damage.
6. These values should be increased whenever compatible with desired design and good molding techniques.
7. Customer-Molder understanding is necessary prior to tooling.

Figure 8.9 Cavity hold tolerances, dimensionally.

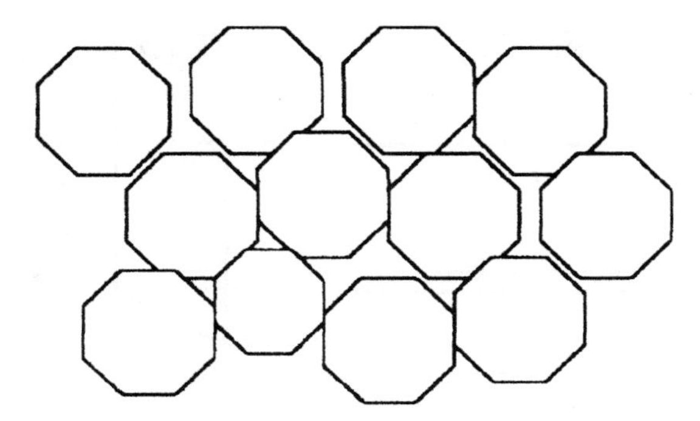

Figure 8.10 Crystal package arrangement (not shown as actual crystal package structure). The more dense crystal packing, the more opaque the material as light is reflected, not passed through the material. Semicrystalline resins have a more open packing structure that make them more translucent than all crystalline resins.

packing structure on cooling, and greater density, which relates to higher material shrinkage. See Figure 8.10 for the packing structure of a crystalline resin and a semicrystalline resin. The material shrinkage for each resin is measured by the supplier under ideal molding conditions and listed as a physical property for each resin.

Amorphous resins do not have a crystalline structure, and their molecular packing is random or less ordered on cooling (see Fig. 8.10), where they have a lower material shrinkage. The noncrystalline structure allows light to pass through, making them transparent even in thick sections up to and including one inch. Some PVC compounds plus acrylic and PC resins have good to excellent optical qualities.

Mold Shrinkage. Mold shrinkage is the same as material shrinkage. Depending on molding conditions, the material's shrinkage can vary from minimum to maximum values as reported by the supplier. The product's mold shrinkage is calculated by the mold designer for the specified resins typical material shrinkage using a shrinkage ratio and past material shrinkage experience based on selected mold temperature, flow into the cavity, gate size and type, number of cavities, packing pressure, and processing factors.

Mold shrinkage is the ratio between the mold cavity dimensions and the molded part's final dimensions after cooling to room temperature expressed in inches per inch or centimeter per centimeter (cm/cm). The mold designer allows for and calculates the anticipated resin's shrinkage on the basis of experience, the material supplier's shrinkage data, gate size, part thickness, mold temperature control, packing pressure, flow direction, and other factors. The mold shrinkage calculation must be as accurate as possible to correctly size the machined mold's cavity dimensions to produce the required part size within tolerance. The mold's part cavity is made larger than the finished product's final dimensions to compensate for material shrinkage as shown in Figure 8.11. A plastic materials mold shrinkage will vary in different directions in the mold cavity. This is due to how the product is layed out in the mold, flow path, and gating plus molding process variables.

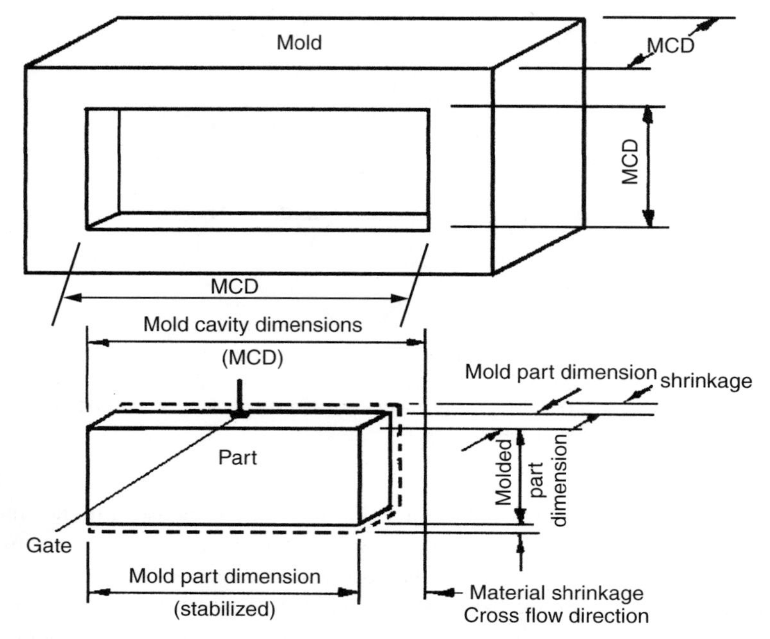

Figure 8.11 Molded part shrinkage (in./in., cm/cm, or percent). (Adapted from Ref. 2.)

Similar relationships apply for anisotropic materials. Transverse to the flow direction, physical properties are lower and material shrinkage higher. Crystalline and fiber-reinforced resins exhibit lower mold shrinkage in the "flow direction" versus the "cross-flow, transverse, direction," and this is true for the amorphous resins only to a lesser degree as shown in Table 8.4 for a variety of generic resins, both reinforced and unreinforced. To minimize property and shrinkage effects, product section thickness should be uniform throughout the product. Section thickness should be only as thick as required to meet product strength and meet flow and fill requirements for the selected resin. Each resin has a minimum section thickness to flow length to fill the product cavity. If these parameters are not compatible, processing problems will result. Always check with your material supplier for the minimum section thickness and flow length for the resin selected for the product.

As section thickness increases, higher mold and material shrinkage will occur in the product and even higher in the transverse direction. Thick parts require the gate to stay open long enough to fully pack out the part and to minimize the differential shrinkage problem. In fiber-reinforced parts, thicker than 0.125 in., the fiber orientation in the product's surface layer is oriented in the flow direction, randomized and in the interior of the part. The thicker the product, the more pronounced this effect. The surface layer may be equal to only 40% of the part's total thickness. In thick parts this produces a more uniform property and shrinkage distribution when fully packed out. The product properties will be higher in the flow direction because of the skin

Table 8.4 Nominal mold shrinkage rates for thermoplastics (in./in.)

Material	Average Rate per ASTM D955		Directional Rates for 0.062-in. Sample	
	0.125 in.	0.250 in.	Flow	Transverse
ABS				
Unreinforced	0.004	0.007	0.005	0.005
30% glass fiber	0.001	0.0015	0.001	0.002
Acetal, copolymer				
Unreinforced	0.017	0.021	0.022	0.018
30% glass fiber	0.003	0.008	0.003	0.016
HDPE, homopolymer				
Unreinforced	0.015	0.030	NA	NA
30% glass fiber	0.003	0.004	0.003	0.009
Nylon 6				
Unreinforced	0.013	0.016	0.014	0.014
30% glass fiber	0.0035	0.0045	0.003	0.004
Nylon 6/6				
Unreinforced	0.016	0.022	0.021	0.021
15% glass fiber + 25% mineral	0.006	0.008	0.006	0.007
15% glass fiber + 25% glass beads	0.006	0.008	0.006	0.008
30% glass fiber	0.005	0.0055	0.003	0.005
PBT polyester				
Unreinforced	0.012	0.018	0.018	0.015
30% glass fiber	0.003	0.0045	0.003	0.007
Polycarbonate				
Unreinforced	0.005	0.007	0.006	0.006
10% glass fiber	0.003	0.004	0.003	0.004
30% glass fiber	0.001	0.002	0.001	0.002
Polyether sulfone				
Unreinforced	0.006	0.007	0.006	0.006
30% glass fiber	0.002	0.003	0.001	0.002
Polyether etherketone				
Unreinforced	0.011	0.013	0.009	0.011
30% glass fiber	0.002	0.003	0.002	0.004
Polyetherimide				
Unreinforced	0.005	0.007	0.006	0.006
30% glass fiber	0.002	0.004	0.001	0.002
Polyphenylene oxide/PS alloy				
Unreinforced	0.005	0.008	0.005	0.005
30% glass fiber	0.001	0.002	0.001	0.002
Polyphenylene sulfide				
Unreinforced	0.011	0.004	0.009	0.011
40% glass fiber	0.002	0.004	0.001	0.003
Polypropylene, homopolymer				
Unreinforced	0.015	0.025	NA	NA
30% glass fiber	0.0035	0.004	0.003	0.009
Polystyrene				
Unreinforced	0.004	0.006	0.005	0.005
30% glass fiber	0.0005	0.001	0.001	0.002

Source: LNP Engineering Plastics, Division of ICI Advanced Materials.

effects and fiber orientation. But randomly oriented fibers in the thick section help to reduce the transverse shrinkage effect and impart more uniform physical property values in the section. Since transverse physical property values are seldom measured, the properties in the transverse direction are assumed to represent only 70% of the actual material property values. These effects should be considered in the decision on where to locate the gate. The mold designer will then adjust the cavity dimensions for estimated shrinkage values as part of their responsibility.

The design team is responsible for determining where the parts gate is located. The gate location has an affect on the physical properties and mold shrinkage of the product. If not located correctly, it can cause weld lines forming in critical sections.

Product dimensions are controlled by processing and mold variables. Any change in the product's section thickness and layout, such as ribs, bosses, and holes, will assist in serving as anchors for some product dimensions. The designer uses ribs to increase the product's strength plus bosses, holes, and major changes in section direction for assisting the mold designer in being able to hold tight dimensional tolerances in the mold cavity as shown in Figure 8.12. All plastics shrink by different amounts on cooling, and the resin needs points to anchor the material to control the shrinkage. The amount of the part's mold shrinkage and direction are considered by the moldmaker to locate the gate and to size the product cavity. Changes in section serve as control points by restricting the material shrinkage in the directions indicated by D, H, and L in Figure 8.12 a,c.

Dimensions not controlled by the mold must have their in-cavity mold dimensions adjusted by the mold designer. These are shown in Figure 8.12b. Each material's anticipated shrinkage value must be adjusted for section thickness, and where the part is gated, points of the material must be injected into the mold cavity. Molding conditions, such as melt temperature; mold temperatures, injection temperatures, and packing pressure, also assist in controlling material shrinkage during manufacture. It is the mold designer's responsibility to analyze the material mold shrinkage values and directions in the mold design to achieve a dimensionally correct molded part. Each products final dimensions are determined on the base resin's shrinkage and whether it is filled or reinforced. Material suppliers provide tables and shrinkage assistance calculations to mold designers to accurately calculate the mold cavity dimensions for a product using their resins. Many mold builders, before finalizing product cavity dimensions, mold sample parts in the new mold to verify their shrinkage calculations against product dimensions. This is important for the designer to understand, as part warpage can occur as a result of fiber orientation and nonuniform shrinkage at ribs, section thickness changes, bosses, and corners. Even fine particle (mineral) reinforced products will exhibit differential mold shrinkage.

A product's mold shrinkage is controlled primarily by the mold's cavity temperature control system. The product cavity gate is also sized to ensure that it remains open long enough until the product cavity is filled and packed out during each molding cycle to obtain the maximum product weight. A higher mold cavity temperature ensures that the gate will remain open to continue to pack out the product cavity for semicrystalline, crystalline, and LCP resins to assist in obtaining maximum material shrinkage in the mold cavity. This makes the product more stable at elevated temperatures and

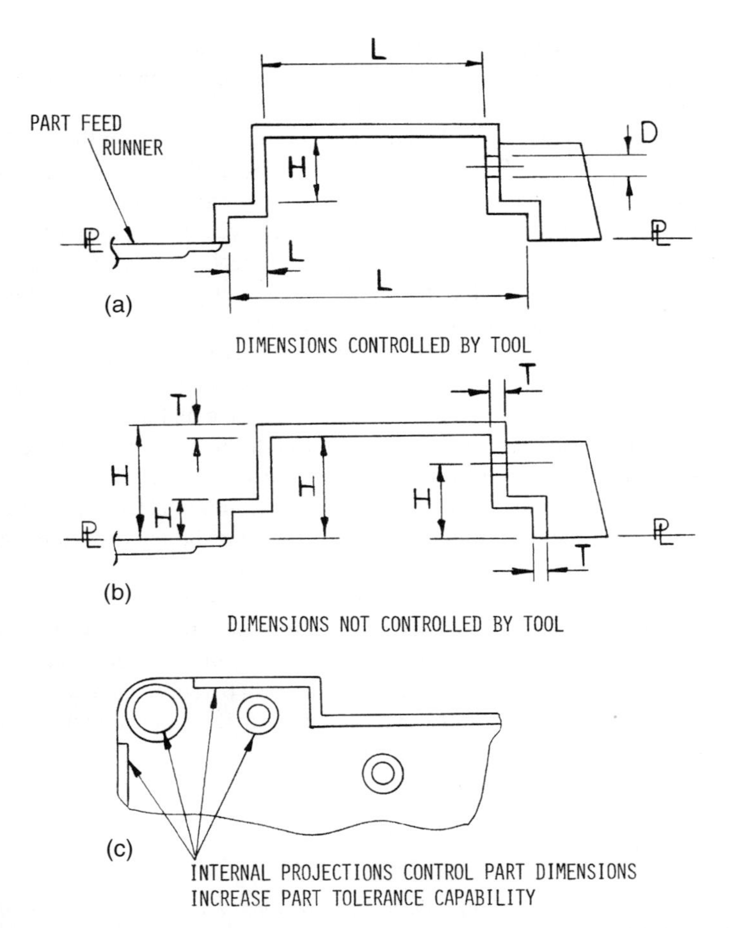

Figure 8.12 The relationship of the mold cavity to part dimension control: (a) dimensions controlled by tool; (b) dimensions not controlled by tool; (c) internal projections control part dimensions increase part tolerance capability. (Adapted from Ref. 3.)

makes it possible to obtain the correct dimensional part tolerances. Crystalline resins are molded at high cavity temperatures above 150°F. Molding crystalline resins at elevated temperatures is used to increase crystallization, obtain the required product size out of the mold, and stabilize and reduce the amount of postmold shrinkage of the product.

Amorphous resins are usually molded in a cold mold, with temperatures ranging from 40 to 100°F. Their in-mold and postmold shrinkage is low, and the cold-temperature mold aids in faster molding cycles. The cold mold extracts heat from the product, that which enhances and expedites product solidification, making the part is rigid enough to be ejected from the mold cavity without distortion caused by ejection pins bending or punching through the part.

Filled and reinforced amorphous resins have even faster cycles as the fillers and reinforcements give the part greater rigidity and strength so that they can be ejected even hotter from the mold cavity without distorting.

Selection of the product's mold cavity temperature is based on the product type of resin gate size, packout and solidification time, and end-use temperature requirements. The molder must be involved in the early product and mold design discussions for the product as decisions made then will affect cycle time and part cost. They can comment on the mold temperature, process control system, machine size, and processing variables to ensure that the product will meet customer requirements.

Other variables are involved in determining the amount of shrinkage that a material experiences in the mold cavity. An important mold cavity variable, besides temperature, and cavity steel involving wear life and heat conductivity, is gate size and its location to fill the part. The product's gate(s) is (are) located at the thickest section of the part. Ideally, when the material flow through the gate enters the cavity, the resin will produce an even melt front to uniformly fill the cavity. This is to ensure an even flow of material into the mold cavity under optimum injection pressure without causing resin shear at the gate. If the gate is undersized and if the injection pressure is too high, gate shear can result, increasing the resin's melt temperature $\geq 100°F$, which can degrade the plastic material's physical properties. The gate size and type must be large enough during fill and then cavity packout to ensure that it is the last section of the part to solidify. This is required to compensate for the material's shrinkage that allows more resin to be forced into the cavity to control part dimensions and eliminate voids and porosity. This also ensures that each part cavity is packed out to a uniform product weight, consistently, cycle to cycle.

REFERENCES

1. L. Leonard, "Design to Minimize Wear and Friction," *Plastics Design Forum* 49–52. (July/Aug. 1990).
2. *General Design Principles—Module II,* H-44600. E. I. Du Pont de Nemours Corp., Wilmington, DE. (Sept. 1999.)
3. "Understanding Tight Tolerance Design," *Plastics Design Forum* 61–71. (March/April 1990).

CHAPTER 9

PRODUCT DESIGN CONSIDERATIONS

EFFECTS OF ELASTICITY, HOMOGENEITY, AND ISOTROPY

Designing a product, any type of product, begins by understanding how the materials considered for the product will behave in the product's end-use application. Also, any material limits or requirements that may have been placed on the designer and design team for the product to be a success must be considered. Plastics are not metals and do not behave like them under all conditions. Metal products are designed as isotropic and homogeneous and with a good degree of elasticity. The design of a plastic product takes advantage of what the plastic material can provide as compared to metal. As long as these advantages and product limitations are considered, a successful product will be produced in plastic.

Designers of plastic products must know and remember that plastics are not linearly isotropic in their behavior under load. Their behavior is nonlinear in response to minor changes in applied load. Temperature and environmental conditions have a greater affect on plastics in terms of how they respond to an applied force. Therefore, the standard metal engineering constants used in design, Young's modulus, and Poisson's ratio must be adjusted for the plastic material. The material constants will vary with the end-use conditions that the product may experience, and these constants are used to describe the mechanical response of the plastic material. These engineering constants change with the material, temperature, and environment and the designer must adjust them to suit the conditions during the design review.

Only for elastic, isotropic materials that respond to loads linearly (i.e., load is proportional to deformation) can these two constants be used to correctly analyze a product under load. Only if the material is homogeneous can these values be used throughout the product design as with metals.

STRUCTURAL PRODUCT ANALYSIS

In structural product analysis the engineer assumes that the material is linearly elastic, homogeneous, and isotropic. This is required for the correct use of the engineering design equations for structural analysis, such as tension, bending, and torsion, which are based on the response of isotropic materials. This is not true for plastics, especially glass-reinforced and liquid crystalline polymers, which are highly anisotropic. Therefore, as the degree of anisotropy increases, the values for these physical property constants, including tensile strength, modulus, and Poisson's ratio, must be adjusted for conditions by the designer to correctly describe the product's material response when subject to external forces. Ensuring that the product's material response is accurately analyzed requires a plastic product, of critical or high liability use, to be thoroughly end-use-tested. This is accomplished preferably by building or molding a prototype in the selected material for testing. This is required to ensure that any variability of a plastic's material properties is considered in the design and use of isotropic analysis techniques using the design equations for structural analysis of the product.

The design team should always consult with their material suppliers to obtain as close to actual end-use material response data as possible for the material's physical property evaluation. This is to ensure that the equation values replicate the material's response for the product's design conditions. Material suppliers develop this product-specific data on their generic resins for customers, but since there are so many product variations, only their standard major products are tested. Since all product conditions vary, the data may only approximate the conditions required for a design situation. In these situations the design team must use a safety factor to ensure that the design analysis, material properties, and processing variables do not affect the product's end-use performance. With the increased use of finite-element analysis (FEA) for complex structures, the behavior and property values of anisotropic materials becomes much more important for correct product analysis.

Factors affecting the design of the product are

1. Product geometry
2. Forces and their location on the product
3. Rate of loading
4. Temperature
5. Environment
6. Material variations
7. Processing, assembly, decorating, shipping, and installation variables

Product Geometry

The product geometry or shape is developed by the designer to fulfill an end-use function. Plastic products are easily contoured to meet the customer's requirements, reduce weight, and add additional functions such as multifunctional and less expensive assembly features. Plastic can replace an existing metal product where the design can now be less expensive and more functional and/or aesthetically pleasing to the user. The design team decides how the product is to be manufactured; injection-molded, extruded, thermoformed, or otherwise processed. The product should be designed for ease and repeatability of manufacture and economics.

Stress–Strain Analysis of the Forces on a Product

Force is defined in physics as any cause that changes, destroys, or produces motion. Therefore, a force is a push, pull, or impact on an object in a static or dynamic mode. Force (F) is expressed as mass (m) times acceleration (a): $F = ma$.

A product design is initially evaluated in static, at-rest, mode for ease of analysis. Dynamic designs involving gears, fans, bushings, pulleys, wheels, and other devices are analyzed as static situations with the dynamic, acceleration, and motion effects considered separately but additive to the static design's analysis. Plastics have an advantage over many conventional materials as they have material properties that translate into helpful physical attributes during a product design. These include energy-absorbing characteristics, natural lubricity, and resistance to chemicals, combined with their capability to withstand high loading forces. Plastic products under a force can also deflect, transferring the forces or distributing them throughout the product without causing a failure at the site of the applied force. This gives the designer an added built-in safety factor in many product applications. This is a very important factor to always consider, as it involves strain rate and creep. Plastics are sensitive to strain rate as force translates into stress on a product in their viscoelastic–viscoplastic behavior. The rate, type, and duration of forces on a body determine whether creep or impact is a factor in the product's mechanical response. The forces rate, magnitude, duration, and type of mechanical loading—tension, compression, flexure, or shear—plus temperature determine what stresses the product must distribute within its structure to remain at rest.

A three-dimensional body subjected to external forces, such as F_1–F_5 in Figure 9.1, distributes these forces internally for the body to remain at rest. To remain at rest, a body must distribute internal reaction forces to counteract all externally applied forces. If the part is sectioned at an arbitrary cross section, a system of forces, called *stresses,* act on the sectioned surface to balance the external forces to keep the part at rest. These stresses act throughout the part and are expressed with a direction and magnitude. Stress will vary throughout the part depending on where the part is sectioned and are called *force* (stress) *vectors*. As an analysis in two dimensions, consider stress, (S), at the cut section 9.1b, with two reaction vectors or components. One stress vector acts perpendicular to the surface and is a normal or direct stress. The second is parallel or in the plane of the cut surface and is a shear stress ($\Im$). Shear stress will be discussed independently.

Direct Stress. A simple tension test (Fig. 9.2) best illustrates this for force, deflection, and stress. Direct stress is the ratio of the applied force to the original

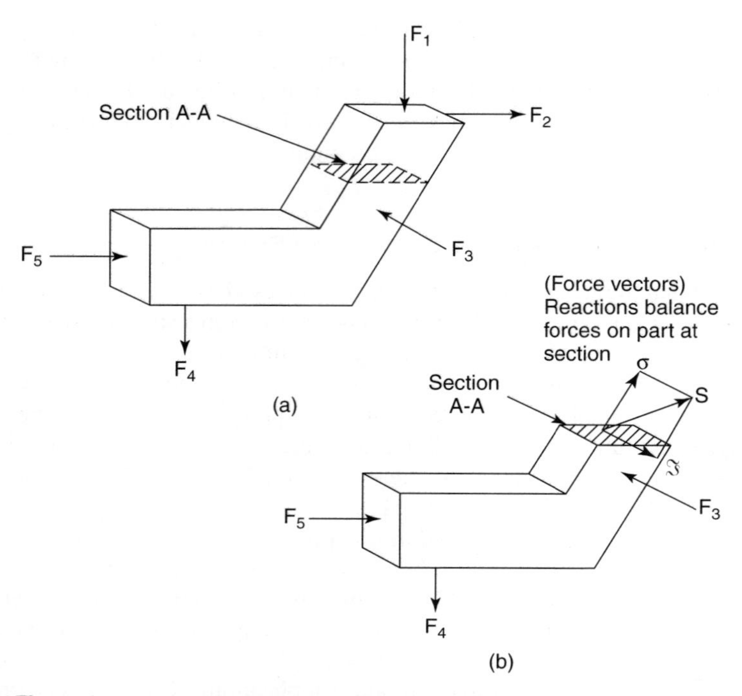

Figure 9.1 Internal forces/stresses acting in a part. (Adapted from Ref. 3.)

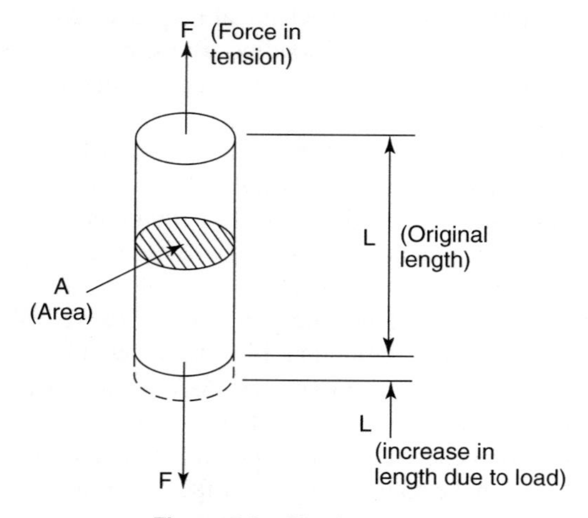

Figure 9.2 Simple tension.

cross-sectional area, expressed in pounds per square inch (lb/in.^{2} or psi). In the metric, International System of Units (SI system), stress is expressed as Newton's per square meter, or Pascal's (Pa):

$$\text{Stress} = \frac{\text{force}}{\text{area}} \quad \text{or} \quad \sigma = \frac{F}{A}$$

The force as applied on the part is said to be in tension and reversed in compression.

Direct Strain. When a force is exerted on a part, a stress results, and the part changes in length, as shown in Figure 9.2. This change from the part's original length (L) is denoted as the change in length ΔL. Strain is defined as a percent change and is dimensionless, that is, is not expressed in measurment units.

$$\text{Strain} = \frac{\text{change in length}}{\text{original length}} \qquad \epsilon = \Delta L/L$$

Material suppliers develop stress–strain material curves for their materials using an Instron tensile testing machine. Material suppliers generate material property data under laboratory-controlled test conditions at anticipated end-use conditions using varying temperature, humidity, and strain rates. Strain rate is the velocity or rate of separation of the test heads when a force is applied to the test sample. The American Society for Testing Materials (ASTM) specifies the rate of test head separation in increments of movement per unit of time, to generate the stress–strain curves for plastic materials. When a slow pull rate or separation is used, such as one inch per minute, the material will be able to absorb more energy before failing than when a higher pull rate is used, such as 5 in. per minute. The material's energy absorption capability, typical stress–strain curve in tension, is shown in Figure 9.3 represented in the slope and area under the curve generated on a graph of force over elongation.

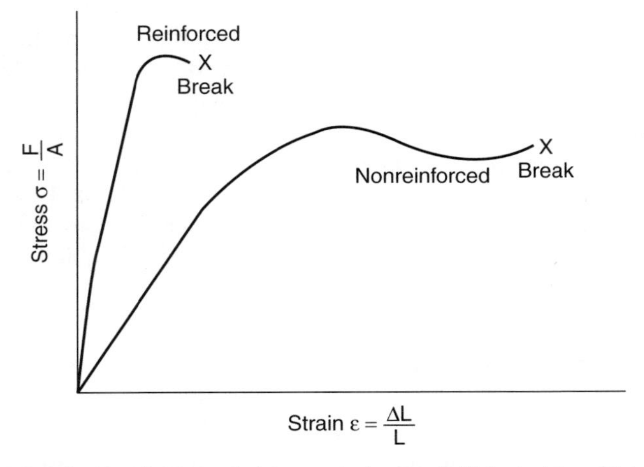

Figure 9.3 Tensile strength (stress–strain curve). (Adapted from Ref. 2.)

Maximum absorption of energy is shown in the material's curve as it reaches the material's yield point, flows, and then fails. An unreinforced plastic will, after reaching its yield point, exhibit a falloff in stress and often a gradual rate of climb in the curve as the material work-hardens and continues to elongate. Reinforced and low-elongation plastics will have a steeper rate of climb, slope, or a more vertical rise to the materials yield point with little yielding and then failure with less area under the curve.

Viscoelastic Behavior

Viscoelastic materials have the ability to return to their original shape after being deformed. These are rubber, thermoelastomers, and highly plasticized or impact-modified thermoplastics. Plastics may exhibit viscoelastic properties at low ($<0.1\%$) strain rates, which allows a return to their original shape without permanent distortion when an intermittent force is removed. But with extended time under a continuous force and at a constant or fluctuating temperature, the material will eventually creep.

Viscoelasticity is a safety factor, seldom considered in the design of a product. Viscoelasticity for nonrepeating intermittent applied forces can allow a product to withstand substantial nonlinear deflections not anticipated during normal use. A *nonlinear deflection* is defined as a deflection in a product section due to a force that causes a deflection greater than the wall thickness of the section.

Deflections of a product section are difficult to analyze. Standard stress analysis theory concentrates and magnifies stresses at the point where a force is acting. Plastics have the ability to absorb and redistribute the stresses to other areas in the product by deflecting and stress transfer through the material. A metal absorbs the main force locally, with only a small redistribution of stress throughout the part. This can result in local fracture or buckling and/or denting a metal product.

Viscoelastic Design Considerations. A product's design analysis is based on a number of assumptions, including magnitude and direction of forces and their steady-state and/or frequency of application on the part. The product is designed for anticipated and/or known forces and end-use environment conditions plus any agency or code requirements. For electrical power tool product testing, the UL drop-impact test is a very severe destructive test that the product must pass for the UL consumer seal of approval. To pass this test, the designer must select a material with multipurpose properties, such as a material with a high impact resistance or viscoelastic properties in a dry-as-molded (DAM) condition, plus tests for strength and rigidity for high temperature operation and heat absorption in use. The product design and material must be capable of passing all test conditions. To aid the product material in absorbing impact energy, the designer may add to the tool housing energy-absorbing external vanes and internal ribs to strengthen and redistribute the stress on impact to stronger sections in the tool housing.

HYGROSCOPIC MATERIAL EFFECTS

Hygroscopic materials, nylon, absorb moisture and become tougher or more elastic. However, a poorly designed nylon product could fail in the material's DAM state.

Table 9.1 Stress types for design analysis

Stress Type	Design Property
Bending	Flexural modulus
Compression	Compressive modulus
Tension	Tensile modulus
Torsion	Shear modulus

The designer must design the product to pass the most severe conditions. When the product is designed using DAM material physical property values, the product will have a high probability of always meeting the product's customer requirements. End-use testing of the molded product is always required to prove material and design before being released to production.

DESIGN EQUATIONS

The standard mechanical design equations used for calculating stress, strain, and deflection were developed for isotropic, homogeneous materials made into simple shapes with stresses under elastic conditions. Plastics do not always behave in this manner as they respond to their end-use environment. Therefore, the standard design equations produce answers not truly realistic for all conditions. The standard design equations assume that all materials follow Hooke's law, specifically, that the strain produced is directly proportional to the amount of applied stress in the part. This is for both tension and compression force. Thermoplastics have a higher tensile modulus than do their compression moduli. With specific thermosets, this is reversed; for instance, phenolic resins have higher compressive strength than tensile strength.

In all design analysis the type of force on the product must be identified so that the correct material property value is used as shown in Table 9.1. The appropriate stress type must be identified to select the correct material property data.

ANALYSIS OF STRESS

When a plastic product is subjected to a simple or complex force loading, the product section reacts to the forces, producing internal stresses in the body to resist the forces. How the stress is distributed in the body sections must be carefully analyzed to determine the reactions of the bodies' stresses to the forces in each section. The theory of superposition allows these forces to be broken down into individual force and vector components so that structural analysis is performed as if only the individual component force were acting. This stress analysis technique allows the individual and different types of stresses to be added to produce a combined loading stress factor in the product. To do this implies a linear relationship between stress and strain that is valid for metals. However, with plastics the relationship is nonlinear. For plastic product design simplification, use the assumption of linearity to facilitate analysis.

Often more than one type of "direct stress" can act on an element of the product. This is often in the same plane and can be a direct tension, compression bending, and/or a torsion force. It is important in these cases to know the maximum direct stress and/or torsion shear stress that occurs in an element's plane of the part. Designers use Mohr's circle to obtain a graphical representation for determining the major stress vectors and their direction. As an aid in determining all forces acting on a product, the designer can use the design checklist, checklist 2 in Appendix A, for assisting in developing the force/stress analysis of the product.

SELECTING THE MATERIAL DESIGN STRESS

Once an analysis of the forces on the product is completed with their individual magnitude and type determined, an initial material selection can be made. In selecting the initial material candidate, more than one generic plastic family may fit the application. According to the type of product, end-use requirements, forces, and anticipated cost, many plastic materials will be likely candidates. The designer's material selection is based on the product requirement manufacture and cost. Assistance from material suppliers' case studies of their products and reference data is very helpful in making the initial selection. After a material candidate(s) is(are) selected, the design analysis for the material and product section thickness begins. The specific material's physical property values, at the product's operating conditions, are selected to size the product's section thickness and shape.

The material's physical property values selected for the design analysis must ultimately ensure that the materials sections thickness will meet the end-use product requirements. For example, if the materials modulus of elasticity (E) were not adjusted lower for higher-temperature exposure, the product could deflect and fail while experiencing a high temperature under load. If too low a material modulus value is used when not required, the product is overdesigned with too thick a section thickness with reference to the selected modulus value. Material is wasted; product weight increases and could cause a molding problem, voids, and porosity in the thick section. A wide range of plastic resin recommended section thicknesses are listed in Table 9.2 for design reference when material thickness is calculated to ensure compatibility with processing and material utilization.

MODES OF PRODUCT FAILURE

Understanding the ways that plastic products can fail should also be understood during the product's material selection process. The design team early in the products design analysis can do a failure mode and effect analysis (FMEA) study to anticipate how and where the product could fail from use and possible misuse. The FMEA is also used to determine whether there are any problem areas in the manufacturing process.

Products fail primarily for mechanical and physical property reasons, followed by chemical, electrical, and environmental problems in elastic and inelastic deformation and fracture.

Table 9.2 Recommended wall thickness for thermoplastic molding materials

Thermoplastic Materials	Minimum (in.)	Maximum (in.)
ABS	0.045	0.140
Acetal	0.015	0.125
Acrylic	0.025	0.150
Cellulosics	0.025	0.187
FEP fluoroplastic	0.010	0.500
Long-strand-reinforced resin	0.075	1.000
LCP liquid crystal polymer	0.008	0.120
Nylon	0.010	0.125
Polyarylate	0.045	0.160
Polycarbonate	0.040	0.375
Polyester	0.025	0.125
Polyethylene (LD)	0.020	0.250
Polyethylene (HD)	0.035	0.250
Ethylene vinyl acetate	0.020	0.125
Polypropylene	0.025	0.300
Noryl[a] (modified PPO)	0.030	0.375
Polystyrene	0.030	0.250
PVC (rigid)	0.040	0.375
Polyurethane	0.025	1.500
Surlyn	0.025	0.750

[a]Registered General Electric material name.

Source: Adapted from Ref. 1.

Elastic Deformation

A product fails in elastic deflection when forces on the body produce too large a deflection that the material, design and physical properties can support. Since most products are composed of more than one individual part, a failure in one can affect the entire product performance. For example, if a molded-in power tool trigger return spring fails, the tool will not operate correctly. Repeating dynamic deflections in parts can result in flexural fatigue failures. These failures may be preceded by audible sounds and visible vibration as with a bearing and gear chatter, knocking, or squeaking. Load-bearing structures, such as a worktable support leg not designed to limit deflection, may fail catastrophically when overloaded. The reason for failure may be attributed to forces on the product exceeding the leg's stiffness, flexure, and stability limit, resulting in the column buckling under the load. Selecting the correct property value for a material's flexural modulus at end-use conditions is critical for rigidity and strength of a product section. The product's internal design, ribs, and section thickness are dependent on the flexural modulus value used in the product's design analysis. Plastic products are usually designed to have a small amount of flexibility within the product design limits. They are rarely designed as rigid, nonyielding structures. It is important to determine whether any deflection is permitted in the product and the allowable amount. Therefore, deflection calculations are used to determine whether any product section deflection results from the applied forces. Use only the plastic materials modulus of elasticity for these calculations, never tensile modulus. Deflection can be controlled in a product section by using a series of ribs for strength and stability. When more rigidity is required, use cross-ribbing to tie the ribs together for added stiffness and strength. Do not try to control deflection by only increasing section thickness. The designer's use of controlled minimum thickness ribs will add more strength with correct material utilization in the product design and only where and when required for product strength and stability.

Inelastic Deformation

A force on the product that the plastic material could not withstand results in creep and material flow known as *inelastic deformation*. This is noticed after the force is removed as the material is permanently deformed, by cold flow. This deformation exceeds the material's elastic limit as the product section under the force does not return to its original shape. If the force on the product is constant, material creep continues until the section is weakened sufficiently by stretching to crack and fail. Products subjected to short-time internal or momentary high loading should use the material's proportional limit or maximum allowable shear stress in the design calculation. Elevated temperatures accelerate this failure mode by lowering the material's elastic properties, resulting in accelerated elongation.

Fracture

Product fracture is caused by a force that exceeded the material's physical property strength. Fracture failure can be catastrophic and instantaneous. Fracture failure develops as a small crack in the product surface or at an internal contamination site and progresses over time through the section of material until total rupture of the material causes failure. Fracture can occur from an impact load, an unexpected high force that instantly exceeds the materials tensile or shear strength.

Fracture also occurs from steady-state, long-term high loading over a period of time (creep rupture). Environmental conditions surface (crazing), or fatigue and vibration by alternating forces can affect the material, resulting in failure. Low temperatures can accelerate material cracking due to lowering the elongation of the material, known as *low-temperature brittle fracture*. The product's maximum principal stresses must be thoroughly analyzed to eliminate fracture failure. If the product could be subjected to impact forces, testing could be conducted to determine its effects on the product. Therefore, it is critical to always radius a corner to reduce stress concentrations as shown in Figure 9.4.

STRESS–STRAIN CURVES

A plastic material working stress is determined from the tested material's stress–strain curves at a specified loading rate in tension or compression. Stress–strain curves are

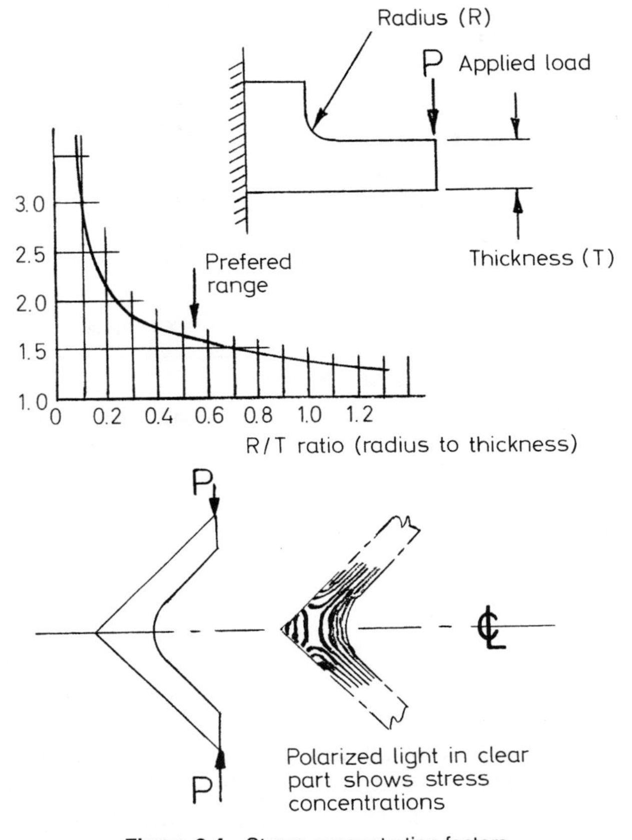

Figure 9.4 Stress concentration factors.

developed for resins in tension, torsion (shear), compression, and flexure as shown in Figure 9.5.

Isotropic materials (metals) have a deformation that is proportional to the applied forces over a range of increasing force. Since stress is proportional to force and strain is proportional to deformation, this implies stress is proportional to strain. Hooke's law states this proportionality as

$$\frac{\text{Stress}}{\text{Strain}} = \text{constant } (E)$$

The constant (E) is called the modulus of elasticity. It is also called *Young's modulus* or *tensile modulus*. For the round bar in Figure 9.2, the tensile modulus is

$$E = \frac{F/A}{\Delta L/L} = \frac{FL}{A\Delta L}$$

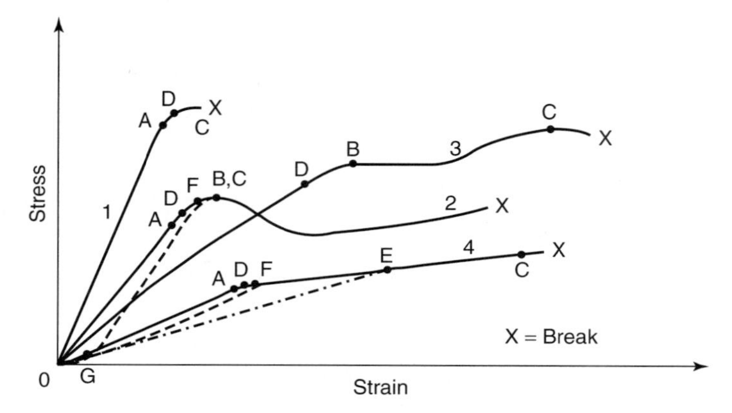

Figure 9.5 Typical stress–strain curves for various plastics. (Adapted from Ref. 5.)

where F = force
ΔL = change in length
A = area of the section
L = length under analysis

Modulus is the initial slope of the stress–strain curve, A–O in Figure 9.5. Materials obeying Hooke's law are elastic; this means that as the force is released, the material returns to its original dimensions.

However, thermoplastics do not follow Hooke's law exactly, except in conditions of less than 0.05% strain. Other materials, such as rubbers and elastomers, also known as *elastic,* will almost always return to their original state, where the stress is proportional to the strain. All materials have variable or fixed properties and may not be linear in their response to stresses. Most metals are linear to their proportional limit on their stress–strain curves.

For plastics, the slope of the curve, the straight portion of the stress–strain curve, is difficult to construct since the material's yield strength is not linear and the generated curve has a gradual bend, not a straight line. It requires some adjustment to construct a straight-line tangent to the initial portion of the curve to measure the slope to obtain the material "modulus." The modulus obtained from the curve is termed the "initial modulus." This is misleading, since the material is elastically nonlinear. Therefore, many material suppliers provide the designer with the 1% secant modulus, which is a better representation of the materials behavior. The secant modulus is a straight line drawn from the 1% strain point, which closely follows the slope of the stress–strain curve. The designer should obtain from the material supplier their stress–strain curves for the material's at-test conditions, namely, with the temperature, moisture, and other levels that you need. If not available in curves, ask what value is available and at what temperature, which is termed, *Young's modulus,* an *initial modulus,* or the *secant modulus*. If you are unsure, use a safety factor to compensate for variation of material properties. Understanding stress–strain curve data for plastics is important when selecting the material's physical property values from environmentally generated stress–strain curves as shown in Figure 9.5.

Proportional Limit. Plastic materials exhibit a point (Fig. 9.5, point *A*) on the curve where typically the slope changes and linearity ends. This is the point where the material begins to creep. Up to point *A* the material reacts linearly, without deviating from the proportionality of stress to strain. Point *A* on curves 1, 2, and 4, on the release of the applied force, return to their original starting points. Some materials retain this proportionality for large values of stress and strain while other have no or little proportionality (Fig. 9.5, curve 3). These materials begin to yield or creep under small amounts of force (stress) and do not have a true proportional limit. Values calculated for *E*, modulus of elasticity, from these curves are expressed in pounds per square inch or pascals.

Yield Point. The yield point (point *B*, Fig. 9.5) occurs on the stress–strain curve at zero slope or when increases in strain occur without an increase in stress. Some materials do not exhibit a true yield point (Fig. 9.5, curves 1 and 4).

Ultimate Strength. This (point *C*, Fig. 9.5) is the maximum stress a material can withstand before failing by breaking (fracture) as the material cold flows until rupture (Fig. 9.5, curves 1 and 4, point *C*). When cold flow occurs, the stress decreases as the material begins to elongate without any increase in stress until rupture just beyond point *C* in Figure 9.5, curve 3. The material may work-harden (a metal term for flow orientation of the molecules in the material which still elongate but require additional force to make it happen). The strain rate remains constant, and as the material work-hardens, which increases the stress in the sample finally and overcomes the material's properties, the material then elongates to fracture and failure. Figure 9.5, curve 2 illustrates how the ultimate strength point was reached and with sustained force how the material work-hardens, rising slightly, until failure.

Elastic Limit. The elastic limit (point *D*, Fig. 9.5) point on the stress–strain curve is where the material has permanently deformed, creep occurred, when the force is removed. The elastic limit occurs just after the proportional limit is reached. Some materials, such as metals, can be stressed beyond their proportional limit (*A*) and have zero strain, with no elongation in the test sample, after the force is removed (Fig. 9.5, curves 1, 2, and 4), but not a thermoplastic.

Secant Modulus. The secant modulus (point *E*, Fig. 9.5) is the ratio of stress to the corresponding strain at any point on the stress–strain curve. It is used to approximate the slope of the stress–strain curve, with or without an offset in the strain value. The secant modulus on Figure 9.5, curve 4 at point *E* is the slope of the line *OE*.

Yield Strength. For materials not exhibiting a yield point, it is desirable to establish yield strength (point *F*, Fig. 9.5) by selecting a stress point on the stress–strain curve just beyond the elastic limit (Fig. 9.5, point *F*, curve 4). This value was developed for materials that do not exhibit a true yield point. Yield strength is for materials with high strain values and high elongation, and is used to

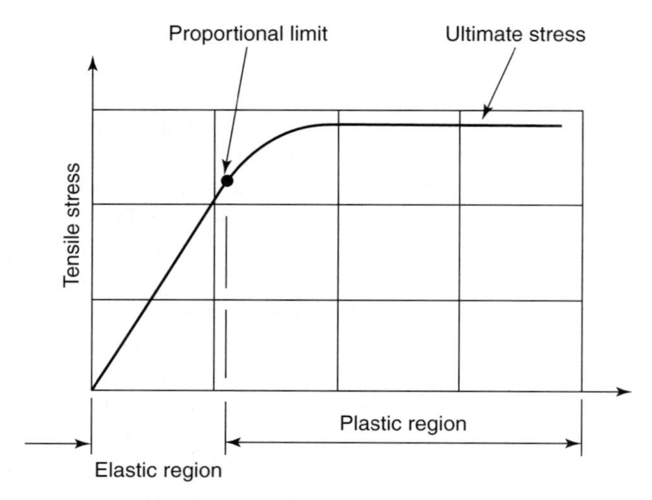

Figure 9.6 Generalized stress–strain curve for plastics.

obtain a realistic yield strength value for the material. Yield strength is usually established by selecting a strain offset of 1–2%, point *G*. From this point a line parallel to points *OA* is drawn to intersect the stress–strain curve at point *F*. The yield strength is then the stress value at this point, denoted as the "yield strength at the selected 2% strain offset."

Selecting the Design Stress

A material's design stress for a product is developed from stress–strain curves at a specific loading state. This value is at or offset from the proportional limit point on the stress–strain curve. This value can then be reduced by a specific percentage, known as a *safety factor*, to be used as the material's end-use allowable design stress. Safety factors will be discussed later in this chapter. When designing in metal, the stress–strain curve is linear up to the proportional limit, where the metal's behavior changes from elastic to plastic. Just beyond this stress point, the elastic limit (yield point) occurs, and the metal flows and a permanent deformation results in the material. It is customary to keep the loading on a part within the elastic region.

With plastics the transition is more gradual. Except for filled and reinforced resins, with low elongation, the curve slowly levels off as in Figure 9.6. For product with calculated section stresses in the plastic region, before the material yields or the elastic limit is reached by flowing, the plastic will recover without any permanent deformation when the force is released. This occurs when the force is applied for a short time period and with no resulting material creep. Figure 9.7 shows thermoplastic, unreinforced polyester, subjected to increasing forces for short-time loading indicating the degree of recovery and percentage of resultant creep. This graph indicates that plastics can be designed for higher stress levels in the material's plastic region and still recover with

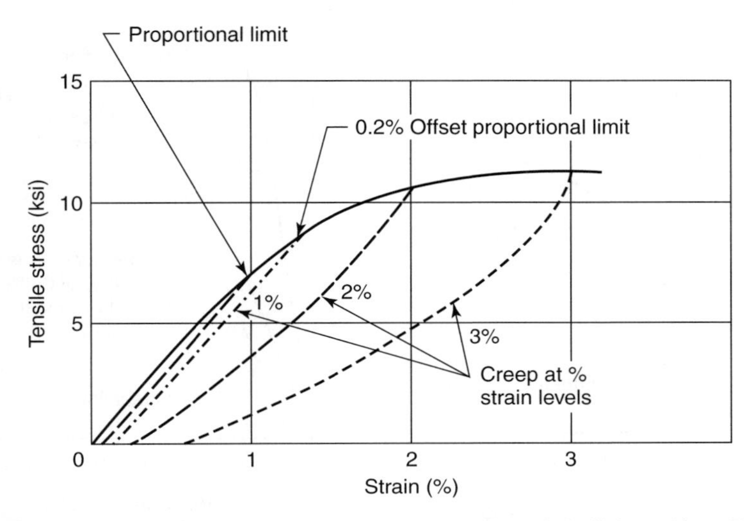

Figure 9.7 Short-duration creep effects beyond proportional limit. (Adapted from Ref. 6.)

minimum or no aftereffects for a short loading time. The material behaves elastically with no measurable creep.

Creep data are very useful in designing a product under high loads with the time varying under load. A problem that the designer encounters is that the proportional limit for thermoplastics is difficult to identify. This is due to the gradual transition from linear behavior to plastic flow. To assist in identifying the proportional limit on the curve, the designer can use a 0.2% strain adjusted offset to locate this point. Using the 0.2% strain offset, a line is drawn parallel to the linear portion of the stress–strain curve. The point at which this line intersects the knee of the curve is the 0.2% offset proportional limit. Using the 0.2% offset value, the material's design stress is adjusted down, decreased, by the safety factor to a value that will suit the part's end-use design requirements. Examples are shown in Figure 9.8a for unreinforced and Figure 9.8b for reinforced thermoplastic resins.

Other material physical property factors to be considered once the material design properties are selected for the products design requirements are described in the following sections.

Material Creep/Elongation

Creep must always be considered during a product's design analysis. Creep is a predictable deformation of a material defined as increasing strain over time in the presence of a constant stress. All plastics creep; the force and time under the load determine the amount of creep. The force can be intermittent or continuous in how it is applied to the product. Whenever possible, always try to limit the products time under a force to reduce the effects of creep. The crystalline plastics have lower creep rates than do amorphous resins. This is due to their strong molecular bonds

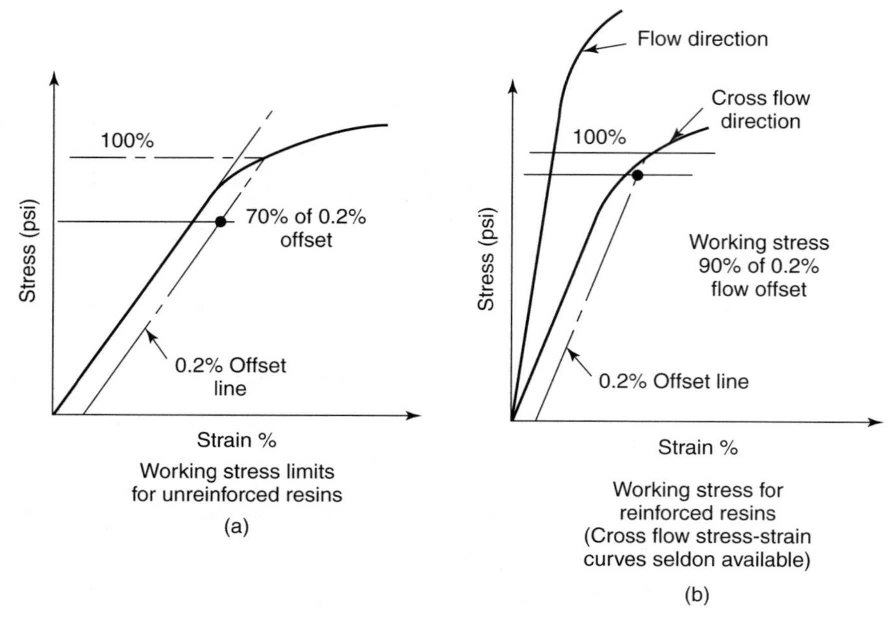

Figure 9.8 Working stress limits: (a) unreinforced resins; (b) reinforced resins (cross-flow stress–strain curves seldom available).

that produce high-strength physical properties in a material. Reinforcements and fibers, especially glass fibers, dramatically improve the creep resistance of a plastic. Thermoset crosslinks on curing (bonds form at sites where molecular chains touch) are more creep-resistant under all physical conditions, loading, time, and temperatures than are thermoplastics. They exhibit even higher properties when filled or reinforced. When designing a product for long-time continuous loading with deflection critical, the material's creep behavior must be carefully analyzed.

Suppliers develop material creep data for their products in various test modes with different tension, compression, and flexure values at various loadings, typically at constant temperature. Under constant load, the sample's deflection or elongation is recorded at regular intervals of hours, days, weeks, or other time periods. Materials are usually tested at four or more stress levels. The stress–strain data recorded against the variable, time, are presented as creep curves of strain versus log time as shown in Figure 9.9, for products under long-time stress conditions. Isochronous stress–strain curves at constant strain yield isometric stress–log time curves, sections taken through the creep curves at constant time. These data are used to analyze the product's material reactions under the various conditions.

Apparent or Creep Modulus

In designing a product under a continuous load, creep considerations are critical. If the modulus of elasticity (E) were used to determine the deflection of a section

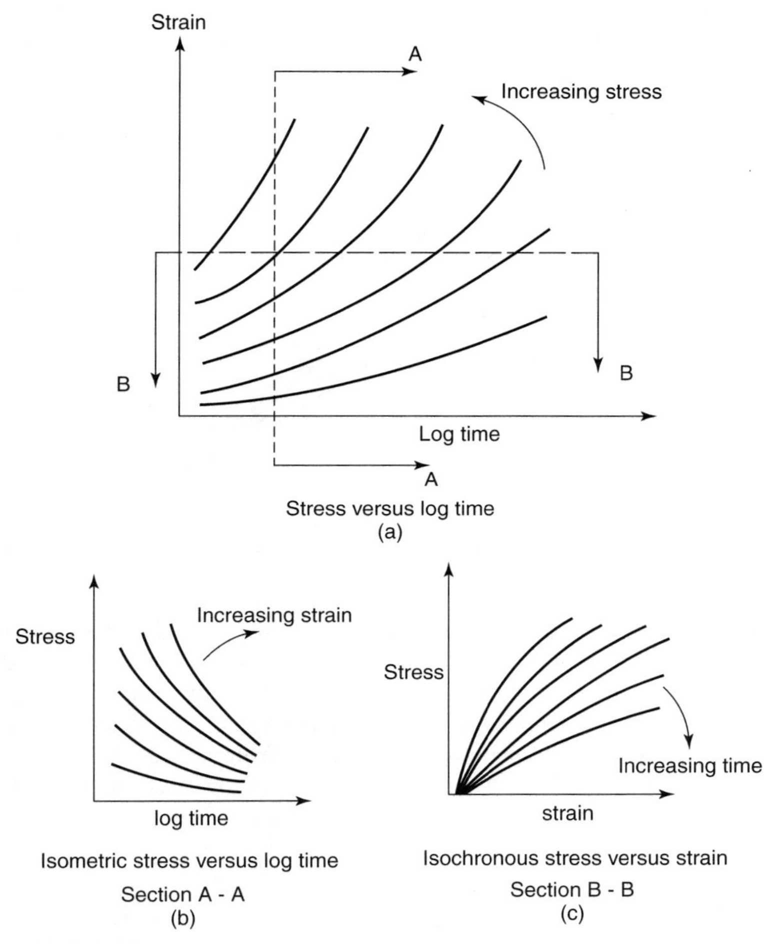

Figure 9.9 Typical creep data curves for plastics: (a) stress versus log time; (b) isometric stress versus log time; (c) isochronous stress versus strain.

under load, error would result as if creep were not considered. When the stress level, temperature, and environmental conditions are known or estimated, and the material suppliers creep curves are used, the modulus (E_{app}), is calculated and used for the material's design stress. The value of E_{app} is obtained from a material's creep modulus–log time curve (Fig. 9.10).

Using the creep curves

$$E_{\text{app}} = \frac{\sigma}{\epsilon_c}$$

where σ equals the calculated stress level and ϵ_c is the strain from the creep curve at anticipated time and temperature. The value, E_{app}, is used instead of E to predict the

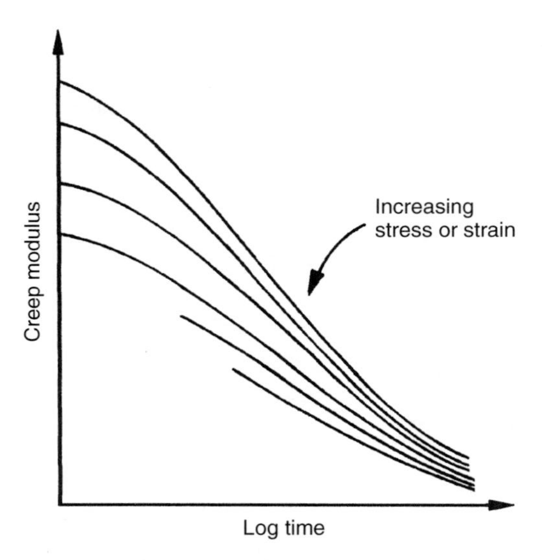

Figure 9.10 Creep modulus.

maximum deflection using the standard engineering design equations for mechanical–structural analysis. Curves of creep modulus or log creep modulus versus log time at either constant stress or strain are usually available from the material suppliers for their major products. These data may also be available in tables at constant stress and temperature for various time periods. If the specific material data are not available, the data of a material closest to the material's physical properties can be substituted. But you should adjust the calculated value using a factor of safety.

Hoop Stress

Stress–strain material data are available for pressure vessels under constant internal pressure in the form of material hoop stress graphs. Hoop stress is the stress component perpendicular to the tensile stress component, on the main axis, in the shell of the pressure vessel. Figure 9.11 is a force–time hoop stress graph for a nylon resin at three test temperatures. The nylon was fully moisture-conditioned to a maximum moisture level or saturated conditions. The graph shows the material's maximum allowable design stress when used to calculate a pressure vessel or piping section thickness at the material's maximum working stress.

Hoop stress is similar to creep rupture data. When a part under constant or alternating forces exceeds the material's allowable elongation, rupture occurs. Failure can be either brittle or ductile, preceded by necking, a thinning of the thickness that precedes rupture. Rupture occurs at the highest stress point usually catastrophically.

Creep rupture data are obtained similar to creep data with measurements taken at timed intervals. Changes in dimension versus time at constant pressure are recorded to material failure. Strain, material elongation, is seldom recorded in creep rupture tests. Strain measurement of a pressure vessel may be misleading since a local imperfection

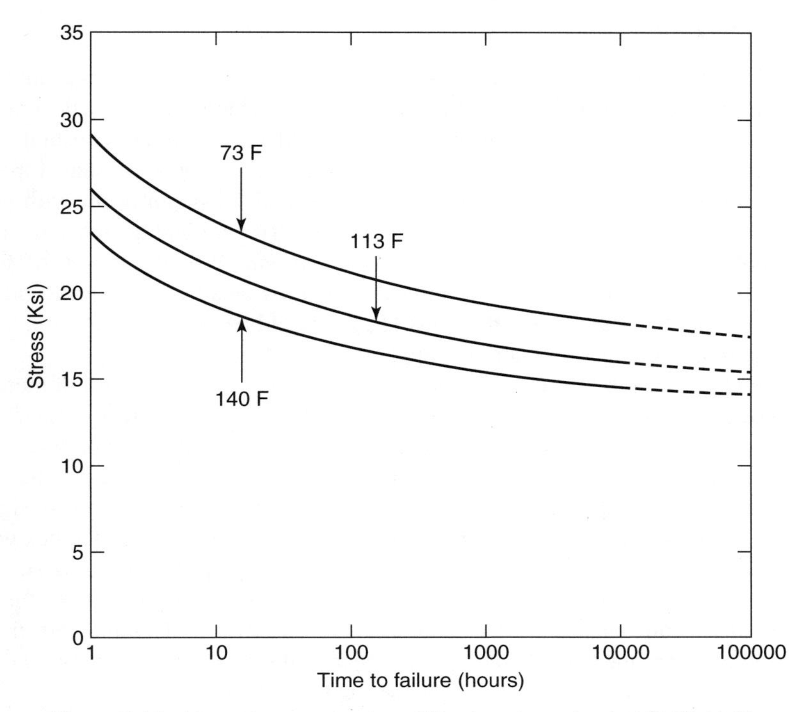

Figure 9.11 Hoop stress curve, type 6/6 nylon pipe saturated (8.5% H_2O).

in the material may elongate more than other sections, causing failure. However, when measured, the vessel's total expansion at set time intervals is equated to changes in the total circumference, and the elongation for the whole is assumed equal in all sections.

Strain, when generated, is plotted as log stress versus log time as shown in Figure 9.12.

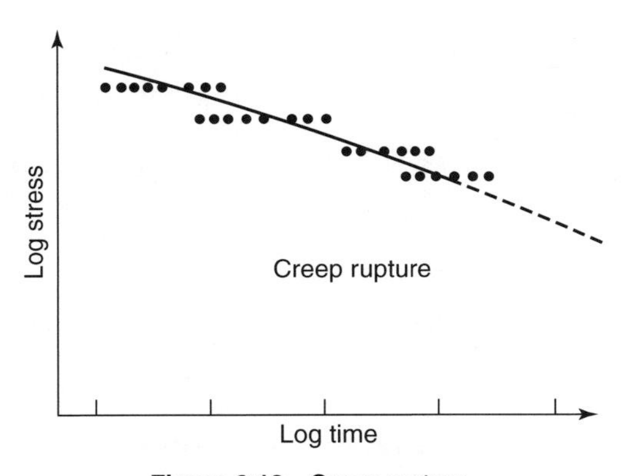

Figure 9.12 Creep rupture.

Stress Relaxation

Stress relaxation is the lowering of stress in a product, usually under a static and fixed force resulting from creep of the material under load. It is identifiable as a decrease in the applied force after a time lapse. A plastic spring depressed under constant force experiences a decreased spring rate. When a product is under a constant force, it will creep when the material's elastic limit is exceeded. This can occur with press fits, mechanical assemblies, and spring applications. The gradual decrease in stress at constant strain, over time, is called *stress relaxation*. When retention of force is important as for a press fit, a mechanical preloaded closure, or a spring ratchet mechanism, the material's creep modulus (E_{app}) should be used to approximate the part's response to elongation under the applied force.

Plastic springs should never be under constant load because of the risk of material creep resulting in stress relaxation. The spring force of plastic springs under load will decay over time. An example is a detent spring on an antifreeze bottle requires a spring force to make it difficult for a child to open the bottle. The spring must be held down while the cap is turned past a detent on the bottle. The spring is designed at rest, not under load, during the time when not in use and only required to flex when needed by the design. Another example is the unscrewing spring ratchet design for automotive gas caps. The spring design is used to prevent overtightening of the gas cap. The spring must also be stiff enough to act as a beam against a molded-in stop opposite the ratchet mechanism on the inner threaded seal assembly, to stop the cap from turning, when removing the cap for fueling (Fig. 9.13).

Stress relaxation data are not always available from material suppliers but can be developed by applying a fixed strain or deflection to replicate a sample and measuring

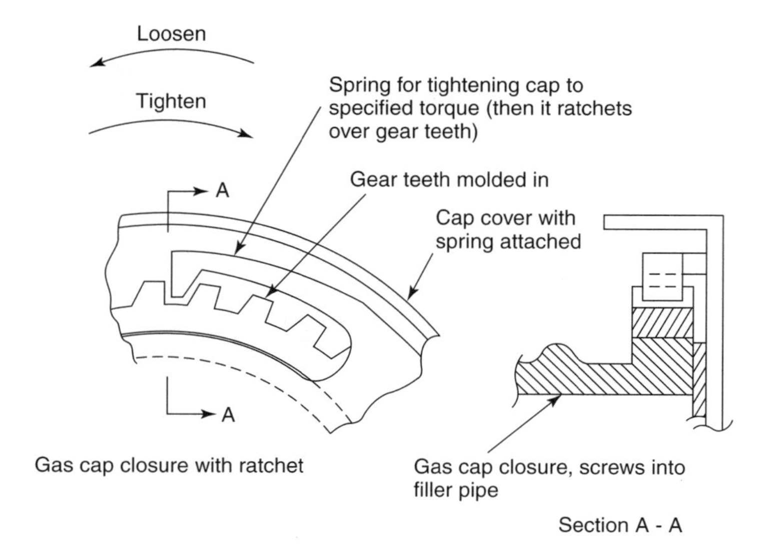

Figure 9.13 Gas cap fuel closure.

the decrease of force or holding power over time. The data can then be presented as a series of curves similar to the isometric stress curves in Figure 9.9. Plastic assemblies requiring fixed preload forces should use molded-in metal inserts to counter the effects of stress relaxation at holddown points. Products can fail over time or during assembly by overstressing or straining the assembly feature, screw, snap in, or press fit. Examples are press-fit and snap-fit components for thread-forming screws as bosses and tubing expanded for a press fit over a metal shaft. There is unfortunately no "relaxation rupture" corollary to creep rupture. Failure can occur immediately as a result of low material elongation and too rapid an assembly by not allowing the material to absorb the energy slowly enough. Failure can also occur over time by creep rupture after the product is in service. The designer must ensure that the stress levels are low and material elongation sufficient to prevent this problem.

Modulus of Elasticity Estimation for High-Strain Applications

A material's modulus of elasticity should be reduced for high-strain applications. Use a strain limit of 20% of the strain at the yield point or yield strength for high-elongation (30%) plastics. For low-elongation materials, less than 5% in the brittle zone and without a yield point, use 20% of the material's elongation at break. If it becomes necessary to extrapolate creep data for times greater than data that are available, extrapolation should not exceed one logarithmic unit of time as shown in Figure 9.12 and only the 20% strain limit of the yield or ultimate strength should be used.

It is recommended that prototype parts be built and thoroughly tested at end-use conditions. If time is critical, increase the temperature by 25% to accelerate the aging process to confirm the design and plastic resin for the product.

Drop-Weight Testing

Drop weight, Gardner impact, or the falling-dart impact test can provide very important information for product analysis when impact on the product is anticipated. Where an impact load strikes the product is important and designing for impact very desirable to product life. Notched Izod or Charpy impact test data are useful, but results from these tests are often misleading and not always representative of the product application. Testing is performed on molded-in or machined notches in test bars that can produce conflicting test results. Since these are areas of high material stress in the samples, end-use testing is always recommended and impact testing required of the product if it is subjected to these types of end-use forces. These notches are areas of high stress concentrations in the material.

To reduce the effects of an impact force, always use a generous radius at internal corners and at changes in section thickness. Even a small radius can reduce the stress concentration factor at these high stress points by 3–4 times using a radius as shown in Figure 9.4. The desired ratio of radius to product section thickness is 0.6.

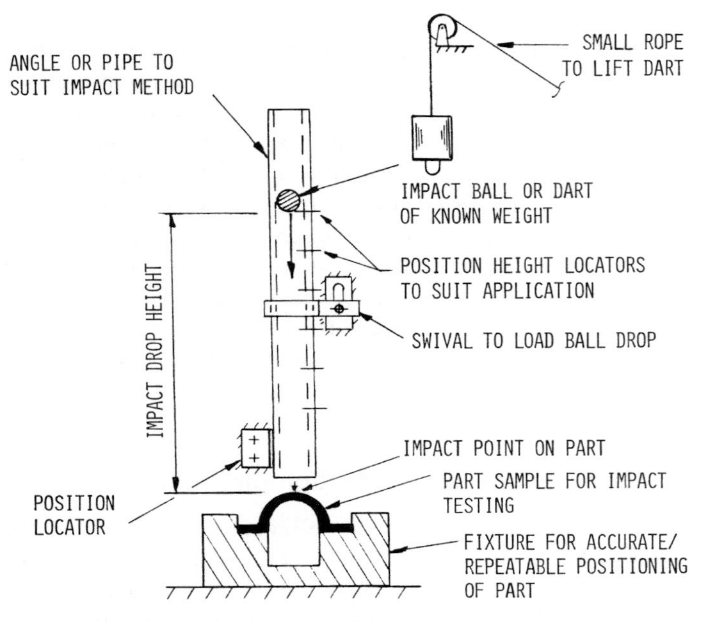

Figure 9.14 Drop weight Gardner impact test apparatus.

Impact Forces

Impact is the rapid application of force concentrated to usually a small area on the part. The greater the contact area of impact, the lower the effect on the product. A large impact area will spread the force out more evenly than will a single point contact. The impact may be a single blow or a series of repeated blows at some frequency over time. Plastics are typically able to withstand impact forces better than are metals without permanent deformation. A plastic can absorb and dissipate the energy by deflecting and returning to their original shape. The product deflection transfers the forces throughout the part by transmitting the force by absorption; deflection, and load transfer throughout the section. Material test data are often reported only for notched Izod impact data. Notched Izod test data do not represent the real impact strength of a material and should be used only to rank material candidates for a product. Izod impact data only examine the notch sensitivity of a plastic resin. Drop-weight (Gardner impact) data on a flat molded plate are shown in Figure 9.14 and are sometimes available. These data are not typical for a molded part, but are the closest representation of how a material behaves under an impact. The ability of a part and material to absorb energy is determined by its shape, thickness, material elongation, and design. Lack of a radius at internal corners is a major contributing reason why plastic products fail in service.

Stress Concentration Factor

The stress concentration factor (SCF) is a unitless number used to estimate the anticipated stress versus the calculated stress in a product section. It is calculated by dividing the selected radius by the section thickness. Sharp (nonradiused) interior corners are areas of high stress concentration. A sharp corner can multiply the stress at an interior corner by a SCF of 2–4 times the actual calculated stress. Therefore, always specify a radius at a corner, either internal or exterior. During molding, sharp corners can cause other problems that affect the part, such as poor flow patterns, increased tool wear, and higher molded-in stresses as the material shrinks in the corner, which result in lower product mechanical properties. As the corner radius increases, the stress concentration factor (the number multiplied by the calculated corner stress, per Fig. 9.4) is lowered. The acceptable value for design and avoiding processing problems is a stress concentration factor of 1.5. An internal corner radius of one-half the wall thickness is acceptable. A minimum radius of 0.020 in. is used for a part section subjected to moderate force and 0.005 in. minimum radius for relatively stress-free sections. External corners can be radiused at a dimension equal to the inside radius plus wall thickness for increasing stress and increasing material flow. For thermosets and low-elongation (<2%) thermoplastics, the inside radius should be increased to equal the wall thickness of the section.

Safety Factors

Safety factors are used to provide a margin of product safety. Safety factors should be used whenever product liability is an issue should the product fail and cause harm to the user. Safety factors are used to ensure that any unforeseen force or end-use situation will not result in the product failing. Safety factors compensate for the nonlinearity of plastic materials and the use of standard design equations. The safety factor selected depends on the product's end-use requirements, such as those related to safety, impact load, and product liability. When a safety factor is used, the product must be thoroughly tested to ensure that it meets all requirements under actual environmental and end-use conditions. Safety and product liability areas should also be evaluated before the product is released to industry or consumers.

The possibility of poor manufacturing quality should be evaluated. Often a destructive impact test is used to evaluate the product versus processing quality. A good manufactured product will pass; others will fail by cracking. One or two products are tested over a shift's production time to prove that processing and resin variables were within limits. Variables monitored are resin variability, molding conditions, assembly, decoration, and agency testing requirements.

Designers ask what degree of safety (safety factors) should be used in a product's "initial" design for selecting materials and determining structural requirements of the part such as wall thickness, ribs, rib locations, and boss size. A range of percentages of strength values are available if only data sheet physical property values are used as presented in Table 9.3. For more demanding applications, using data sheet values, only this percentage of value numbers should be used for the design of the product.

A factor of safety is selected for critical applications where conditions, forces, and the environment are not always fully known. Safety factors allow and compensate for variations in material, processing, and design to ensure that the product performs as required. The safety factor for unreinforced resins was selected at 70% of the material's design stress and 90% for reinforced material.

Table 9.3 Design strength (safety factors) for plastic part design[a]

	Failure (%)	
	Noncritical	Critical
Intermittent (nonfatigue) loading	25–50	10–25
Continuous loading	10–25	5–10

[a]These values represent are only a percentage of strength values based on marketing data product sheets, based in turn on stress and maximum temperature, only for preliminary design analysis, not to replace thorough end-use testing. Contact your material suppliers for their recommendations.

In Figure 9.8b for a reinforced material, stress/strain curves were generated for both the flow and cross-flow directions of a fiber-reinforced test bar and combined on the same stress–strain diagram. Fiber-reinforced materials have different physical properties in their flow and cross-flow directions. Materials tested in the flow direction, fibers lining up, have higher physical property values versus transverse direction. This is due to the fibers with a length–width orientation lining up parallel in the flow direction. This orientation results in lower properties in the cross-flow direction. Therefore, the designer must consider the lower material physical properties in the cross-flow direction. Unfortunately, physical property material test values are usually reported only in the flow direction, so the design stress values must be adjusted accordingly. A design rule to follow for analysis of unreinforced cross-flow material properties is that the cross-flow values are typically only 70% of the value recorded in the flow direction. As wall thickness increases over 0.10 in. the material properties orientation will be more random, resulting in more isotropic property values. In the part section, 0.25 in. thicks, the safety factor can be 80% of the calculated value.

Safety factors for reinforced resins use 75% of the material's properties in the flow direction for thickness of ≤ 0.10 in. and less and 85% for thickness >0.25 in.

Failure Analysis

If failure occurs during preproduction testing, the product's design and material can be adjusted. Failure analysis is assisted with the use of problem-solving analysis. Failure analysis and solutions to a problem are often elusive. The design team must be careful to avoid moving the current problem into another section of the part or create a new one by changing the design or material too quickly. Failure analysis consists of documenting all variables that could contribute to failure. The Ishikawa or fishbone diagram analysis method is used to collect information by outlining all variables involved in an operation. An example is shown in Figure 9.15 for diagramming process and other variables to be considered as contributing or involved in the product failure.

Another failure analysis technique is the Taguchi design of experiments (DOE) analysis. DOE is a technique used to evaluate a series of variables or manufacturing methods selected as possibly causing or contributing to the problem. These selected interacting variables are used in the analysis. The variables' extreme, high and low, values, which may have significant effects on the problem, are evaluated with the

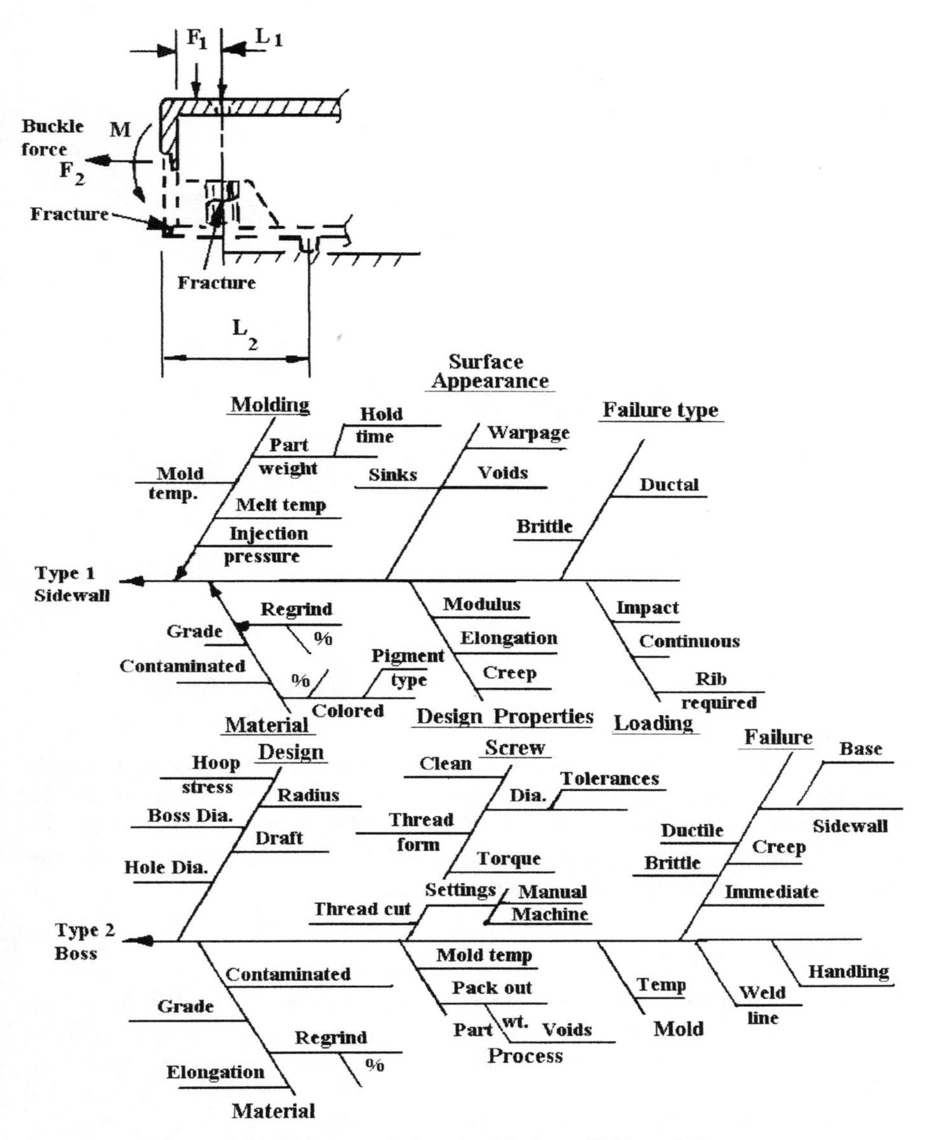

Figure 9.15 Failure analysis using Ishakawa "fishbone" diagrams.

other variables in a randomized matrix analysis. DOE analysis randomizes the set of variables' values and compares their high–low value interactions in a selected number of experimental runs, which quickly evaluates their effects on the problem and leads toward a suggestive solution. DOE reduces the normal trial–error analysis by over 500-fold in time and effort. DOE reduces the hundreds of individual trial possibilities to a minimum number depending on the number of variables to be evaluated, yielding

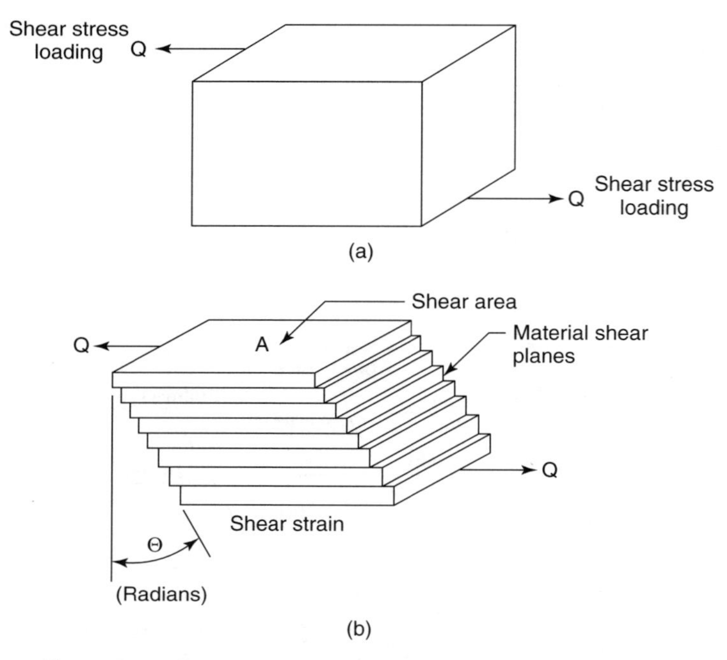

Figure 9.16 Shear stress–strain in tension. (Adapted from Ref. 8.)

good results. For very complex product analysis the designer can choose to use the maximum shear (also known as *Coulomb,* or *Tresca*) theory. It is recommended that the shear strength (not the data sheet value) used for a design equal one-half the material's elastic limit strength. If not, then use the published data sheet shear strength or one-half the tensile strength, whichever is lower.

Shear Stress

All materials subjected to a force in tension, compression, torsion, and flexure are also subjected to internal, equal and opposite shear forces (Q). The analysis evaluates a material as a set of infinitesimally thin-layered sections, as shown in Figure 9.16. When a material is subjected to a force, each thin layer has a potential to slide over its adjoining layers. This is shear, a sliding of parallel planes, due to a deforming off-center force. Each thin plane has forces acting on it that try to tear each plane apart. To resist the force and remain at rest produces a shear force component (Q) acting on all planes of the body.

A shaft in torsion and a beam at its anchor point or at a change in section has shear components acting to tear the material apart at these planar sections. Shear is a very important design consideration for a product. Shear stress is defined as being equal to $\Im$:

$$\Im = \frac{\text{shear force}}{\text{area resisting shear}} = \frac{Q}{A}$$

Shear stress is always tangent to the area on which it acts. Shear strain, measured in radians, is the angle of deformation θ. Shear modulus (G), according to Hooke's law, is proportional to the shear stress:

$$\text{Shear modulus} = \frac{\text{shear stress}}{\text{shear strain}} = \frac{\Im}{\theta} = \text{constant} = G$$

Shear modulus is also called the *modulus of rigidity* (G), and is directly comparable to the modulus of elasticity (E).

Poisson's Ratio

Poisson's ratio is the material constant used to characterize a material assumed to be linearly elastic, homogenous, and isotropic. Plastics are inherently nonlinear and anisotropic in their behavior. Therefore, material suppliers for use with the standard design equations have estimated Poisson's ratio.

Poisson's ratio is a constant required for stress and deflection analysis for the design of plates, shells (pressure vessels), and rotating disks (fans and pulleys). Poisson's ratio is the ratio of a body's lateral strain to longitudinal strain within the material's elastic range and designated by the Greek letter γ.

When a bar is subjected to a tensile force, as in Figure 9.17, it is elongated by an amount ΔL, resulting in a longitudinal strain, ϵ:

$$\epsilon = \frac{\Delta L}{L}$$

As a result of the force the bar experiences lateral changes, a decrease in its width and thickness due to stretching. These changes are opposite in sign [contracting (−) vs. stretching (+)] to the longitudinal strain and expressed as:

$$\epsilon_{\text{lateral}} = \frac{\Delta b}{b} = \frac{\Delta d}{d}$$

$$\Delta b = b - b' \text{ width}$$

$$\Delta d = d - d' \text{ thickness}$$

Provided the deformation of the material is in the elastic range, the ratio of lateral to longitudinal strain is always a constant, γ, called Poisson's ratio:

$$\gamma = \frac{\text{lateral strain}}{\text{longitudinal strain}} = \text{a constant}$$

$$= \frac{\Delta d/d}{\Delta L/L}$$

Poisson's ratio for plastic materials is not a constant and is affected by factors such as the intensity of the force, time underload, creep, temperature, environment, and

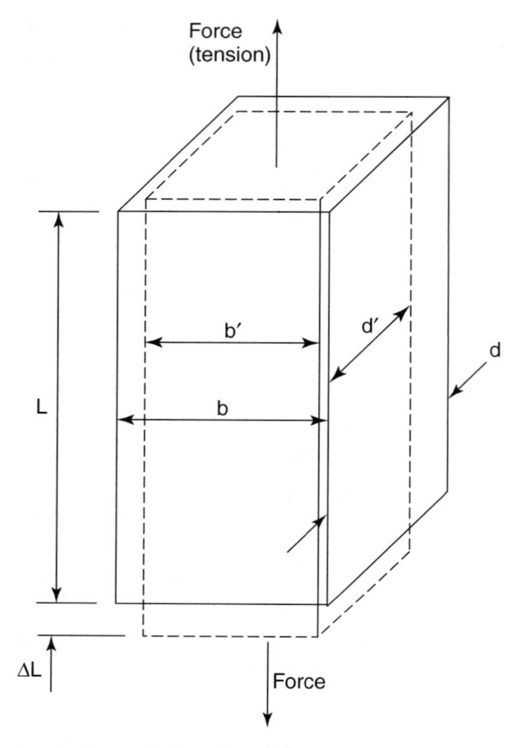

Figure 9.17 Dimensional change in length, width, and thickness from a tensile force on an object. (Adapted from Ref. 9.)

material composition. Classically, (γ) is between zero (no lateral contraction) and (0.5) (constant volume deformation). Table 9.4 shows a range of (γ) values for various materials.

The designer must contact the material supplier to obtain the value of Poisson's ratio for the resin they are considering for the design of the product. If Poisson's ratio is not available, a default value of 0.35 is usually representative for use in a design.

Relationship between Material Constants

The three material constants (E), modulus (G), shear, and (γ), Poisson's ratio, are based on elasticity concepts, and their relationship is expressed as

$$\frac{E}{G} = 2(1 + \gamma)$$

This relationship holds for most metals and is generally applied to injection-molded thermoplastics and thermoset.

Table 9.4 Poisson's ratio[a] for typical thermoplastics and other materials

Material	Range of Poisson's Ratio
Aluminum	0.33
Carbon steel	0.29
Rubber	0.50
Rigid thermoplastics	
Unfilled	0.20–0.45
Filled or reinforced	0.10–0.45
Rigid thermosets	
Unfilled	0.20–0.40
Filled or reinforced	0.20–0.40
Structural foam	0.30–0.40

[a] For specific Poisson ratio values, contact your material supplier as the value will vary with temperature and other environmental conditions.

Source: Adapted from Ref. [2].

Measures of Strength and Modulus

The stress–strain curves for plastic materials in tension and compression are similar. At high stress levels the compressive strain is less than tensile strain. This is illustrated in tensile loading that results in a defined failure; the test specimen elongates and fails. A test block under compression as shown in Figure 9.18 produces a slow finite

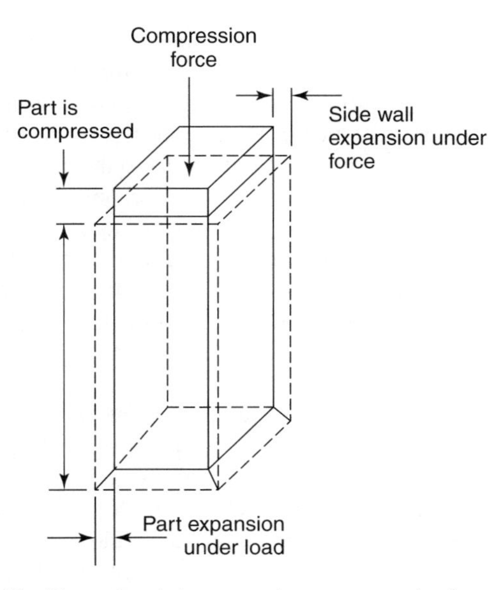

Figure 9.18 Dimensional change under a compressive force on a part.

yielding. Failure can occur by material buckling, sides pushing out and shearing, loss of height contacting another part, or another mode of failure such as a slender column that buckles in bending. Compressive strength is therefore the stress expressed in pounds per square inch (pascals) [psi (Pa)] required to deform a standard test block to a specified strain. Compression modulus is reported as an initial modulus. This means that at a specified stress, a resulting strain occurs that is equivalent to reporting a secant or a point modulus, anywhere on the stress–strain curve. If an injection- or compression-molded plastic product had a uniform density in section thickness, shear stress consideration would replicate an isotropic material, such as metal. But an injection- or compression-molded plastic product's wall sections exhibit a nonuniform density or thickness. The section's outer skin is typically denser than the inner core thickness. This is visible when the part is sectioned and the inner core is of a lighter shade than that of the outer skin surfaces. For wall sections less than 0.10 in. thick, the section density is considered uniform. In wall sections greater than 0.10 in. the skin effect applies. This means that the section has two skins, inner and outer, with a less dense core. In thicker-walled parts the inner core density due to material shrinkage is lower. This is because this area is the last to solidify and is termed "material-starved" unless during molding it is fully packed out before the part's gate freezes. Ideally the outer skin and inner core sections should be uniform in density, but this never occurs except in very thin ($\leq$0.100-in.) wall sections.

With reinforced resins, the surface skin effect is more apparent. For injection-molded reinforced products the material contacting the walls of the mold cavity quickly solidify with the reinforcement oriented in the flow direction. As the mold cavity fills, the pressure against the molten resin and solidifying product surface skin, against the mold cavity surface, becomes greater, forming an even denser surface skin. The center core section's material flows over the newly laid-down skin and continues to fill the mold cavity with fiber direction usually more random in orientation depending on the product's section thickness. The thicker the section thickness, the more random the glass fiber and/or mineral particles will be in the section. When the cavity is full, the final packing pressure continues to push more hot resin into the cavity as the material solidifies and shrinks from the cavity walls to the center core where the last hot resin is introduced. During this time the skin becomes even denser because of the packing pressure on the molten core being pushed against the mold's cavity wall surfaces. If more hot resin is not packed into the core of the wall before it solidifies, the section on cooling and shrinking is less dense. In thick-walled (>0.10-in.) parts, the orientation of the reinforcement in the skin layer is in the flow direction with the core reinforcement randomly oriented. This results in a lower section strength in the cross-flow direction, as shown in Figure 9.8b.

Because of the random orientation of reinforcement, only 75% of the reinforced material's strength properties are used for design purposes in the product's cross-flow direction. It is very important early in the product's design to determine where the part will be gated to anticipate the product's strength properties in the part based on the direction of material flow to fill the product's mold cavity. Check with your material supplier to get their product design mold gate location recommendations before finalizing the design.

Bending Strength and Flexural Modulus

Material suppliers provide flexural data on their materials at specified loading rates (application of force with time) test conditions. The rate at which a force is applied on a material will generate a specific stress–strain curve. A material subjected to bending forces should use the stress–strain curve for materials that are developed at a similar loading rate, and this curve should reflect the service conditions in temperature and environment. Materials subjected to different loading rates (as in tensile elongation) will produce different responses and different stress–strain curves. The slower the loading rate (≤ 1 in. deformation per minute), the more energy the material will be able to absorb until its' elastic limit is reached and it yields. A fast loading rate (>2–3 in./min) in elongation will cause a ductile material to fail in a brittle-type versus elastic failure. The material will not be able to absorb all the energy of elongation created by the tensile pull forces and will rupture prematurely. Always ask your material supplier if the strain rate is available for the material tested. Usually a standard ASTM strain rate is specified for each test procedure. If not specified, end-use testing is mandatory.

When a test beam (usually rectangular in shape) is bent in flexure to develop a stress–strain physical property data curve (Fig. 9.19a), the beam's internal stress distribution is as shown in Figure 9.19b. At the beam's neutral axis a line of zero stress results with the upper region in compression and the lower region in tension. Simple beam theory assumes the following assumptions:

1. The beam is initially straight, unstressed, and symmetric.
2. Beam material is homogenous, isotropic, and linearly elastic.
3. The material's proportional limit is not exceeded.
4. Young's modulus is the same in tension and compression for the material.
5. Deflections are small, resulting in all planar cross-sections remaining planar before and after bending, with no shear.

Typical beam formulas and beam section properties yield the following relationships:

$$\text{Bending stress } \sigma = 3FL/2bh^2$$

$$\text{Bending or flexural modulus } E = FL^3/4bh^3Y$$

where Y is deflection at the load point.

Flexural modulus is the initial modulus from the bending load deflection curve. When designing parts subjected to bending, use only the flexural modulus of the material, E. Do not use the tensile stress–strain curve value, which is used only for tension loading conditions.

Vibration and Fatigue Resistance

Vibration, either irregular or harmonic, causes flexing in a section of the product. An alternating force on a body that causes the product to vibrate or resonate at

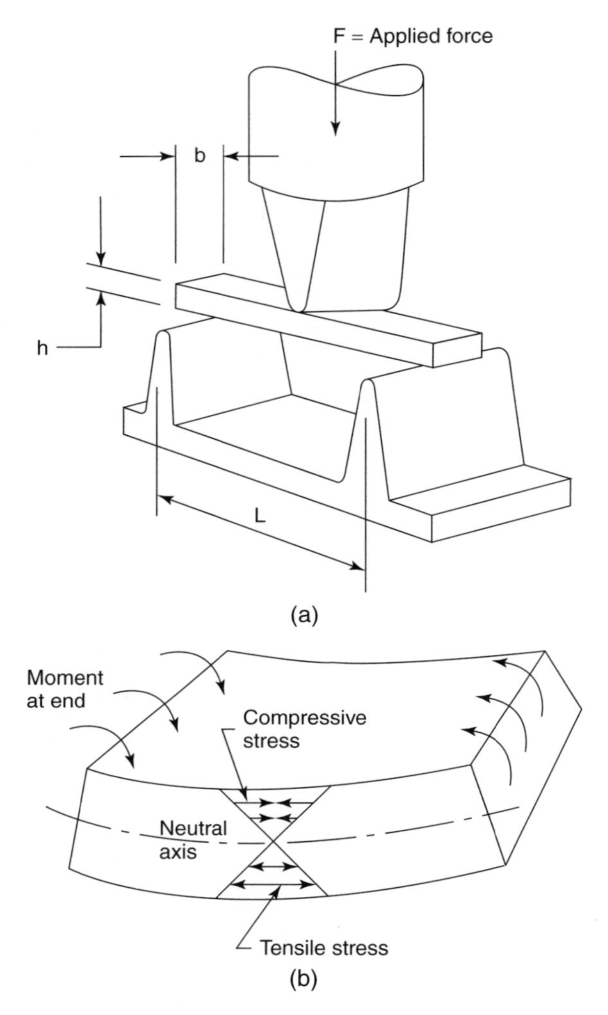

Figure 9.19 Simple beam in bending.

frequencies that can cause fatigue failures causes vibrations. Plastic products can be firmly anchored or mounted on an energy-absorbing mounting device to damp out resonate energy. Vibration can cause a body to resonate at a critical material frequency that can build up heat in a section and cause fatigue failure over time.

Most products are subjected to vibration forces of varying magnitudes. Products exposed to vibration are prime candidates for fatigue failure if severe or longlasting. A simple snap-action latch, intermittently operated, can be subjected to fatigue failure if repeatably flexed. Depending on the rate of flexing, frequency, or external vibration encountered, the type and elongation or flexural motion of a part, fatigue is produced in the section of the product leading to failure. Tests have shown that plastics subjected

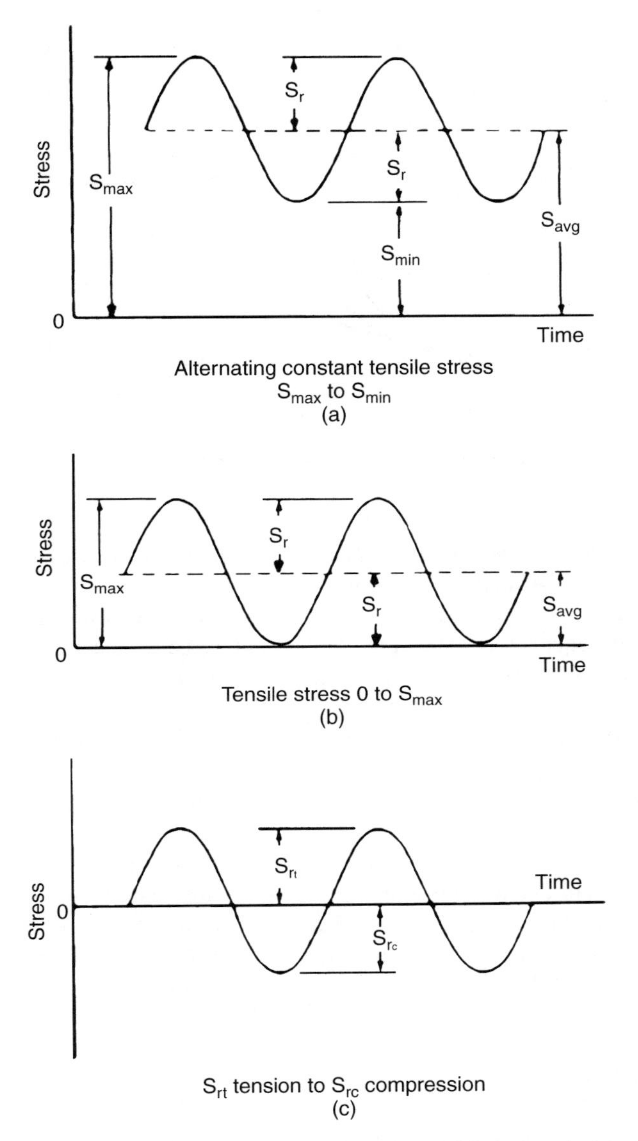

Figure 9.20 Fatigue stress patterns: (a) fluctuating stress; (b) repeated stress; (c) reversed stress. (Adapted from Ref. 7.)

to cyclic tension, compression, or both (see Fig. 9.20) at high-speed cyclic stresses fail at stress levels far below their tensile and/or compressive strength. This type of failure is called *fatigue failure*.

The worst-case scenario for fatigue failure is cyclic tension and compression loading per Figure 9.20c. Testing shows that, at or below 1800 cpm (cycles per minute),

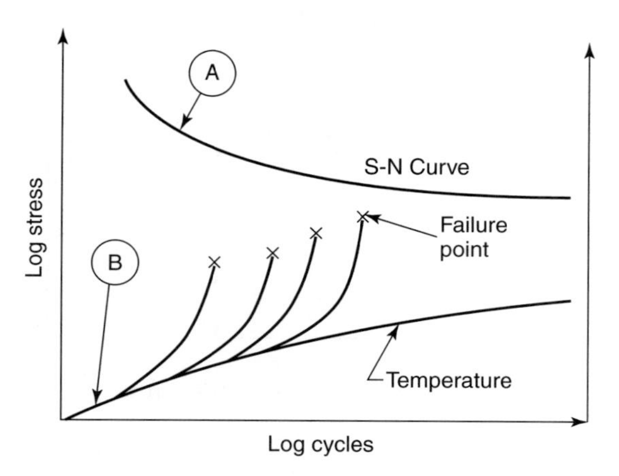

Figure 9.21 Fatigure testing effects on plastics.

frequency of loading does not have an effect on the number of cycles to failure at a given stress level. A material's fatigue resistance is obtained by flexing a standard test specimen at varying stress levels on a Sonntag-Universal testing machine. The typical testing frequency is 1800 cpm. A plastics fatigue resistance is the stress level at which it passes one million cycles without failure. The data is recorded using an *S–N* curve plotting (log stress over log cycles) is shown in Figure 9.21, curve *A*. The cycle loading rate profile and test temperature should be reported on the curve. Heat buildup, leading to thermal failure within the part, also contributes to fatigue failure as shown in Figure 9.21, curve *B*. Thermal failure is amplified by increasing the frequency of the cycle stress.

Testing at different frequencies, mean stress, waveforms, and test methods, namely, tension rather than cyclic tension/compression flexing, can generate different *S–N* curves. For critical products, dynamic shakers are used with temperature chambers. Whenever possible end-use-test the part under actual conditions to determine the true failure endurance of the product when subjected to significant cyclic loading.

REFERENCES

1. R. D. Beck, *Plastic Product Design,* Van Nostrand Reinhold, New York; 1970.
2. *A Guide for Designing with Engineering Plastics,* E. I. Du Pont de Nemours Corp., Wilmington, DE, 1990.
3. *Designing with Plastics, Design Handbook,* Hoechst Celanese (TDM-1) HCER-92-313/10M692: 3-3, Chatham, NJ.
4. P. Richards, "Push Plastics to Their Design Limits," *Plastics World* (Directory), 381–385 (1988).
5. *Designing with Plastics, Design Handbook,* Hoechst Celanese (TDM-1) HCER-92-313/10M692: 3-8, Chatham, NJ.

6. *Designing with Plastics, Design Handbook,* Hoechst Celanese (TDM-1) HCER-92-313/10M692: 3-9, Chatham, NJ.
7. *General Design Principles—Module 1,* E-80920-2, E. I. Du Pont de Nemours Corp., Wilmington, DE.
8. *Designing with Plastics, Design Handbook,* Hoechst Celanese (TDM-1) HCER-92-313/10M692: 7-3, Chatham, NJ.
9. *Designing with Plastics, Design Handbook,* Hoechst Celanese (TDM-1) HCER-92-313/10M692: 3-4, Chatham, NJ.

CHAPTER 10

STRUCTURAL PRODUCT ANALYSIS

Structural analysis of a product is used to determine how forces act on the product. This information is then used in the design of the product and for selecting the material with which it will be manufactured. The design analysis determines how the product's shape and material will distribute the forces acting on the product. The analysis also ensures that the selected material for the product has the necessary physical properties to perform its end-use function without failing. Plastics are used in structurally demanding product applications for varying lengths of time under severe and changing end-use conditions. These products require that their design and materials of construction meet or exceed their end-use design requirements in many demanding applications without prematurely failing.

The designer's responsibility is to analyze these forces and, using the appropriate design equations, determine the stress–strain curve, torsion, and moments acting on the product. Drawing a force diagram (Fig. 10.1) of the proposed product helps develop this information. To assist in analyzing these forces, the designer should use the product design, product development, and material checklists (see checklists 2, 4, and 7 in Appendix A). These lists of questions will guide the designer and the design team in determining the product's shape, strength, and properties and in selecting the plastic product's material. With this information, a preliminary list of plastic materials can be selected for consideration for the product. The materials selected are based on the product end-use function, designer's knowledge of plastic resins, data from the checklists, and product material suppliers data sheets.

Once the forces acting on the part are translated into stress and strain values, the product's material selection is determined. When designing a metal part, there are standard available material sizes for selection for manufacture of the product. These include sheet thickness and structural shapes, angles, and channels, which can be selected from suppliers. The design team using plastic is not restricted to standard

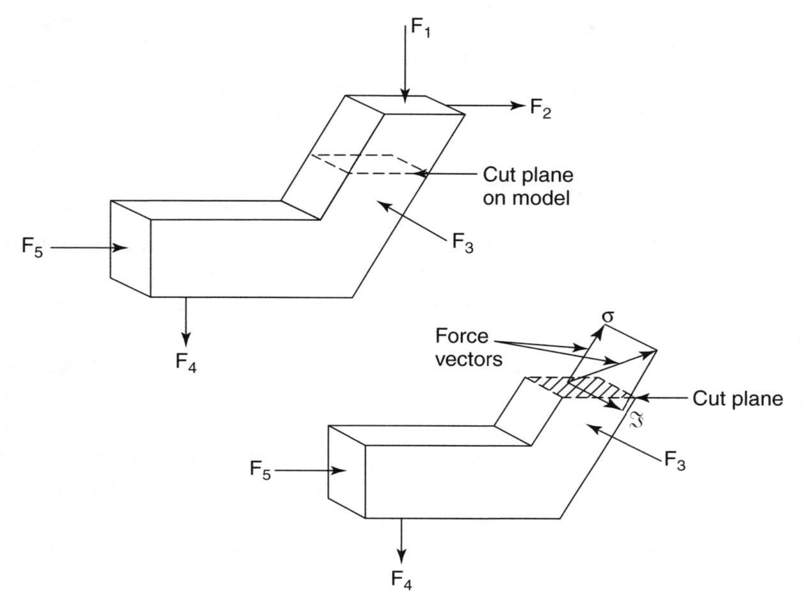

Figure 10.1 Forces on a body and resulting internal stresses.

shapes and can select a material and thickness to size and shape the product to meet the requirements for the application.

One of the designer's main considerations is to determine how the product is to be manufactured. This involves selecting materials that can be used for the manufacturing method to be chosen for manufacture of the product. This in turn has a direct effect on the tooling and corresponding machinery to be used for the manufacture of the product. Each has a direct correlation to the final product cost, design, assembly, and method of manufacture.

During the initial design phase using the checklists, the design team should consider including multifunction and product capabilities in their design analysis in an attempt to reduce the number of parts. This is not always possible using metal but is with plastics. Multifunctional capability includes press, molded-in, and snap-fit assemblies versus screws, internal/integral springs, index cams, bosses for assembling the product, alignment slots, and bearings to reduce product components and add value to the product. These can be incorporated into the product at a slight added up-front tooling cost.

To keep creep and stress relaxation to a minimum, always design with plastic to limit the product's time under load. This is not always possible but is the goal. This can be accomplished when good design principles are considered and used. Design the product so that after assembly it is under minimum load by the use of built-in stops or load transfer points in the part to transfer forces not supported elsewhere. This will assist in reducing creep, deflection, or torque, which can extend a product's life and service reliability.

When a plastic product is designed for continuous loading, the material selection, part shape, creep factors, and design must be carefully analyzed. This involves the use of appropriate material safety factors when the product is designed. This is often the opposite of designing a metal product, which is historically overdesigned. The goal in designing a plastic product is to minimize the section thickness of the product and when additional structural strength is needed, add ribbing. This is easier to do with plastics than with metals. The ribbing can be added in the tool for a low initial tooling added cost. This strengthens the product, makes it easy to manufacture, remove the product from the tool, and in some product shapes resists warpage.

Products must be designed to ensure that the external and internal assembly forces do not exceed the material's plastic or elastic limit. Higher loading may be sustained but for only a limited time of 10–20 min maximum. The intensity of any high short-time loading must be considered as the product material may creep if overstressed. Under these conditions the standard design equations, based on Hooke's law, can be used with reasonable accuracy. Most plastic products fall into this category, meaning that no excessive, long-term forces are applied and product deflections are minimal ($\leq$0.100 in.). Long-term loads can be accommodated when correct product design and material are selected. This means selecting section thickness and profiles that can withstand these forces with minimal deflection.

DEFINING STRUCTURAL REQUIREMENTS

The allowable maximum stress on a product should be limited to 75% of the plastic material's physical properties at operating conditions. Never use the material's maximum allowable physical properties or design the product to use the maximum properties of the plastic material as reported by the material suppliers. Any unforeseen variables in the product's manufacture or material could produce products that do not meet these maximum properties. Engineers old and new to plastic product design must also consider the stresses involved with product assembly, forces, decorating, and shipping. These include assembly operations using screws, snap and press fits, decorating requirements and their stresses, solvents, painting, and rough handling and extreme temperatures experienced during shipping. These are seldom considered until a problem occurs requiring a redesign or material change that can be very expensive in time and money.

With the proliferation of new high-strength resin compounds, the designer has an opportunity to effectively reduce part weight, lower cost, and increase product reliability. The designer now has access to computer design software to analyze products of higher complexity and multiload requirements with very precise stress analysis to tailor the material in the product design for increased consumer safety and reliability. The computer-aided design (CAD) software programs provide an accurate, fast, and detailed analysis using finite-element analysis (FEA). The first step in a new design or redesign of a product is to sketch out how the finished product will look. Next determine the type, magnitude, and duration of the forces and how and where they act on the product.

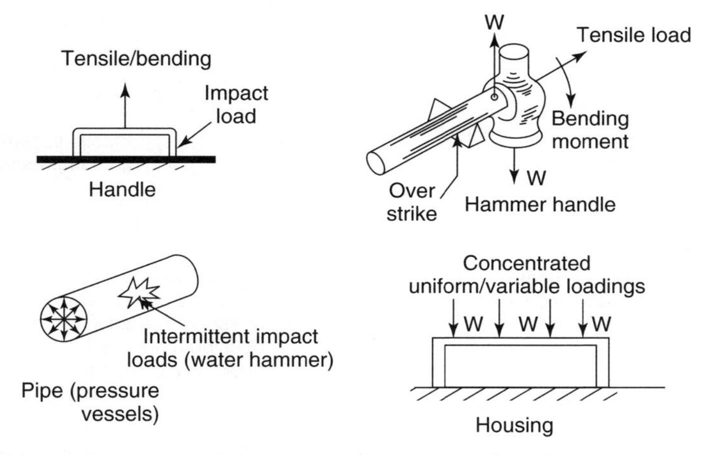

Figure 10.2 Forces on products that can vary and exceed anticipated forces.

Direct forces can be concentrated at a point, a line, or a defined boundary or distributed over an entire area. Their line of action and where the forces act on the product are diagrammed on the sketch. If not known, they should be estimated.

With an existing metal part considered for redesign in plastic, testing under end-use conditions can yield this information. Be sure to include the weight of large parts and internal components in the analysis. Examples of direct applied loads are shown in Figure 10.2. Indirect or reaction forces occur as a result of the direct loading. These occur as moments around anchor points or as a result of the forces and their reactions trying to keep the body in equilibrium.

Products are also subjected to strain-induced forces when deflection occurs in a section of the product. The forces that act on a product from a direct physical reaction developed by applied strain are shown in Figure 10.3. Strain-induced loading, compressive or hoop, is dependent on the modulus of elasticity of a material and the deflection forces on the part section. Depending on the method of strain-induced force; the stress in the part will decrease over time as a result of material creep. This occurs when a pin is press-inserted into a small hole in a boss, or with a plastic coupling under an applied compressive force. Plastics under a sustained force will creep, and the designer must ensure that the stress does not exceed the material's ultimate elongation, or failure can occur. Press-fit assembly force will decrease over time as a result of creep, resulting in stress relaxation in the press-fit assembly. If the boss for the press fit is restrained from creeping, using a metal retaining ring securing the material of the boss, material stress relaxation and press-fit joint retention strength will be maximized. If the material of the boss has too high an elongation, it will creep, causing overtime, a reduction in the press-fit joint strength. In cases where the interference is too great and the material's elongation too low, the boss may fail by fracture. Poor assembly design and thermal strain-induced loading in service can aggravate these situations.

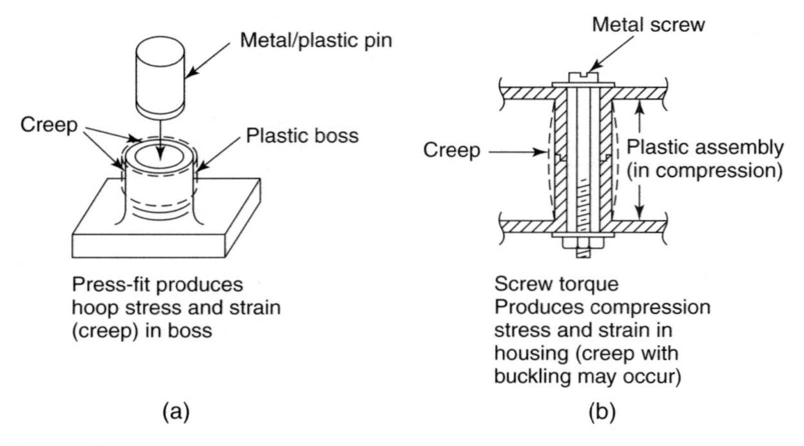

Figure 10.3 Stress-induced loading with resultant strain (creep): (a) press fit produces hoop stress and strain (creep) in boss; (b) screw torque produces compression stress and strain in housing (creep with buckling may occur. (Adapted from Ref. 1.)

SUPPORT CONDITIONS

When forces are applied to a product, equal and opposite forces in the product are developed to maintain equilibrium of the product in service. These balancing or reaction forces in the product occur at the product support locations. These support conditions, shown in Figure 10.4 for structural analysis, are defined as

1. *Unsupported* (free)—allows the edge of a body to freely rotate or move in any direction.

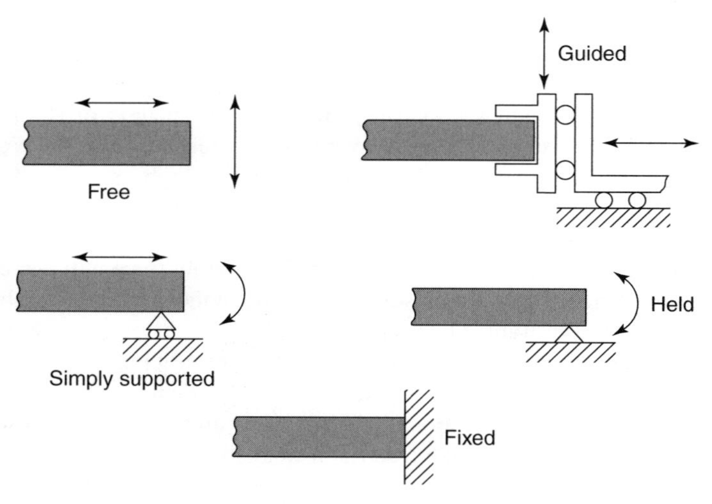

Figure 10.4 Typical beam end connections. (Adapted from Ref. 6.)

2. *Guided*—condition is similar to that of the unsupported condition except that rotation is restricted.
3. *Simply supported*—restricts transverse movement in one direction but allows rotation.
4. *Pinned* (*held*)—similar to simply supported, allowing rotation at the pin.
5. *Fixed* (*clamped* or *built- in*)—this support condition at the edge or end of a beam or plate firmly restricts all end motion. With plastic supports, unless the support is firmly attached to a fixed, rigid solid surface, there is usually some deflection at the attachment point.

DESIGN ASSUMPTIONS AND SIMPLIFICATIONS

The most common support conditions are free, simply supported, and fixed. The following assumptions and simplifications can be used for plastic product design:

1. The product under analysis can be separated and analyzed as one or more simple structures, such as beam, plate, column, or pressure vessel.
2. The use of the standard design equations assumes the material to be linearly elastic, homogeneous, and isotropic.
3. To use the standard design equations, it is assumed that the force is a simple concentrated or evenly distributed static load. The force is applied slowly for a short period of time and then removed. For continuous loading conditions, resulting in creep, relaxation, or fatigue, the appropriate material modulus and/or rupture strength data must be used in the equations for the appropriate end-use

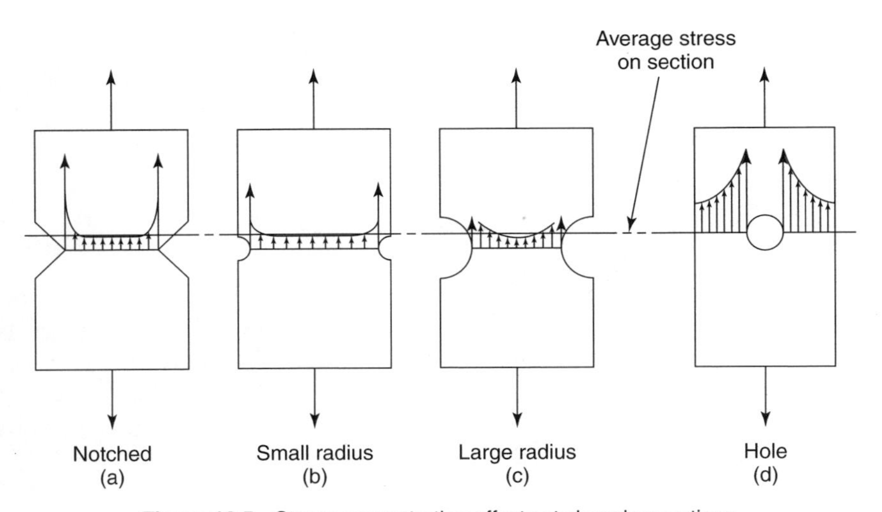

Figure 10.5 Stress concentration effects at changing sections.

conditions. These data are obtained from the material supplier for each material tested under the specific conditions.

4. The product under analysis has no molded-in or residual material stresses. These non-force-related stresses can appreciably reduce the allowable forces that a product and material can withstand. All plastic products have these induced manufacturing stresses, and they can be minimized when a good product and tool design and manufacturing practices are followed using their material supplier's recommendations.
5. The design equations apply to all sections of the structure. This includes the point of applied load or any other sections of the part. Where part section thickness changes at corners, bosses, and holes, the stress concentration factors at these points must be calculated and considered in the design of the product. The stress at a hole is typically 2 times the calculated stress in the product's section as shown in Figure 10.5d.

STRESS CONCENTRATION FACTORS

If there are section changes in a product at a rib, corner, or where the wall thickness increases to withstand a greater force acting on the product, higher stresses occur in these sections. If the product is subjected to an impact, these internal stresses are amplified as shown in the stress amplification curves in Figure 10.6. The curves were developed to show the amplification factor resulting from impact loading at these sections.

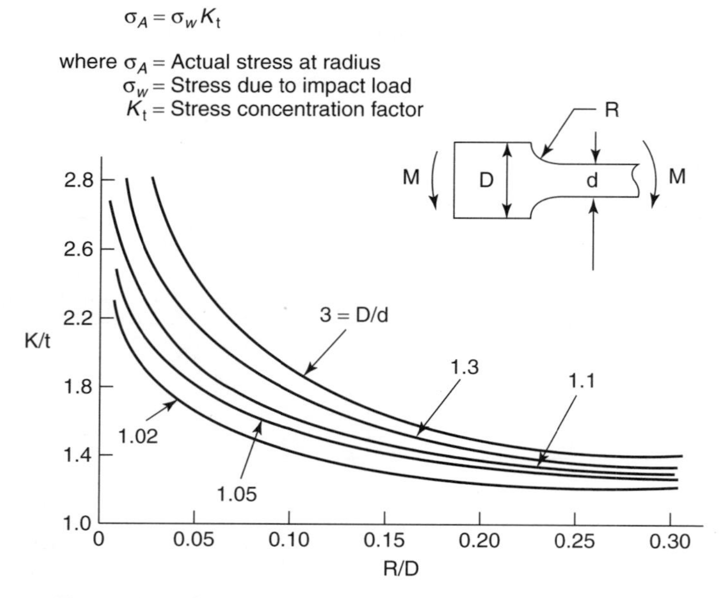

Figure 10.6 Stress concentration factor. (Adapted from Ref. 10.)

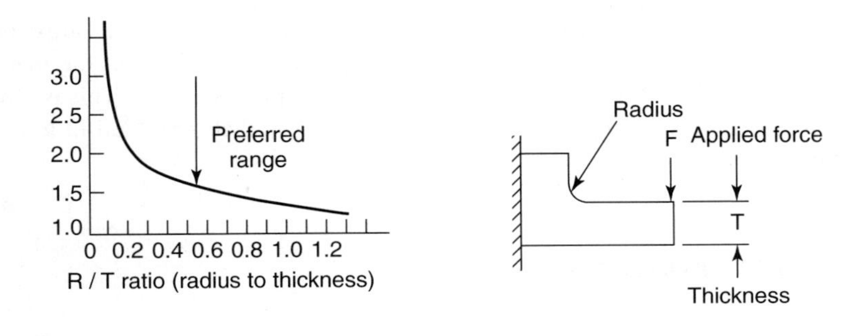

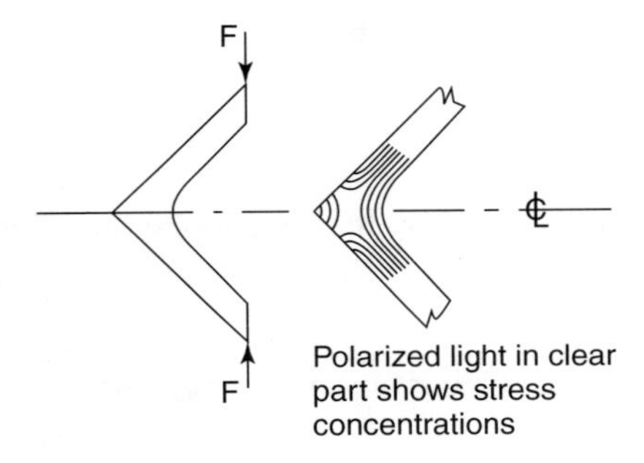

Figure 10.7 Stress concentration factor.

After the impact stress is calculated at these sections, the total section stress at these points can be calculated using the following equation:

$$\sigma_A = \sigma_w K_t$$

where $\sigma_A =$ actual stress at radius
$\sigma_w =$ stress at impact point
$K_t =$ stress concentration factor

Using this analysis for parts subjected to impact loading, the corners and changes in the part section can be analyzed for strength and material suitability.

The optimum stress concentration factor at a corner is 0.5–0.6. This is the ratio of the interior radius of the corner to part thickness, per the graph in, Figure 10.7. The stress at a corner increases as the radius is decreased, which results in a notch effect. Figure 10.5a–c depicts the average stress versus the stress intensity of zero radiuses to more liberal radii. Products with a hole (Fig. 10.5d) and at a change in section thickness also have higher stress concentrated at these points. The stresses at the edges of a hole are amplified as at a sharp notch in Figure 10.5a.

During manufacture, as the material flows around an obstruction or where a core pin forms an opening, high residual material stresses are formed at these points. The designer can attempt to modify the product at these points to reduce the stress or go to a secondary operation. The design team can specify machining out the material after manufacture that will lower the stress level at this point by 80%. This is always a possible solution if testing or the FEA analysis shows that a molded-in hole is a potential site for failure.

STRUCTURAL ANALYSIS

Using the standard design equations based on Hooke's law, design analysis for short-term loading on a plastic part is presented.

Hooke's law states that the stress (S) is equal to the modulus of elasticity E times the strain (ϵ): $S = E\epsilon$

Tensile stress and bending stress are defined as follows:

1. *Tensile stress*—in tension the force divided by the section area develops the maximum stress:

$$\text{Tensile stress } S = F/A = \frac{\text{force}}{\text{area}} \quad \text{lb/in.}^2 \qquad \text{or} \qquad \text{kg/cm}^2$$

2. *Bending stress*—in bending the maximum stress is calculated as

$$\text{Bending stress (lb)} = S = \frac{My}{I} = \frac{M}{Z} = \frac{\text{moment}}{\text{section modulus}}$$

where M = moment (lb.)
$M = FL$ (in · lb. pound or m · kg) (force times lever arm at the applied load point)
$Z = I/y$ = section modulus ($\text{in}^3 \cdot \text{cm}^3$).
y = distance from neutral axis to extreme outer fibers of the section
I = moment of inertia for the section profile under analysis.

The calculations used to determine the I and y values of typical sections are shown in Table 10.1 for nonsymmetric sections as the short-legged I beam. Channel and tee sections, with y, the distance from the neutral axis to their outer surfaces, are not symmetric.

Beams

Beams can be supported in many ways and subjected to forces in tension, compression, and torsion. To determine the force reactions on them, use the equations for various beam conditions as shown in Table 10.2.

Table 10.1 Section properties

Form of Section	Area A	Distance from Centroid to Extremities of Section y_1, y_2	Moments of Inertia I_1 and I_2 about Principal Central Axes 1 and 2	Radii of Gyration, r_1 and r_2, about Principal Central Axes
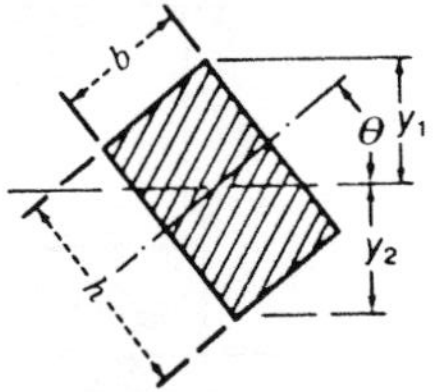	$A = bh$	$y_1 = y_2 = \dfrac{h\cos\theta + b\sin\theta}{2}$	$I_1 = \dfrac{bh}{12}(h^2\cos^2\theta + b^2\sin^2\theta)$	$r_1 = \sqrt{\dfrac{h^2\cos^2\theta + b^2\sin^2\theta}{12}}$
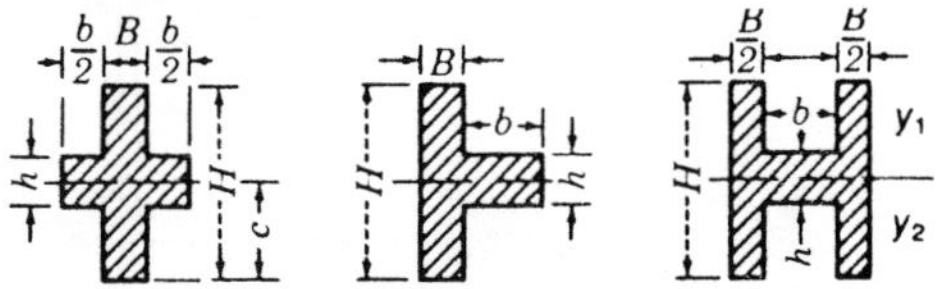	$A = BH + bh$	$y_1 = y_2 = \dfrac{H}{2}$	$I_1 = \dfrac{BH^3 + bh^3}{12}$	$r_1 = \sqrt{\dfrac{BH^3 + bh^3}{12(BH + bh)}}$
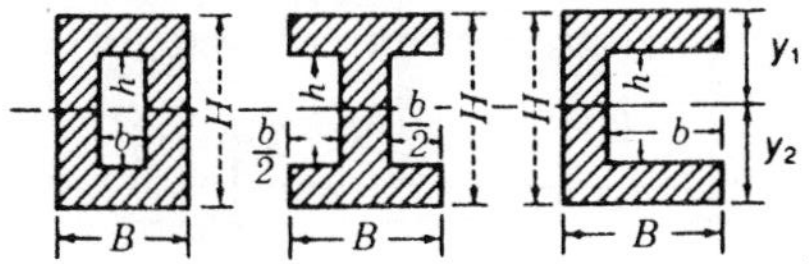	$A = BH - bh$	$y_1 = y_2 = \dfrac{H}{2}$	$I_1 = \dfrac{BH^3 - bh^3}{12}$	$r_1 = \sqrt{\dfrac{BH^3 - bh^3}{12(BH - bh)}}$
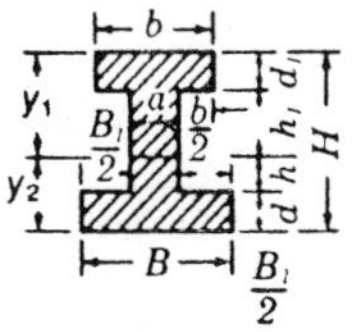	$A = bd_1 + Bd + H(h + h_1)$	$y_1 = H - y_2$ $y_2 = \dfrac{1}{2}\dfrac{aH^2 + B_1d^2 + b_1d_1(2H - d_1)}{aH + B_1d + b_1d_1}$	$I_1 = \frac{1}{2}\left(By_2^3 - B_1h^3 + by_1^3 - b_1h_1^3\right)$	$r_1 = \sqrt{\dfrac{1}{(Bd + bd_1) + a(h + h_1)}}$

Table 10.1 (*Continued*)

Form of Section	Area A	Distance from Centroid to Extremities of Section y_1, y_2	Moments of Inertia I_1 and I_2 about Principal Central Axes 1 and 2	Radii of Gyration, r_1 and r_2, about Principal Central Axes
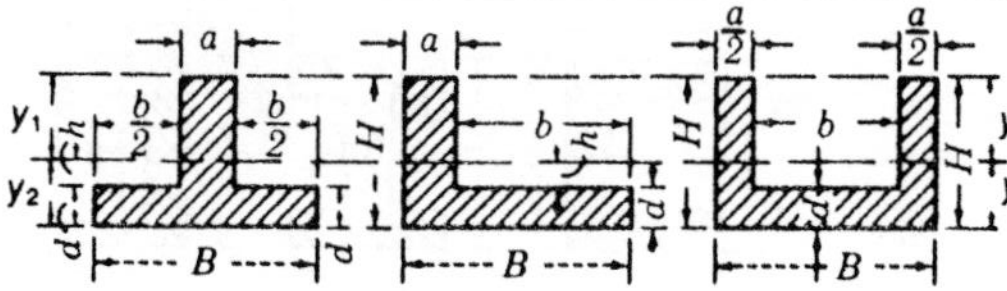	$A = BH - b(H - d)$	$y_1 = H - y_2$ $y_2 = \frac{1}{2}\frac{aH^2 + bd^2}{(aH + bd)}$	$I_1 = \frac{1}{2}\left(By_2^3 - bh^3 + ay_1^3\right)$	$r_1 = \sqrt{\frac{I}{(Bd + a(H - d)}}$
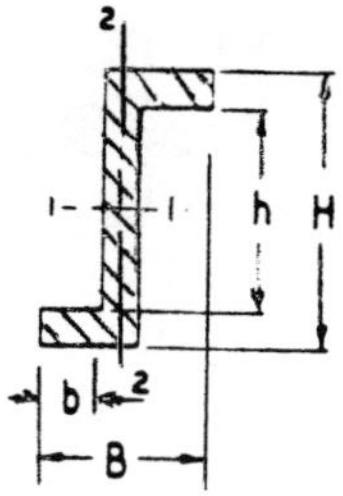	$A = B(H - h) + h(B - 2b)$	$y_1 = \frac{H}{2}$ $y_2 = \frac{B - 2b}{2} + b$	$I_{t-1} = \frac{(B - b)H^3 - b[H - 2(B - 2b)]^3}{12}$ $I_{t-2} = \frac{H(B^3 - 2b^3) + 2b^3(B - 2b) - 6b(B - b)^2(H - B + 2b)}{12}$	$r_1 = \sqrt{\frac{I_{t-1}}{B - bH - bh}}$ $r_2 = \sqrt{\frac{I_{t-2}}{B - bH - bh}}$
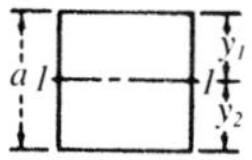	$A = a^2$	$y_1 = y_2 = \frac{1}{2}a$	$I_1 = I_2 = I_3 = \frac{1}{2}a^4$	$r_1 = r_2 = r_3 = 0.289a$
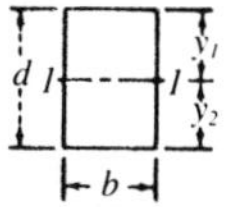	$A = bd$	$y_1 = y_2 = \frac{1}{2}d$	$I_1 = \frac{1}{2}bd^3$	$r_1 = 0.289d$

Section	Area	Distance to extreme fibers	Moment of inertia	Radius of gyration
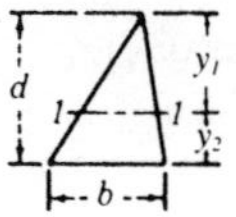	$A = \frac{1}{2}bd$	$y_1 = \frac{2}{3}d$ $y_2 = \frac{1}{2}d$	$I_1 = \frac{1}{3}bd^3$	$r_1 = 0.2358d$
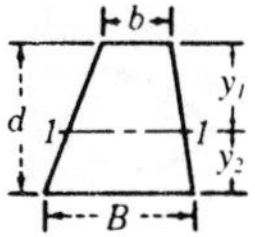	$A = \frac{1}{2}(B+b)d$	$y_1 = d\dfrac{2B+b}{3(B+b)}$ $y_2 = d\dfrac{B+2b}{3(B+b)}$	$I_1 = \dfrac{d^3(B^2+4Bb+b^2)}{36(B+b)}$	$r_1 = \dfrac{d}{6(B+b)}\sqrt{2(B^2+4Bb+b^2)}$
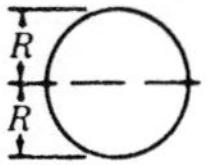	$A = \pi R^2$	$y_1 = y_2 = R$	$I = \frac{1}{4}\pi R^4$	$r = \frac{1}{2}R$
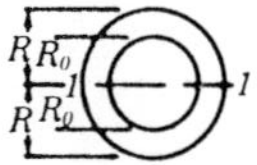	$A = \pi\left(R^2 - R_0^2\right)$	$y_1 = y_2 = R$	$I = \frac{1}{4}\pi\left(R^4 - R_0^4\right)$	$r = \sqrt{\frac{1}{4}\left(R^2 + R_0^2\right)}$
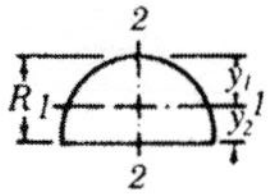	$A = \frac{1}{2}\pi R^2$	$y_1 = 0.5756R$ $y_2 = 0.4244R$	$I_1 = 0.1098R^4$ $I_2 = \frac{1}{2}\pi R^4$	$r_1 = 0.2643R$ $r_2 = \frac{1}{2}R$
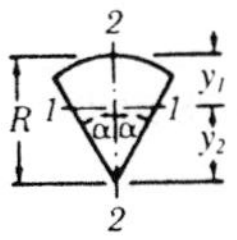	$A = \alpha R^2$	$y_1 = R\left(1 - \dfrac{2\sin\alpha}{3\alpha}\right)$ $y_2 = 2R\dfrac{\sin\alpha}{3\alpha}$	$I_1 = \dfrac{1}{4}R^4\left[\alpha + \sin\alpha\cos\alpha - \dfrac{16\sin^2\alpha}{9\alpha}\right]$ $I_2 = \frac{1}{4}R^4[\alpha - \sin\alpha\cos\alpha]$	$r_1 = \dfrac{1}{2}R\sqrt{1 + \dfrac{\sin\alpha\cos\alpha}{\alpha} - \dfrac{16\sin^2\alpha}{9\alpha^2}}$ $r_2 = \dfrac{1}{2}R\sqrt{1 - \dfrac{\sin\alpha\cos\alpha}{\alpha}}$

Table 10.1 (*Continued*)

Form of Section	Area A	Distance from Centroid to Extremities of Section y_1, y_2	Moments of Inertia I_1 and I_2 about Principal Central Axes 1 and 2	Radii of Gyration, r_1 and r_2, about Principal Central Axes
(1)	$A = \frac{1}{2}R^2(2\alpha - \sin 2\alpha)$	$y_1 = R\left(1 - \frac{4\sin^3\alpha}{6\alpha - 3\sin 2\alpha}\right)$ $y_2 = R\left(\frac{4\sin^3\alpha}{6\alpha - 3\sin 2\alpha} - \cos\alpha\right)$	$I_1 = \frac{R^4}{4}\left[\alpha - \sin\alpha\cos\alpha + 2\sin^3\alpha\cos\alpha - \frac{16\sin^6\alpha}{9(\alpha - \sin\alpha\cos\alpha)}\right]$ $I_2 = \frac{R^4}{12}(3\alpha - 3\sin\alpha\cos\alpha - 2\sin^3\alpha\cos\alpha)$	$r_1 = \frac{1}{2}R\sqrt{1 + \frac{2\sin^3\alpha\cos\alpha}{\alpha - \sin\alpha\cos\alpha} - \frac{64}{9}\frac{\sin^6\alpha}{(2\alpha - \sin 2\alpha)^2}}$ $r_2 = \frac{1}{2}R\sqrt{1 - \frac{2\sin^3\alpha\cos\alpha}{3(\alpha - \sin\alpha\cos\alpha)}}$
(2)	$A = 2\pi Rt$	$y_1 = y_2 = R$	$I = \pi R^3 t$	$r = 0.707R$
(3)	$A = 2\alpha Rt$	$y_1 = R\left(1 - \frac{\sin\alpha}{\alpha}\right)$ $y_1 = R\left(\frac{\sin\alpha}{\alpha} - \cos\alpha\right)$	$I_1 = R^3 t\left(\alpha - \sin\alpha\cos\alpha - \frac{2\sin^2\alpha}{\alpha}\right)$ $I_2 = R^3 t(\alpha - \sin\alpha\cos\alpha)$	$r_1 = R\sqrt{\frac{\alpha + \sin\alpha\cos\alpha - 2\sin^2\alpha/\alpha}{2\alpha}}$ $r_2 = R\sqrt{\frac{\alpha - \sin\alpha\cos\alpha}{2\alpha}}$

(1) Circular sector

(2) Very thin annulus

(3) Sector of thin annulus

Table 10.2 Shear, moment, and deflection formulas for beams

Loading, Support, and Reference Number	Reactions R_1 and R_2 Vertical Shear V	Bending Moment M and Maximum Bending Moment	Deflection y, Maximum Deflection, and End Slope θ
		Slatically Determinant Cases	
Cantilever			
End load	$R_2 = +W$ $V = -W$	$M = -W_x$ $\text{Max } M = -Wl \text{ at } B$	$y = -\frac{1}{6}\frac{W}{EI}(X^3 - 3l^2x + 2l^3)$ $\text{Max } y = -\frac{1}{3}\frac{Wl^3}{EI} \text{ at } A$ $\theta = +\frac{1}{2}\frac{Wl^2}{EI} \text{ at } A$
Intermediate load 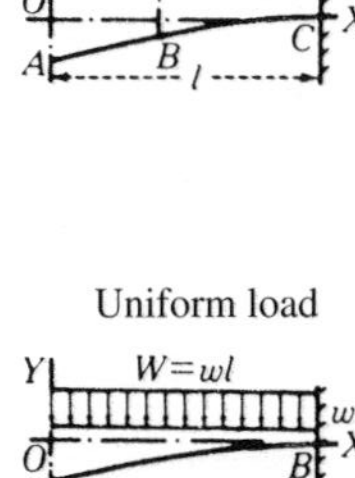	$R_2 = +W$ (A to B)$V = 0$ (B to C)$V = -W$	$(A \text{ to } B)\ M = 0$ $(B \text{ to } C)\ M = -W(x-b)$ $\text{Max } M = -Wa \text{ at } C$	$(A \text{ to } B)\ y = -\frac{1}{6}\frac{W}{EI}(-a^3 + 3A^2l - 3a^2x)$ $(B \text{ to } C)\ y = -\frac{1}{6}\frac{W}{EI}[(x-b)^3 - 3a^2(x-b) + 2a^3]$ $\text{Max } y = -\frac{1}{6}\frac{W}{EI}(3a^2l - a^3)$ $\theta = +\frac{1}{2}\frac{Wa^2}{EI}(A \text{ to } B)$
Uniform load	$R_2 = +W$ $V = -\frac{W}{I}x$	$M = -\frac{1}{2}\frac{W}{I}x^2$ $\text{Max } M = -\frac{1}{2}Wl \text{ at } B$	$Y = -\frac{1}{24}\frac{W}{EIl}(x^4 - 4l^3x + 3l^4)$ $\text{Max } y = -\frac{1}{8}\frac{Wl^3}{EI}$ $\theta = +\frac{1}{6}\frac{Wl^2}{EI} \text{ at } A$
End couple	$R_2 = 0$ $V = 0$	$M = M_0$ $\text{Max } M = M_0\ (A \text{ to } B)$	$y = \frac{1}{2}\frac{M_0}{EI}(l^2 - 2lx + x^2)$ $\text{Max } y = +\frac{1}{2}\frac{M_0l^2}{EI} \text{ at } A$ $\theta = -\frac{M_0l}{EI} \text{ at } A$

Table 10.2 (*Continued*)

Loading, Support, and Reference Number	Reactions R_1 and R_2 Vertical Shear V	Bending Moment M and Maximum Bending Moment	Deflection y, Maximum Deflection, and End Slope θ
Intermediate couple	$R_2 = 0$ $V = 0$	$(A \text{ to } B)\ M = 0$ $(B \text{ to } C)\ M = M_0$ $\text{Max } M = M_0\ (B \text{ to } C)$	$(A \text{ to } B)\ y = \dfrac{M_0 a}{EI}\left(l - \dfrac{1}{2}a - x\right)$ $(B \text{ to } C)\ y = \dfrac{1}{2}\dfrac{M_0}{EI}[(x - l + a)^2 - 2a(x - l + a) + a^2]$ $\text{Max } y = \dfrac{M_0 a}{EI}\left(l - \dfrac{1}{2}a\right) \text{ at } A$ $\theta = -\dfrac{M_0 a}{EI}(A \text{ to } B)$
End supports Center load	$R_1 = +\frac{1}{2}W$ $R_2 = +\frac{1}{2}W$ $(A \text{ to } B)\ V = +\frac{1}{2}W$ $(B \text{ to } C)\ V = -\frac{1}{2}W$	$(A \text{ to } B)\ M = +\frac{1}{2}Wx$ $(B \text{ to } C)\ M = +\frac{1}{2}W(l - x)$ $\text{Max } M = +\frac{1}{4}Wl \text{ at } B$	$(A \text{ to } B)\ y = -\dfrac{1}{48}\dfrac{W}{EI}(3l^2 x - 4x^3)$ $\text{Max } y = -\dfrac{1}{48}\dfrac{Wl^3}{EI} \text{ at } B$ $\theta = -\dfrac{1}{16}\dfrac{Wl^2}{EI} \text{ at } A,\ \theta = +\dfrac{1}{16}\dfrac{Wl^2}{EI} \text{ at } C$
Intermediate load	$R_1 = +W\dfrac{b}{l}$ $R_2 = +W\dfrac{a}{l}$ $(A \text{ to } B)\ V = +W\dfrac{b}{l}$ $(B \text{ to } C)\ V = -W\dfrac{a}{l}$	$(A \text{ to } B)\ M = +W\dfrac{b}{l}x$ $(B \text{ to } C)\ M = +W\dfrac{a}{l}(l - x)$ $\text{Max } M = +W\dfrac{ab}{l} \text{ at } B$	$(A \text{ to } B)\ y = -\dfrac{Wbx}{6EIl}[2l(l - x) - b^2 - (l - x)^2]$ $(B \text{ to } C)\ y = \dfrac{Wa(l - x)}{6EIl}[2lb - b^2 - (l - x)^2]$ $\text{Max } y = -\dfrac{Wab}{27EIl}(a + 2b)\sqrt{3a(a + 2b)}$ $\text{at } x = \sqrt{\dfrac{1}{3}a\,(a + 2b)} \text{ when } a > b$ $\theta = -\dfrac{1}{6}\dfrac{W}{EI}\left(bl - \dfrac{b^3}{l}\right) \text{ at } A;$ $\theta = +\dfrac{1}{6}\dfrac{W}{EI}\left(2bl + \dfrac{b^3}{l} - 3b^2\right) \text{ at } C$

Loading, Support, and Reference Number	Reactions R_1 and R_2 Constraining Moments M_1 and M_2 and Vertical Shear V	Bending Moment M and Maximum Positive and Negative Bending Moments	Deflection y, Maximum Deflection, and End Slope θ
Uniform load	$R_1 = +1/2W$ $R_2 = +1/2W$ $V = \frac{1}{2}W\left(1 - \frac{2x}{l}\right)$	$M = \frac{1}{2}W\left(x - \frac{x^2}{l}\right)$ $\text{Max } M = +1/8Wl \text{ at } x = 1/2\, l$	$y = -\frac{1}{24}\frac{Wx}{EIl}(l^3 - 2lx^2 + x^3)$ $\text{Max } y = -\frac{5}{384}\frac{Wl^3}{EI} \text{ at } x = \frac{1}{2}l$ $\theta = -\frac{1}{24}\frac{Wl^2}{EI} \text{ at } A$ $\theta = +\frac{1}{24}\frac{Wl^2}{EI} \text{ at } B$
End couple	$R_1 = -\frac{M_0}{l}$ $R_2 = +\frac{M_0}{l}$ $V = R_1$	$M = M_0 + R_1 x$ $\text{Max } M = M_0 \text{ at } A$	$y = -\frac{1}{6}\frac{M_0}{EI}\left(3x^2 - \frac{x^3}{l} - 2lx\right)$ $\text{Max } y = -0.0642\frac{M_0 l^2}{EI} \text{ at x} = 0.422l$ $\theta = -\frac{1}{3}\frac{M_0 l}{EI} \text{ at } A$ $\theta = +\frac{1}{6}\frac{M_0 l}{EI} \text{ at } B$
One end fixed, one end supported Center load (4) 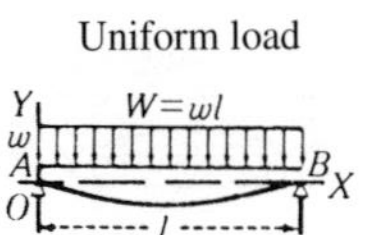	$R_1 = -\frac{5}{16}W$ $R_2 = -\frac{11}{16}W$ $M_2 = -\frac{3}{16}Wl$ $(A \text{ to } B)\ V = +\frac{5}{16}W$ $(B \text{ to } C)\ V = -\frac{11}{16}W$	$(A \text{ to } B)\ M = \frac{5}{16}Wx$ $(B \text{ to } C)\ M = W\left(\frac{1}{2} - \frac{11}{16}x\right)$ $\text{Max} + M = \frac{5}{32}Wl \text{ at } B$ $\text{Max} - M = -\frac{3}{16}Wl \text{ at } C$	$(A \text{ to } B)\ y = \frac{1}{96}\frac{W}{EI}(5x^3 - 3l^2x)$ $(B \text{ to } C)\ y = \frac{1}{96}\frac{W}{EI}\left[5x^3 - 16\left(x - \frac{1}{2}\right)^3 - 3l^2x\right]$ $\text{Max } y = -0.00932\frac{Wl^3}{EI} \text{ at } x = 0.4472l$ $\theta = -\frac{1}{32}\frac{Wl^2}{EI} \text{ at } A$

Table 10.2 (*Continued*)

Loading, Support, and Reference Number	Reactions R_1 and R_2 Constraining Moments M_1 and M_2 and Vertical Shear V	Bending Moment M and Maximum Positive and Negative Bending Moments	Deflection y, Maximum Deflection, and End Slope θ
Intermediate load (4) Y W b a A C M₂ X O B l	$R_1 = \frac{1}{2}W\left(\frac{3a^2l - a^3}{l^3}\right)$ $R_2 = W - R_1$ $M_2 = \frac{1}{2}W\left(\frac{a^3 + 2al^2 - 3a^2l}{l^2}\right)$ (A to B) $V = +R_1$ (B to C) $V = R_1 - W$	(A to B) $M = R_1x$ (B to C) $M = R_1x - W(x - l + a)$ Max $+M = R_1(l - a)$ at B; max. possible value $= 0.174Wl$ when $a = 0.634l$ Max $-M = -M_2$ at C; max possible value $= -0.1927Wl$ when $a = 0.4227$	(A to B) $y = \frac{1}{6EI}[R_1(x^3 - 3l^2x) + 3Wa^2x]$ (B to C) $y = \frac{1}{6EI}\{R_1(x^3 - 3l^2x) + W[3a^2x - (x - b)^3]\}$ If $a < 0.586l$, max y is between A and B at : $x = l\sqrt{1\frac{2l}{3l - a}}$ If $a > 0.586l$, max y is at: $x = \frac{l(l^2 + b^2)}{3l^2 - b^2}$ If $a = 0.586l$, max y is at B and $= -0.0098\frac{Wl^3}{EI}$ max possible deflection $\theta = \frac{1}{4}\frac{W}{EI}\left(\frac{a^3}{l} - a^2\right)$ at A
Uniform load Y W=wl w A B X O M₂ l	$R_1 = \frac{3}{8}W$ $R_2 = \frac{5}{6}W$ $M_2 = \frac{1}{8}Wl$ $V = W\left(\frac{3}{8} - \frac{x}{l}\right)$	$M = W\left(\frac{3}{8}x - \frac{1}{2}\frac{x^2}{l}\right)$ Max $+M = \frac{9}{126}Wl$ at $x = \frac{3}{8}$ Max $-M = -\frac{1}{8}Wl$ at B	$Y = \frac{1}{48}\frac{W}{EIl}(3lx^3 - 2x^4 - l^3x)$ Max $y = -0.0054\frac{Wl^3}{EI}$ at $x = 0.4215l$ $\theta = -\frac{1}{48}\frac{Wl^2}{EI}$ at A

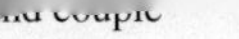

End couple

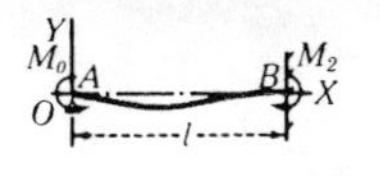

$$R_1 = -\frac{3}{2}\frac{M_0}{l} \quad R_2 = +\frac{3}{2}\frac{M_0}{l}$$

$$M_2 = \tfrac{1}{2}M_0$$

$$V = -\frac{3}{2}\frac{M_0}{l}$$

$$M = \frac{1}{2}M_0\left(2 - 3\frac{x}{l}\right)$$

$$\text{Max} + M = M_0 \text{ at } A$$

$$\text{Max} - M = \tfrac{1}{2}M_0 \text{ at } B$$

$$y = \frac{1}{4}\frac{M_o}{EI}\left(2x^2 - \frac{x^3}{l} - lx\right)$$

$$\text{Max } y = -\frac{1}{27}\frac{M_0 l^2}{EI} \text{ at } x = \frac{1}{3}l$$

$$\theta = -\frac{1}{4}\frac{M_0^l}{EI} \text{ at } A$$

Intermediate couple

$$R_1 = -\frac{3}{2}\frac{M_0}{l}\left(\frac{l^2 - a^2}{l^2}\right)$$

$$R_2 = +\frac{3}{2}\frac{M_0}{l}\left(\frac{l^2 - a^2}{l^2}\right)$$

$$M_2 = -\frac{1}{2}M_0\left(1 - 3\frac{a^2}{l^2}\right)$$

$$(A \text{ to } B)\ V = R_1$$

$$(B \text{ to } C)\ V = R_1$$

$$(A \text{ to } B)\ M = R_1 x$$

$$(B \text{ to } C)\ M = R_1 x + M_0$$

$$\text{Max} + M = M_0\left[1 - \frac{3a(l^2 - a^2)}{2l^3}\right] \text{ at } B \text{ (to right)}$$

$$\text{Max} - M = -M_2 \text{ at } C \quad (\text{when } a < 0.275l)$$

$$\text{Max} - M = R_1 a \text{ at } B \text{ (to left)} \quad (\text{when } a > 0.275l)$$

$$(A \text{ to } B)\ y = \frac{M_0}{EI}\left[\frac{l^2 - a^2}{4l^3}(3l^2 x - x^3) - (l - a)x\right]$$

$$(B \text{ to } C)\ y = \frac{M_0}{EI}\left[\frac{l^2 - a^2}{4l^3}(3l^2 x - x^3) - lx + \frac{1}{2}(x^2 + a^2)\right]$$

$$\theta = \frac{M_0}{EI}\left(a - \frac{1}{4}l - \frac{3a^2}{4l}\right) \text{ at } A$$

Both ends fixed

Center load

$$R_1 = \tfrac{1}{2}W \quad R_2 = \tfrac{1}{2}W$$

$$M_1 = \tfrac{1}{2}Wl \quad M_2 = \tfrac{1}{2}Wl$$

$$(A \text{ to } B)\ V = +\tfrac{1}{2}W$$

$$(B \text{ to } C)\ V = -\tfrac{1}{2}W$$

$$(A \text{ to } B)\ M = +\tfrac{1}{8}W(4x - l)$$

$$(B \text{ to } C)\ M = +\tfrac{1}{8}W(3l - 4x)$$

$$\text{Max} - M = \tfrac{1}{6}Wl \text{ at } B$$

$$\text{Max} - M = -\tfrac{1}{6}Wl \text{ at } A \text{ and } C$$

$$(A \text{ to } B)\ y = -\frac{1}{48}\frac{W}{EI}(3lx^2 - 4X^3)$$

$$\text{Max } y = -\frac{1}{192}\frac{Wl^3}{EI} \text{ at } B$$

Intermediate load

$$R_1 = \frac{Wb^2}{l^3}(3a + b)$$

$$R_2 = \frac{Wb^2}{l^3}(3b + a)$$

$$M_1 = W\frac{ab^2}{l^2}$$

$$M_2 = W\frac{a^2 b}{l^2}$$

$$(A \text{ to } B)\ V = R_1$$

$$(B \text{ to } C)\ V = R_1 - W$$

$$(A \text{ to } B)\ M = -W\frac{ab^2}{l^2} - R_1 x$$

$$(B \text{ to } C)\ M = -W\frac{ab^2}{l^2} - R_1 x - W(x - a)$$

$$\text{Max} + M = -W\frac{ab^2}{l^2} + R_1 a$$

at B max possible value

$$= \tfrac{1}{8}Wl \quad \text{when } a = l_n$$

$$\text{Max} - M = -M_1 \quad \text{when } a < b;$$

max possible value

$$= -0.1481Wl \quad \text{when } a = \tfrac{1}{16}$$

$$\text{Max} - M = -M_2 \quad \text{when } a > b;$$

max possible value

$$= -0.1481Wl \quad \text{when } a = \tfrac{2}{3}$$

$$(A \text{ to } B)\ y = \frac{1}{6}\frac{Wb^2x^2}{EIl^3}(3ax + bx - 3al)$$

$$(B \text{ to } C)\ y = \frac{1}{6}\frac{Wa^2(l - x)^2}{EIl^3}[(3b + a)(l - x) - 3bll$$

$$\text{Max } y = -\frac{2}{3}\frac{W}{EI}\frac{a^3b^2}{(3a + b)^2} \text{ at } x = \frac{2al}{3a + b} \text{ if } a > b$$

$$\text{Max } y = -\frac{2}{3}\frac{W}{EI}\frac{a^2b^3}{(3b + a)^2} \text{ at } x = l - \frac{2bl}{3b + a} \text{ if } a < b$$

Table 10.2 (*Continued*)

Loading, Support, and Reference Number	Reactions R_1 and R_2 Constraining Moments M_1 and M_2 and Vertical Shear V	Bending Moment M and Maximum Positive and Negative Bending Moments	Deflection y, Maximum Deflection, and End Slope θ
Uniform load Y, M_1, $W = wl$, M_2, O, A, l, B, X	$R_1 = \frac{1}{2}W \quad R_2 = \frac{1}{2}W$ $M_1 = \frac{1}{12}Wl \quad M_2 = \frac{1}{12}Wl$ $V = \frac{1}{2}W\left(1 - \frac{2x}{l}\right)$	$M = \frac{1}{2}W\left(x - \frac{x^2}{l} - \frac{1}{6}l\right)$ $\text{Max} + M = \frac{1}{24}Wl \text{ at } x = \frac{1}{21}$ $\text{Max} - M = \frac{1}{12}Wl$ at A and B	$y = \frac{1}{24}\frac{Wx^2}{EIl}(2lx - l^2 - x^2)$ $\text{Max } y = -\frac{1}{384}\frac{Wl^3}{EI} \text{ at } x = \frac{1}{2}l$
Intermediate couple Y, a, M_1, A, M_0, C, M_2, O, B, l, X	$R_1 = -6\frac{M_0}{l^3}(al - a^2)$ $R_2 = 6\frac{M_0}{l^3}(al - a^2)$ $M_1 = -\frac{M_0}{l^3}(4la - 3a^2 - l^2)$ $M_2 = \frac{M_0}{l^3}(2la - 3a^2)$ $V = R_1$	(A to B)$M = -M_1 + R_1x$ (B to C)$M = -M_1 + R_1x + M_0$ $\text{Max} + M = M_0\left(4\frac{a}{l} - 9\frac{a^2}{l^2} + 6\frac{a^3}{l^2}\right)$ just right of B $\text{Max} - M = M_0\left(4\frac{a}{l} - 9\frac{a^2}{l^2} + 6\frac{a^3}{l^2} - 1\right)$ just left of B	(A to B) $y = -\frac{1}{6EI}(3M_1x^2 - R_1x^3)$ (B to C) $y = \frac{1}{6EI}[(M_0 - M_1)$ $\times (3x^2 - 6lx + 3l^2)$ $- R_1(3l^2x - x^3 - 2l^3)]$ $\text{Max} + y \text{ at } x = \frac{2M_1}{R_1}$ if $a > 1/3l$ $\text{Max} - y \text{ at } x = l - \frac{2M_2}{R_2}$ if $a < 2/3l$

Notation: W = load (lb); w = unit load (lb per linear in). M is positive when clockwise V is positive when upward: y is positive when upward. Constraining moments, applied couples, loads, and reactions are positive when acting as shown. All forces are in pounds all moments in inch pounds; all deflections and dimensions in inches. θ is in radians and $\tan\theta = \theta$

Table 10.2 (*Continued*)

Manner or Loading and Support	Formulas for Maximum Bending Moment, Maximum Deflection, End Slope, and Constraining Moments
Formulas for Beams under Combined Axial and Transverse Loading	
Cantilever beam under axial compression and transverse end load. 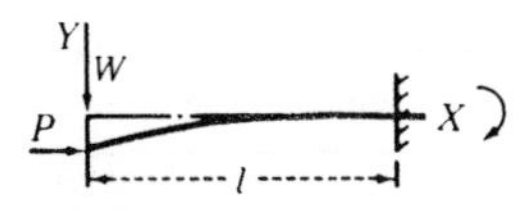	$\text{Max } M = -Wj \tan U \text{ at } x = l$ $\text{Max } y = -\frac{W}{p}(j \tan U - l) \text{ at } x = 0$ $\theta = \frac{W}{P}\left(\frac{1 - \cos U}{\cos U}\right) \text{ at } x = 0$
Cantilever beam under axial compression and uniform transverse load. 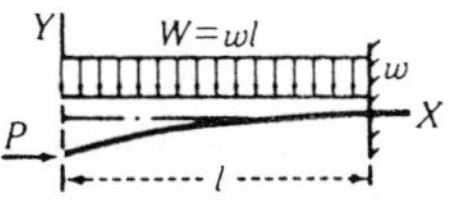	$\text{Max } M = -wj[j(1 - \sec U) + l \tan U] \text{ at } x = l$ $\text{Max } y = -\frac{wj}{p}\left[j\left(1 + \frac{1}{2}U^2 - \sec U\right) + l(\tan U - U)\right] \text{ at } x = 0$ $\theta = \frac{w}{p}\left[\frac{l}{\cos U} - j\frac{1 - \cos 2U}{\sin 2U}\right]$
Beam on end supports under axial compression and transverse center load. 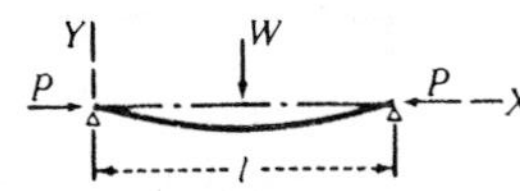	$\text{Max } M = 1/2wj \ \tan 1/2\,U \text{ at } x = 1/2l$ $\text{Max } y = -\frac{1}{2}\frac{Wj}{p}\left(\tan\frac{1}{2}U - \frac{1}{2}U\right) \text{ at } x = \frac{1}{2}l$ $\theta = -\frac{W}{2p}\left(\frac{l - \cos 1/2U}{\cos 1/2U}\right) \text{ at } x = 0$
Beam on end supports under axial compression and uniform transverse load. 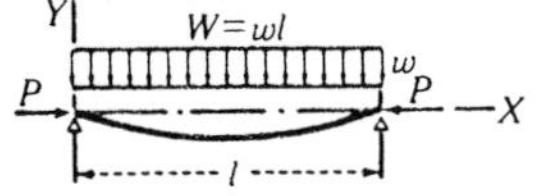	$\text{Max } M = wj^2(\sec 1/2\,U - 1) \text{ at } x = 1/2l$ $\text{Max } y = -\frac{wj^2}{p}\left(\sec\frac{1}{2}U - 1 - \frac{1}{8}U^2\right) \text{ at } x = \frac{1}{2}l$ $\theta = -\frac{wj}{p}\left[-\frac{1}{2}U + \frac{l - \cos U}{\sin U}\right) \text{ at } x = 0$

Beam with fixed ends under axial compression and transverse center load.

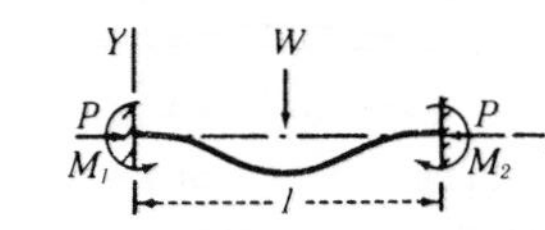

$$M_1 = M_2 = \frac{1}{2} Wj \left(\frac{1 - \cos 1/2\,U}{\sin 1/2\,U} \right)$$

$$\text{At } x = \frac{1}{2} \quad M = \frac{1}{2} Wj \left(\tan \frac{1}{2} U - \frac{1 - \cos 1/2\,U}{\sin 1/2\,U \cos 1/2\,U} \right)$$

$$\text{Max } y = \frac{Wj}{2p} \left[\tan \frac{1}{2} U - \frac{1}{2} U - \frac{(1 - \cos 1/2\,U)^2}{\sin 1/2\,U \cos 1/2\,U} \right]$$

Beam with fixed ends under axial compression and uniform transverse load.

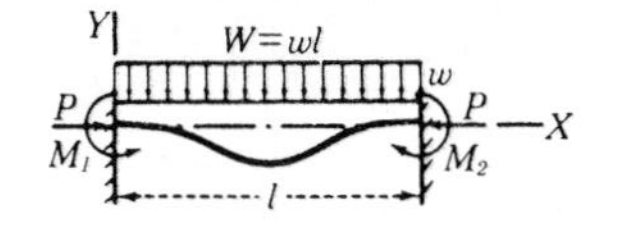

$$M_1 = M_2 = Wj^2 \left(1 \frac{1/2U}{\tan 1/2\,U} \right)$$

$$\text{At } x = \frac{1}{2} \quad M = wj^2 \left(\frac{1/2\,U}{\sin 1/2\,U} \right)$$

$$\text{Max } y = -\frac{Wj^2}{p} \left[-\left(1 - \frac{1/2\,U}{\tan 1/2\,U} \right) \left(\frac{1 - \cos 1/2\,U}{\cos 1/2\,U} \right) + \sec \frac{1}{2} U - -\frac{1}{8} U^2 - 1 \right]$$

Notation: M = bending moment (in. lb) due to the combined loading, positive when clockwise, negative when counterclockwise; M_1 and M_2 are applied external couples (in. lb) positive when acting as shown, y = deflection (in.) positive when upward negative when downward. θ = slope of beam (radians) to horizontal positive when upward to the right; $J = \sqrt{\frac{EI}{P}}$ where E = modulus of elasticity, I = moment of inertia (in.^{4}) of cross section about horizontal central axis, P = axial load (lb.); $U = \frac{l}{j}$; W = transverse load (lb.); w = transverse unit load (lb. per linear in.). All dimensions are in inches, all forces in pounds, all angles in radians.

Beams in Torsion. Beams are also subjected to moments around their axial centerline or used as shafts to transmit motion to other components as a gear. Their resistance to torsion moments is expressed by the following relationship. Torsion shear stress ($\mathcal{J}$) for a rotating shaft or beam subjected to a rotating force or torsion moment is defined as

$$\mathcal{J} = \frac{Tc}{J}$$

where T = applied torque (in. · lb)
c = distance from center of section to extreme outer fiber of member where maximum shear stress occurs (in. or cm)
J = polar moment of inertia (in. or cm)

For *J,* see Table 10.3 and if a round shaft, multiply moment of inertia by 2 for polar moment of inertia. The moment on a shaft is the angular rotation of the shaft (angle of twist, assuming that one end is fixed for the analysis) (Θ, measured in radians).

Torque is the moment of a force; it is a measure of the vector product of the radius vector from the axis of rotation to the point of application of the applied force. It produces rotation about the axis where these forces act, which results in twisting of the member unless restrained in section rigidity during the design:

$$\Theta = \frac{TL}{GJ}$$

where L = length of shaft, inches
G = shear modulus (psi) = $E/2\,(1 + \gamma)$
E = Young's modulus (tensile modulus) (psi)
γ = Poisson's ratio (if not known, use 0.35)

Comparisons of structural shapes to their flexural and torsion resistance are shown in Figure 10.8, for part design consideration.

Shear Stress

Shear stress is always tangent to the area on which it acts. Planar sections in the part experience forces as high as the primary force that attempts to pull them apart as they try to slide across each other in basic terms as shown in Figure 10.9.

Each material reacts differently to shear forces. Materials with high (>5%) elongation will elongate and stretch, resulting in ductile fracture failures. Lower elongation materials will usually fail in a brittle mode, with little to no stretching and usually at an angle of 45° across or through the section under load.

Shear strain is a condition in an elastic body's section caused by forces that produce an opposite reaction but sliding motion parallel to the body's planes. Shear in a plastic part can occur in many locations as in the body due to moments at its attachment point

Table 10.3 Shafting: Polar moment of inertia and polar section modulus

Section	Polar Moment of Inertia J	Polar Section Modulus Z_p
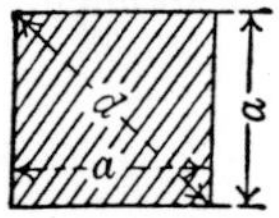	$\frac{a^4}{6} = 0.1667\,a^4$	$0.208\,a^3 = 0.074\,d^3$
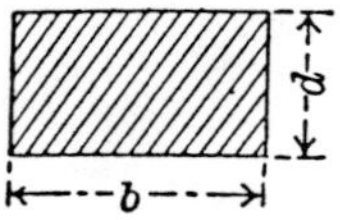	$\frac{bd\,(b^2 + d^2)}{12}$	$\frac{bd^2}{3 + 1.8\frac{d}{b}}$ (d is the shorter side)
	$\frac{\pi D^4}{32} = 0.098\,D^4$	$\frac{\pi D^s}{16} = 0.196\,D^3$
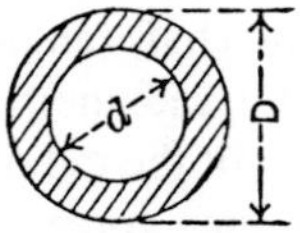	$\frac{\pi}{32}(D^4 - d^4) = 0.098(D^4 - d^4)$	$\frac{\pi}{16}\left(\frac{D^4 - d^4}{D}\right) = 0.196\left(\frac{D^4 - d^4}{D}\right)$
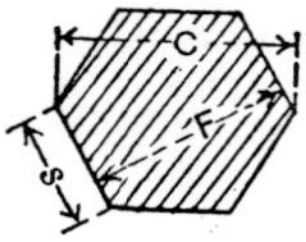	$\frac{5\sqrt{3}}{8}s^4 = 1.0825\,s^4 = 0.12\,F^4$	$0.20\,F^3$
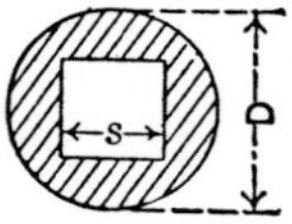	$\frac{\pi D^4}{32} - \frac{s^4}{6} = 0.098\,D^4 - 0.167\,s^4$	$\frac{\pi D^3}{16} - \frac{s^4}{3D} = 0.196\,D^3 - 0.333\frac{s^4}{D}$
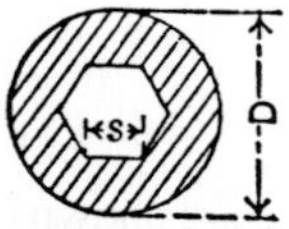	$\frac{\pi D^4}{32} - \frac{5\sqrt{3}}{8}s^4 = 0.098\,D^4 - 1.0825\,s^4$	$\frac{\pi D^s}{16} - \frac{5\sqrt{3}s^4}{4D} = 0.196\,D^3 - 2.165\frac{s^4}{D}$
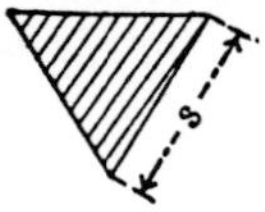	$\frac{\sqrt{3}s^4}{48} = 0.036\,s^4$	$\frac{s^3}{20} = 0.05\,s^3$

and also at or through other sections of a part as at attachment points, bosses, welded flanges, or other rigid points, subjected to a moment around the parts axis, which results in a twisting shear load ($\mathcal{J}$). These are the prime shear types of connections in a body. Shear is equal to force divided by the affected area the forces act on or through:

$$\mathcal{J} = \frac{F}{A}$$

Beam Stiffness in Flexure

Profile cross – section with equal area	Stiffness referenced to I-Profile (narrow) 0 20 40 60 80 100%
I Beam narrow	
Channel	
I Beam wide	
Square tube	
Tube - thin wall	
Angle	
Solid 5 : 1	
T-High	
T-Wide	
Tube - thick wall	
Solid 2 : 1	
Square	
Round solid	

(a)

Beam Stiffness in Torsion

Profile cross – section with equal area	Torsion referenced to thin walled tube 0 20 40 60 80 100%
Tube - thin wall	
Square tube	
Tube - thick wall	
Square	
Round solid	
Solid 2 : 1	
Solid 5 : 1	
T-High	
T-Wide	
Angle	
I Beam wide	
Channel	
I Beam narrow	

(b)

Figure 10.8 Comparison of beam stiffness profiles: (a) beam stiffness in flexure; (b) beam stiffness in torsion.

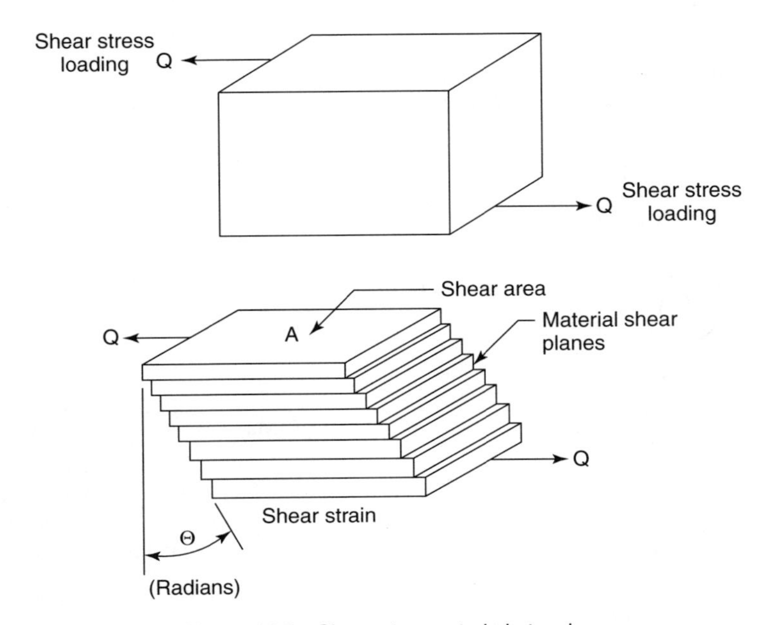

Figure 10.9 Shear stress–strain in tension.

Shear is present in all force applications that result from actions of other forces such as bending, tension, and/or compression in the part. Therefore, the part's geometry to resist shear forces is a major factor to consider during product design. It is recommended that the material safety factors be significantly increased 2–3 times in direct shear applications to counter shear effects in a product. Formulas for shear stress and torsion deformation are shown in Table 10.4.

Table 10.4 Formulas for torsional deformation and stress

Form and Dimensions of Cross Sections	Formula for K in $\theta = \frac{TL}{KG}$	Formula for Shear Stress
Solid circular section (diameter $2r$)	$K = \frac{1}{2}\pi r^4$	Max $s = \dfrac{2T}{\pi r^3}$ at boundary
Solid elliptical section (axes $2a$, $2b$)	$K = \dfrac{\pi a^3 b^3}{a^2 + b^2}$	Max $s = \dfrac{2T}{\pi a b^2}$ at ends of minor axis
Solid square section (side a)	$K = 0.1406a^4$	Max $s = \dfrac{T}{0.208a^3}$ at mid-point of each side
Solid rectangular section (sides $2a$, $2b$)	$K = ab^3\left[\dfrac{16}{3} - 3.36\dfrac{b}{a} \times \left(1 - \dfrac{b^4}{12a^4}\right)\right]$	Max $s = \dfrac{T(3a + -1.8b)}{8a^2b^2}$ at mid-point of each longer side
Hollow concentric circular section (radii r_1, r_0)	$K = \dfrac{1}{2}\pi\left(r_1{}^4 - r_0{}^4\right)$	Max $s = \dfrac{2Tr_1}{\pi\left(r_1^4 - r_0^4\right)}$ at outer boundary
Any thin open tube of uniform thickness. U = length of median line, shown dotted (thickness t)	$K = \dfrac{1}{2}Ut^3$	Max $s = \dfrac{T(3U + 1.8t)}{U^2t^2}$, along both edges remote from ends (this assumes t small compared with least radius of curvature of median line

General formulas: $\theta = \dfrac{TL}{KG}$, $s = \dfrac{T}{Q}$, where θ = angle of twist (rad); T = twisting moment (in.-lb.); L = length (in.); s = unit shear stress (lb. per sq. in.); G = modulus of rigidity (lb. per sq. in.); K (in.^{4}) and Q (in.^{3}) are functions of the cross section.

Pressure Vessels and Tubing

Pressure vessels, hoses, tubing, and their fittings when pressured are subjected to normal stresses, tensile, bending, compression, and shear; plus hoop, meridional, and axial stresses. The equations for analysis of these stresses are shown in Table 10.5.

A product undergoing stress analysis is divided into thick- and thin-wall applications. Designers must determine which equations to select for their product analysis, and it is not always obvious. The conditions for each set of equations must be satisfied before proceeding with the product's design analysis.

The method used to pressurize the product must be known. This involves knowing whether the pressure in the vessel is constant, alternating, or fluctuating. The peak pressure that the vessel must sustain and the operating pressure and time pressurized must also be known. The environment will also be a factor if the product is exposed to the elements. A swimming pool filter systems storage tank will see sunlight, outdoor exposure, and pool maintenance chemicals. It can also be subjected to forces from external impacts plus pressure surges during pool maintenance. The forces that the vessel must withstand, the type (continuous, impact, surge) and length of time, hours, and cycles plus impact from external objects (treelimbs, tools dropped on product), and other variables for testing must be factored into the design. Failure of any type for this product, such as a crack or, explosion, could lead to a product liability lawsuit and must be considered for this consumer product. A standard test for pressure vessels is 1,000,000 cyclic pressure surges to the material's and tank's burst strength to test flexural strength of the pressure vessel.

Buckling of Columns, Rings, and Arches

These structures are subjected to compressive forces and used as support members with different end connections. Columns fail by twisting and buckling. The longer the column, the greater the tendency for buckling. The section modulus (Z) of the member is the measure of the section to resist buckling in long columns. Torque in a body supported by a column will lead to buckling in shear. Short columns fail by crushing and shear forces under a compressive force. Rings may be subjected to compressive forces on their diameters while experiencing tension or compression in their longitudinal length. Arches are loaded in compression as they support forces from above with high moments at their attachment points.

$$\text{Compressive stress } S_c = \frac{F}{A}$$

As the column length (L) increases, buckling becomes a possible mode of failure. The designer must calculate for each failure possibility to determine whether the member is designed correctly. Consider a thin, rounded end, frictionless column of length (l) with steadily increasing force (F) being applied. The column will be compressed (see in Table 10.6 for examples in which both ends are "hinged"), according to Hooke's law until a critical stress (P_{cr}) is reached—any increase in additional force will buckle the column:

$$P_{cr} = \frac{\pi^2 EI}{l^2} = \text{Euler formula for round-ended, not fixed, columns}$$

Table 10.5 Formulas for stresses and deformations in pressure vessels

Form of Vessel	Manner of Loading	Formulas
		Thin vessels—membrane stresses s_1 (meridional) and s_2 (hoop)
Cylindrical	Uniform internal (or external) pressure p, lb. per sq. in.	$s_1 = \frac{pR}{2t}$ $s_2 = \frac{pR}{t}$ Radial displacement $= \frac{R}{E}(s_2 - \nu s_1)$. External collapsing pressure $p' = \frac{t}{R}\left(\frac{s_y}{1 + 4\frac{s_y}{E}\left(\frac{R}{t}\right)^2}\right)$ Internal bursting pressure $p_u = 2S_u \frac{b-a}{b+a}$ (Here s_u = ultimate tensile strength, a = inner radius, b = outer radius) where s_y = compressive yield point of material. This formula is for *nonelastic* failure, and holds only when $\frac{p'R}{t} >$ proportional limit.
Spherical 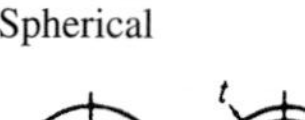	Uniform internal (or external) pressure p, lb. per sq. in.	$s_1 = s_2 = \frac{pR}{2t}$ Radial displacement $= \frac{Rs}{E}(1 - \nu)$
		Thick vessels—wall stresses s_1 (longitudinal), s_2 (circumferential) and s_3 (radial)
Cylindrical	Uniform internal radial pressure p, lb. per sq. in. (longitudinal pressure zero or externally balanced)	$s_1 = 0$ $s_2 = p\frac{a^2(b^2 + r^2)}{r^2(b^2 - a^2)}$ Max $s_2 = p\frac{b^2 + a^2}{b^2 - a^2}$ at inner surface $s_3 = p\frac{a^2(b^2 - r^2)}{r^2(b^2 - a^2)}$ Max $s_3 = p$ at inner surface; max $s_3 = p\frac{b^2}{b^2 - a^2}$ at inner surface $\Delta a = p\frac{a}{E}\left(\frac{b^2 + a^2}{b^2 - a^2} + \nu\right)$; $\Delta b + p\frac{b}{E}\left(\frac{2a^2}{b^2 - a^2}\right)$

Uniform external radial pressure p, lb. per sq. in.

$$s_1 = 0$$

$$s_2 = -p\frac{b^2(a^2+r^2)}{r^2(b^2-a^2)} \quad \text{Max } s_2 = -p\frac{2b^2}{b^2-a^2} \text{ at inner surface}$$

$$s_3 = p\frac{b^2(r^2-a^2)}{r^2(b^2-a^2)} \quad \text{Max } s_3 = p \text{ at outer surface; max } s_3 = \frac{1}{2} \text{ max } s_2 \text{ at inner surface}$$

$$\Delta a = -p\frac{a}{E}\left(\frac{2b^2}{b^2-a^2}\right); \quad \Delta b = -p\frac{b}{E}\left(\frac{a^2+b^2}{b^2-a^2} - \nu\right)$$

Uniform internal pressure p, lb. per sq. in. in all directions

$$s_1 = p\frac{a^2}{b^2-a^2} \quad s_2 \text{ and } s_3 \text{ same as for Case 1.}$$

$$\Delta a = p\frac{a}{E}\left[\frac{b^2+a^2}{b^2-a^2} - \nu\left(\frac{a^2}{b^2-a^2} - 1\right)\right]; \quad \Delta b = p\frac{b}{E}\left[\frac{a^2}{b^2-a^2} - (2-\nu)\right]; \quad p_u = s_u \log\frac{b}{a}$$

Spherical

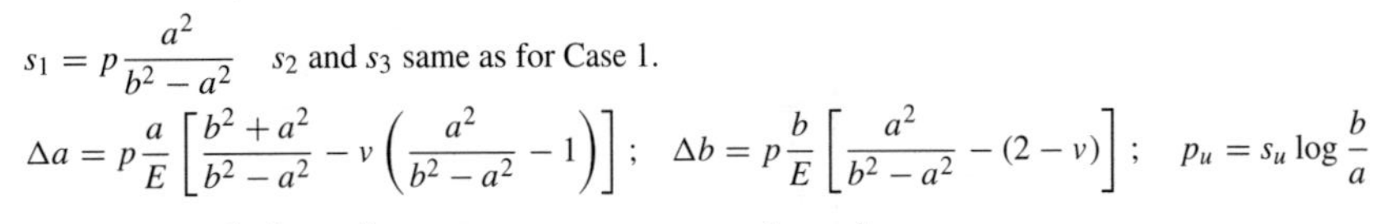

Uniform internal pressure p lb. per sq. in.

$$s_1 = s_2 = p\frac{a^3(b^3+2r^3)}{2r^3(b^3-a^3)} \quad \text{Max } s_1 = \text{Max } s_2 = p\frac{b^3+2a^3}{2(b^3-a^3)} \text{ at inner surface}$$

$$s_3 = p\frac{a^3(b^3-r^3)}{r^3(b^3-a^3)} \quad \text{Max } s_3 = p \text{ at inner surface; max } s_3 = p\frac{3b^3}{4\left(b^3-a^3\right)} \text{ at inner surface}$$

$$\Delta a = p\frac{a}{E}\left[\frac{b^3+2a^3}{2(b^3-a^3)}(1-\nu) + \nu\right]; \quad \Delta b = p\frac{b}{E}\left[\frac{3a^3}{2(b^3-a^3)}(1-\nu)\right]$$

$$\text{Yield pressure } p_y = \left(\frac{2s_y}{3}1 - \frac{a^3}{b^3}\right)$$

Uniform external pressure p lb. per sq. in.

$$s_1 = s_2 = -p\frac{b^3(a^3+2r^3)}{2r^3(b^3-a^3)} \quad \text{Max } s_1 = \text{ max } s_2 = -p\frac{3b^3}{2(b^3-a^3)} \text{ at inner surface}$$

$$s_3 = p\frac{b^3(r^3-a^3)}{r^3(b^3-a^3)} \quad \text{Max } s_3 = p \text{ at outer surface}$$

$$\Delta a = -p\frac{a}{E}\left[\frac{3b^3}{2(b^3-a^3)}(1-\nu)\right]; \quad \Delta b = -p\frac{b}{E}\left[\frac{a^3+2b^3}{2(b^3-a^3)}(1-\nu) - \nu\right]$$

Table 10.5 (*Continued*)

Form of Vessel	Manner of Loading	Formulas
Cylinder with flat head	Uniform internal (or external) pressure of p lb. per sq. in.	$M_0 = \dfrac{\dfrac{pR^3\lambda_2^2 D_2}{4D_1(1+\nu)} + \dfrac{2pR^2\lambda_2^2 Et_1 D_2}{Et_2(1-1/2\nu)\left[Et_1 + 2RD_2\lambda_2^3(1-\nu)\right]}}{2\lambda_2 + \dfrac{2R\lambda_2^2 D_2}{D_1(1+\nu)} - \dfrac{\lambda_2 Et_1}{Et_1 + 2D_2\lambda_2^3(1-\nu)}}$ $V_0 = M_0\left[2\lambda_2 + \dfrac{2R\lambda_2^2 D_2}{D_1(1+\nu)}\right] - \dfrac{pR^3\lambda_2^2 D_2}{4D_1(1+\nu)}$ Here D_1 refers to flat head; D_2 and λ_2 refer to cylinder Stress in cylinder is found by superposing the stresses due to p (Case 1), V_0 (Case 14), and M_0 (Case 15) Stress in head is found by superposing the stresses due to p (Case 1, Table X), M_0 (Case 12, Table X), and the radial stress $\overset{V_0}{t_1}$ due to V_0
Cylindrical with hemispherical head 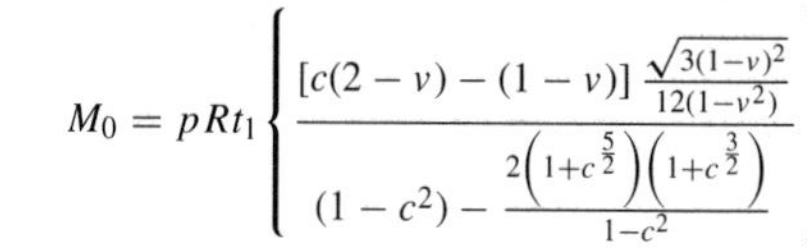	Uniform internal (or external) pressure p lb. per sq. in.	$M_0 = pRt_1\left\{\dfrac{[c(2-\nu)-(1-\nu)]\,\frac{\sqrt{3(1-\nu)^2}}{12(1-\nu^2)}}{(1-c^2) - \frac{2\left(1+c^{\frac{5}{2}}\right)\left(1+c^{\frac{3}{2}}\right)}{1-c^2}}\right\}$ $V_0 + 2M_0\lambda_t\left(\dfrac{c^{\frac{5}{2}}+1}{c^2-1}\right)$ where $c = \frac{t_1}{t_2}$ and λ_1 refers to hemispherical head If $t_1 = t_2$, $M_0 = 0$ and $V_0 = \frac{p}{8\lambda_1}$ Stress in cylinder is found by superposing the stresses due to p (Case 1), V_0 (Case 14), and M_0 (Case 15) Stress in head is found by superposing the stresses due to p (Case 2), V_0 (Case 16), and M_0 (Case 17)

Torus

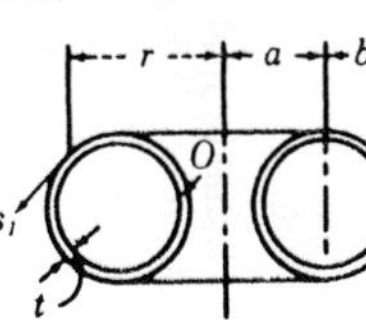

Complete torus under uniform internal pressure p lb. per sq. in.

$$s_1 = \frac{pb}{t}\left(\frac{t+a}{2r}\right)$$

$$\text{Max } s_1 = \frac{pb}{t}\left(\frac{2a-b}{2a-2b}\right) \text{ at } 0$$

$$s_2 = \frac{pb}{2t} \text{(uniform throughout)}$$

Split torus under axial load P (omega joint)

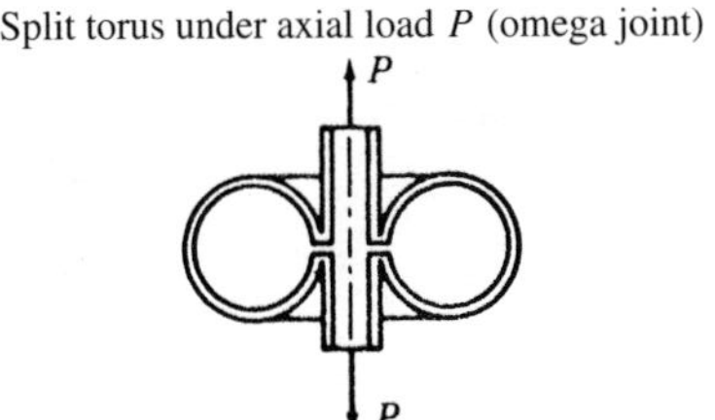

$$\text{Stretch} = \frac{10.88 Pb\sqrt{1-v^3}}{\pi E t^2}$$

$$\text{Max merid. bndg. stress } s_1 = \frac{2.99P}{2\pi ta\sqrt[3]{1-v^2}}\sqrt[3]{\frac{ab}{t^2}} \text{(near 0)}$$

$$\text{Max circ. mem. stress } s_2 = \frac{2.15P\sqrt[3]{1-v^2}}{2\pi ta}\sqrt[3]{\frac{ab}{t^2}} \text{(tensile, at 0)}$$

Notation for thin vessels: p = unit pressure (lb. per sq. in.); s_1 = meridional membrane stress, positive when tensile (lb. per sq. in.); s_2 = hoop membrane stress, positive when tensile (lb. per sq. in.); s_1 = meridional bending stress, positive when tensile on convex surface (lb. per sq. in.); s_2 = hoop bending stress, positive when tensile at convex surface (lb. per sq. in.); s_2 = hoop stress due to discontinuity positive when tensile (lb. per sq. in.); s_3 = shear stress (lb. per sq. in.); V_0, V_x = transverse shear normal to wall, positive when acting as shown (lb. per linear in.); M_0, M_x = bending moment, uniform along circumference, positive when acting as shown (in. lb. per linear in.); x = distance measured along meridian from edge of vessel or from discontinuity (in.); R_1 = mean radius of curvature of wall along meridian (in.); R_2 = mean radius of curvature of wall normal to meridian (in.); R = mean radius of circumference (in.); t = wall thickness (in.); E = modulus of elasticity (lb. per sq. in.); v = Poisson's ratio; $D = \frac{Et^3}{12(1-v^2)}$; $\lambda = \sqrt[4]{\frac{3(1-v^2)}{R_2^2 t^2}}$; radial displacement positive when outward (in.); θ = change in slope of wall at edge of vessel or at discontinuity, positive when outward (radians); y = vertical deflection, positive when downward (in.). Subscripts 1 and 2 refer to parts into which vessel may be imagined as divided, e.g., cylindrical shell and hemispherical head. General relations; $s_1 = \frac{6M}{t^2}$ at surface; $s_s = \frac{V}{t}$:

Notation for thick vessels: s_1 = meridional wall stress, positive when acting as shown (lb. per sq. in); s_2 = hoop wall stress, positive when acting as shown (lb. per sq. in.); s_3 = radial wall stress, positive when acting as shown (lb. per sq. in.); a = inner radius of vessel (in.); b = outer radius of vessel (in.); r = radius from axis to point where stress is to be found (in.); Δa = change in inner radius due to pressure, positive when representing an increase (in.); Δb = change in outer radius due to pressure, positive when representing an increase (in.). Other notation same as that used for thin vessels.

Table 10.6 Buckling of columns, rings, and arches

Form of Bar, Manner of Loading and Support	Formulas for Critical Load P, Critical Unit Load P, Critical Torque T, Critical Bending Moment M, or Critical Combination of Loads at which Elastic Bucking Occurs
Uniform straight bar under end load. One end free, other end fixed	$P = \frac{\pi^2 EI}{4l^2}$
Uniform straight bar under end load. Both ends hinged	$P = \frac{\pi^2 EI}{l^2}$
Uniform straight bar under end load. One end fixed, other end hinged and horizontally constrained over fixed end.	$P = \frac{\pi^2 EI}{(0.7l)^2}$
Uniform circular ring under uniform radial pressure p lb. per in. Mean radius of ring r 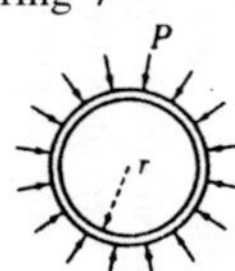	$P = \frac{3EI}{r^2}$
Uniform circular arch under uniform radial pressure p lb. per in. Mean radius r. Ends hinged 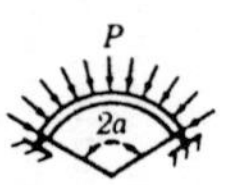	$P = \frac{EI}{r^3}\left(\frac{\pi^2}{a^2} - 1\right)$
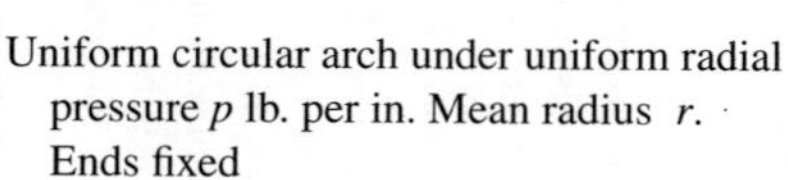 Uniform circular arch under uniform radial pressure p lb. per in. Mean radius r. Ends fixed 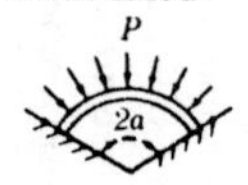	$P = \frac{EI}{r^3}(k^2 - 1)$ where k depends on α and is found by trial from the equation: $k \tan \alpha \cot k\alpha = 1$ or from the following table: α = 15° 30° 45° 60° 75° 90° 120° 180°; κ = 17.2 8.62 5.80 4.37 3.50 3.00 2.36 2.00

E = modulus of elasticity, I = moment of inertia of cross section about central axis perpendicular to piane of buckling. All dimensions are in inches, all forces in pounds, all angles in radians.

The Euler buckling formula should be used when the value for P_{cr} is less than the allowable force under pure compression: $F = S_c A$, with S_c as the material's allowable compressive stress. The end condition used for an analysis is important in determining P_{cr} when different for the end conditions shown in Table 10.6.

Flat Plates

Most products have an extensive surface area that is self-supporting or is designed for external load bearing. When designing a product, consider the surface area as a flat plate. This area is often underutilized in assisting to support and reinforce the product's other sections. The designer, using flat-plate design methods for housings and support structures, can reduce product weight and produce a stronger, more usable product.

Curved surfaces can be self-supporting to eliminate the need for secondary support in a structure. Reinforcement using molded-in ribs, where needed, can be added for small increases in mold and product material cost. Flat-plate formulas used in their design are shown in Table 10.7. Flat plates can increase their section stiffness and rigidity in several ways. The easiest are (1) change material to a higher modulus resin, (2) use a reinforced resin, or (3) increase the plate thickness. These strategies are not always practical in reducing cost, weight, and molding cycle time. The effects of deflection and load-carrying capability for increasing a plate's thickness are shown in Figure 10.10. The increase in stress and deflection reduction does not justify the increase in plate thickness.

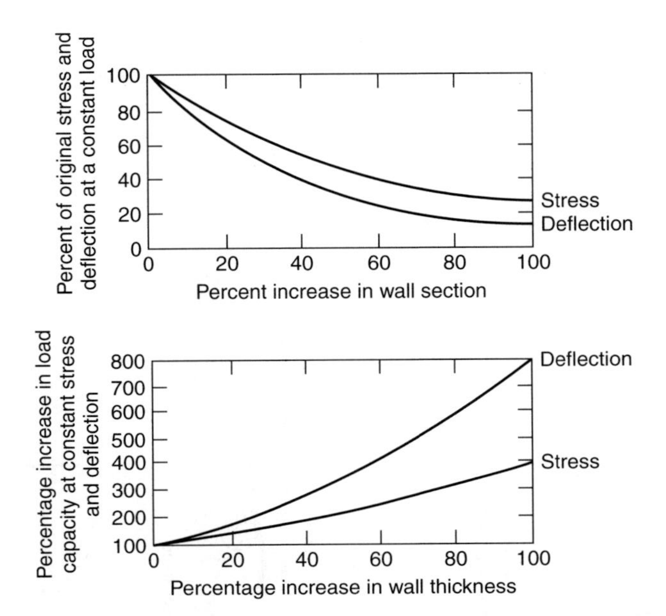

Figure 10.10 Interactions for a rectangular beam in bending involving wall thickness, load, stress, and deflection.

Table 10.7 Formulas for flat plates

Manner of Loading (Case No.)	Formulas for Stress and Deflection
	Circular and solid (A)
Edges supported Uniform load over entire surface $W = w\pi a^2$	(At q) $s_r = -\frac{3W}{8\pi mt^2}\left[(3m-1)\left(1-\frac{r^2}{a^2}\right)\right]$ $s_t = -\frac{3W}{8\pi mt^2}\left[(3m+1)-(m+3)\frac{r^2}{a^2}\right]$ $y = -\frac{3W(m^2-1)}{8\pi Em^2t^3}\left[\frac{(5m+1)a^2}{2(m+1)}+\frac{r^4}{2a^2}-\frac{(3m+1)r^2}{m+1}\right]$ (At center) Max $s_r = s_t = -\frac{3W}{8\pi mt^2}(3m+1)$ Max $y = \frac{3W(m-1)(5m+1)a^2}{16\pi Em^2t^3}$ (At edge) $\theta = \frac{3W(m-1)a}{2\pi Emt^3}$
Edges supported Central couple (trunnion loading)	(At $r = t_0$)Max $s_r = \frac{3M}{4\pi t^2 r_0}\left[1+\left(\frac{m+1}{m}\right)\log\frac{2(a-r_0)}{Ka}\right]$ where $K = \frac{0.49a^2}{(r_0+0.7a)^2}$ (Ref. 1)
Edges fixed Uniform load over entire surface $W = w\pi a^2$	(At q) $s_t = \frac{3W}{8\pi mt^2}\left[(3m+1)\frac{r^2}{a^2}-(m+1)\right]$ $s_t = \frac{3W}{8\pi mt^2}\left[(m+3)\frac{r^2}{a^2}-(m+1)\right]$ $y = -\frac{3W(m^2-1)}{16\pi Em^2t^3}\left[\frac{(a^2-r^2)^2}{a^2}\right]$ (At edge) Max $s_t = \frac{3W}{4\pi t^2}$ $s_t = \frac{3W}{4\pi mt^2}$ (At center) $s_r = s_t = -\frac{3W(m+1)}{8\pi mt^2}$ Max $y = -\frac{3W(m^2-1)a^2}{16\pi Em^2t^3}$

Edges fixed

Central couple (trunnion loading)

(At $r = r_0$) $\text{Max } s_r = \dfrac{3M}{4\pi t^2 r_0}\left[1 + \left(\dfrac{m+1}{m}\right)\log \dfrac{2(0.45a - r_0)}{0.45ka}\right]$

$$\text{where } k = \frac{0.1a^2}{(r_0 + 0.28a)^2}$$

Circular, with concentric circular hole (circular flange)

a
b

(B)

Outer edge supported

Uniform load over entire actual surface

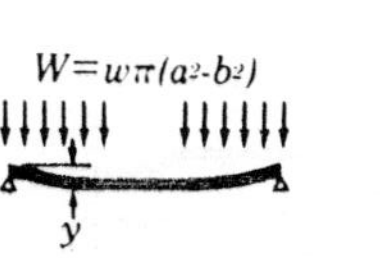

(At inner edge) $\text{Max } s = s_t = -\dfrac{3W}{4mt^2(a^2-b^2)}\left[a^4(3m+1) + b^4(m-1) - 4ma^2b^2 - 4(m+1)a^2b^2 \log \dfrac{a}{b}\right]$

when b is very small, $\text{Max } s = s_t = -\dfrac{3wa^2(3m+1)}{4mt^2}$

$$\text{Max } y = -\frac{3w(m^2-1)}{2m^2Et^3}\left[\frac{a^4(5m+1)}{8(m+1)} + \frac{b^4(7m+3)}{8(m+1)} - \frac{a^2b^2(3m+1)}{2(m+1)} + \frac{a^2b^2(3m+1)}{2(m-1)}\log\frac{a}{b} - \frac{2a^2b^4(m+1)}{(a^2-b^2)(m-1)}\left(\log\frac{a}{b}\right)^2\right]$$

Outer edge supported

Uniform load along inner edge

(At inner edge) $\text{Max } s = s_t = -\dfrac{3W}{2\pi mt^2}\left[\dfrac{2a^2(m+1)}{a^2-b^2}\log\dfrac{a}{b} + (m-1)\right]$

$$\text{Max } y = -\frac{3W(m^2-1)}{4\pi Em^2t^3}\left[\frac{(a^2-b^2)(3m+1)}{(m+1)} + \frac{4a^2b^2(m+1)}{(m-1)(a^2-b^2)}\left(\log\frac{a}{b}\right)^2\right]$$

Table 10.7 (*Continued*)

Manner of Loading (Case No.)	Formulas for Stress and Deflection
Outer edge fixed and supported Uniform load over entire actual surface $W = w\pi(a^2-b^2)$	(At outer edge) $\text{Max } s_r = \frac{3w}{4t^2}\left[a^2 - 2b^2 + \frac{b^4(m-1) - 4b^4(m+1)\log\frac{a}{b} + a^2b^2(m+1)}{a^2(m-1)+b^2(m+1)}\right] = \text{Max } s$ (At inner edge) $\text{Max } s_t = -\frac{3w(m^2-1)}{4mt^2}\left[\frac{a^4 - b^4 - 4a^2b^2\log\frac{a}{b}}{a^2(m-1)+b^2(m+1)}\right]$ $\text{Max } y = -\frac{3w(m^2-1)}{16m^2Et^3} \times \left\{a^4 + 5b^4 - 6a^2b^2 + 8b^4\log\frac{a}{b} - \frac{[8b^6(m+1) - 4a^2b^4(3m+1) - 4a^4b^2(m+1)]\log\frac{a}{b} + 16a^2b^4(m+1)\left(\log\frac{a}{b}\right)^2 - 4a^2b^4 + 2a^4b^2(m+1) - 2b^6(m-1)}{a^2(m-1)+b^2(m+1)}\right\}$
Outer edge fixed and supported Uniform load along inner edge W	(At outer edge) $\text{Max } s_r = \frac{3w}{2\pi t^2}\left[1 - \frac{2mb^2 - 2b^2(m+1)\log\frac{a}{b}}{a^2(m-1)+b^2(m+1)}\right] = \text{Max } s \text{ when } \frac{a}{b} < 2.4$ (At inner edge) $\text{Max } s_t = \frac{3w}{2\pi mt^2}\left[1 + \frac{ma^2(m-1) - mb^2(m+1) - 2(m^2-1)a^2\log\frac{a}{b}}{a^2(m-1)+b^2(m+1)}\right] \text{Max } s \text{ when } \frac{a}{b} > 2.4$ $\text{Max } y = -\frac{3w(m^2-1)}{4\pi m^2Et^3}\left\{a^2 - b^2 + \frac{2mb^2(a^2-b^2) - 8ma^2b^2\log\frac{a}{b} + 4a^2b^2(m+1)(\log\frac{a}{b})^2}{a^2(m-1)+b^2(m+1)}\right\}$

Inner edge fixed and supported Uniform load along outer edge

(At inner edge) Max $s_r = \frac{3w}{2\pi t^2}\left[\frac{2a^2(m+1)\log\frac{a}{b} + a^2(m-1) - b^2(m-1)}{a^2(m+1)+b^2(m-1)}\right]$

(At outer edge) Max $y = \frac{3w(m^2-1)}{4m^2\pi E t^3}\left[\frac{a^4(3m+1) - b^4(m-1) - 2a^2b^2(m+1) - 8ma^2b^2\log\frac{a}{b} - 4a^2b^2(m+1)(\log\frac{a}{b})^2}{a^2(m+1)+b^2(m-1)}\right]$

Concentrated load applied at outer edge

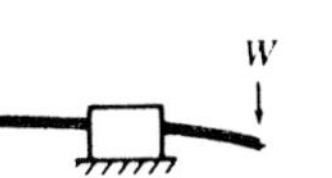

(At inner edge) Max $s_r = \beta\frac{W}{t^2}$ where β may be found following table:

$\frac{a}{b}$	1.25	1.50	2	3	4	5
β	3.7	4.25	5.2	6.7	7.9	8.8

(values for $\nu = 0.3$)

Square, solid

Edge supported above and below (Corners held down) Uniform load over entire surface

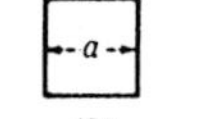

(D)

(At center) $s_a = -\frac{0.2208wa^2(m+1)}{mt^2} = \text{Max } s \quad \text{Max } y = \frac{0.0487wa^4(m^2-1)}{m^2Et^2}$

(Formulas due to Prescott [1])

Table 10.7 (*Continued*)

Manner of Loading (Case No.)	Formulas for Stress and Deflection
All edges fixed Uniform load over entire surface	(At center of each edge) $s_a = \dfrac{0.308wa^2}{t^2} = \text{Max } s$ (At center) $s = -\dfrac{6w(m+1)a^2}{47mt^2}$ (Ref. 5). Max $y = -\dfrac{0.0138wa^4}{Et^2}$ (other formulas based on coefficients given by Timoshenko [2] $v = 0.3$)
	Equiltateral triangle, solid
Edges supported Distributed load of intensity w over entire surface	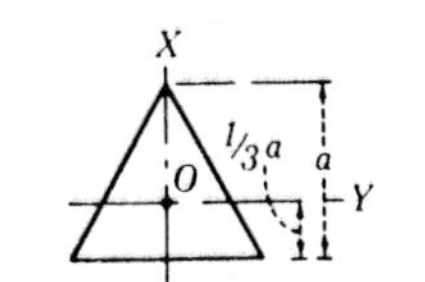 Max $s_x = 0.1488\dfrac{wa^2}{t^2}$ at $y = 0$, $x = -0.062a$ Max $s_y = 0.1554\dfrac{wa^2}{t^2}$ at $y = 0, x = 0.129a$ (Values for $v = 0.3$) Max $y = \dfrac{wa^4(m^2-1)}{81Et^3m^2}$ at 0.
	Circular sector, solid
Edges supported Distributed load of intensity w over entire surface	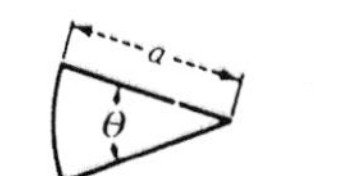 Max $s_r = \beta\dfrac{wa^2}{t^2}$ Max $s_1 = \beta_1\dfrac{wa^2}{t^2}$ Max $y = \alpha\dfrac{wa^4}{Et^2}$

Solid semicircular plate

Uniform load w, all edges fixed

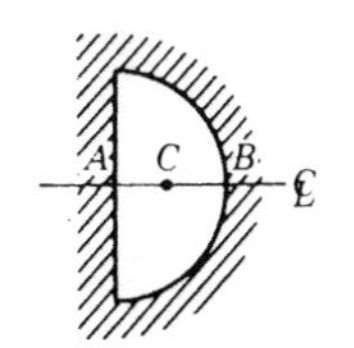

(values for $v = 0.3$)

θ	45°	60°	90°	180°
β	0.102	0.147	0.240	0.522
β_1	0.114	0.155	0.216	0.312
α	0.0054	0.0105	0.0250	0.0870

$$\text{Max } s = s_r \text{ at } A = \frac{0.42wa^2}{t^2}$$

$$s_r \text{ at } B = \frac{0.36wa^2}{t^2}$$

$$\text{Max } s_t = \frac{0.21wa^2}{t^2} \text{ at } C$$

Notation: W = total applied load (lb.); w = unit applied load (lb. per sq. in.); t = thickness of plate (in.); s = unit stress at surface of plate (lb. per sq. in.); y = vertical deflection of plate from original position (in.); θ = slope of plate measured from horizontal (rad); E = modulus of elasticity; m = reciprocal of v, Poisson's ratio. q denotes any given point on the surface of plate; r denotes the distance of q from the center of a circular plate. Other dimensions and corresponding symbols are indicated on figures. Positive sign for s indicates tension at upper surface and equal compression at lower surface; negative sign indicates reverse condition. Positive sign for y indicates, upward deflection negative sign downward deflection. Subscripts r, t, a, and b used with s denote respectively radial direction, tangential direction, direction of dimension a, and direction of dimension b. All dimensions are in inches. All logarithms are to the base e ($\log_e x = 2.3026(\log_{10} x)$.

Source: [1] Prescott, J., "Applied Elasticity," Longmans, Green & Company, 1924. [2] Timoshenko, S., and J. M. Lessells: " Applied Elasticity," Westinghouse Technical Night School Press, 1925.

Using a material with higher physical properties can maintain the same plate thickness, as the stress increases, but may not justify the increase in weight and material cost. When the design requires a stronger plate, consider a ribbed plate to reduce the section thickness of the original material. The internal ribs will increase the plate's stiffness, that is, the moment of inertia, and will increase the section strength and rigidity with minimal material usage when designed correctly.

Ribbed Plate Design

The use of internal ribs for reinforcing a plastic product surface must be carefully analyzed from two viewpoints. If the external aesthetics or show surface of the product must be smooth without any noticeable surface blemishes or warping, the ribs must be carefully designed to eliminate any postmolding surface problems. If the ribs are not correctly designed, the resulting surface effects can be sink marks and/or an outline of the ribbing pattern visible to the user. These effects can be minimized with good rib design as shown in Figure 10.11 or disguised with surface treatments in various ways. A textured viewing surface is one method often used when permitted. With highly glossy viewing surface, it is much harder to hide the effects of internal ribs, but with good rib design, it is possible. The effects on a highly polished surface with

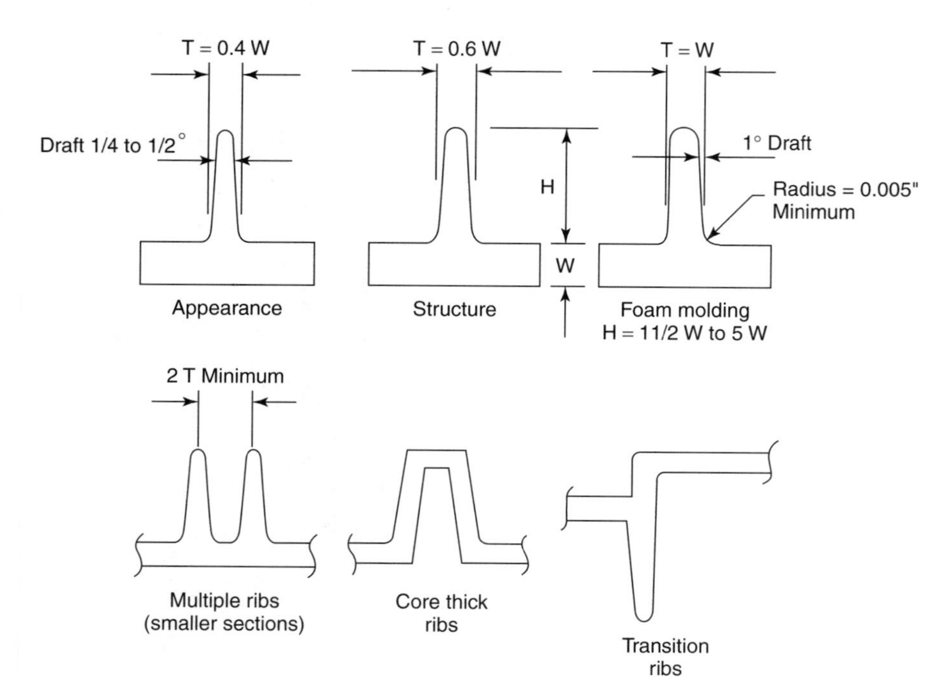

Figure 10.11 Rib designs.

Table 10.8 Recommended wall thickness for thermoplastic molding materials

Thermoplastic Materials	Minimum (in.)	Maximum (in.)
ABS	0.045	0.140
Acetal	0.015	0.125
Acrylic	0.025	0.150
Cellulosics	0.025	0.187
FEP fluoroplastic	0.010	0.500
Long-strand-reinforced resin	0.075	1.000
LCP liquid crystal polymer	0.008	0.120
Nylon	0.010	0.125
Polyarylate	0.045	0.160
Polycarbonate	0.040	0.375
Polyester	0.025	0.125
Polyethylene (LD)	0.020	0.250
Polyethlene (HD)	0.035	0.250
Ethylene vinyl acetate	0.020	0.125
Polypropylene	0.025	0.300
Nory[a] (modified PPO)	0.030	0.375
Polystyrene	0.030	0.250
PVC (rigid)	0.040	0.375
Polyurethane	0.025	1.500
Surlyn	0.025	0.750

[a]Registered General Electric material name.

Source: Adapted from Ref. 7.

well-designed ribbing can be minimized when the mold temperature control and molding variables are tightly controlled. Crystalline and reinforced resins have a greater problem eliminating rib surface effects than do amorphous and thermoset resins. This is due to their higher mold shrinkage. The appearance and warpage problem is less for the lower-shrinkage amorphous resins and thermosets Ribs structurally increase rigidity and load-bearing strength. Their use can dramatically reduce product weight, resulting in thinner plate sections and shortened molding cycle time. Consider ribs when section thickness exceeds the material's recommended wall thickness as shown in Table 10.8, or when the product's deflection must be eliminated. End-use product testing will show if the rib design is adequate. It is initially best to design with small ribs as they can always be increased in size. This is relatively easy as metal removal in the mold is involve only in increasing the ribbing.

A rib's size, refering to Figure 10.11, is based on the product's wall thickness (W), where the rib's thickness (T) is a percentage of W. Single rib height is a function of structural requirements, with $1.5W$ to $5W$ typically maximum. Examples of correct rib proportions are shown in Figure 10.12.

There are many forms of ribbing; the majority are unidirectional, with cross-ribbing (boxing). Rib proportions are determined by the structural requirements of

2" × 1/4" Beam-Effect of 1/8" thick rib on proper

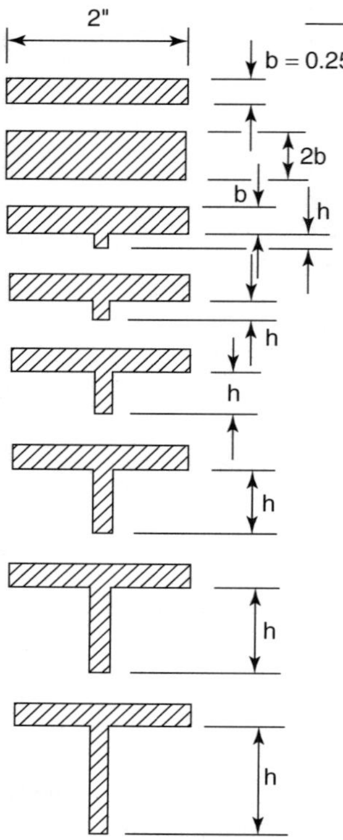

Section	Rib Size (h)	h/b	Percent Increase weight	Percent Increase stiffness	Percent Stiffness weight
2"	b = 0.25"	—	—	—	—
	2b	—	+100	+ 700	7.0
	h	1/2	+3.12	+ 23	7.3
		1/1	+6.25	+ 77	12.3
	h	2/1	+12.5	+ 349	28.0
	h	3/1	+19.0	+ 925	49.3
	h	4/1	+25.0	+1901	76.0
	h	5/1	+31.0	+3352	107.2

Figure 10.12 Rib effects on a beam. (Adapted from Ref. 9.)

the product. But if the ribs are too large, then molding and aesthetic requirements become more important. It is best to use several smaller well-proportioned ribs rather than one large rib to satisfy both structural and aesthetic requirements.

Unidirectional Ribbing

Single or multiple unidirectional ribs affect a section's structural efficiency by increasing load-bearing strength while maintaining minimal wall thickness and product weight. The effective use of a rib to increase the original plate's strength is illustrated in Figure 10.12. Doubling the section thickness increases the product weight. But using a rib height to thickness ratio of 3:1 with only a 19% weight increase yields greater stiffness and strength with a stiffness percentage increase of 925% versus 700% for an increase in stiffness.

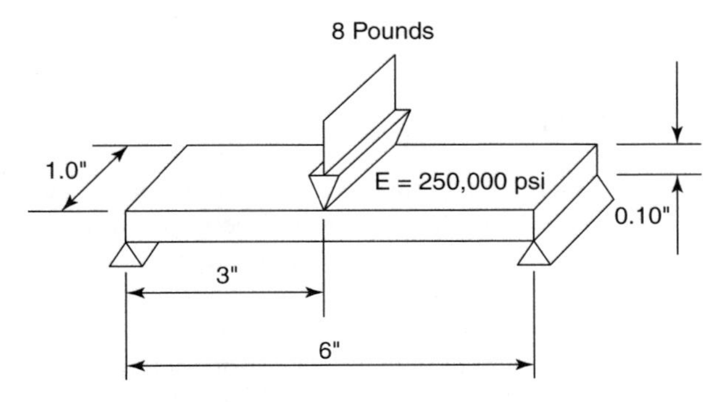

Figure 10.13 Beam loading.

Example 10.1. Flat-Plate Design Design a machine housing cover at room temperature with a concentrated load of 8 lb. For analysis, represent it as a uniform beam segment cut from the product with a constant wall thickness of 0.100 in. as shown in Figure 10.13. The maximum stress, using standard beam equations, is calculated at 7207 psi, and the maximum deflection, assuming a flexural modulus at room temperature of 250,000 psi, is 1.728 in. For the material selected, deflection and stress are unacceptable. To increase rigidity, reduce deflection and stress and a select single rib section. The rib's dimensions are initially selected at 0.04 in. wide by 0.50 inc. high, as the opposite side is a show surface. A draft of 0.5° on the sides of the rib is selected for easy release from the mold and to minimize material as shown in Table 10.9. Using the new section dimensions to calculate the new product stress and deflection, stress is decreased to 3000 psi and deflection reduced to 0.072 in. This is acceptable

Table 10.9 Effects of rib and cross-sectional changes on a part

Section	Section Geometry (Values in Inches)	Cross Section (Area)	Maximum Stress (psi)	Maximum Deflection (in.)
A	1.0" × 0.10"	0.100	7207	1.728
B	0.040" rib, 0.50"	0.120	3000	0.072
C	1.0" × 0.287"	0.287	871	0.0727

with increases in part weight of only 25%. If increasing the plate thickness, without a rib, were considered, a plate thickness of 0.287 in. would be required with an increase of 240 % in product weight.

If temperature were increased in this example, the resin modulus E would be reduced and a thicker section or higher rib would be required to meet the new requirements. The designer could also consider using several smaller ribs to solve the increase temperature requirement. There are always other alternatives to increase the product strength and stiffness when conditions require improvements.

When section structural strength is the main consideration, rib thickness can be increased by up to 75% or 100% of wall thickness. Rib sections approaching material thickness should be used sparingly as they can create product and manufacturing

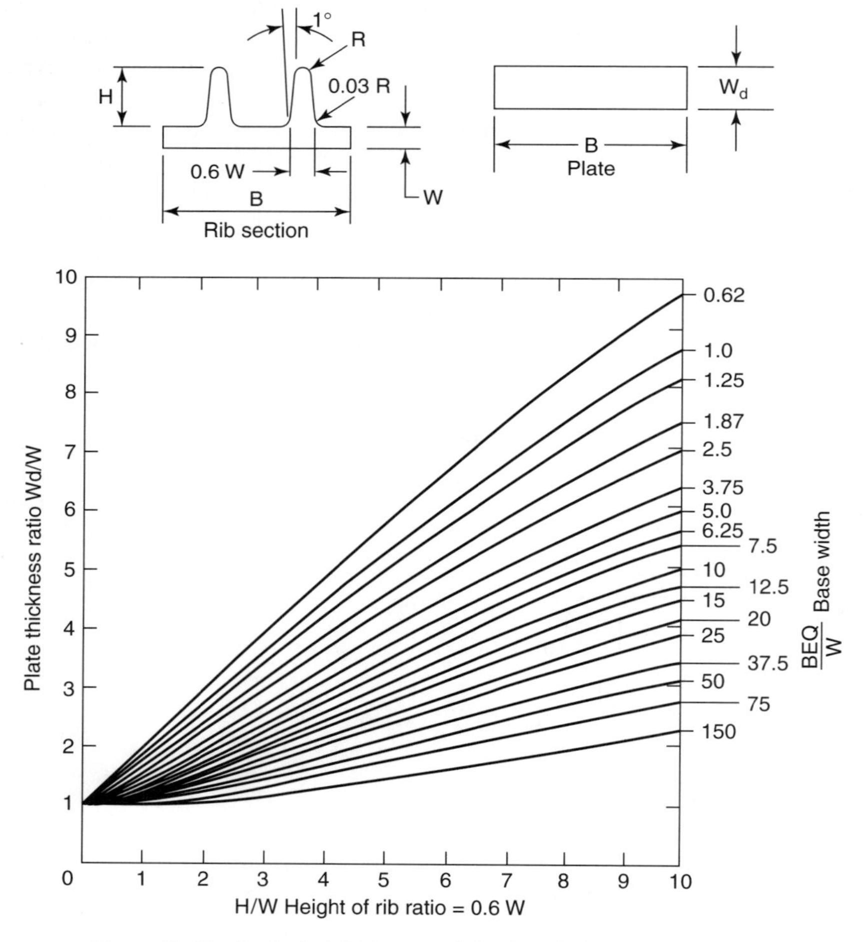

Figure 10.14 Equivalent thickness—deflection. (Adapted from Ref. 10.)

problems such as voids, warping, and longer molding cycle times. Ribs should be tailored to meet stress requirements in the product and should be designed to match the stress distribution in each section. Ribs should have a minimum of 0.25–0.50° of draft for ease of removal from the mold. A minimum radius of 0.01 in. is recommended at the rib base with no sharp corners. This reduces the stress concentrations at these critical load transfer points. When a single rib becomes too big or thick or aesthetically unacceptable, smaller, evenly spaced ribbing is used.

Multiribbing Considerations

To assist the designer with multirib plate design, dimensionless ratio curves were developed. These curves compare flat-plate geometry to unidirectional multiribbed structures of equal stiffness and strength. These ribbed design plate curves are shown in Figures 10.14–10.17 for plates with a rib thickness of 60% and 75% of the plate thickness. A flat plate is analyzed by dividing it into small equal sections, and the

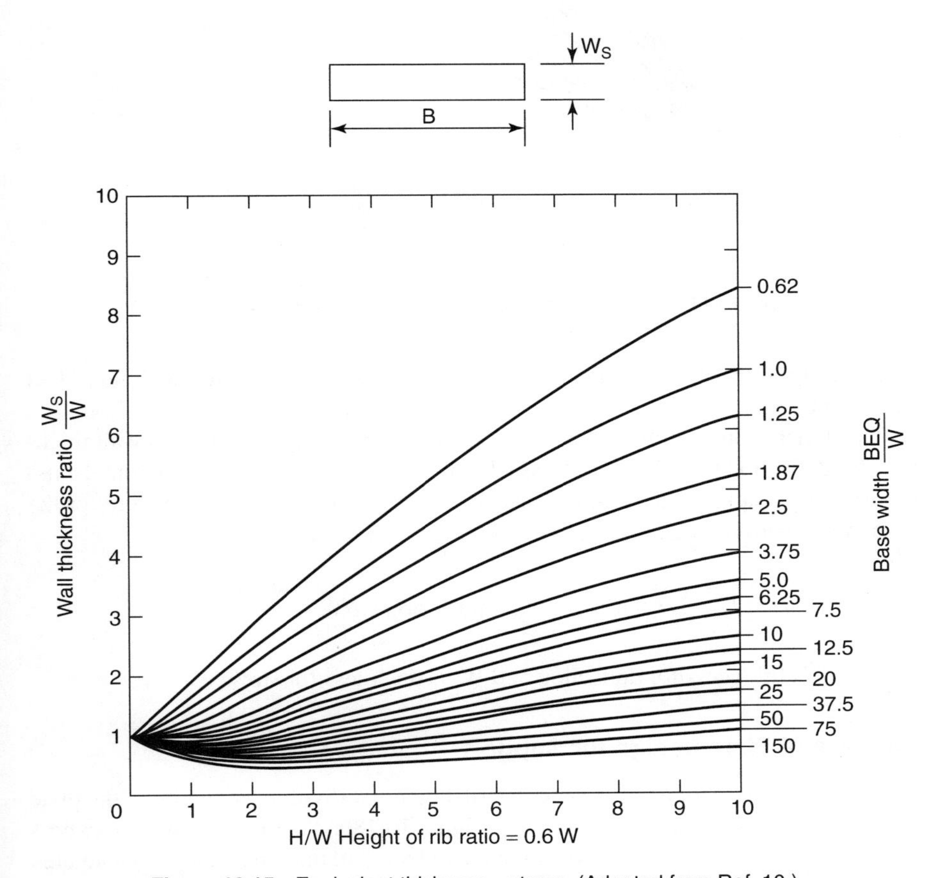

Figure 10.15 Equivalent thickness—stress. (Adapted from Ref. 10.)

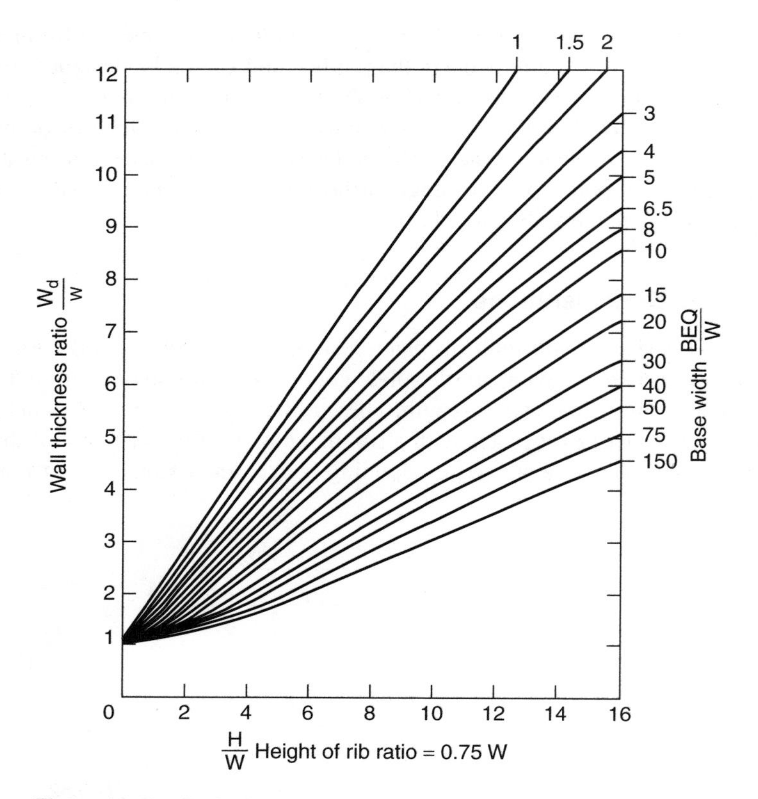

Figure 10.16 Equivalent thickness—deflection. (Adapted from Ref. 10.)

moment of inertia of the smaller section is calculated and compared with its ribbed equivalent.

The dimensionless ratio curves were developed in terms of the plate's wall thickness for deflection (W_d/W) and stress (W_s/W), with the abscissa (X, horizontal axis) representing a ratio of rib height to plate thickness (H/W). The nomenclature for the rib and plate cross sections is shown in Figure 10.18 with

$$t = T - 2H \tan \propto; \qquad A(\text{area})\text{ rib} = BW + \frac{H(T+t)}{2}$$

where W_d is thickness for deflection and W_s is thickness for stress.

Analysis of Plate Design for Support

The design analysis for a flat plate with a rib begins by dividing the width of the plate into small equal sections as illustrated in Figure 10.18b. An estimated wall thickness is chosen (W) (Fig. 10.18b), and the moment of inertia of this single plate is calculated and then compared with a ribbed equivalent plate (Fig. 10.18a).

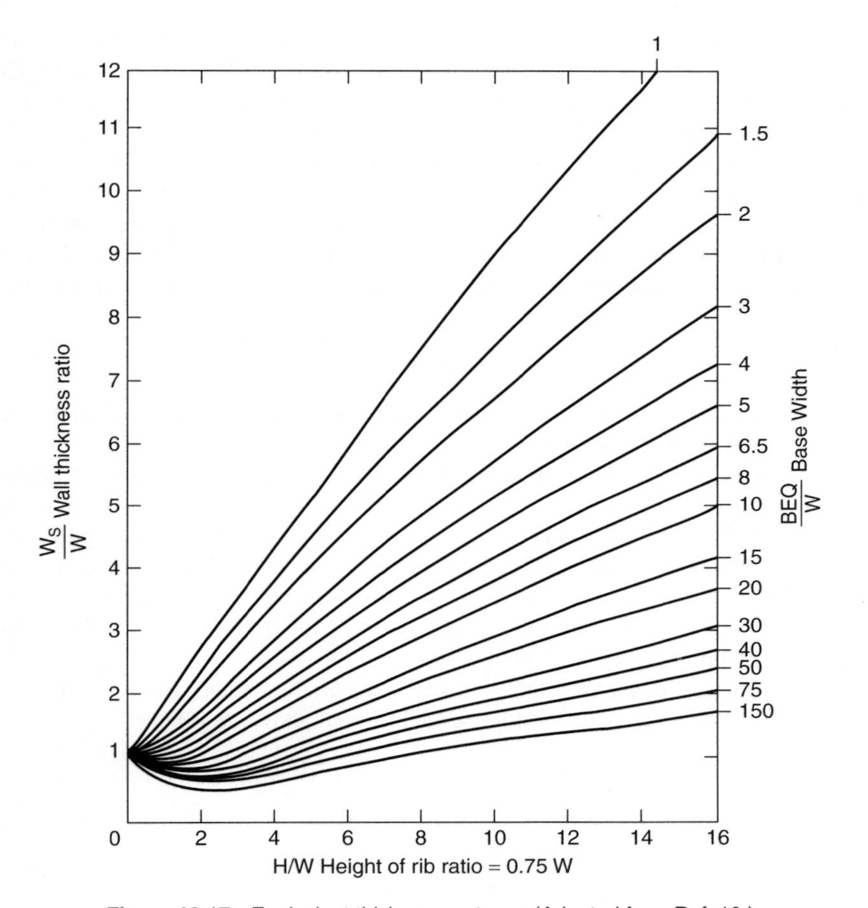

Figure 10.17 Equivalent thickness—stress. (Adapted from Ref. 10.)

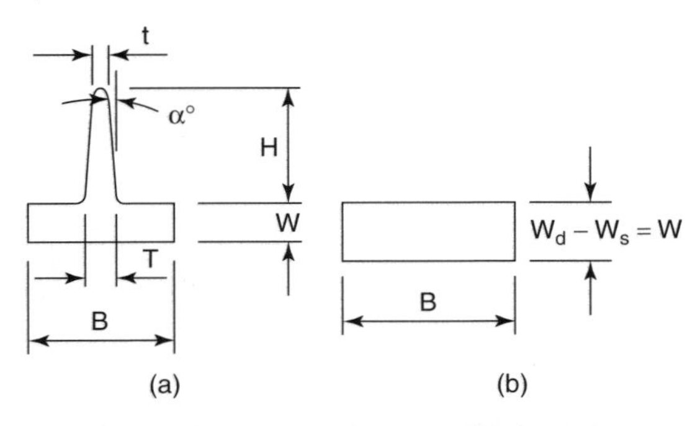

Figure 10.18 Housing cover—Example 10.2.

The deflection and stress curves were developed using a single rib section of width B and an equal width (B) flat plate. The curves developed are plotting ratios of the plate thickness versus the ribbed plate, rib height, and plate thickness, with the resulting curves equal to the total width (B) of the plate under study, in Figure 10.14 and 10.15, using their moment of inertia equations divided by the number of ribs, then divided by the ribbed plate thickness. These curves were calculated on the section's plate thickness for deflection (W_d/W) or plate thickness for stress (W_s/W) curves plotted on the Y axis (ordinate) against the ratio of rib height to plate thickness (H/W) on the X axis.

To define a single or smaller section of the whole structure, the term BEQ is used:

$$BEQ = \frac{\text{total width of section}(B)}{\text{number of ribs}(n)} = \frac{B}{n}$$

These curves simplify plate design, in contrast to considerable trial and error using deflection and stress calculations to determine rib height.

Example 10.2 will illustrate the simplicity these curves have for the design analysis.

Example 10.2 An aluminum enclosure is used to support auxiliary equipment that is being replaced with plastic. It is attached (fixed) at one end and is loaded on the top surface with a uniform 40 lb over its entire surface. The selected material for replacement is a molded modified polyphenylene oxide (PPO) or Noryl®resin.* The design is at room temperature conditions.

The designer is to calculate the enclosure's material thickness and if too thick, an equivalent ribbed section is used for the enclosure as shown in Figure 10.19:

$$\text{Aluminium flex modulus } E_A = 10.3 \times 10^6 \text{psi}$$

$$\text{PPO flex modulus } E_p = 3.6 \times 10^5 \text{ psi at room temperature}$$

$$\text{Thickness PPO wall} = W_d \text{ (unknown)}$$

$$\text{Thickness aluminum wall } W_A = 0.15\text{in.}$$

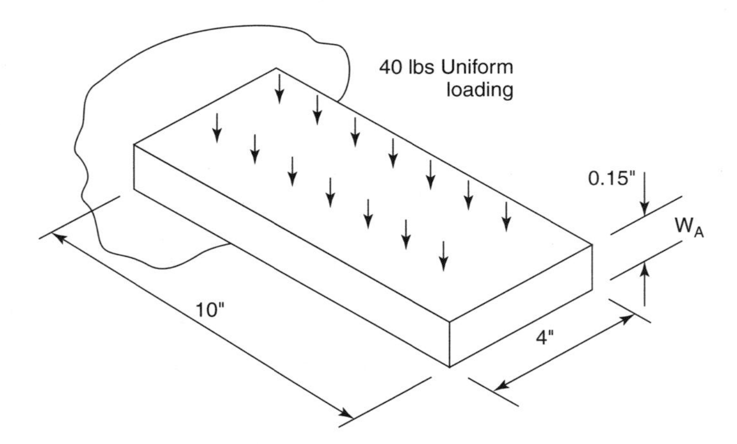

Figure 10.19 Housing cover—Example 10.2.

*Noryl is a registered General Electric trade name.

The plate thickness for the PPO cover with equivalent stiffness is determined by equating the product of the modulus and the cube of the plate thickness for each material:

$$E_A \times W_A^3 = E_p W_p^3$$
$$(10.3 \times 10^6) \times (0.15)^3 = 3.6 \times 10^5 W_d^3$$
$$W_d^3 = 28.6 \times 0.003375$$
$$W_d = {}^3\sqrt{0.0965}$$
$$W_d = 0.453 \text{ in. plate thickness in PPO}$$

This is too thick a plate for material utilization and a commercial molding cycle. Therefore, consider a ribbed plate to reduce the wall thickness. Using the aluminum enclosure cover thickness of 0.15 in., divide the cover width (an assumption) into 10 (n) equally spaced rib sections to begin the analysis. Calculation of the values required for use with deflection curve yields:

$$\frac{W_d}{W} = \frac{0.453}{0.150} = 3.02 \qquad \text{Base} = 4 \quad \text{in.}; n = \text{number of ribs} = 10$$
$$BEQ = \frac{B}{n} = \frac{4}{10} = 0.40 \qquad \frac{BEQ}{W} = \frac{0.40}{0.150} = 2.66$$

Using the rib thickness equal to $0.6W$ of wall thickness deflection curve in Figure 10.14. Height (H)/width (W) is H/W = 3.25, where $H = 3.25 \times 0.15 = 0.49$. Use a 0.50-in.-high rib.

From the stress graph in Figure 10.15, *H/W* is equal to 3.25 and *BEQ/W* equal to 2.66:

$$\frac{W_s}{W} = 2.2; \qquad W_s = 2.2 \times 0.150 \text{ in.} = 0.33 \text{ in.}$$

Now, calculate the moment of inertia and section modulus for the ribbed area plate:

$$I = \frac{B(W_d^3)^3}{12} = \frac{4 \times (0.453)^3}{12} = 0.031 \text{ in.}^4$$
$$Z = \frac{B(W_s^2)}{6} = \frac{4 \times (0.33)^2}{6} = 0.073 \text{ in.}^3$$

Maximum deflection (Y) at the free end is

$$Y_{max} = \frac{FL^3}{8EI} = \frac{40 \times 10^3}{8.36 \times 10^5 \times 0.031} = 0.45 \text{ in.}$$

Maximum stress (σ) at the fixed end is:

$$\sigma_{\max} = FL/2Z = 40 \times 10^3/2 \times 0.073 = 2739 \text{ psi}$$

The Noryl (high-glass-transition-temperature material) has an average maximum tensile strength of 8000 psi; therefore, a safety factor of almost 3 is realized. The designer should compare the stress loading level with the material's creep curves as the part is under constant load to ensure that it does not fail by creep rupture. However, the cover is not a true cantilever section and is supported at the free end.

Example 10.3 The deflection and stress curves can also be used to analyze an existing structure for material substitutions once the maximum stress and deflection are calculated for the structure. Then, if end-use conditions such as temperature, humidity, and loading, should vary, the respective stress–strain curves for each material can be compared to see if the material will perform as required. Assume that the following structure is to be analyzed as shown in Figure 10.20. Using the data from the beam and loading conditions for a 30% glass-reinforced polyester (PET) resin with a flexural modulus (E) of 1,500,000 psi at room temperature, we obtain

$$BEQ = B/n = \frac{3}{5} = 0.6$$

$$\frac{BEQ}{W} = \frac{0.6}{0.2} = 3$$

$$H = 0.75 - 0.2 = 0.55$$

$$\frac{H}{W} = \frac{0.55}{0.2} = 2.75$$

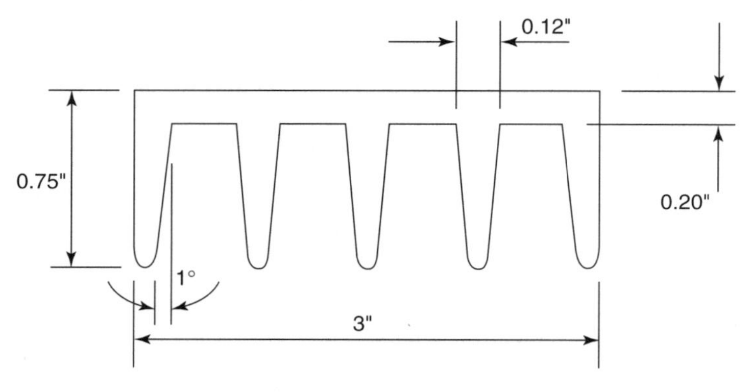

Figure 10.20 Channel ribs.

Using the deflection and stress curves, $W_d + W_s$ are:

$$W_d/W = 2.7; \qquad W_d = 2.7 \times 0.2 = 0.54$$

$$W_s/W = 1.8; \qquad W_s = 1.82 \times 0.2 = 0.36$$

$$I = \frac{B(W_d^3)}{12} = \frac{3 \times (0.54)^3}{12} = 0.039 \text{ in.}^4$$

$$Z = \frac{B(W_s^2)}{6} = \frac{3 \times (0.36)^2}{6} = 0.065 \text{ in.}^3$$

Note: $(F = W)$ continuous loading on beam from the equations

$$Y_{max} = \frac{5FL^3}{384EI} = \frac{5 \times 100 \times (18)^3}{384 \times 1.5 \times 10^6 \times 0.039}$$

$$= 0.13 \text{ in.}$$

$$\sigma_{max} = \frac{FL}{8Z} = \frac{100 \times 18}{8 \times 0.065} = 3461 \text{ psi}$$

The resin tensile strength is 19,500 psi, which yields a safety factor of 5. This section is overdesigned for the application, and a lower reinforcement (try 15% glass reinforcement) loading may be selected or the structure modified by decreasing the number or ribs or adjusting their geometry. A few trial calculations will determine the best way to proceed. Do not forget temperature increases or other environmental considerations that may affect the final product.

Cross-Ribbing

With the versatility that plastic materials offer, equipment housing can be designed to support and perform more functions. The housing can be designed with higher rigidity to support additional forces on their external and internal sections. External brackets can be added to make them self-standing and supporting (standalone structures). Internally they support heavy power supplies and other electronic equipment, which emit considerable heat even when cooled with internal fans during operation. When unsupported product sections need stiffening, the design structure in Figure 10.21 becomes practical, and easily manufactured by injection molding.

A section's rigidity is directly proportional to the moment of inertia of the cross section. On the basis of this relationship, a graph was developed (Fig. 10.22) to simplify the analysis of using a cross-ribbed section for a product. Using the relationship between flat plates and cross-ribbed plates, ribs in the plane of the plate, and running side to side, the graph uses the identical moment of inertia equation to describe their dimensional relationship. The graph was developed using unity, or one, where L designates as the section length and width. The baseline of the graph shows values from 0 to 0.2 for the product of the nonribbed wall thickness t_A. The number of cross-ribs per inch is designated n, which is divided by the width of the plate L, which is one or

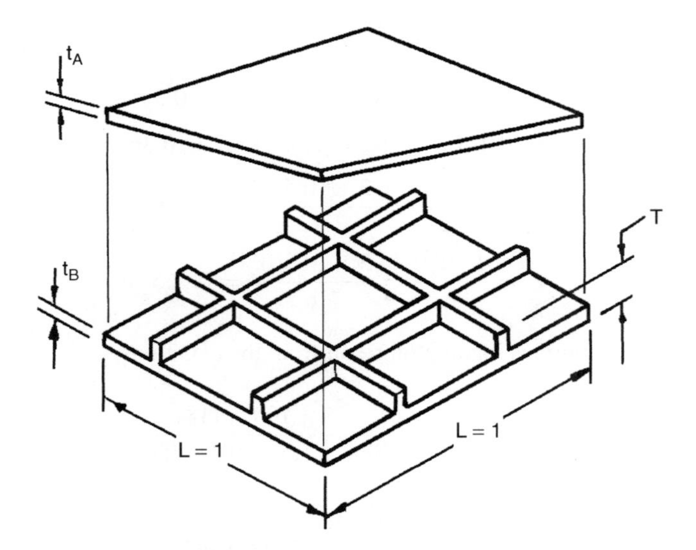

Figure 10.21 Ribbed wall design.

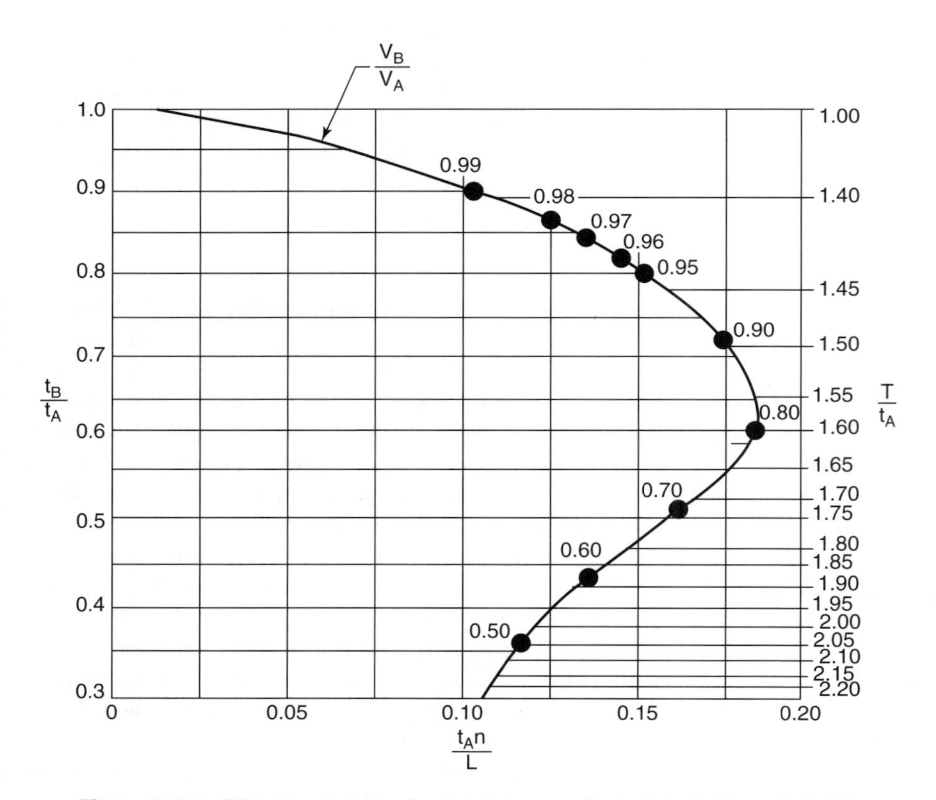

Figure 10.22 Ribbed wall design (material conservation). (Adapted from Ref. 10.)

unity. In the graph, the rib thickness is equal to the plate thickness t_B. As explained in Example 10.3, this is not desirable. Therefore, a thinner rib can be calculated to reduce thickness and material weight while increasing strength. This is done using the dimensions and spacing from the graph in Figure 10.22.

The left-hand ordinate (t_B/t_A) is the value for the ratio of the ribbed wall thickness (t_B) to the nonribbed wall thickness (t_A). The right-hand ordinate gives the values for the ratio of the overall thickness of the ribbed part (T) to the nonribbed wall thickness (t_A). The curve on the graph is the ratio of the volume of the cross-ribbed plate (V_B) to the volume of the new to be designed plate (V_A). The new plate (V_A) will have cross-ribs added to the section to reduce cross-ribbed plate (V_B) thickness. The volume ratios, (V_B/V_A) will equate the minimum volume of material required to provide the same structural stiffness as the original cross-ribbed plate (V_B).

The following examples explain how this is accomplished.

Example 10.4 As shown with the comparison of a typical flat plate to a flat plate with a longitudinal rib design, plate thickness can be reduced while maintaining the same or better strength properties and reducing part weight. The objective involves reducing the sections thickness (t_B). Product material utilization (reducing material weight and molding considerations for the material selected) tends to favor a wall thickness (t_B) of ≤ 0.25 in. versus the plate thickness (t_A) of 0.55 in. as shown in Figure 10.23. Known: $t_A = 0.55$ in. Required wall thickness $t_B = 0.25$ in:

$$\frac{t_B}{t_A} = \frac{0.25}{0.55} = 0.45$$

From the curve in Figure 10.22, we obtain

$$\frac{T}{t_A} = 1.84 \qquad \text{or} \qquad T = (1.84)(0.55) = 1.012 \text{ in.}$$

From the intersection on the curve, $V_B/V_A = 0.62$ and

$$(t_A)\frac{n}{W} = 0.14. \qquad \text{or} \qquad n = \frac{(0.14)(1)}{0.55} = .25 \text{ ribs/in.}$$

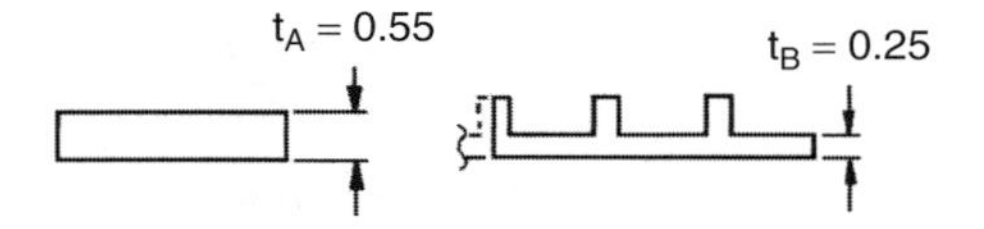

Figure 10.23 Ribbed wall design.

Analysis:

$$\text{Wall thickness} = 0.25\text{ in.}$$

$$\text{Rib height} = T - t_B = 1.012 - 0.25 = 0.76\text{ in.}$$

$$\text{Rib spacing} = 0.25\text{ ribs per inch or 1 rib every 4 in.}$$

$$\text{Material savings} = 38\%\text{ over original plate.}$$

Example 10.5 Reduce part weight with cross-ribbed design for a plate with thickness

$$t_A = 0.180\text{ in.}$$

Required: material reduction of 35%.

$$\frac{V_B}{V_A} = 0.65\text{ from the curve Figure 10.22.}$$

$$\frac{(t_A)n}{W} = 0.145 \quad\text{or}\quad n = 0.145 \times 1/0.180 = 0.80\text{, rounded}$$

to one rib every 1.25 in.

$$\frac{t_B}{t_A} = 0.475 \quad\text{or}\quad t_B = 0.475 \times 0.180 = 0.0855\text{-in.-thick wall}$$

$$\frac{T}{t_A} = 1.79 \quad\text{or}\quad T = 1.79 \times 0.180 = 0.322\text{ in.}$$

$$\text{Rib height} = T - t_B = 0.322 - 0.085 = 0.237\text{ in.}$$

$$\text{Wall thickness} = 0.085\text{ in.}$$

$$\text{Rib height} = 0.237\text{ in.}$$

$$\text{Rib spacing} = 1\text{ rib on 1.25 in. centers}$$

$$\text{Material savings} = 35\%$$

Example 10.6 Other cases may occur as a result of packaging requirements as limiting rib height dimensions. Required: Reduce part weight with rib height (T) restricted to 0.400 in. Known: wall thickness (t_A) is 0.275 in.:

$$\frac{T}{t_A} = \frac{0.400}{0.275} = 1.45\text{ from curve in Figure 10.22}$$

$$\frac{(t_A)n}{W} = 0.160 \quad\text{or}\quad n = \frac{0.160 \times 1}{0.275} = 0.58\text{ rib/in.}$$

$$\frac{t_B}{t_A} = 0.78 \quad\text{or}\quad t_B = (0.78)(0.275) = 0.214\text{ in.}$$

$$\frac{V_B}{V_A} = 0.94\text{ or }1 - 0.94 = 0.06\text{, a 6\% savings in material}$$

There is a small material reduction of only 6%, but if the part volume is large, this equates to a considerable material savings. The rib spacing will be on 1.2-in. centers. If thinner ribs are required for appearance or functional reasons, the same strength can be obtained by holding the product of the number of ribs and the rib thickness constant. If the rib thickness selected was 50% of (t_B) or 0.107 in., the number of ribs must be increased and on 0.60-in. centers.

Example 10.7 When the number of ribs is limited because of product shape, cutouts, or locations of internal components, or must match rib spacing with a adjoining part or decorative element, the designer may still be able to obtain a volume reduction by specifying rib spacing. Known: present wall thickness (t_A) = 0.065 in. Required: rib spacing per inch (n) = 2. With a base (L) of unity

$$\frac{(t_A)(n)}{W} = \frac{(0.065)(2)}{1} = 0.13$$

From the curve in Figure 10.22, we obtain

$$\frac{t_B}{t_A} = 0.43 \qquad \text{or} \qquad t_B = (0.43)(0.065) = 0.028 \text{ in.}$$

$$\frac{T}{t_A} = 1.90 \qquad \text{or} \qquad T = (1.90)(0.065) = 0.124 \text{ in.}$$

$$\frac{V_B}{V_A} = 0.58 = 1 - 0.58 = 42\% \text{ material savings}$$

The new design is

$$\text{Wall thickness } t_B = 0.028 \text{ in.}$$

$$\text{Rib height}(T - t_B) = 0.096 \text{ in.}$$

$$\text{Rib spacing} = 2 \text{ per inch (2 in.)}$$

(*Note:* The reduced wall thickness may cause a molding problem. If the material selected does not have good flow characteristics and the flow length is long, the section may not fill. The designer must know the material's minimum product or wall thickness:flow length ratio, presented in Table 10.8.)

Similar flat-plate studies have been conducted using round gussets at the corners of the ribs plus multiribbed patterns. These plates were tested and ranked against a single flat plate of equivalent stiffness for bending, torsion, and strength:weight ratio. They were also joined in a sandwich assembly by hot-plate or vibration welding. They were tested for comparative values with the results presented in Table 10.9. The five plate designs in Figure 10.24 show the amplification factor in Table 10.10 obtained in comparison to a flat-plate design.

The starlike gusset design for plate 5 adds very little to plate rigidity in bending and a very minimal amount in torsion from the small star ribs. Also, the bond between the

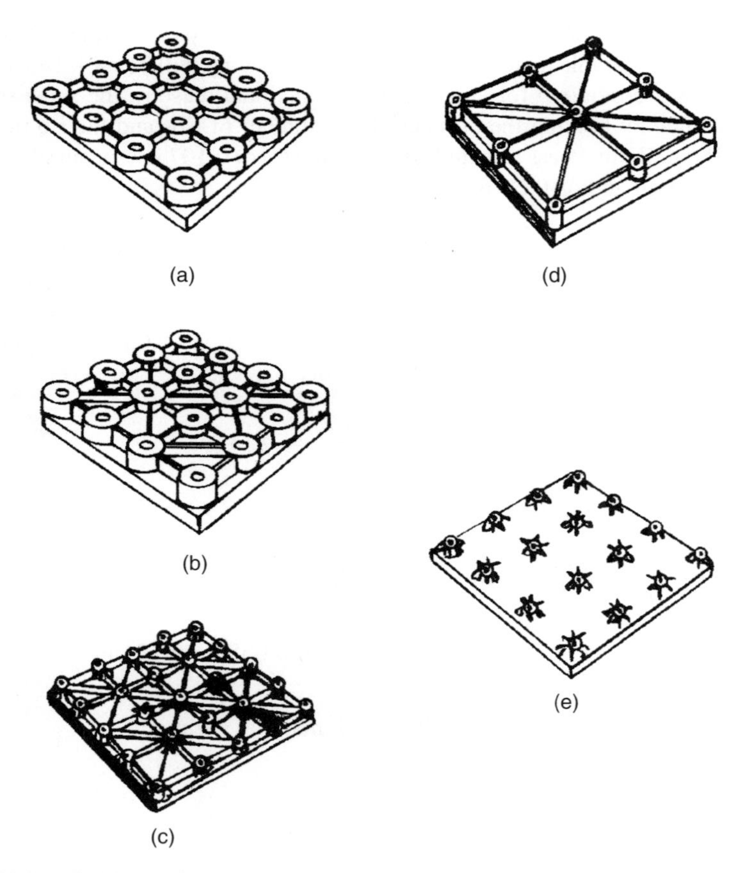

Figure 10.24 Reinforced flat plates: (a) round gussets and straight ribs; (b) round gussets and straight/diagonal ribs; (c) starlike gussets and radial ribs (close spacing); (d) starlike gussets and radial ribs (for spacing); (e) starlike gussets (no ribs).

plates in a sandwich construction of reinforced or nonreinforced material will only be as strong as the parent resin and the method of assembly. No reinforced materials intermix to significantly strengthen the fusion weld at the joint interface.

These plate reinforcement technique layouts (unidirectional, cross-ribbing, gusset designs) are proven methods to reduce material product cost and weight while achieving the product's property values. For all pattern designs, extensive end-use testing is required to verify whether product objectives are achieved.

DESIGNING PLASTIC SPRINGS

Plastics are ideally suited for performing multifunctional applications. The physical properties of most plastics allow incorporation of many design features in a single

Table 10.10 Reinforced plate property amplification factor for gussets

Load	Application			Plate Number			
				1	2	3	4
Torsion							
Single			3.5	33	11.3	10.6	1.3
Sandwich			29	44	48	24	4.8
Bending (single)	*X* axis		10	12.5	16.3	8.8	1.0
	Y axis		10	12.5	16.3	8.8	1.0
Bending (sandwich)	*X* axis		11	16	32	3.6	3.6
	Y axis		11	16	32	3.6	3.6
Strength: weigth ratio	Single torsion		2.1	18	7.3	6.8	1.0
	Single bending	*X*	5.8	6.7	9.1	5.7	0.74
		Y	5.8	6.7	9.1	5.7	0.74
	Sandwich torsion		23	32	33	1.0	40
	Sandwich bending	*X*	8.5	11	21	2.6	2.6
		Y	8.5	11	21	2.6	2.6

Source: Adapted from Ref. 4.

product. One of these is the successful use of specific plastic materials in spring applications. Plastic springs should be used only for limited intermittent spring action. Plastic springs should never be kept under a constant load. Springs under constant load creep and stress-relax. This causes the spring's rate, to decrease quickly over time. Spring rate loss accelerates with temperature and environment effects, such as exposure to chemicals, moisture, and sunlight, so only specific resins may fit specific applications. Any spring required for constant loading should be designed in a suitable spring material.

Plastic springs used for intermittent actions can be accurately designed to function in many applications. Depending on the frequency of operation, plastic springs are limited only in their design by the amount of deflection and potential fatigue-induced stress of multiple, rapid operations. Reinforced plastics are used for spring applications as long as the requirements do not exceed the material's physical properties, especially deflection and elongation. High spring forces can be obtained using reinforced materials for only intermittent-use applications. Acetal and nylon are the most frequently used plastics for springs. They have a high modulus of elasticity, strength, creep resistance, fatigue resistance, and resiliency and also good environmental resistance to oils and solvents. Nylon, which picks up moisture, will experience a decrease in modulus of elasticity over time in wet and humid environments, but it will dry out when removed from this wet environment. Then, over time, the spring material will dry out and regain its nearly original spring constant. Using Acetal, a designer can inexpensively mold in integral, light spring actions that can function for many thousands of operations. Other plastics are used for spring applications in closures and in snap-fit and spring applications. The material selected will depend on the design of the application, deflection, force required, and number of operations. The typical plastic spring design is a cantilever beam. Other spring-related design shapes can be

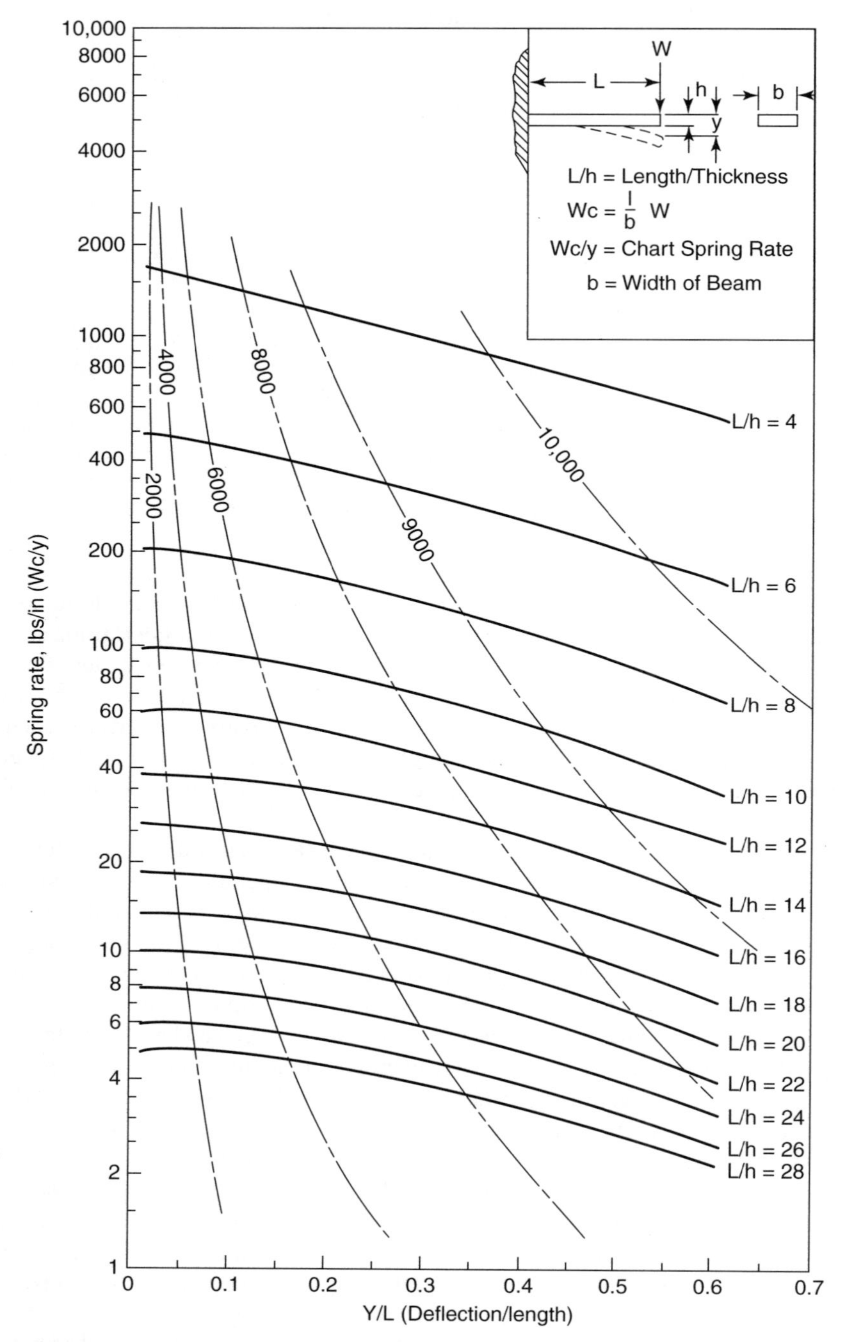

Figure 10.25 Spring rate chart for cantilever springs of acetal homopolymer resin at room temperature [73°F (23°C)]. (Adapted from Ref. 10.)

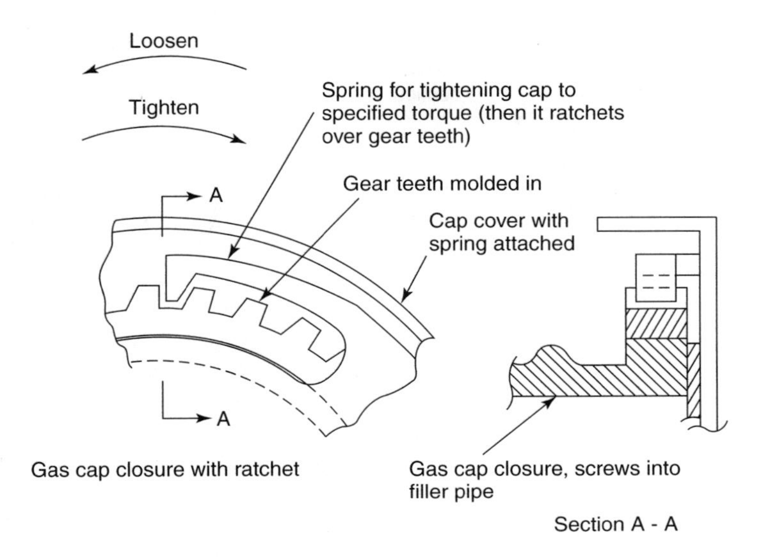

Figure 10.26 Beam loading.

used, and suppliers of Acetal resins have developed spring rate charts that are used for various types of spring geometry.

The spring rate chart in Figure 10.25 was developed for cantilever springs with a spring width (beam width b) of 1 in. The spring chart can be used for cantilever spring designs, operated intermittently, at room temperature (73°F) for a spring thickness up to 0.25 in. thick. (Springs thicker than 0.25 in. will have accuracy diminished using this chart.) Plastic springs should not be used to store energy, kept under constant load or deflection. They should be designed in an at-rest position and operated only when their action is required to perform an operation.

One of the longest uses of Acetal in a spring application is the automobile two-piece gas cap, as shown in Figure 10.26. Molded as separate components before assembly, the cap and Acetal spring's screw thread nozzle closure are snap-fit together. The cap is designed to seal with a leakproof closure and to prevent overtightening on closure. The spring tension was designed to seal the fill pipe to a specified sealing force or tightening torque. When the person putting on the cap reaches this torque, the spring deflects. This deflection of the spring arm causes the cap to ratchet or turn with respect to the seal surface and therefore prevents the closure from being overtightened. When the predetermined seal force on the rounded cap threads is obtained, the Acetal springs flex over the molded-in ratchets in the cap. This spring flexing stops the tightening action by the cap and prevents the threads from being overtorqued. To remove the cap, the spring's cantilever arms act as a beam that transfers the cap's removal torque to the backside of the cap's ratchet mechanism, which is molded as a 90° corner, which the cap pushes on to turn and loosen the filler neck closure.

Other automotive uses are in the mileage odometers. They are designed with a single trigger odometer resetting mechanism and spring detents for the numbers showing the mileage traveled. Electrical connectors and housing covers use spring designs for

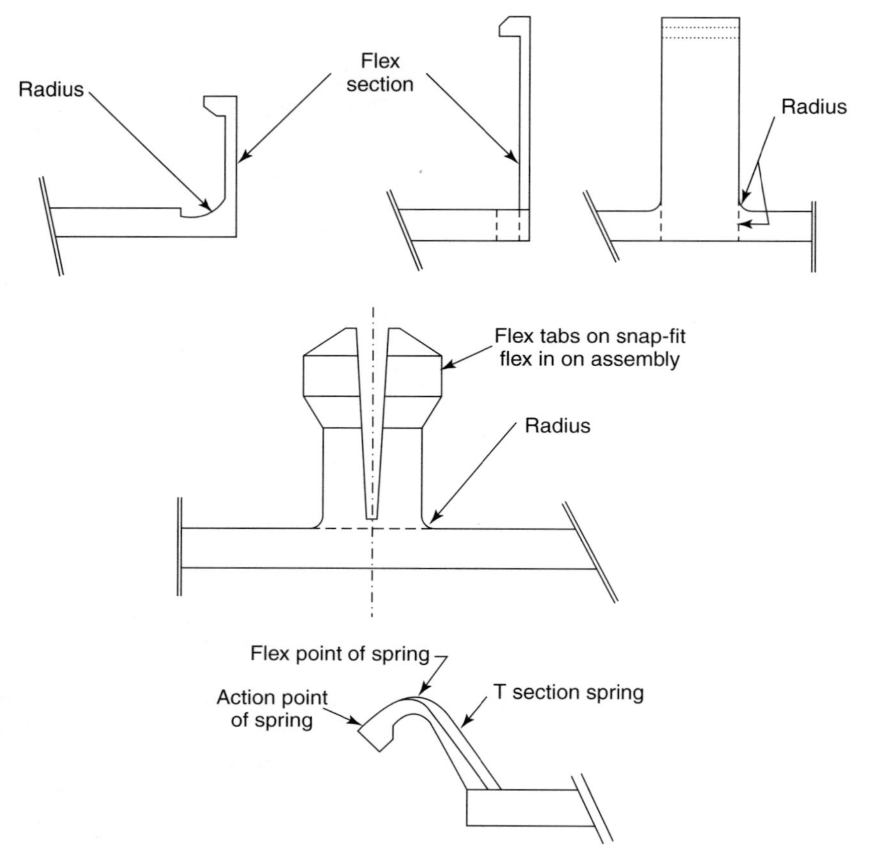

Figure 10.27 Snap fits.

their snap assemblies. These built-in snap assemblies are easily operated when the designer builds in finger grips for their disassembly for service or replacement.

When other spring design shapes, not cantilever, are required, it is recommended that machined or molded prototypes be made and tested for their design suitability. Many different spring shapes can be used. The many shapes possible are shown in Figure 10.27. The standard design equations for complex shapes, "U" and "C" sections, can be used for initial calculations and part sizing. Then end-use testing will verify their action and any force limitations. When modeling a spring from bar stock, the spring rate will vary and be lower. Stock shapes, specifically extruded bar stock, have an even less dense center that is not accurately representative of a molded shape. Therefore, testing a molded prototype product versus a model made from bar stock is recommended to verify design calculations. Also, the performance of a molded spring sample will be about 25–50% better in terms of spring rate and flexural strength than will a machined part cut from a stock shape.

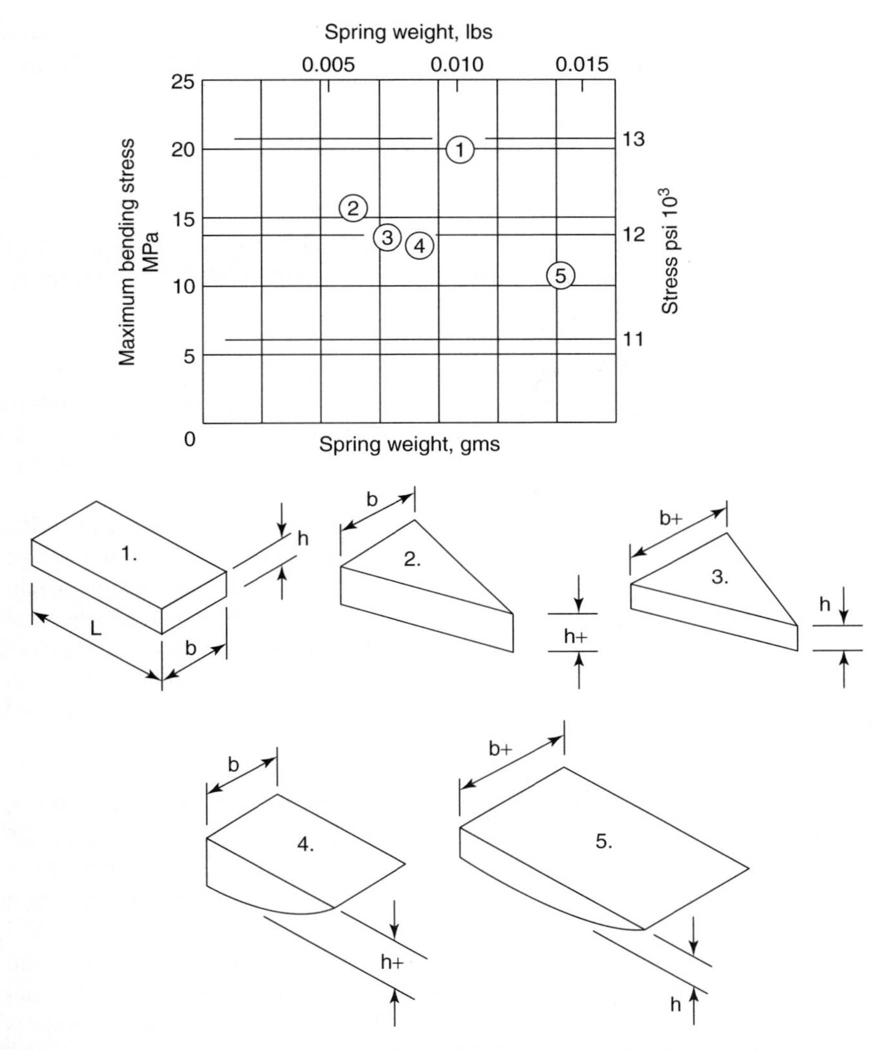

Figure 10.28 Bending stress versus spring weight for various spring designs at room temperature [73°F (23°C)].

Plastic springs should be designed to operate at low stress levels to avoid exceeding their yield stress, which would result in cold flow or creep and lower spring rate. Spring beam shapes have been analyzed by material suppliers to produce an equivalent spring rate and are shown in Figure 10.28. Most spring designs use the constant strength beam formula for a given spring rate and part weight. All spring designs are easily molded and minimize material except number 5. With high-volume parts, material savings to minimize product cost is very important.

The most widely used spring shape is number 1, with a constant rectangular cross section. The initial spring rate is calculated using the cantilever beam deflecting formula

$$\frac{W}{y} = \frac{3EI}{L^3}$$

where W/y is equal to the spring rate in pounds per inch for a one-inch-wide spring from the spring rate chart in Figure 10.25. The following examples use the spring rate graph shown in Figure 10.25.

Example 10.8 When designing a spring, several trial calculations are usually required. The first decision is to select a material that has spring physical properties to handle the requirements. The questions answered in the design and materials checklists (checklists 2 and 4 in Appendix A) will assist. Also, decide whether the spring will be an integral part of the original design or a separate item.

The selected material for the example is acetal homopolymer resin. The spring section will be a rectangular shape (see Fig. 10.28) and number 1. The spring will be integrally molded in a squeeze-type trigger handle with a nonloaded flexible mechanism to produce a spring rate (W/y) of 6 lb/in. This force is produced when the handle is deflected (y) by 0.45 in. Spring dimensions are limited by the confines of the enclosure, whose interior dimensions include a (L) length maximum of 1.875 in. and a width (b) maximum of 0.50 in. The design problem is to calculate the required thickness (h) of the spring.

Solution Select a spring width value (b) of 0.375 in. to fit the enclosure. When selecting trial spring dimensions, remember that the spring rate graph in Figure 10.25 was developed for a 1-in. spring width (b). To adjust the selected spring width to use the graph, (W_c/y) must be multiplied by $1/b$ [<> 1] to obtain the adjusted graph value. In this example, multiply the spring rate (W_c/y) or 6 lb by the ratio of 1/0.375 to adjust for the narrower spring width to fit inside the enclosure. The spring width could have been selected wider or narrower, which has a direct relationship to the final spring's thickness. When space allows, use the maximum width to reduce thickness and stress in the material. Continuing in the analysis

$$W_c/y = (6)\frac{1}{0.375} = 16 \text{ lb/in. spring rate.}$$

Using (16) as the (W_c/y) vertical value from the graph and extending a horizontal line across the graph, there are now multiple choices of (y/L) and (L/h) values on this line that will yield a solution.

Next, assume that L (spring length) = 1.75 in., based on spring dimension clearance inside the housing. Then $y/L = 0.45/1.75 = 0.257$. On the horizontal y/L axis at 0.257, a vertical line intersects the (W_c/y) curve equal to 16 lb./in. just above the (L/h), which equals an 18-curve line. Interpolating between (L/h) = 18 and (L/h) = 16 yields a value of (L/h) = 17.5. Now h can be determined: $L/h = 17.5$,

$h = L/17.5 = 1.75/17.5 = 0.10$-in.-thick section for a 0.375-in.-wide by 1.75-in.-long spring.

Determining the material stress level by interpolation from the graph yields a value of 6400 psi. This value is satisfactory for acetal, since the spring operation is intermittent. If spring operation requires 60 operations per minute, a lower stress level is required to limit the flexural stress on the spring. Since the stress in the spring is almost at the maximum stress (7800 psi) allowable for the material, stress in the spring material can be decreased in several ways:

1. Decreasing (y/L)
2. Decreasing (W_c/y)
3. Increasing (L/h).

The options available are

1. Increase L to 1.825 -in. maximum length, with same clearance.
2. Increase b to 0.490 -in. maximum spring width.

The new spring width decreases (W_c/y) to $6/0.490 = 12$ lb/in. spring rate from the graph. The (y/L) value is adjusted down to

$$\frac{y}{L} = \frac{0.45}{1.825} = 0.246$$

This new vertical line, which intersects very near the (L/h) line, equals 20. Solving for (h), = we obtain $L/20 = 1.825/20 = 0.091$ in. spring thickness. The reduced stress at this point on the chart (W_c/y) and (L/h) is 6000 psi. This is a more acceptable value for this spring's design.

Example 10.9 Determine the force needed to deflect an existing acetal rectangular shaped spring $(y) = 0.300$ in. with the following dimensions:

$$L = 0.750\,\text{in.}$$

$$h = 0.050\,\text{in.}$$

$$b = 0.500\,\text{in.}$$

Solution

$$\frac{L}{h} = 0.750/0.050 = 15$$

$$\frac{y}{L} = 0.300/0.750 = 0.400$$

The spring rate (W_c/y) is at the intersection of (L/h), and (y/L) by interpolating is 18 $(W_c/y) = 18$ lb/in. Therefore, W_c = spring rate times deflection equals (18)

(0.30) = 5.4 lb of force. Since the graph was developed for a 1-in.-wide spring, the actual load to deflect the spring 0.300 in. is $(W_c/y) = 1/lb = 5.4 \times 2 = 10.8$, where $W_c = 1/0.500 = 10.8$ lb of force.

The stress level from the graph is approximately 8000 psi. This is too high a stress level and must be reduced. Reducing the (y/L) factor or using a modified spring shape, numbers 2 through 4 in Figure 10.28, can best accomplish this. Try reducing $(y/L) : y = 0.15$, $y/L = 0.2$, then Wc/y = 33 lb/in. and $W_c = 33 \times 0.15 = 4.95$ lb. Adjusting the spring width, W_c = 1/*b* × spring rate:

$$W_c = \frac{1}{0.50(4.95)} = 2.475 \text{ lb of spring force.}$$

The spring stress level from the graph is 6200 psi, which is acceptable for intermittent operation. Analysis: If the deflection cannot be reduced, try decreasing the length of the spring (L).

Snap-Fit Design Modifications

Snap-fit designs are variations of spring designs. They are one-time flexural assemblies or multiple assemblies for access to repair components inside depending on the product functions.

Conventional snap-fit design assumes that all deflection occurs in the cantilever spring arm. This has placed restrictions on some plastics, especially the reinforced grades, from being considered for some part designs using this method of closure. Studies have proved that in some design situations the part's sidewall will deflect slightly, which reducing the stress level on the snap-fit arm as shown in Figure 10.29. The following information can be used for these and other resins when the snap-fit requirements are more strenuous.

Cantilever snap- fits are a spring design that can operate once or in multiple operations, usually at product assembly or when serviced. This is as simple as a child's lunchbox cover, an infant's safety harness for a car, swing and stroller seats, or a male/female electrical connector snap-fit assembly for an automotive wire harness. Snap fits are subjected to flexural stresses and tensile loading after assembly to hold parts together. The undercut depth (y), of the snap-fit (Fig. 10.30) should be evaluated for sufficient cross section to withstand the holding forces after assembly. The desired assembly force for noncritical applications is just enough to hold the parts together while avoiding creep.

For critical applications such as child safety seat harness buckles, the interference (Y) depth of undercut and lever arm must be capable of withstanding sudden tensile impact forces. The snap-fit lug and arms cannot fail in a crash situation with a small child buckled into the safety seat. In these applications end-use testing is mandatory. No sharp internal corners should be used on the snap-fit assembly.

The lever arms must be enclosed within the snap fit housing to avoid bending of the arms in a high-stress situation. However, the snap-in forces must be sufficient to hold the closure tightly secure yet flexible enough for parents to release the child snap-in

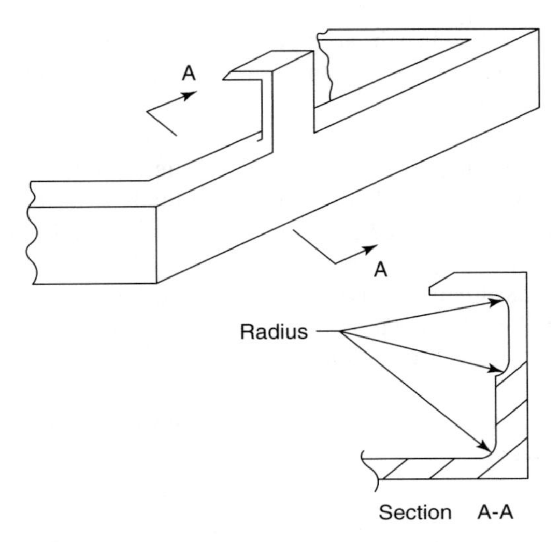

Figure 10.29 Sidewall bending to assist in stress reduction at snap fit.

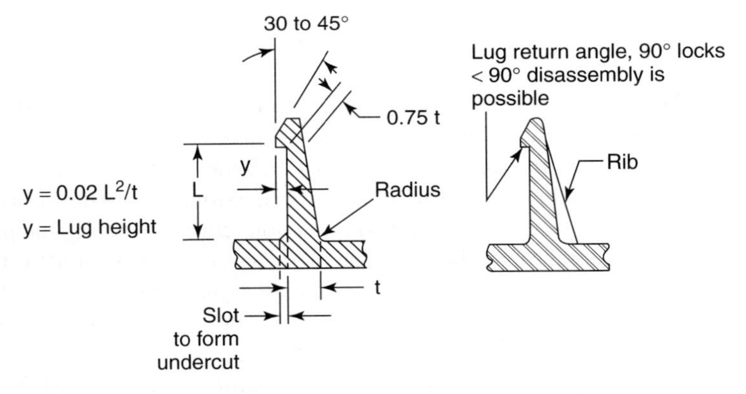

Figure 10.30 Locking lugs. (Adapted from Ref. 3)

harness without an extreme amount of force. This assembly will experience a high degree of use, and an audible snap, in click sound is desired to let the person buckling the snap know that it is secure in a darkened automobile. Plastics can provide many security features when the part is properly designed to use these features.

Therefore, there are combinations of forces acting on a snap-fit cantilever lug, both during and after assembly. The best design is to have no load on the snap-fit arms except tension after assembly. Assembly forces can often exceed the designed material's properties depending on the speed of automatic or manual assembly. If the assembly is too rapid for a low-elongation material, the snap fit can fail. This often occurs because the material is overstressed too rapidly and not able to absorb the energy created during the rapid flexing of the assembly on the lever arm tab. These forces must be considered when the snap fit is designed. Reinforced and nonreinforced plastics are used as long as the snap-fit design does not exceed the resin's physical properties during assembly.

Snap fits can be designed as tamperproof or for multiple open and closings by the design angle of the snap-fit undercut, as shown in Figure 10.30. Snap fits not on the edge of a part should use as liberal a radius at the base of the lug as possible to reduce stress. Even parts at an edge can be designed with radii or modifications to limit the amount of deflection as shown in Figure 10.30. Snap fits can be designed in many ways; the most common types are shown in Figure 10.30. Many variations are possible, from slotted circular, hollow pin designs, and ribbed lugs as shown for achieving a constant stress distribution over the entire length of the lug and for greater holding power.

Snap-fit design uses the cantilever beam equation for a beam mounted on a solid, nondeflecting surface:

$$\text{Stress } S = \frac{FLy}{I}$$

$$\text{Deflection } y = \frac{FL^3}{3EI}$$

Conventional design for a snap fit uses the strain equations ($\epsilon = S/E$) to determine a material's suitability. This standard analysis can give misleading results and not consider the potential flexing of the sidewall to which the snap fit is attached, which can flex during the assembly of the snap fit. When the maximum stress during assembly is below the material's yield stress, the flexing section returns to its original position. When the stress exceeds the material's yield strength, it causes the material to elongate. This is often recognized visually as stress whitening at the base of the snap fit. This results in creep of the material caused by the deflection (y), with the material yielding, and, if too great, fracture results. Therefore, use the calculated strain as an indicator for determining whether the material's elastic properties will be exceeded. When available, use the material's lower dynamic strain limit as the value for E.

To design a snap fit, the beam's dimensions are selected based in accordance with the requirements. These dimensions are used to calculate the stress and strain values as shown in Figure 10.31.

If a tapered beam is selected as in Figure 10.32, use the proportionality constant (K) to calculate the beam cross-section's strain value. This constant (K) should also be used when a rib is required to stiffen the beam as shown in Figure 10.30.

The snap-fit sections must also be correctly aligned during assembly. Misalignment of the parts can cause the mating parts, the beam, to have a greater deflection (y). This can create a larger strain in the section that may exceed the material elastic strain limit, resulting in creep and/or failure.

To assist in assembly and reduce the assembly force, the leading edge of the snap fit should be tapered. This improves the alignment of the snap fit and also assists in reducing the assembly force and stress on the snap fit.

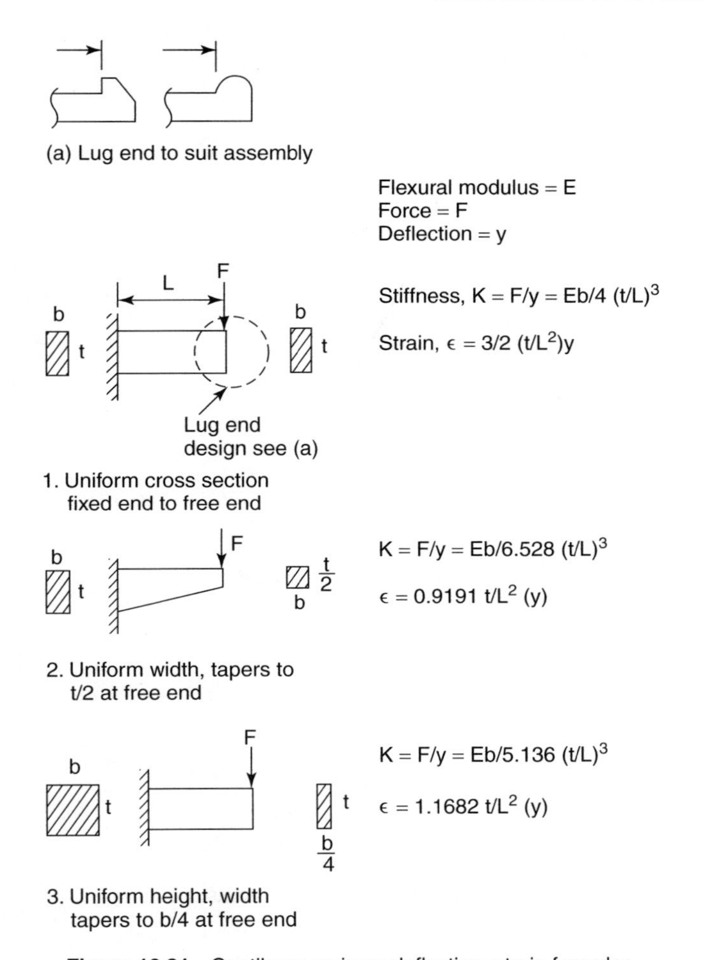

Figure 10.31 Cantilever springs: deflection–strain formulas.

Snap-Fit Assembly Speed

The speed of making the assembly should also be considered. This is important for high-modulus reinforced materials with low elongation. Automatic assembly, usually lasting for only a fraction of a second, may not allow these materials enough time to absorb the energy of assembly, and they may crack and fail. A test should be conducted to determine the nominal speed of assembly to avoid failure at assembly. Parts should always be fixtured for correct alignment and assembly speed selected to ensure that the material is not overstressed during assembly. Testing will show the fixtures required and how rapidly they can be snapped together.

The higher the material's modulus of elasticity, the longer the length (L) desired for a deflecting beam. The beam length is determined by clearance and part shape to allow the length (L) to be increased to keep the material's stress and strain low.

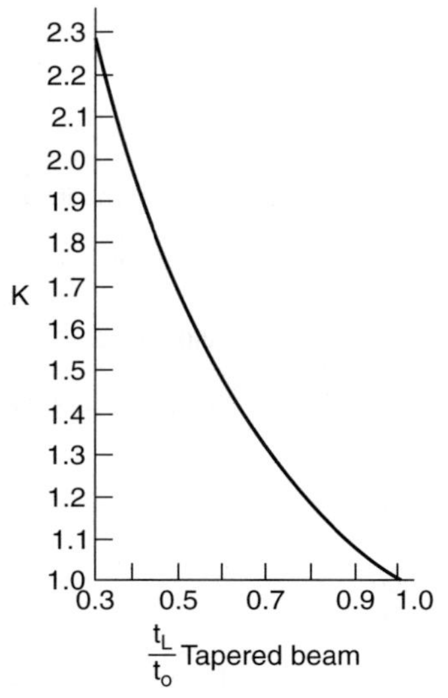

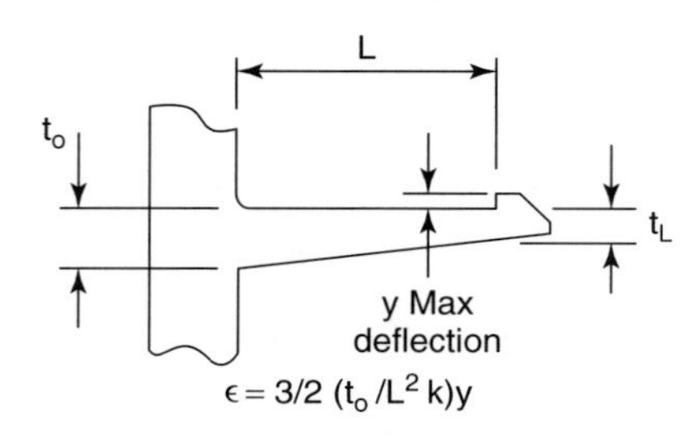

Figure 10.32 Tapered beams.

New Snap-Fit Stress Deflection Analysis

Reevaluation of snap-fit design analysis has produced the following results. This analysis assumed that the sidewall of the attached snap-fit lever arm would also deflect by a small amount during the snap assembly. The sidewall deflection increases the actual snap-fit lever arm dimension that was used to calculate the deflection stress. The longer lever arm for the snap-fit reduces the deflection dimension of the lever arm, resulting in lower strain of the snap fit at the point of attachment. The new beam analysis, based on tests showing the ratio of a beam's snap-fit arm length (L), and thickness (t), predicts a deflection for snap-fit beams greater than that for standard beam end-condition equations. A series of curves (Fig. 10.33) were developed based on test data and are known as the "deflection magnification factor curves." The curves are for flat-plate and wall sections that predict a material's ability to stay within its allowable strain values. The set of curves were developed (see Fig. 10.34): (1–3) for a beam on a solid wall and (4–5) for a beam on a channel or flexible wall.

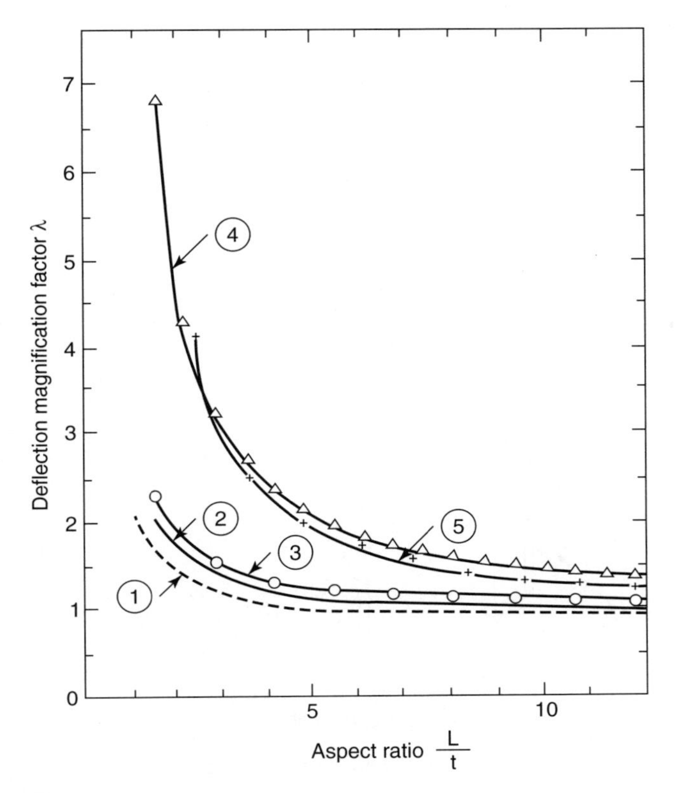

Figure 10.33 Deflection factor. Curve numbers correspond to lug position numbers as shown in Figure 10.34.

When the standard design equations show that the stress or strain is exceeded and conditions require the selected dimensions, evaluate the new deflection equation. Use the respective curve for the design of the snap-fit based on its position on the plate, solid wall or channel, and flat plate, in Figure 10.34. The new deflection formula for the snap-fit is

$$y = \frac{FL^3}{3EI\alpha}$$

The allowable deflection for a rectangular cross section beam is

$$y = \left(\frac{0.67\epsilon L^2}{t}\right)\alpha$$

$$F = \left(\frac{bt^2}{6}\right)\left(\frac{E\epsilon}{L}\right)\left(\frac{1}{\alpha}\right)$$

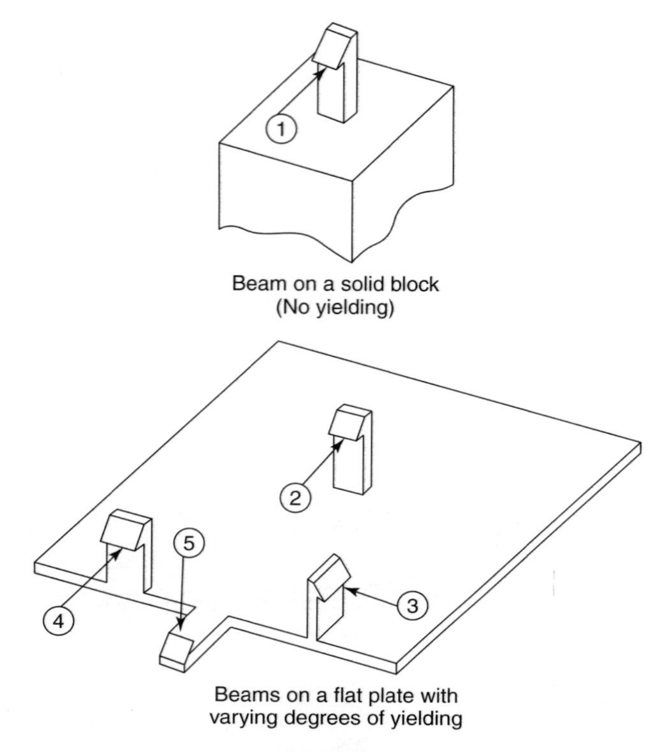

Figure 10.34 Snap-fit beam configuration.

The allowable stress and strain for a rectangular cross section are

$$\text{Stress} = S = \frac{1.5tyE}{L^2\alpha}$$

$$\text{Strain} = \epsilon = \frac{1.5ty}{L^2\alpha}$$

The assembly force is calculated:

$$W = \frac{F\mu + \tan \propto}{1 - \mu \tan \propto}$$

where F = applied force
L = beam length to undercut
E = Young's modulus of material
I = area moment of inertia
α = deflection magnification factor

Table 10.11 Recommended allowable strain values of plastic materials at room temperature of 73°F

Material	Unreinforced (%)	Reinforced (%), 30% Short Glass Fiber
ABS	6–7	—
Acetal	1.5	—
Nylon type 6, DAM	8	2.1
Polycarbonate	\$–9.2	—
PBT	8.8	–
PBT	—	1.5
PBT	9.8	—

[a]Recommended: 70% of tensile yield strain value of the material.

Source: Adapted from Ref. 3.

y = maximum/minimum allowable beam deflection
b = beam width
t = beam thickness
ϵ = maximum strain at base of cantilever (see Table 10.11 for typical plastic allowable strain values; if not listed for your material, ask your material supplier for the values)
S = maximum bending stress
$\propto$ = snap-fit mating angle
μ = coefficient of friction
W = assembly force

Example 10.10 Determine the maximum deflection (Y), snap-fit locking force (F), and assembly force for an edge-mounted, type 4, snap-fit lug based on the following information and shown in Figure 10.35: Material = acetal homopolymer

$$\text{Beam length } L = 0.50 \text{ in.}$$

$$\text{Beam width } b = 0.375 \text{ in.}$$

$$\text{Beam thickness } t = 0.125 \text{ in.}$$

$$\text{Mating angle } \propto = 45^\circ$$

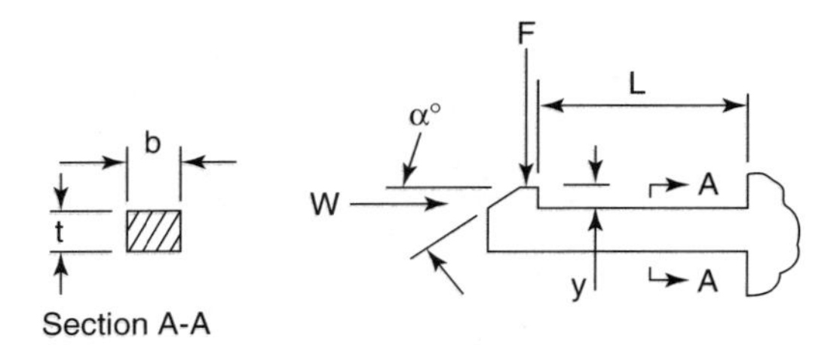

Figure 10.35 New deflection formula.

Solution Allowable strain = 1.5% based on 70% of tensile yield strain. The standard formula method is as follows:

$$Y = \frac{(0.67(\epsilon L^2))}{t} = \frac{0.67(0.015)(0.5)^2}{0.125} = 0.020 \text{ in.}$$

The deflection force is

$$\begin{aligned} F &= \frac{(3EI)}{L^2} = \frac{(bt^2 E)}{6L} \\ &= \frac{((0.375)(0.125)^2(4.1 \times 10^5)}{6(0.5)} \\ &= \frac{24,023}{3} = 8008 \text{ lb,} \end{aligned}$$

which is an unrealistic value! This is an unworkable snap-fit. Using the new method with the deflection magnification factor (α). Figure 10.33 yields

$$\frac{L}{t} = 0.50/0.125 = 4 \qquad \text{(using curve number 2, } \alpha = 2.6\text{)}$$

$$Y' = Y\alpha = Y' = 0.020 \times 2.6 = 0.052 \text{ in.} \qquad \text{(allowable deflection)}$$

The deflection factor takes into account the fact that the snap-fit sidewall will also deflect, which permits double the allowable deflection or interference for the snap-fit detent. The deflection force (F) at the deflection end is

$$\begin{aligned} F &= \frac{3EI}{L^2\alpha} = \frac{bt^2 E\epsilon}{6L\alpha} \\ &= \frac{(0.375)(0.125)^2(4.1 \times 10^5)(0.015)}{6(0.50)(2.6)} \\ &= 4.61 \text{ lb} \end{aligned}$$

Assembly force W is determined as follows: For acetal, the coefficient of friction (u) is 0.20 static against steel for this example:

$$\begin{aligned} W &= \frac{F(\mu + \tan \propto)}{1 - \mu \tan \propto} \\ &= \frac{4.61(0.20) + \tan 45^\circ)}{(1 - (0.20)(\tan 45^\circ)} \\ &= \frac{4.61(1.20)}{1 - 0.20} \\ &= 6.92\text{lb} \end{aligned}$$

Example 10.11 A short 30% glass-reinforced type 6 nylon snap-fit lug type 2 or 3 (see Fig. 10.34) is used to attach a mating part cover of a housing. Determine whether the material and snap-fit lug dimensions will work:

$$\begin{aligned}\text{Beam length } L &= 0.150 \text{ in.}\\ \text{Beam width } b &= 0.500 \text{ in.}\\ \text{Beam thickness } t &= 0.050 \text{ in.}\\ \text{Mating angle } \propto &= 30^\circ\\ \text{Deflection } y &= 0.025 \text{ in.}\end{aligned}$$

Maximum strain (ϵ) obtained by the standard method is

$$\epsilon = \frac{1.5\,ty}{L^2} = \frac{1.5(0.050)(0.025)}{(0.150)^2} = 0.083\%$$

Using the new deflection magnification factor, curve 1, we obtain

$$\frac{L}{t} = \frac{0.150}{0.050} = 3, \alpha = 1.7$$

with the new strain (ϵ') factor calculated; checking $\epsilon' = \epsilon/\alpha = 0.083/1.7 = 4.9\%$ too high! Modify lug and increase length (L):

$$\begin{aligned}L &= 0.250 \text{ in.}\\ \epsilon &= \frac{1.5(0.050)(0.025)}{(0.25)^2} = 3.0\%\\ \frac{L}{t} &= 4, \qquad \alpha = 1.4\\ \epsilon' &= \frac{\epsilon}{\alpha} = 0.030/1.4 = 2.1\%\end{aligned}$$

This is the maximum allowable strain, and if molding conditions are not controlled, the snap fit can break. Therefore, decreasing beam thickness (t) to an allowable, approximately 1.5% should reduce the snap-fit strain. Do not assume that the nylon will pick up moisture before assembling the product. Always use dry-as-molded (DAM) material properties, or consider an impact-modified grade with higher elongation. The material will be more expensive but will have higher elongation in the DAM state. Also, evaluate the assembly force and consider a slow assembly time to allow the lug to yield, absorbing the assembly force energy, so that it will not break. It is better to redesign the lug's the example illustrated by reducing the lug thickness and/or width.

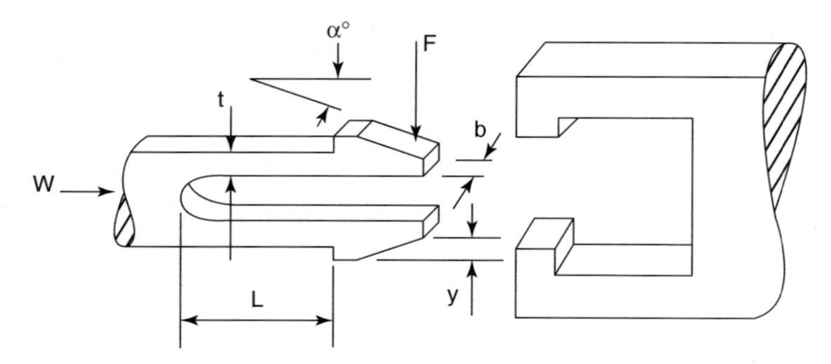

Figure 10.36 Double-cantilever snap fit.

Example 10.12 Refer to Figure 10.36 for dimensions as listed. Double the cantilever snap fit for an acetal connector:

$$L = 0.50\text{ in.} \qquad Y = 0.050\text{ in.}$$
$$t = 0.075\text{ in.} \qquad \propto = 30^\circ$$
$$b = 0.250\text{ in.} \qquad \in = 1.5\%$$

with E = Young's modulus for acetal and coefficient of friction (μ) = 0.30 (acetal to acetal—static). Using the conventional method

$$F = \frac{3EIY}{L^3} = \frac{3(410,000)(8.7 \times 10^6)(0.050)}{(0.50)^3} = \frac{0.535}{0.125}$$

Assembly force $F = 4.28$ lb and

$$\text{Strain } \epsilon \text{ max} = \frac{1.5ty}{L^2} = \frac{1.5(0.075)(0.050)}{(0.50)^2} = \frac{0.0056}{0.25}$$
$$= 2.25\% \text{ strain}$$

Using the deflection magnification factor (α), we obtain

$$\frac{L}{t} = \frac{0.500}{0.075} = 6.66$$

Using curve 2, $\alpha = 1.8$, and strain adjusted for deflection magnification factor $\in' = \in/\alpha$:

$$\in' = \frac{2.25}{1.8} = 1.25\%$$

Thus $\in'$ for acetal = 1.5%, which is acceptable.

Assembly force adjusted for the deflection magnification factor $F' = F/\alpha$ is

$$F' = \frac{4.28}{1.8} = 2.4\ \text{lb deflection per assembly arm}$$

Assembly force for the connector is

$$\begin{aligned} w &= \frac{F'\mu + \tan \propto}{1 - \mu\ \tan \propto} \\ &= 2.4 \times \frac{(0.30) + \tan 30^\circ}{1 - (0.30\ \tan 30^\circ)} \\ &= 1.297/0.827 \\ &= 1.57\ \text{lb/arm} \times (2\ \text{arms}) = 3.17\ \text{lb total assembly force.} \end{aligned}$$

Snap- fits can also be designed to be disassembled if the interior lug return angle is less than 90°. However, the holding power will be lower. The same assembly force equations apply only using the return angle ($\propto$), on the back of the snap-fit arm. There are other forms of snap-fit assemblies besides the cantilever beams. The designers' requirements and moldability only limit them.

Typical Snap-Fit Designs

The typical snap fit designs used for assembling plastic-to-plastics parts and combination materials are shown in Figure 10.37. In all designs the following guidelines need to be followed:

1. Uniform wall thickness throughout, with no stress risers.
2. Freedom to expand and move, and deflect. Molded-in strippable and snap-fit parts are shown in Figure 10.38 and materials strain values, in Table 10.12.
3. A round snap-fit geometry is preferred for molding and assembly. A rectangular snap fit requires special tooling to form and release the undercut snap-fit arm as it must be free of the mold cavity steel when the product is ejected from the mold cavity. A rounded lug design often relies on the material's deflection or being squeezed while being stripped from the mold cavity. It is important for all snap-fit designs that mold cavity steel be removed from or around the snap fit so that it can deflect on ejection, or it may be destroyed.
4. Review the part's mold filling design drawings to be sure where the gate is located so that a weld line will not be formed where the snap fit must flex.

Designing for Round Snap Fits

Pin-style tongue-and-groove and return-angle snap fits are used for assembling and sealing many large and small products. All plastic materials have their recommended

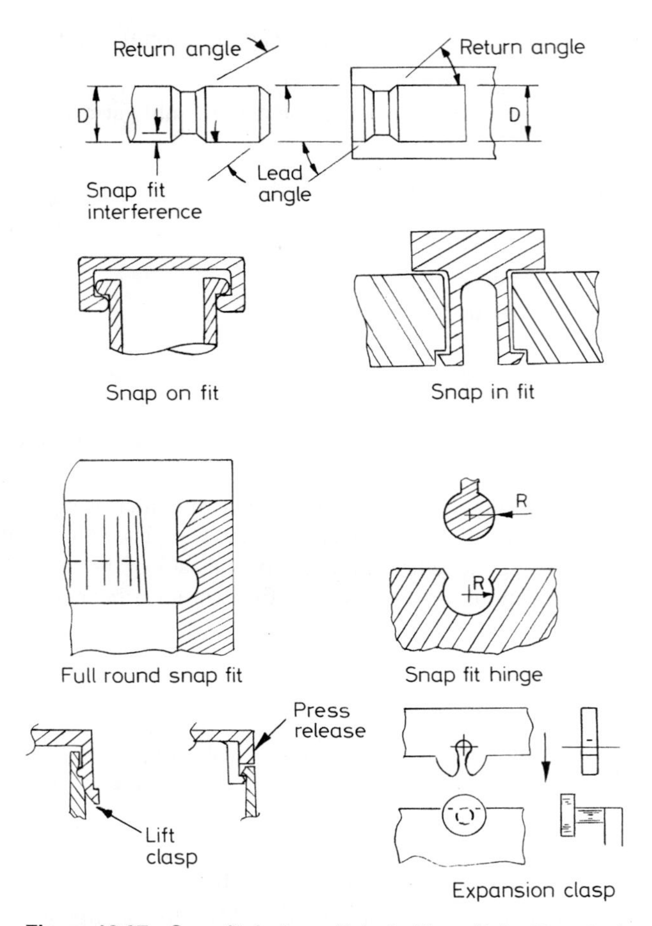

Figure 10.37 Snap-fit designs. (Adapted from Refs. 10 and 7.)

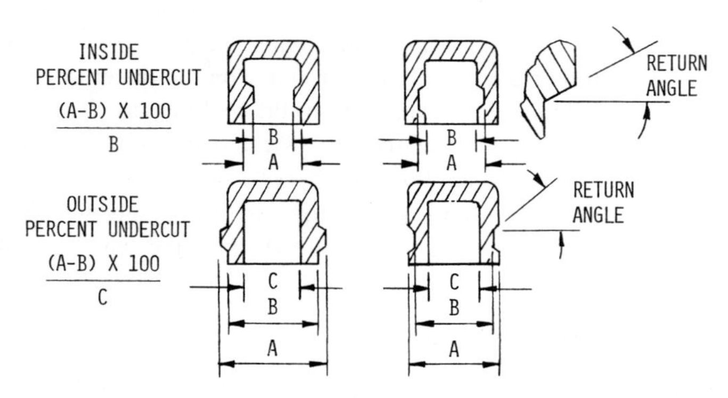

Figure 10.38 Press-fit design. (Adapted from Ref. 6.)

Table 10.12 Mold cavity strippability strain values for a plastic material

Material	Strain at 150°F (%)
ABS	8
SAN	NR[a]
Polystyrene	NR
Acetal	5
Acrylic	4
Nylon	9
Polyethylene (low-density)	21
Polyethylene (high-density)	6
Polypropylene	5
Polyallomer	15
Polycarbonate	NR
Noryl[b]	NR
Surlyn[c]	10

[a] Not recommended.

[b] General Electric Company trademark (PPO/styrene blend).

[c] E. I. Du Pont de Nemours trademark (ionomer).

snap-fit interference dimension for assembly and holding capability. Most round containers use the tongue-and-groove assembly method. Airtight seals are possible with the correct design interference. It is the designer's responsibility to select the dimension to provide the closures required for sealing the product. Dimensions vary with the elongation of resin selected and per the requirements of the closure.

When a metal shaft and plastic hub are assembled using a snap-fit interference assembly, the plastic hub must expand as the pin is inserted, and the amount of expansion is material-dependent. These types of assemblies are shown in Figure 10.38.

The higher the material's elongation, the greater the allowable interference. The percentage of interference varies from 1% for small shafts ($\frac{5}{32}$ inch) up to 2% for $1\,\frac{3}{8}$ in. shafts, such as those using type 6 nylon resin, which has a high DAM material elongation. These closures require a tight seal and considerable force to open and close. Without the small cutout in the snap-fit cover and sealing lip, they would be almost impossible to open. The snap-fit cap has about 4–5% interference. After every opening and closing there will be some degree of relaxation on the assembly as it shrinks back down on the mating sections. The childproof snap-fit bottle closures use polyethylene and PVC materials for these inexpensive but high-product-liability-type products. The closure protects the child from opening the bottle. The assembly force for a circular snap fit is important when determining the method of assembly, whether manual or machine press. The calculations are divided into two sections:

1. The force needed to expand the hub
2. The force needed to overcome friction during assembly

To assist in lining up the assembly, beveled edges on the hub and shaft are recommended where the assembly's beveled edges slide past each other, which occurs at the point of maximum hub expansion and stress on the hub. The equations that these assembly forces develop are as follows. The geometry factor (W) is given as

$$W = \frac{1 + (D_s/D_h)^2}{(1 - (D_s/D_h)^2)}$$

$$I = (S_d D_s / W)[((W + \mu_h)/E_h) + ((1 - \mu_s)/E_s]$$

where D_h = outside diameter of hub
D_s = diameter of shaft
E_h = tensile modulus of elasticity of hub
E_s = modulus of elasticity of shaft
I = diameter interference
μ_h = Poisson's ratio of hub material
μ_s = Poisson's ratio of shaft material
S_d = material design stress
W = geometry factor

The force to overcome friction is given as

$$F_f = \frac{f S_d D_s L_s \pi}{W}$$

where F_f = friction force
f = coefficient of friction
S_d = stress due to interference
D_s = shaft diameter
L_s = length of interference sliding surface

The assembly force required to expand the hub is

$$F_e = \frac{(1 + f)\tan(n) S_d \pi D_s L_h}{W}$$

where F_e = expansion force (lb)
n = angle of beveled surfaces
S_d = stress due to interference (psi)
D_s = shaft diameter (in.)
W = geometry factor
L_h = length of open hub expanded (in.)

For a blind hub, use 2 times the shaft diameter. Therefore, the total assembly force for the snap fit is

$$F \text{ assembly} = F_e + F_f$$

For a locking assembly, a 45° return angle on the undercut section is usually satisfactory. This depends on the materials used and the forces developed during assembly. Any angle on the undercut greater than 45° and up to 90° can be used to permanently lock the assembly together. When the return angle bevel approaches 60°, the pin forming the angle will require a collapsing core in the mold cavity.

Combinations of circular snap fits can be designed using vertically slotted hubs or shafts. This reduces the assembly force and also limits the torque that can be applied to the assembly. The degree of retention force required for the assembly is decreased by the slot size. The disassembly force to the snap fit will be equal to the assembly force if the angles are the same. Remember that the hub will relax after being expanded within a reasonable time period and will develop the calculated retention force. To increase this force, a permanent metal band can be added to thin hubs to ensure good retention of the assembly.

When metal molded-in inserts for screw assembly are used for parts requiring frequent disassembly, the recommended boss wall thickness is as shown in Table 10.13. Inserts should be smooth with no sharp corners and clean and heated when high-shrink

Table 10.13 Minimum wall thickness for molded in inserts

	Diameter of Molded in Inserts					
Plastic Resins	0.125	0.250	0.375	0.500	0.750	1.00
ABS	0.125	0.250	0.375	0.500	0.750	1.00
Acetal	0.062	0.125	0.187	0.250	0.375	0.500
Acrylics	0.093	0.125	0.187	0.250	0.375	0.500
Cellulosics	0.125	0.250	0.375	0.500	0.750	1.00
Ethylene vinyl acetate	0.040	0.085	NR*	NR	NR	NR
FEP (fluorocarbon)	0.025	0.060	NR	NR	NR	NR
Nylon	0.125	0.250	0.375	0.500	0.750	1.00
Noryl (modified PPO)	0.062	0.125	0.187	0.250	0.375	0.500
Polyallomers	0.125	0.250	0.375	0.500	0.750	1.00
Polycarbonate	0.062	0.125	0.187	0.250	0.375	0.500
Polyethylene (high-density)	0.125	0.250	0.375	0.500	0.750	1.00
Polypropylene	0.125	0.250	0.375	0.500	0.750	1.00
Polystyrene	NR	NR	NR	NR	NR	NR
Polysulfone	NR	NR	NR	NR	NR	NR
Surlyn (ionomer)	0.062	0.093	0.125	0.187	0.250	0.312

[a]Not recommended.

Source: Adapted from Ref. 7.

resins are used. This will reduce the stress of the resin molded around the inserts and eliminate stress cracking in service. All designs for product assembly should be molded and tested to ensure that they will perform as required in the end-use application.

REFERENCES

1. *Plastics Design Forum* (Sept./Oct. 1987).
2. C. Lee and A. Dubin, "Improving Snap Fit Design" *Plastics Design Forum,* 65–70, (Nov./Dec. 1987).
3. *A Guide for Designing with Engineering Plastics,* E. I. Du Pont de Nemours Corp., Wilmington, DE, 1990.
4. *General Design Principles—Module 1,* E-80920-2, E. I. Du Pont de Nemours Corp., Wilmington, DE.
5. "Shafting." in *Machinery Handbook,* 21st ed. Industrial Press New York. 1979, p. 453.
6. *Designing with Plastics, Design Handbook,* Hoechst Celanese (TDM-1) HCER-92-313/10M692: 8-6, Chatham, NJ.
7. R. D. Beck, "Plastic Product Design," Van Nostrand Reinhold, New York, 1970.
8. *Designing with Plastics, Design Handbook,* Hoechst Celanese (TDM-1) HCER-92-313/10M692: 8-7, Chatham, NJ.
9. *Designing with Plastics, Design Handbook,* Hoechst Celanese (TDM-1) HCER-92-313/10M692: 9-2, Chatham, NJ.
10. *General Design Principles—Module 1,* E-62617, E. I. Du Pont de Nemours Corp., Wilmington, DE.

CHAPTER 11

DESIGN FOR PRODUCT PERFORMANCE

The key to developing a product that will meet customer satisfaction and end-use performance is teamwork with "total involvement" of the design team. Early selection and collaboration of the team members will assist in ensuring that all pertinent design and manufacturing areas are explored and questions answered. It will also speed production if knowledge of the product's tooling, manufacturing process, assembly, and decoration will be within the program guidelines to meet cost and produce a quality product.

DESIGN FOR ASSEMBLY AND SERVICE

Designing the product beyond the initial end-use requirements is necessary to produce a product that will meet customer satisfaction beyond the requirements. Moderate product cost plus the desire and need by the customer makes a product valuable and assists the producer in making a profit that meets their company goals. Product and function consolidation is achieved with forward thinking on how the product will be used in service by the customer. Also, the ease of manufacture and durability and repair or replacement of a consumable item is a requirement for the success of the product in service. This process begins by analyzing the end-use function of the product and devising methods to combine various functions into a single product whenever possible or what functions could be added without a high additional cost to make the product even more useful and valuable to the customer.

A high-profile OEM (original equipment manufacturer) supplier once told me that this was the one function that they needed to improve in their work with their customers. The cost of making changes, after the design was thought to be complete, was costing them their profit margin. They needed to spend more time with their customers to ensure that any changes were quickly acted on and that the cost was a shared cost, not just a concern of the suppliers when a customer change was made. They also needed to ensure that all their departments were involved with the design of the product to ensure that all questions were answered and considered before the tooling and manufacture started. This early review is initially done to reduce the number of components in an assembly. Consideration is given to the material of manufacture, and it is not always the least expensive material that will do the job. Cost is one item to consider but the versatility of the base resin and what it can offer to the entire product must also be considered. Often a more expensive and versatile resin that will assist in reducing inventory, with fewer molds and purchased products, can ultimately lead to lower production and product cost. Consolidation of products and materials can improve the profitability and lead to fewer customer problems and aid in the success of a program. This process can also give the customer a more useful product at lower cost. Couple this with simplifying assembly, and serviceability will further increase the product's value to a customer.

This technique, coined "designing for manufacture, assembly, and service" (DMAS), encompasses the manufacture, assembly, and service requirements during the product's early design stage that anticipates, eliminates, and solves production and field problems. DMAS brings to the design team the field service repair technicians, who typically are never in the design loop. Their input can assist the design team to make the product more user-friendly, reducing their service warranty costs and field problems. Easily replaced, used-up, or worn-out products can be stocked by the customer and replaced with the same color-coded products, making the product even more user-friendly.

The objectives of DMAS are to increase reliability and quality while reducing the number of components, simplifying assembly and repair, plus reducing design time, service, product liability, and product cost. After product consolidation has been considered, the assembly of the products into the final product is explored for both assembly and customer service and repair if applicable.

COMPUTER-AIDED DESIGN AND MANUFACTURE

All factors in the design process are critical to the program's success. An area that will assist in answering product design questions and providing additional information for material selection is computer-aided design.

Computer-Aided Design

Computer-aided design (CAD), when coupled with computer-aided manufacture (CAM), encompasses more than just the ability to design and display the product in either two- of three-dimensional (2D or 3D) space. The versatility and strengths of the new CAD/CAM software programs now used by designers can assist in saving time for analysis and reach more answers more quickly than ever before. This is in the areas of product and tool design, plus assist manufacturing in knowing before manufacture how each mold and material will perform in the processing phase of manufacture. The software with the right input of information can answer design, tooling, and control of manufacturing questions for all products. It greatly assists the engineers by being more standardized to supply the needs of plastic product design and manufacture. CAD software performs finite-element analysis and solid modeling with form and fit capability for products. By simulating the product's end-use performance in the software analysis, the reactions based on known input data show the effects of forces on the product and illustrate, in very fine detail, the stress concentrations in the product. This simplifies the design process and reduces design time from days to hours.

CAD with support software also assists in designing the tooling with information on cooling and material flow analysis. These software programs can visibly demonstrate material pressure gradients to assist in profiling injection pressure and fill time for filling and packing out the product's mold cavity. Then, once a product's material is selected, the simulation programs can visually show how the molding process will affect the product's tolerances, internal residual stresses, weld lines, packout, and potential warpage from the fill and cooling software program. These programs can evaluate product design; tooling, and processing conditions in a simulation to visualize how over 400 related variables interact when plastic pellets are converted into a high-performance plastic product.

CAD as a designer tool is very high-powered, with the end results only as good as the input information used to guide the analysis. Before using the CAD program, a rough or preliminary product stress analysis should be performed to verify whether the program variables are within anticipated results. Once the structural analysis is completed, prototype testing during the design stage; either modeled or molded, is recommended to prove the reliability of the design to meet product requirements. CAD can assist in these areas to ensure that success for the design of the product is attained.

CAD creates both 2D and 3D design drawings on a computer screen. The 2D system is basically a clean drawing system producing legible product development drawings and notes. It is an excellent training tool for the designer to become familiar with the computer and related software. The 3D system presents products in their true perspective, three-dimensional objects that can be rotated, sectioned, and viewed from any angle. Any change made in one view will automatically be reproduced in every other view. Updating is then instant and continuous until the final product dimensions and design parameters are established.

Any product's orthogonal or auxiliary views can be generated from the 3D model. With the 3D product model completed, stress, thermal, and plastic flow analysis can be performed. Once the model is created, data are transferred into the ASCII (American Standard Code for Information Interchange) format and downloaded into plastic flow and cooling analysis programs for calculating filling, packing, tolerance determination, and cooling stages for injection-molding review. Graphic displays can depict product temperature gradients, including shear rates and material flow velocity profiles occurring during the molding cycle. These will show the design team where the product's gates should be located for mold filling that is affected by wall thickness changes and where weld lines will occur. Volumetric shrinkage, gate freeze-off and internal residual stress areas will also be visualized. Versatility is obtained by performing work in one view while the computer displays and updates multiple views.

Software programs also allow sketching a mold base around a pregenerated cavity layout, estimating mold cost, including cavity and cores, and bill of material generation for the mold. Mold design costs and time can be reduced by up to 50% using the CAD system. Savings can also be achieved using CAM with numerically controlled (NC) machines to produce prototypes. The main advantage is the reproducibility of NC machining to produce identical mold cavities. Using an NC programmer, tool paths can be created in multiaxis orientation from the mold or wire-frame model to accurately machine the product model and assist in machining the mold cavity.

Selecting the CAD System

There are many CAD/CAM/CIM (computer-integrated manufacture) software programs for the designer and manufacturing personnel to select from when designing a product. The software must fit your design and manufacturing requirements, and it is difficult with so many products available to choose from that claim to be compatible with existing software, better, faster, and more user-friendly. A set of questions and answers as to what an engineer should look for in software need to be addressed:

1. What are your co-workers using, and who may be doing a similar job? Will their software do what you want? If so, stop here. You will have a source for training and troubleshooting and ways to increase the productivity of your output. Also ensure compatibility of software with departments who will interface, in real time, with the product design program.
2. Price is not always the best measure of software performance. Price out systems within your range of use do value. Some less expensive programs have as much function capability and ease of use than do more expensive and high-powered programs. Investigate whether software training, on site or at their facility, and system support are included in the cost.
3. How much use will the program get, daily or just weekly? Ease of learning and remembering how it works is very important if not used daily. The more the use, the more sophisticated may be the requirements of the software.
4. Narrow your software selections down to a maximum of 10 or fewer programs. Then explore and catalog the primary function of the programs you find of interest. Also, find out if they were developed for your industry or for metals, assembly, and plastics. Remember that one program may not meet all your needs. Narrow your search to software that meets your specific requirements now and in the near future.
5. Visit computer shows and vendor demonstrations. The Internet and vendors often have trial time periods for you to evaluate and use their full software

Figure 11.1 SLA-250 equipment setup. (Courtesy of 3D Systems, Valencia, CA.)

programs to see if they meet your requirements. This is an excellent time to find out if the software really meets your needs. Also, visit the SciTech International Website at `www.scitechint.com` for an evaluation and assistance.

6. After trying the software, narrow your search to two or three selections. Obtain trial copies of each program, and evaluate it for ease of operation and for fulfilling your design requirements.

CAD-Generated Models

Computer-generated prototypes using finite-element analysis (FEA) with the new model and tooling prototype software programs can answer many design, product layout, and fit questions. Data from these modeling techniques can be used to generate numerically controlled (NC) tapes to fabricate physical models. Models can take on many forms as machined, molded-in prototype, die-cast tooling, or SLA/SLS (stereolithography/selective laser sintering) and other trademark systems, such as fused deposition modeling (FDM), computer-generated as shown in Figures 11.1 and 11.2. The method used and material of the model will determine the answers that they can provide if capable of being tested in the end-use environment.

PROTOTYPE MODELING

Prototyping, a model of the product and/or assembly, will aid the design team and their customers in answering many questions as to design, moldability, tolerances,

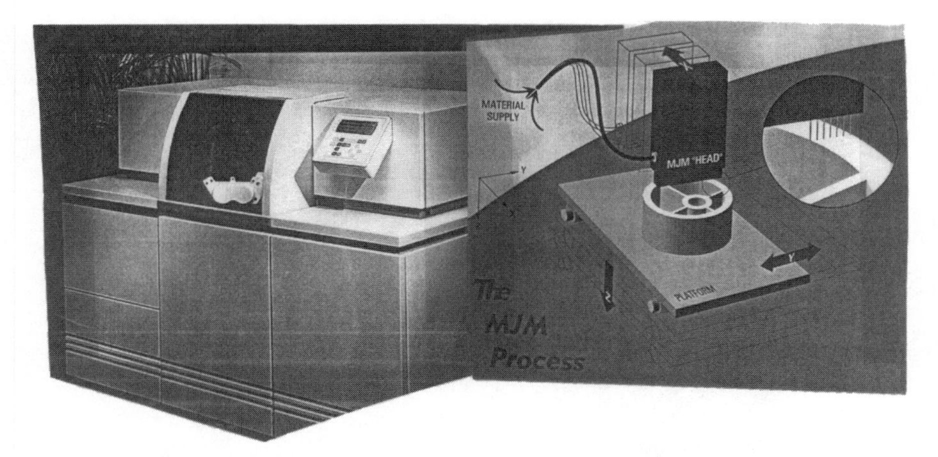

Figure 11.2 FDM equipment setup. (Courtesy of Stratasys, Inc., Eden Prairie, MN.)

decoration, assembly, and end-use suitability. This should be done early in the product's design process once the product's geometry is defined and a material candidate has been selected for the product. Then, depending on the type of prototype selected, computer generated or modeled (machined, molded, SLS, or SLA) and the degree of accuracy required, the cost and the model's ability to generate data to questions needing answers can be determined. Accurate prototypes are required for complex, tight tolerance products with extensive detail and product functions such as internal springs, bearings, and snap fits. They are also needed for product end-use testing of mechanical, electrical, thermal, chemical, and environmental properties. The most accurate model will duplicate the method of manufacture and its material. The following prototype modeling techniques are now discussed for consideration for selecting the type of model to prototype.

SLA/SLS Models

These guidelines will assist the design team in determining the product's mold ability and sections in the product that can be modified to increase the product's quality and end-use properties before cavity steel is cut.

SLA/SLS models use resins to form the model, one layer at a time, usually employing laser sintering. These models are fragile and, depending on the resin used, can experience shrinkage and tolerance problems with time. They are expensive but can be produced in hours rather than days or weeks for machined or molded prototypes. They can serve as a model for the production mold if desired, in the same way as a wood pattern is for metal castings.

SLA/SLS models with the newer resin systems, such as acrylate, epoxy, and newer resins, can withstand high mechanical loads and are used to develop assembly systems, form and fit considerations, and other uses as appropriate for the material used to generate the model. They can also achieve excellent accuracy, dimensional stability, strength, and high heat resistance for testing in their end-use environment. Prototype full size products can be generated using the 3D database. A model can be made using NC machining; SLA (stereolithography), SLS (selective laser sintering), or other modeling technique directly from the computer-generated model. SLA is a model generation processes where thin horizontal slices of the computer model are built up, slice by slice, by a laser using a photosensitive resin. The resin is in a container that holds the liquid resin to make the model and has a movable platform where the three-dimensional model will be reproduced in a holder that contains the liquid resin. When the surface of the resin is exposed to the laser, tracing the shape of the computer-generated section of the product called a "slice," in two dimensions, it solidifies the exposed resin. The model on the platform is then indexed down a small amount into the unreacted liquid resin to just cover the surface of the model for the generation of the next model layer. The process continues slice by slice until a 3D model is completed as shown in Figure 11.3.

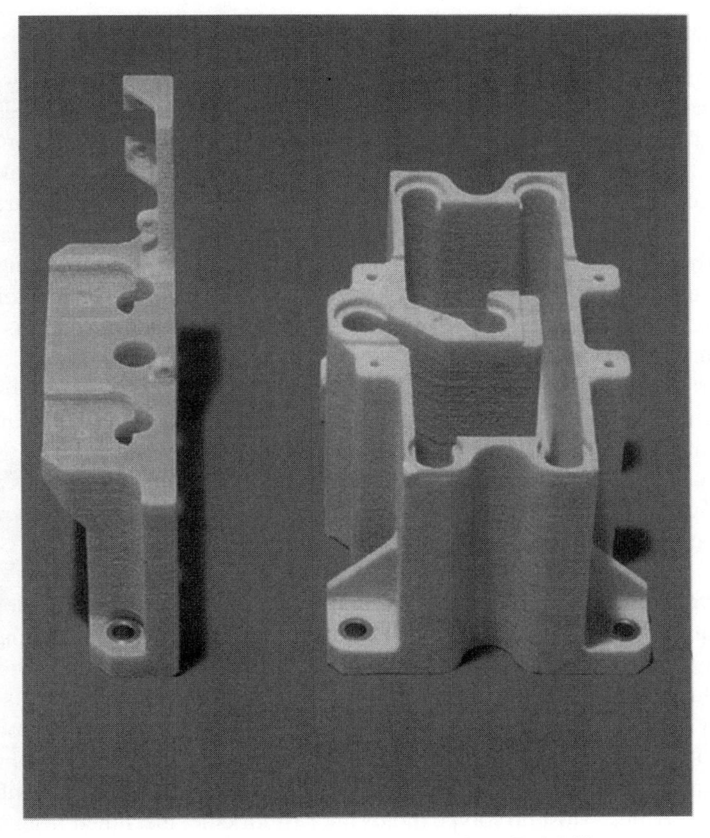

Figure 11.3 SLA model. (Adapted from Ref. [5].)

There are three other model generation techniques available; SLS, fused deposition modeling (FDM), and laminated object manufacture (LOM). The last two, FDM and LOM, are suitable for testing and the models generated must be carefully handled, as they are very fragile. Some earlier resins had postmodel shrinkage problems resulting in dimensional problems after being generated. But new modeling resins have eliminated this problem; check with your suppliers regarding the right SLA/SLS modeling resins to use.

Machined Models

Machined prototypes are made from single or multiple blocks of plastic, preferably the material selected for the product if available. These models will exhibit mechanical, chemical, and thermal properties different from those of a molded product. Specifically, if a different material is used and even with the same material, there will be differences in material density, surface skin effects, flow pattern effects, and fiber orientation when reinforced.

Prototype Molded Models

The plastic injection-molded prototypes should duplicate the design in wall thickness, curvature, fillets, radius draft, and other features such as molded-in-inserts. Any variation can mask assembly or end-use problems. Prototype molding a product for testing will indicate whether the gating is correct to control warpage, shrinkage, fill, locations of weld lines and their strength, fiber orientation effects, gas entrapments, and venting requirements. Any change in runner type, length, or size is a major mold change from the prototype mold to the production mold, and can create changes in the production product that can affect the products end-use performance and tolerances. Therefore, extensive testing of both the prototype and production products is required even if changes are not required.

The production mold will produce products that are different from those of the prototype mold. If more cavities are added, the pressure, fill rate, temperature profile in the cavity, and other parameters can and will change. This is especially true when the mold cavity materials are changed, as will the thermal conductivity rates affecting mold temperature balance. The intent of testing is to make sure that there are no surprises encountered when the product goes into production.

Prototype Mold. The prototype mold, typically a single-cavity mold, should have a runner and gate layout that will replicate the production tool and the cooling system and cavity functions as close to the proposed production mold as possible. This may be difficult to do, as the intent of prototyping is to be cost-conscious but still produce a testable product to evaluate and answer questions an how the product will perform. If all the features were put in the prototype mold, such as cooling mold materials and ejection system, it would be the production mold, less the final number of product cavities. Most prototype molds are cast kirksite, machined from soft steel or aluminum, with minimal cooling (using a single cooling circuit for both cavity and core and with inserts usually hand-loaded). Standard mold frames are used with the mold cavity inserted into a standard mold base opening for support using the ejection system built into the mold base.

A multiutility die (MUD) mold frame is designed to hold single or even multiple cavity inserts with an internal, semistandard ejection system. The MUD frame's system cooling is limited as it is universally designed to hold a cavity insert with minimum capability. The cooling is limited to the frame with only conduction of heat transfer to the MUD frame's cooling system. A prototype mold, if the material selection is still in doubt, is built with standard tolerances to accommodate the different material selections and still be within approximate product size for the product's dimensional requirements.

The major benefit of using a prototype mold is that the material will closely approximate the final product in how the product is packed out and how the material reacts to the forces on the product in its end-use environment. A molded, machined, or SLA/SLS product's physical properties are only 33.3–50% as strong as the molded prototype product. After a prototype-molded product is qualified for the application, the prototype mold can be fine-tuned or made more to the material's selected tolerance requirements. The gating can be recut and relocated to a better location with cooling channels added to the mold cavity for making preproduction test products. This gives the design team the best preproduction test product to finalize the product dimensions and perform actual end use testing.

Regrind Consideration

If product and runner regrind is allowed for use in the product, make sure that the percentage is specified, with molded prototype products molded and tested as with virgin resin. This is often overlooked in the design stage, and depending on the mold, runner layout, and scrap products, the reuse of this material is rarely considered. Most products can use 20% regrind without a noticeable lowering of the product's physical properties. This question of regrind usage must be answered before allowing the molder to use regrind in any percentages.

Molding results from a prototype mold, unless the mold is as close to the proposed production mold as possible, can vary considerably. Many questions on the suitability of the selected material can be answered and a wide variety of resins can be evaluated if shrinkage values are not too different, say, within 0.005 in./in.

PROTOTYPE AND END-USE PRODUCT TESTING

Product and assembly design requirements should guide the test plan for both prototype and production products. The product's test plan must be developed on the basis of end-use requirements. The test plan must duplicate the product requirements in both testing materials that will be approved and specified for the product in the material supplier's laboratory or your facility. This involves using a prototype product, modeled or injection-molded in a prototype mold, usually a single cavity, and often in MUD, and frame type of mold.

Die-Cast Models

Die-cast tooling can be used and converted for producing a very close model for testing at a reasonable cost. It should have uniform wall thickness and preferably less than $\frac{3}{16}$-in.-thick wall sections. Products from die-cast molds can be used for end-use testing and as a product shrinkage guide for the production mold.

The accuracy of the prototype product's test results is based on the quality and reproducibility of the product. Only the design team and management can decide which method of prototyping will be used for the program.

Prototype Molded Product Characteristics

Areas the prototype molded products can answer are

1. Does it meet end-use test requirements for mechanical, thermal, electrical, chemical, environmental, and other criteria (creep, and deformation)?
2. Do decoration techniques work, are they acceptable, and does the color match exactly?
3. Are the dimensional tolerances correct?
4. Are there warp, distortion, sink, flow marks, or other surface or product defects?
5. Are there areas of potential failure, such as weld line?

Assembly

1. Can the product be mated to other products?
2. Can it be assembled as proposed, press/snap-fit, welding, bonding, or similar?
3. Are the fit and tolerances correct?
4. Does the product work in the total assembly?
5. Can the assembly be more efficient?
6. Do purchased products, screws, inserts, and other components fit correctly?
7. Is the assembly procedure correct as written?

Moldability. Questions regarding moldability must be answered early in the design program. Some products initially may not appear to be moldable, and some serious thinking must be done to design the product for manufacture. Molds have a great capability for making even complex products with in-mold operations such as unscrewing cores, camming sections, and in-mold operations occurring before the mold opens. These questions must be answered, and you should consult your mold engineers and manufacturing personnel to assist in these areas plus your material suppliers.

1. Can it be molded as designed?
2. Was shrinkage estimated correctly in critical areas?
3. Is the product in tolerance?
4. Is the heating/cooling system adequate, or does it need to be modified?

5. Does the product release reliably, and are the ejection system and draft adequate?
6. How wide is the processing window to produce good products?
7. Is the estimated cycle within projected cost limits?
8. Can regrind be used, and if so, in what percentage?

Testing of prototype products should be to the original design criteria giving equal weight to all factors being evaluated. Should prototype products vary slightly from the specifications, adjust the product to fit without modifying other products that may already be finalized. If not able to do this, note the deviation and see if it has an effect on test results. Or if major, adjust the prototype model and rerun the products for testing.

Prototype products can answer the questions previously listed for product, assembly, and molding ability. Problems discovered during this initial testing stage can be corrected more easily than when the production mold is completed. Problems or modifications not changed or corrected before the production mold's design is finalized and built may require production mold changes later in the program. But a problem not solved at the start, before the mold and components wear, could cause a severe mold casualty.

Such a problem was experienced with a machine tool housing when the armature bushings were changed after the production tool was completed. This required core modifications, which resulted in weakening the original core strength in the cavity. Months later, as a result of not making a new core, it failed when a section of the core broke out as a result of the weakening of the original core caused by the tool builder shimming some sections versus replacing the core with a new single piece. During the high-pressure injection of material, the core fatigued. Resin eventually worked into an undercut section and caused the section to fail during ejection of a product. This caused major mold and product downtime until a new cavity core could be completed for each mold cavity. The OEM decided to continue running the two-cavity mold as a single-cavity mold. The injection molder modified their molding cycle to ensure that the product tolerances were still within the tolerance limits specified. They blocked off the broken core cavity until two new cores were ready to be installed. This minimized product loss to the assembly line and kept the line running. This was a new product, and marketing would not permit the plant to not supply product to point-of-sale sites.

Major material suppliers have made available their physical property product information data files involving the effects of temperature, creep, environmental exposure, and other parameters in computer files that can be accessed by the CAD system for FEA design analysis.

General Electric Company has their snap-fit program and PETS (plastic education and troubleshooting), which allows their customers to enter their database for design and material assistance when specifying their products.

CAD Systems

CAD systems can generate the following types of product formats with their respective software programs: (1) Wire-frame model, 2D and 3D; (2) surface model, 2D; and

(3) solid model, 3D:

1. Wire-frame modeling is more basic using single lines and curves to form the outline and surfaces of the product. It can be used to generate 2D orthographic drawings, which will show all lines, even hidden ones, unless they are programmed to remain hidden on the screen.
2. Surface models are the most useful capable of being generated from wire frame or solid models. They can be shaded to create a picture and used to create NC tool paths or meshed to create finite-element models for filling, packing, cooling, or warpage analysis.
3. Solid modeling is the most useful, showing the product with thickness, mass, and moments of inertia. It is more expensive than wire-frame modeling and requires more memory due to the number of data points used to construct the model. They can be shaded, thus providing a reasonable picture of the finished product.

FINITE-ELEMENT ANALYSIS

Finite-element analysis (FEA) is a numeric analysis in which the product, a complex problem, is broken down into very small subproblems that can be solved by a series of mathematical equations in a software program. The analysis of the forces acting on a body will vary with the experience of the engineer or designer in defining the forces and the complexity of the product under analysis. Time involved for setup of the matrix for defining the product in free space with the newer software programs can range from 2 or 3 days for less complex loaded plastic products to 10–14 days, which is typical for more complex products. Ninety percent of the FEA time is involved in creating the computer model, and the more complex the loading on the product, the tighter the mesh must be created.

FEA is used to

1. Optimize product design where complex geometries and loading restrict textbook solutions
   a. Determine highly stressed areas of the modeled product, including shear stress, creep, and buckling
   b. Ensure that the product can survive subjected to a given force, impact, or fatigue force
   c. Conduct failure analysis
   d. Redesign a product before committing it to hard tooling
   e. Reduce trial and error and guesswork in producing a product
2. Reduce lead time in product design
3. Minimize material usage and cost
4. Reduce product weight
5. Reduce molding cycle by reducing wall thickness

6. Minimize tooling changes caused by redesign of the product
7. Reduce prototype modeling and testing time
8. Increase design accuracy by pinpointing problem areas in the product and evaluating their solutions

There are three main steps in doing an FEA for a product: (1) preprocessing or modeling, (2) processing, and (3) postprocessing.

Preprocessing Analysis

Preprocessing utilizes the design checklist data plus reasonable assumptions on the product end-use functions and requirements. Determining whether your analysis will be in two or three dimensions is the first step in modeling. There are general guidelines to follow, and these are involved with formulating the problem. The majority of information required would be obtained from the design checklist in Appendix A that should contain

1. The forces on the product and where they act
2. Whether forces are concentrated or distributed and their duration of time acting on the product
3. How the product is held in place and whether it is fixed
4. Whether it can rotate or move in any direction
5. Trial calculations to determine the estimated stress, strains, and deflections and where they act on the product for later comparison with the FEA model
6. The product's strength, stiffness, and dimensional requirements
7. What material physical properties are needed for the product
8. How much deflection can be tolerated or produced by the forces
9. Whether the product is subjected to creep
10. Thermal expansion and contraction effects on material properties and dimensions

The final question involves a decision by the designer and the team's estimation of the product's complexity, type of results needed, and the time needed to set up a 3D model versus a 2D model with product thickness specified to obtain a suitable solution. The type and size of the element selected for analysis determine this.

The final question to be answered is to determine the type and size of elements selected for FEA analysis.

FEA Elements

An "element" in simplistic terms is a two-dimensional planar area section of a product. The element is defined as beginning at a corner called a "node." A line connecting two nodes is called a "boundary." The area enclosed by three or more boundaries is

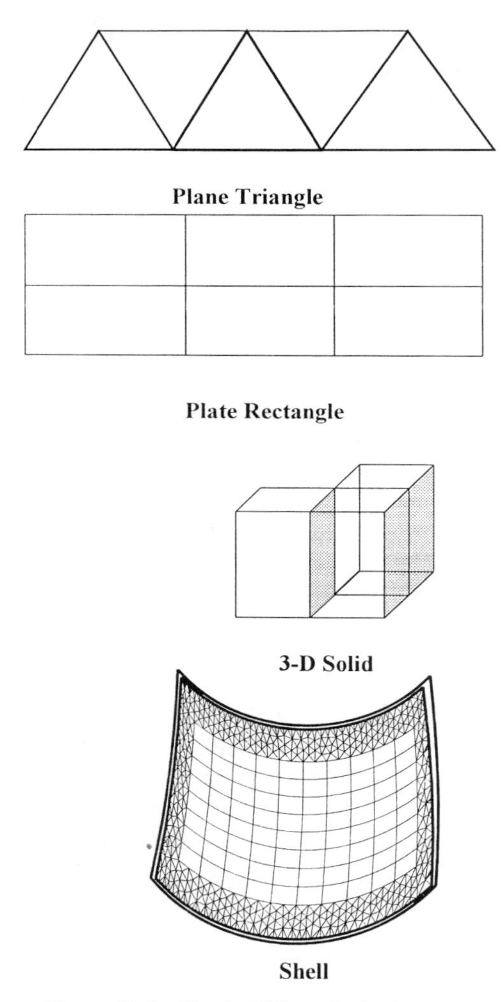

Figure 11.4 Simple CAD analysis elements.

an "element." The simplest of elements is the simple plane triangle, then the plate, and shell and 3D solid as shown in Figure 11.4.

The fewer the number of nodes used for analysis, the less complex the analysis. This means that the computer has fewer equations to solve. Usually for the first analysis this is acceptable to locate the high stress areas. Once these areas are identified, a denser node layout, composed of 4–10 node tetraelements with powerful algorithms, is used to determine the highest stress points in the section. Typically the first analysis is done in 2D layout. A 3D solution is seldom necessary for very thin-walled products (<0.010″ in. thick). But this also means that the final analysis is not as accurate. The rule to use is if an element's total thickness is not greater than 20% of its length, a 3D solid element is not used.

Plate and Shell Elements

Determining the shape of the element is at the discretion of the designer and the shape of the product under analysis. Plate elements are used for flat structures such as laminates and sheets. For thin curved surfaces the shell element is used and is the most common for plastics. The nodes are usually located at midplane, and a thickness is specified. The thickness can vary, and the procedure requires more time to set up because of a three-dimensional analysis of defining the model points in 3D space.

A three-dimensional solid such as a beam or cube is described with eight nodes. It requires more time to set up and calculate but is more accurate and is used to determine internal stresses and temperature gradients in the product such as during mold fill and cooling analysis.

There are almost 100 different types of elements available; even nonstandard ones can be created, for which modeling and processing will take more time. The element is defined by giving each node geometric coordinates in a specified order. When the element is selected, it is then oriented in space using a coordinate system.

The most common is the Cartesian (X, Y, Z) coordinate system, but polar coordinates can be specified, spherical, or elliptical. Once the coordinate system is selected, the product is divided into elements. Determining how many elements to divide the product into depends on the degree of accuracy desired and setup and processing time. The elements linked together form a mesh, and the finer the mesh, and the more nodes used, the higher the accuracy within the defined area as shown in Figure 11.5.

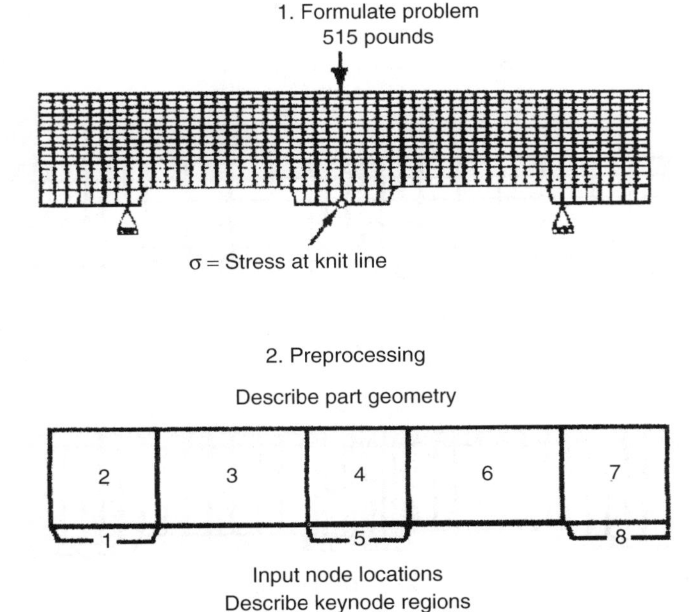

Figure 11.5 Analysis of problem and setting up the node density.

The designer locates the suspected high-stress areas of the section with the first analytical stress analysis performed. Then the fine-mesh elements are specified at suspected high-stress areas, known as *key node regions* (KNRs). Fine meshes are also used at holes, corners, and product discontinuities. The purpose of the KNR is to determine as accurately as possible the stress in these critical areas. The KNR with a finer number of meshes will yield a higher accuracy in stress determination for selecting the product material.

Mesh Density Setup

The FEA analysis begins with the surface of the product drawn in two- or three-dimensional space. Software support systems simplify this by importing standard shapes to assist in building the product model. Once the rough outline and shape of the product is built, the designer then smooths the surfaces to suit the finished product's shape. At this time coordinates are entered into the program and sections divided up into key node regions. This is shown in Figure 11.6. The KNRs are positioned in areas of probable high stress points from the analytical analysis previously conducted on a trial basis and the designer's experience from prior product analysis. The designer can wait until the first trial stress analysis is completed before locating the key node high-stress mesh system. This method is preferred, as the first stress analysis distribution will locate the high stress points in the product's sections. Using a larger mesh density in noncritical regions and concentrating a lower mesh density where the stress is known to be higher will reduce the computer program calculation time. This method concentrates the analysis in the anticipated high-stress areas as shown in Figure 11.7.

For example, a support fixture can be broken down into 200 elements. Then, on the basis of the initial computer stress analysis, use finer meshes to determine

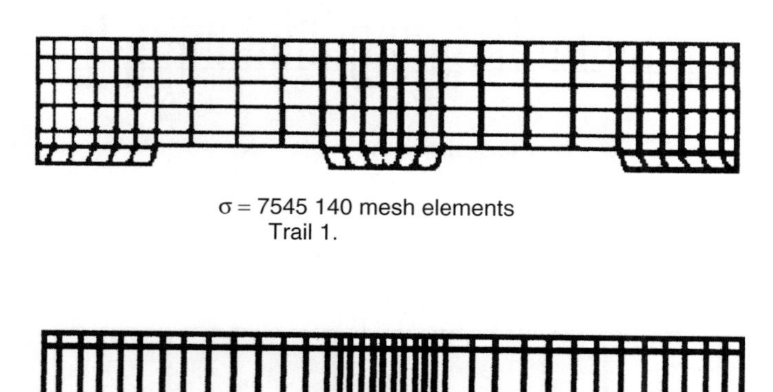

σ = 7545 140 mesh elements
Trail 1.

σ = 7839 327 mesh elements
Trail 2.

Figure 11.6 Finite-element analysis—preprocessing.

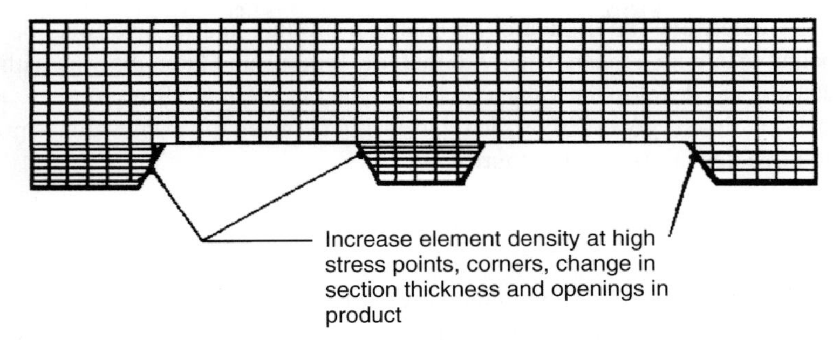

Figure 11.7 Finite-element analysis—final analysis.

the actual stress at likely high stress points. According to the force diagram of the product initially described, these could be impact or support points where forces can concentrate in the product.

Symmetry

If there is symmetry to the product, then only one half of those areas need to be analyzed, with a note input to the computer that the line of symmetry is not a free surface. As an example, only a quarter of a circle need be analyzed when symmetric on a product surface, but the forces on each half or quarter must be equivalent.

Boundary and Constraint Conditions

The designer must also specify constraints or boundary conditions, which describe how specific nodes are constrained or able to move. Each node has 6 degrees of freedom; it can move in the X, Y, or Z direction, or it can rotate about the X, Y, or Z axes. These nodes are typically defined in the Cartesian coordinate system.

Examples of this are a node at a fixed boundary that cannot move in the X, Y, or Z direction. Rotation at the node must also be specified; for instance, it may not be able to rotate about the Y and Z axis, but can about the X axis.

Processing Time and Node Density

How well the designer initially analyzed the products stress requirements on the basis of developed analytical results and knowledge of product design and how the product is expected to react will determine the mesh density initially set up for the product's analysis. This can be learned only from experience, but understanding the parameters of FEA analysis can save the designer both setup and processing time, or the "number crunching" that the computer performs, as it is commonly called. Processing time is dependent on the number and complexity of the elements, or node density. Computer time may be hours, days or even weeks for very complex analyses.

To initially determine the high stress points in a product and to save time, begin as described above, with a mesh density to map the regions and later analysis with finer mesh densities to develop the values on the product at these KNR regions.

Processing time is governed by the relationship between time (t), which is proportional to node density (n), to the fourth power:

$$t \propto n^4$$

Speed is also a function of the computer used for the analysis and the software requirements for its operation. During each mesh analysis cycle, the computer calculates the data and progresses to the adjacent elements equal to

$$\frac{n(n+1)}{2}$$

This is the triangular sum formula, where n is the number of elements. Since one additional element is processed each time, the growth of this factor can be treated as varying between $(n(n+1)/2)$ and n^2. For large values of (n) to be calculated, it can be treated as n^2. Because n is proportional to the node density squared, the total time for processing is proportional to the fourth power of the node density. On some computers using 2D triangular mesh, the time to complete a 30-cm^2 mesh at 0.78 nodes/cm is approximately 24 min. Increasing this to 1.18 nodes/cm or 1.5 times increases processing time by a factor of 5, or 120 min. As computer processors become faster, the processing time decreases. Depending on the computer program, it may be required to input the basic equilibrium equations for the computer to perform the analysis. But in the majority of FEA software programs this is included.

There are CAD/FEA interfaces available that allow the CAD-generated surface to be divided into elements, saving preprocessing time. Inputting the trial material's physical property data involves entering the material's modulus of elasticity (E) and Poisson's ratio (γ). For plastics a single curve or series of curves for E are entered because of the materials' nonelastic behavior at different temperatures and strain rates. The selection of the curves (Fig. 11.8) input must account for the conditions that the product will experience in service, temperature, humidity, dryness, and other factors as well as the data available from the material supplier. The computer has the ability to interpolate these data into a curve for the desired temperature and strain rate.

The designer can also tailor the material's physical property data by placing the product reinforcement rib's thicker sections and other properties only where needed. The designer can enter both resin type and reinforcement data into the computer, specifying type, quantity, location, and orientation of the reinforcement. This implies (using FEA) that the strength that a product requires can be specifically located to yield the best results when converting to plastic from a metal product. FEA allows the designer to input the plastic material's entire physical property data with increased accuracy, thereby reducing the safety factor question by accurately determining the stress at key points using the material's actual properties for the end-use conditions.

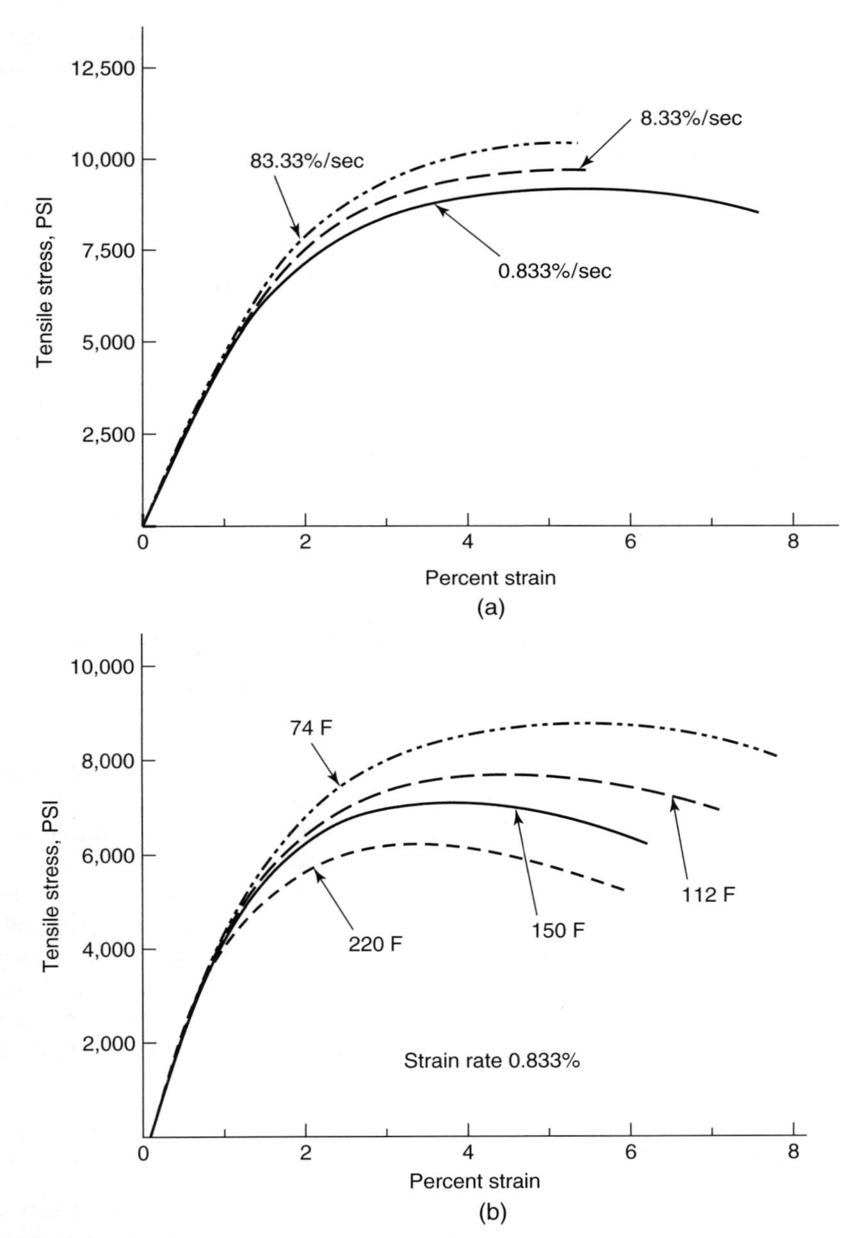

Figure 11.8 Effects of strain rate and temperature on PPO modulus values: (a) entire field showing stress concentrations in 1000 psi; (b) center section, high-stress area of part (numbers in 1000 psi). (Adapted from Ref. 19.)

When doing the calculation by hand, only a single value of E is used, requiring a larger safety factor. Now the computer can calculate the stiffness between nodes and the "stiffness matrix" of the entire model.

Once the product model is described with node density layout, the next step involves inputting the forces into the geometric model. Each force's magnitude, direction, and point of application must be specified.

These forces include

1. Continuous or distributed loads on an element boundary
2. Concentrated loads on a node
3. Pressure
4. Gravitational forces
5. Centrifugal forces
6. Dynamic forces, which vary as a function of time, such as wind
7. Impact forces
8. Vibration and other forces
9. Displacements
10. Moments
11. Thermal stresses, caused by thermal variation across the product

Processing the FEA Model

The computer now begins the analysis by solving the simultaneous equations for the node layout. Each element has a matrix of equations generated that are combined into a global set of matrix equations for analysis. FEA solves either matrix or differential equations giving approximate solutions based on mesh densities.

This is the most time-consuming step of the analysis, often requiring hours for mesh density control. After the calculations are completed, the data are graphically displayed on the product, as isoquants of various stress levels, like a topographic map using colors to designate and identify stress levels as shown in Figure 11.9a. Highly stressed areas can be enlarged for more detailed analysis as shown in Figure 11.9b. The output is more graphic when color defines the stress areas with topographical boundaries.

PostProcessing

After the first analysis is completed, it will usually be necessary to make changes in the mesh density. This is done to ensure that accurate predictions of the products stress are calculated and to optimize the mesh density of each key node region to ensure that the maximum stress values are determined. However, from the initial results, it must be determined whether the results are as anticipated and within the expected levels. The FEA results can be compared with the initial, single-point analysis, hand-generated. If the results for magnitude and location of the stress fields appear where they are

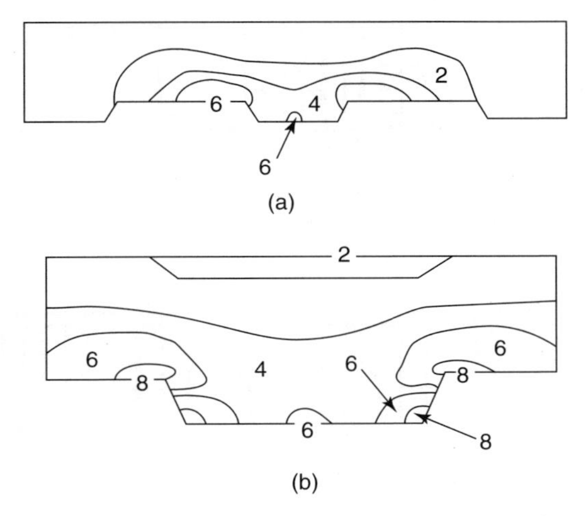

Figure 11.9 Finite-element analysis stress field displays (processing). (Adapted from Ref. 10.)

anticipated, the nodal placement of the forces and their action fields are probably correct. If not, then the inputted data should be rechecked for any errors. This should include material property data, curves, and constraints placed on the product or data not considered applicable but a factor in the products function, such as restraining forces if securely anchored in place or to another product.

The software program rotates the body showing where the deflections and shear forces act on the product to assist in the analysis. This simulation will show where additional ribs, thicker sections, and so on can be located to control excessive deflections and twist in the product if greater than anticipated.

Adding reinforcement and stiffening to the section and then rerunning the analysis in the same mesh density will show the effects and indicate other weak areas for more detailed analysis. In cases of deflecting numbers, an analysis can be run to determine the allowable deflection permitted in a situation before the section will fail. These data can be used to determine product tolerance allowance on critical areas and rule out noncritical areas.

Retrial Calculations. The numbers of elements selected, including mesh density, should be doubled with each successive analysis in the key element or node areas to determine how the stress values change. This will develop the maximum section stress values in critical sections. A point will be reached, after several analyses with increasing mesh densities, where the increase in calculated stress stabilizes, or is maximized for the time and accuracy desired for the section. When the last change to the previous analysis increasing stress is 50 psi or less, the next analysis will show marginal increase in stress level. The designer can then be assured that the calculated stress is near the peak value for the section. The relationship of increasing mesh density to predictable results is shown in Figure 11.10.

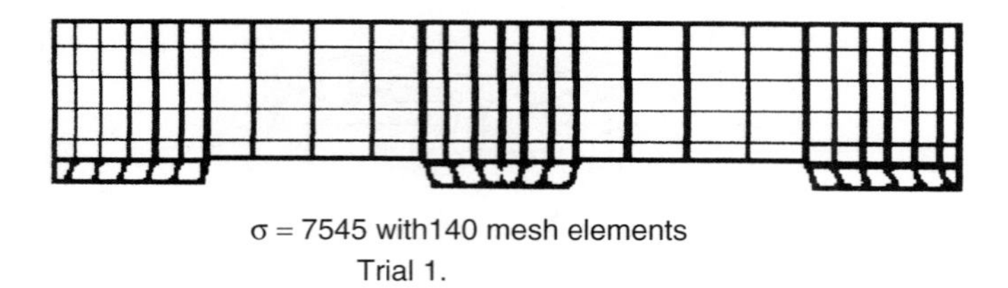

σ = 7545 with140 mesh elements
Trial 1.

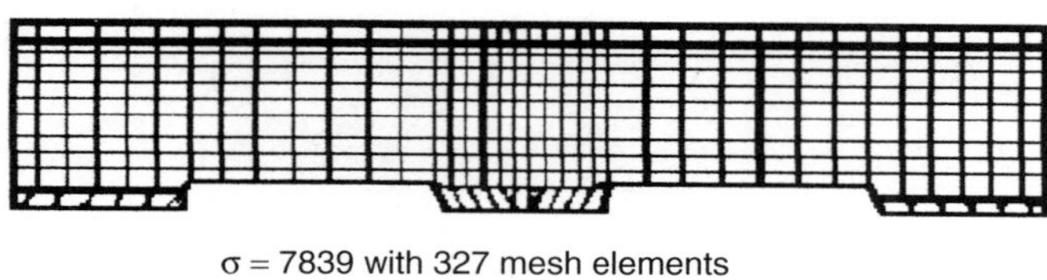

σ = 7839 with 327 mesh elements
Trial 2.

Figure 11.10 Accuracy increases as the number of nodes increases.

Failure Analysis

FEA is used for failure analysis should a product fail in service, and an analysis can also be simulated prior to design completion using conditions that may impact the product in service but that were not originally considered in the analysis. Alternate materials can also be evaluated as potential material candidates for cost savings and future consideration for a redesign of the product prior to prototyping. Any forces for a nontypical situation can be set up and analyzed. This information is used to predict where failure can occur or if it already occurred, where the product could be better designed or laid out in the mold considering areas such as gating, weld lines, and material section. This will allow for modifications to prevent future problems. Different material candidates can be evaluated to assist in selecting the best and most economical resin candidates. Additional FEA analysis techniques, include mold fill and cool analysis using the same nodal mesh density model generated for FEA or used for analyzing resin flow, heat transfer, and cooling.

FEA reduces the product's design time and increases product reliability and customer confidence, especially for high-product-liability items. FEA pinpoints areas of high stress as shown in Figure 11.11 that can be compensated for in designing the product correctly the first time. This will allow the design team to tailor the design to the material's properties. It will also allow lower safety factors and often point out material properties not always obtained by the typical property testing methods.

For example, materials exhibiting high compressive strength will exhibit shear strength values higher than they actually are and should be considered when using data sheet shear strength values in actual product design. As earlier discussed, data sheet values should not be used in product designs, only end-use property data from the testing of materials. The FEA nodal mesh density also provides the mold fill and cool

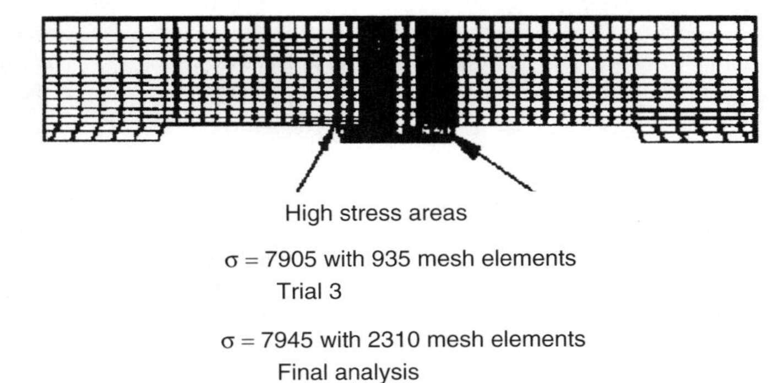

Figure 11.11 Fine meshing highlights high stress areas, and doubling mesh number achieves maximum stress levels in the critical areas.

analysis used for analyzing resin flow, heat transfer, and cooling when designing the production mold. FEA software suppliers often offer this capability in their programs, which are compatible with the ANSI setup and notation for using the same data points as the stress analysis.

Mold Design Considerations

The design of the production mold is crucial to obtaining a high-performance and quality product. The design team's responsibility continues until their customer accepts the product. Therefore, how a material will process in the mold must be considered. There are several hundred interactive materials, molds, and processing variables for viscoelastic nonisotropic plastic resins. The processing variables control the quality of the products manufacture using the best designed and best built mold that the program will allow within its cost limitations. The mold must be built according to the product's design requirements and required tolerances to obtain the specified dimensions, tolerances, volume of products, and cycle-to-cycle product reproducibility to meet the customer's quality specifications.

Table 11.1 lists basic mold design information for the design of a good mold. Use the mold and processing checklists to answer the questions necessary for designing a mold to produce quality products. All questions regarding the processing and manufacturing variables should be known prior to designing the production mold. Then, when manufacturing begins, fine-tuning the molding operation is all that will be necessary by the molding team. The design team can determine how the product will process in the mold even before it is built by using the FEA model information in a software program for mold filling and cooling analysis. Then, if any product revisions are necessary to improve performance or molding quality, they can be incorporated into the production mold.

FEA analysis will save both time and money to keep the program on schedule as computerized manufacturing steps, shown in Figure 11.12. After the product design

Table 11.1 Mold design guidelines

Part size and volume	$\leq$100 in.^{2} of projected area; $\geq$1 million parts	>160 in.^{2} of projected area; $\leq$1 million pieces	Same as previous columns; $\leq$100,000 pieces(or preproduction)	Prototype (facsimile)
Mold design	Complete details of all part form dimensions in the mold design by moldmaker; all inserts detailed except standard purchased items (injector pins, bushings, etc.)	Same as previous column	General concept layout of mold only; details to be supplied by mold maker optional	Same as previous column
Cavity/core steel	Hardened AISI H13, S7, 420 stainless steel A2	Prehardened steel, P20, 414 stainless steel	P20 cold rolled steel, cast steel	Aluminum or any of the previous columns
Base steel	Prehardened steel 280–320, BHN, 4140/P20	Same as previous column, integral base and/or cavity core	Any steel	Any steel or aluminum
Gating	Determined by molder to meet part design	Same as previous column	Same as previous column	Same as previous column
Base construction	Standardized A (cavity plate) or B (ejection half) series-style bases guided ejection; tapered parting line locks (locators with zero clearance)	A or B series integral cavity and/or core could be cut in solid base instead of inserts; guided ejection; parting line locks	Whatever is necessary to hold cavity and core (doweled location on top of plate; universal mold base system)	Same as previous column
Cooling	In cavity and core block; also in inserts to maintain best possible temperature control (use heat pipes)	Same as previous column	Whatever is necessary to simulate production parts (possible sacrifice; longer cycle); typically water in mold base only	Same as previous column
Area requiring moving components (not in line of draw)	Slides: mechanical or hydraulic; lifters angled or straight; all components to be hardened and made of dissimilar material or hardness to eliminate galling	Same as previous columns	Possible short cut, but must simulate production parts	Could be loose pieces (manual inserts etc.) that come out with part and are reinserted for next cycle
Special details	Fragile area of core or cavity inserted for ease of maintenance; bar ejection; blade ejection	Same as previous columns	Not necessary	Not necessary

Source: Adapted from Ref. 6.

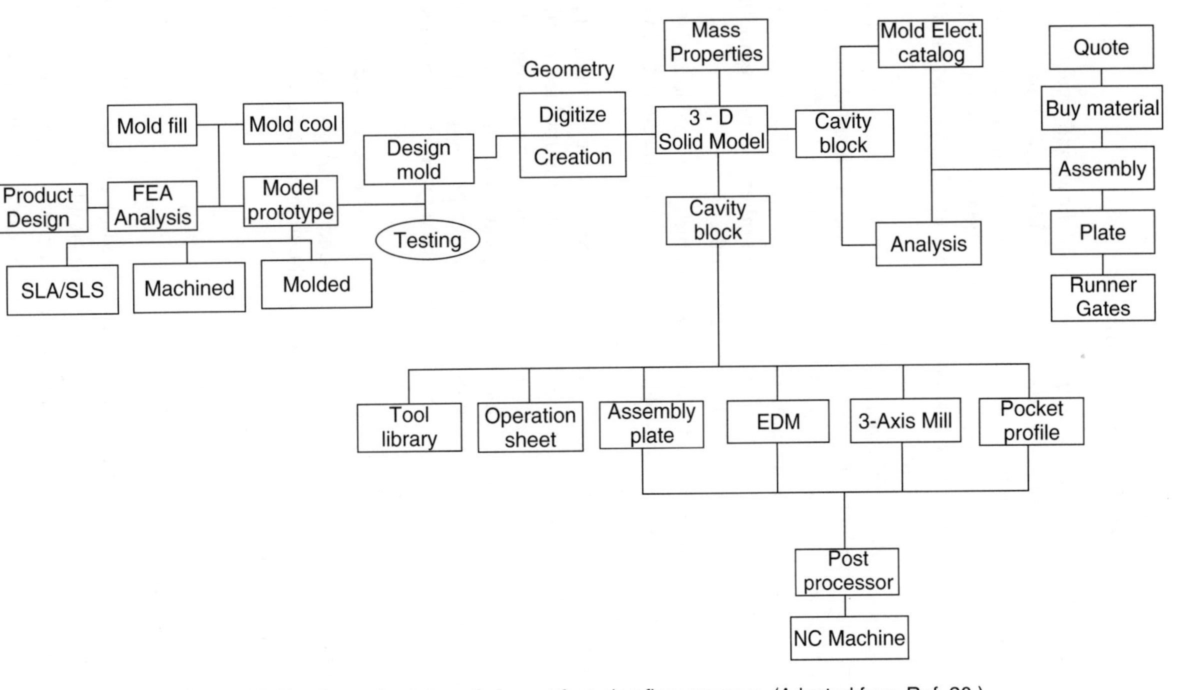

Figure 11.12 Computer-integrated manufacturing flow program. (Adapted from Ref. 20.)

analysis and model generation and testing, the mold and processing variables are evaluated. If FEA analysis is not used, other mold and processing analysis methods can be used to evaluate the system.

Alternate Mold Process Analysis Methods

1. Resin characteristic analysis: melt index, spiral flow, and material shrinkage in the mold based on standard processing conditions using published data and flow test versus product dimensional and tolerance requirements. *Analysis:* Results obtained are standard, typical results from nonproduction or standardized test conditions, material test samples, and/or molds. Poor translation to actual production mold and products.
2. Sampling in a similar mold. *Analysis:* Popular method but actual product results can vary when product geometry changes at a critical location or mold layout, cooling and gate and runner design variables are changed.
3. Sampling in the production mold. *Analysis:* Can be expensive if incorrect gating, cooling, runner system, cavity dimensions, and other features, combined with processing variables, do not attain desired results.
4. Use of oversimplified FEA 3D software yielding questionable results used to develop item 3. *Analysis:* Variables selected do not represent actual mold, material, and processing conditions.
5. Sampling in a prototype mold. *Analysis:* Used in major product programs. Products from the production mold rarely replicate the molded results of the prototype-molded products. Material flow, cavity temperature control, number of cavities, and final production machine settings are never matched. Production product results can and do vary extensively from prototype products from a single-cavity prototype mold. But it can produce a product for end-use testing that will be close to finished product properties.
6. Reliable 2D or 3D moldability analysis (fill, cool, shrinkage and tolerances).
7. Combining product/process analysis items 5 and 6 will give the design team the best calculated and observable picture obtainable before the product and production mold designs are finalized.

Mold Filling Analysis (Simplified)

Plastic material flow into a mold cavity, either heated or cooled, is nonisothermal. When the molten plastic comes in contact with the mold surfaces, the runner and mold cavity becomes fixed in place and immobile. The skin of the product freezes in successive layers as the material cools and solidifies when injected in a continuous flow, under pressure, into the mold cavity. The molten resin flows like lava from a volcano as shown in Figure 11.13. The melt front flows from the molten center section onto the cavity surfaces by flowing over the first material laid down on the cavity walls. The core material flowing over the resin just laid down on the cooler

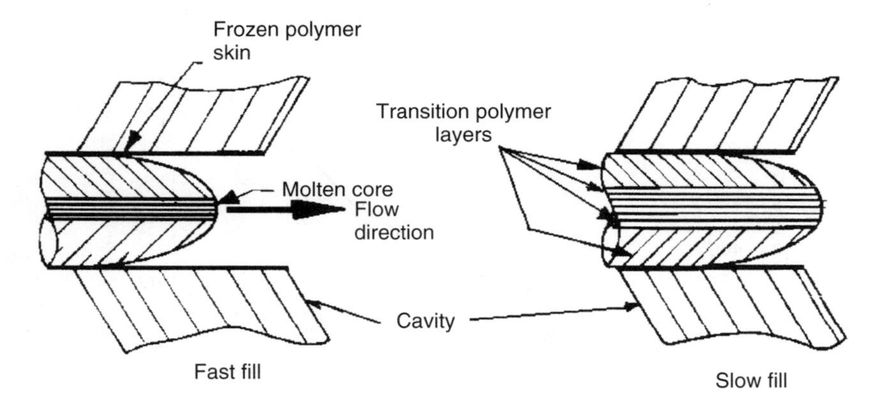

Figure 11.13 Effects of injection rates on cavity filling.

mold surface adheres and solidifies. This boundary layer of initial resin laid down on the cavity wall restricts the material's flow channel. In the mold cavity this is the product's section thickness through which the follow on molten resin in the barrel in front of the screw must flow. The buildup of resin, known as the "skin effect," also insulates the runner and cavity sections during filling, which helps to keep the material hot and fluid until the cavity is filled. Then additional packing pressure densifies the core, under packing pressure with more resin forced into the mold cavity, as the resin cools and shrinks. These narrowing sections require high fill pressure to continue moving and packing the resin into the mold cavity, which affects the dynamics of the liquid resin. If the pressure and injection rates are too high, this will increase the material's shear heat, which will thin the melt viscosity, affect material properties, and possibly cause flashing of the mold parting lines. The fill analysis run on the FEA model can answer questions as to the correct fill rate and pressure profile to use during the molding operation. Some thin sections may require higher pressure and a faster fill rate to fill and pack out the product in higher-viscosity resins. This is an area that the design team can rely on for answers from their production/processing personnel. The material's skin or boundary-layer thickness can be controlled by injection rate and cavity surface temperature. Typically the hotter the boundary layer, the denser it will become as packing pressure is increased. Each variable affects the product's cycle time, based on the resin used, product geometry, wall thickness, and the mold's cooling design and temperature dynamics. Each material and product will have an optimum fill rate to fill the product cavity without creating excessive gate shear heat produced in the resin. Cavity fill and packout must be accomplished without creating excessive molded-in material stresses, which can create warping and lowering of physical property values in some product geometry's. In multicavity tooling a rule used for approximating fill rates, in grams per second, is shown in Figure 11.14 for reference purposes. The design team must be aware of these material molding dynamics to correctly locate the cavity gate to ensure that maximum fill and flow are attained, especially for long flow and thin product sections. If the mold is not gated

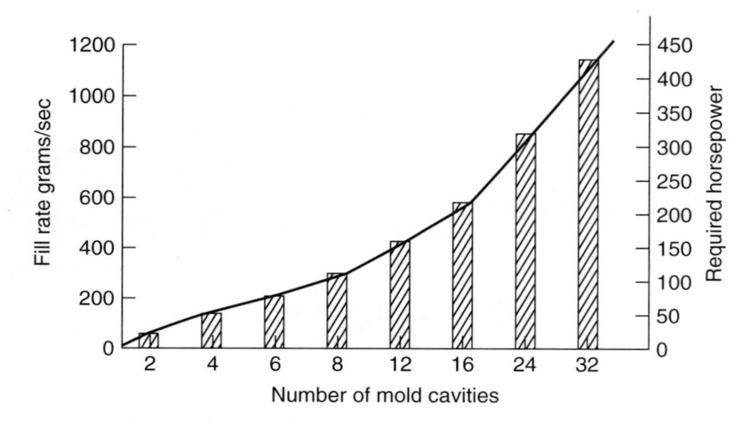

Figure 11.14 Mold cavity fill rates.

correctly, the molder will have to adjust their molding conditions, melt temperature, fill rate, mold temperature, cycle time, and pressure to produce acceptable molded products. The plastic resins must always be kept within the supplier's recommended processing temperature range and not overheated.

If filling the mold cavity is a problem, it can be related to runner and/or gate size or temperature. Often increasing the mold cavity surface temperature can increase flow by extending the resin's freeze-off time. This often increases cycle time but allows the cavity to be filled and packed out without creating excessive shear heat in the material at the gate, which can degrade the material. The product cavity must be uniformly packed out to optimize the material and product's physical properties. The design team must consider these items before the final product section, thickness, and production mold is finalized.

Mold Filling and Cooling Analysis

A majority of plastic support software companies have developed the software for CAE/CAD/CAM systems to both improve product and process analysis and speed product development programs. These programs can accurately analyze the requirements for filling and cooling the mold cavity to produce quality products. New mold software programs assist in sizing and selecting mold components, mold base, support plates, ejection system, and other elements for reducing the time required for designing and building the mold. This is accomplished with software programs that can generate numerically controlled (NC) data to machine mold cavities using NC machining methods.

This involves designing in 3D for the layout and location of the product cavities, the size and type of runner system to use, plus gating and layout of the mold cavity cooling lines. The program will also assist in calculating the temperature profile as the cavity fills and the venting as the cavity is filled for efficient and uniform cavity temperature control. All of these factors, coupled with the product's geometry, material, and Injection-molding machine conditions, must be integrated together to produce a product to meet the "customers satisfaction."

The desire and need to speed new products to market and reduce product liability can be enhanced with these new techniques. Engineers and designers are moving away from using the traditional product design methods, where decisions were made continuously in the program and not finalized until a prototype testing of a modeled product was completed. This was costly and time-consuming.

Simultaneous or Concurrent Engineering Concept

Enter the "simultaneous" or "concurrent engineering" concept, which places the responsibility for the product's design and manufacture early in the program. Using the existing simulation software systems, the variables for all phases of design, manufacture, and even through secondary finishing operations can now be decided early in the program.

This involves more initial time spent in the areas often overlooked or not considered until prototyping showed a problem existed. The design team approach using all the companies and supplier assistance in the initial stage of product design is showing positive results. It has reduced product revision costs in the programs by the elimination of engineering, tooling, and processing changes before the product is in production. The old method of design, build, test, redesign, design, build, test, and redesign has been eliminated.

This reduces the traditional method with higher costs versus the "concurrent engineering" method. The concurrent method requires more time early in the program, but the important decisions are made early, which will dramatically reduce cost and produce a world-class product which is the basis for the ISO 9000 manufacturing quality system.

Some companies will take the traditional approach of designing a product and having a mold built without answering all the design and mold questions such as filling and cooling. Then, when products are produced, they are not to size or are warped, or other problems may have occurred as sink marks on a show surface. This also includes processing problem areas such as venting, draft, and ejection points. These companies often relied on their in-house or outside mold builders to know what the product must do in its end-use environment and produce a mold to accomplish these items. The resulting mold and processing parameters were often found incapable of producing acceptable products no matter how hard the molder tried.

This often led to product and tooling changes, gate redesign, cooling layout inadequate for temperature control of the cavity, and product dimensions that left the molder with a mold totally incapable of producing good products. Often the mold had so many revisions that it would fail early in production, resulting in additional costs, lost production, and an unsatisfied customer.

To solve these problems, companies developed software programs to simulate early in the design phase a model of the product for analysis. The standard CAE software now used is very versatile and becoming easier to use. Product development

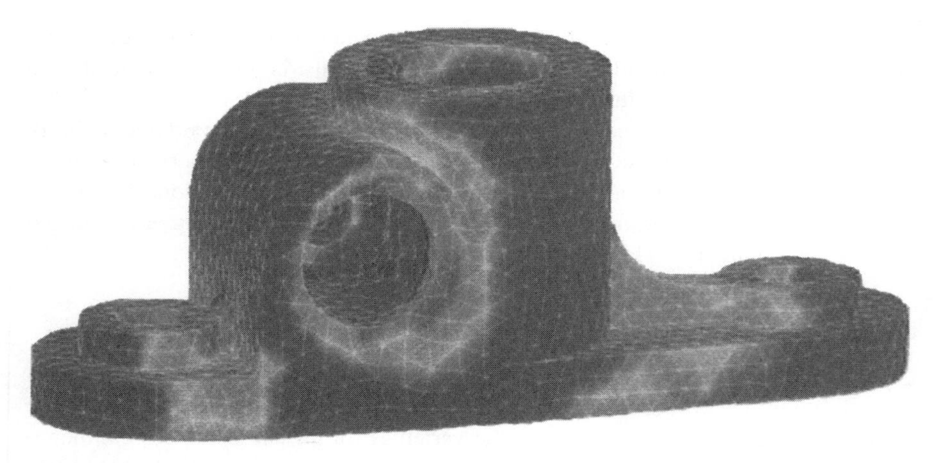

Figure 11.15 3D solid finite-element analysis mesh model for analysis. (Adapted from Ref. 3.)

begins with developing a solid computer-generated model of the product as shown in Figure 11.15. This model presents a 3D image of the product used to verify the form, fit, and function of the finished product. It also provides weight, volume, and inertia properties that can be used for cost and performance evaluation of the product for the end use environment.

Using FEA a mesh model, superimposed over the solid model, Figure 11.15 is generated to analyze performance and process simulation of the product. The mesh density of the model is selected to yield the desired results on the basis of initial structural design calculations. The mesh model is used to analyze the product's static and dynamic behavior properties using material data and time–temperature curve values supplied by the material supplier for end-use operating conditions. Then, if changes were required, a new material or section change could be made and a reanalysis of the product run until acceptable.

These relatively inexpensive changes were not possible before when using a model or molded prototype product and testing it under end-use conditions. As a result, many decisions made early in the program had to be changed in accordance with the test results. The cost and time were high, and results often necessitated a material or product design change that required a new prototype. If the new material's shrinkage was very different, it could require extensive prototype mold changes. Now, once the initial design is accepted, the manufacturing engineers for the mold design can then use the FEA mesh model. They can use the model to simulate mold filling, cooling, venting, and other process parameters. These include pressure, temperature, melt rheology, mold temperature, product section thickness, and time considerations to produce a good product. The software can produce color contour plots of flow velocities, pressures, temperature shear rate and stress, flow fronts with respect to time, and weld line formation locations identified with respect to gate locations. By optimizing the mold and process parameters, one can now produce a saving in cycle time and improve product quality.

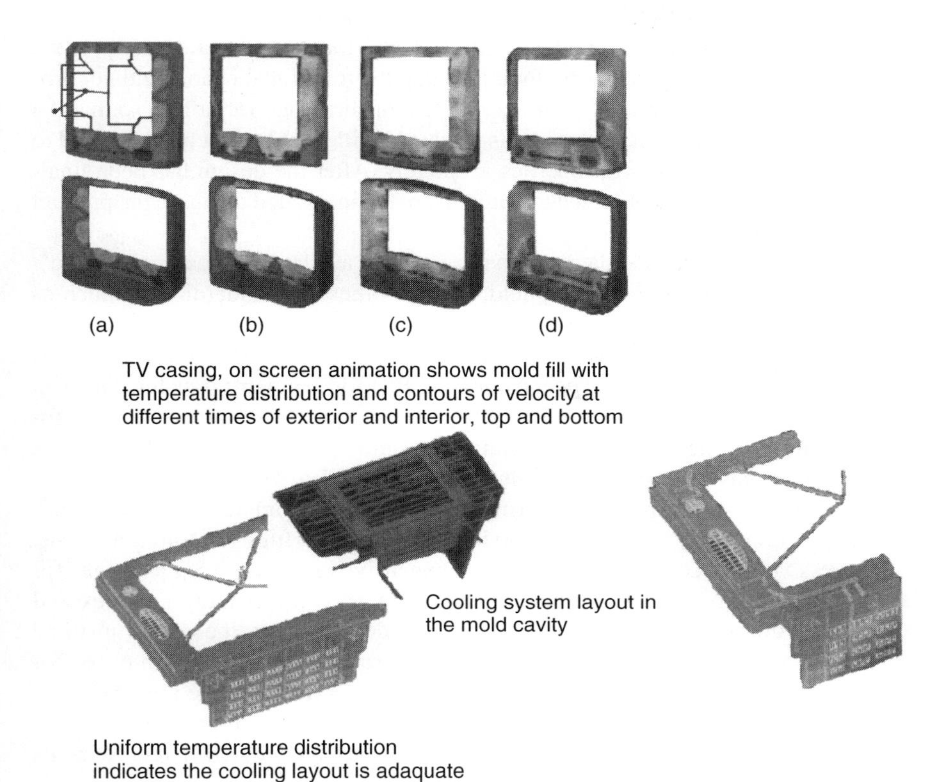

Figure 11.16 Mold fill and temperature profiles assist in molding. (Adapted from Ref. 4.)

By knowing ahead of time the fill pattern, temperature profile in the cavity of the product and cavity surface (see Fig. 11.16), the time required to reach minimum product ejection temperature, deformation due to thermal induced warpage can be eliminated. These data can be used to maximize coolant flow and pressure as temperature rises in the mold during filling and temperatures through each element thickness to correctly size and route the optimum cooling line layouts.

Product filling and solidification analysis uses the same FEA model, utilizing data from each to improve results. This has also led to warp and distortion analysis improvements. Information from material suppliers on their resin rheology is becoming more available. But with the proliferation of compounds, this still needs to be improved. They customarily have data only on their most common grades, and this information database needs to be increased for all of their grades.

Optimizing a product's performance and processing variables must go together in the analysis. Thick sections required for fill must be evaluated for weight and ribbed construction considered that add strength and flow directors for a product. Optimizing product thickness and strength for fill and warpage elimination can lead to a less costly and better-designed and more easily produced product.

Also, other resins can be simulated in the model for cost reductions or product improvements by adding additional functions that the traditional resin could not provide. This eliminates the prototype mold problem of shrinkage variations, so that if a prototype mold is later built, the products produced will be able to fit and be tested to simulate the end-use product properties accurately. After the design has been finalized, the CAD package can be used to output from the integrated database the product and assembly drawings.

There are also mold base data software programs available to design the mold's components as from DME in Southfield, MI and others to standardize as much as possible on the mold components.

The data stored in the CAE solid model system, supplied by most vendors, support the *Initial Graphics Exchange Specification* (IGES) for geometry transfer. This can be used to produce complex 3-to-5 axis sculptured surface NC program tapes for cavity machining. The use of NC machining can reduce mold lead times by 25–33% and also ensure cavity-to-cavity reproducibility for multicavity molds.

The use of the concurrent engineering with design team interaction plus the use of reliable engineering methods for product design, mold filling, cooling analysis, and process variable selection can yield less expensive products. CAD packages for data transfer are now under $500.00, but the FEA software is more complex and expensive. However, when it is used for product development, the cost is justified as the user gains proficiency and realizes that the savings can be well in excess of the software initial cost and learning time curve to effectively utilize the programs. Additional benefits are having the documentation in "real time" for all phases of the program and the analytical results documented to meet the ISO 9001 requirements and meeting the customer's product "satisfaction."

Analysis Utilizing FEA for Molding Assistance

The FEA design analysis determines the major stress points in a product by studying the applied forces on the product. The same mesh layout is used to analyze how the cavity fills. The variables involved are injection and packing pressure and shear of the resin during injection at the gate, which affects the product's packout, porosity, warpage, and surface appearance, including blush and sink marks. Tests have shown that the volatility of a contour parameter, a sharp corner, is a measure of its sensitivity to relative, rather than absolute, mesh density. Geoffrey Engelstein, GR Technical Services, Mountainside, NJ, conducted tests on mold filling analysis (MFA), using melt temperature, injection pressure, and material shear stress at the center of a 5-cm square for a range of node densities. His results are shown in Figures 11.17 relating the variables of shear stress versus pressure and temperature to fill time.

The analysis concluded that a 5-cm square with 5 nodes/cm shows the same accuracy as a 25 cm square with 1 node/cm. The most reactive in variable analysis were shear, shear stress, pressure, and resin density, and temperature was the least reactive.

Most of the variable discrepancies were noted at the beginning or end of the fill cycle with the points, somewhat random, but converging on a curve with minimum variation for analysis. When more precise values are required, such as for blush or

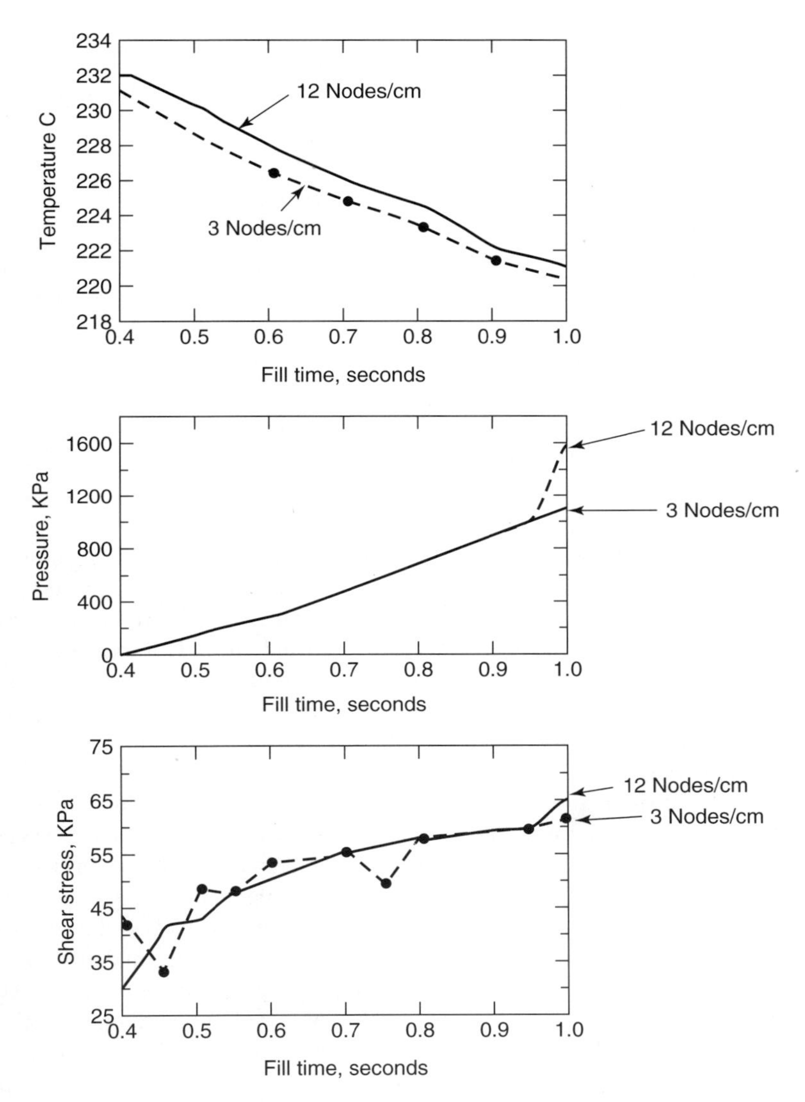

Figure 11.17 Property effects based on node density. (Adapted from Ref. 9.)

optical clarity, using the variable shear stress, a higher mesh density is recommended. The same is true for stress and deflection accuracy as shown in Figure 11.18.

A similar fill time analysis is used to determine where weld lines may occur. When two converging melt fronts meet at the edge of a mesh element, or if unvented air is trapped in a blind pocket of the cavity or blocked from escaping from the mold cavity by too narrow a vent, MFA could identify a weld line forming at a single node position.

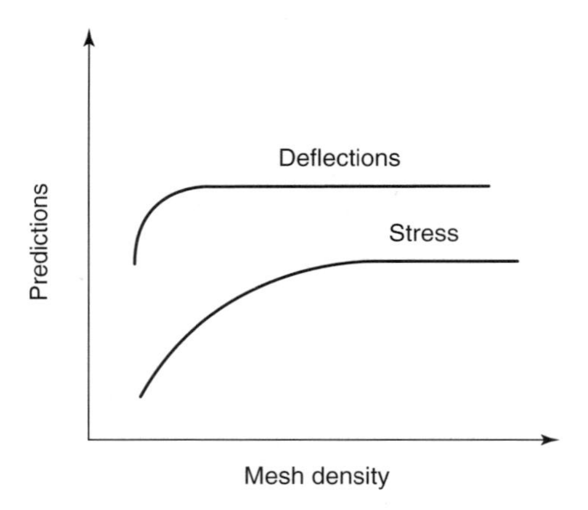

Figure 11.18 Mesh density increases reliability.

Weld lines form wherever melt fronts meet after separating, as flowing around an obstruction, boss, or hole in the product. They also form when multiple product gating is used and the converging melt fronts meet within the interior of the product. The flow through an element that is considered flat is smooth and linear. However, in a section where the path changes at curves or obstructions, the mesh density must be increased for more accurate analysis. When beginning an FEA fill analysis, select 30–50 nodes along the length of the product regardless of its overall size. This allows each node in the mesh analysis to have a calculated fill time associated with it to determine when the described area of the product is filled and packed out 100%. Within any element, any point along the edge or at an interior section will have a fill time that is the average of the element's node geometry, whether triangular, rectangular, cubic, or of another shape. A weld line formed by trapped air will form at the node that is filled last and has the highest calculated fill time value. Accuracy of the analysis is then related to node density with a 0.05-cm accuracy of location using a 2-node/cm density.

Users of these software analysis systems have to rely on experience when setting up the initial mesh density. This is to avoid chasing a weld line around a section of the product. Once a weld line is identified, a higher density is specified in the area. Some software programs perform this modification automatically, known as *adaptive meshing*.

The design team using FMA realize that the precise locations of the calculated values are not absolute. The location and value, minimum or maximum, are always on a node. A detailed location can be obtained by increasing mesh density in any suspected high stress area. Using midplane, 2D analysis linking numeric thickness, further considerations are the effects of bosses, curves, fillets, and rib sections. If insufficient meshing, low node density is used in these areas, it can affect the product's filling analysis, due to decreasing fill pressure. Pressure drops occur whenever the resin is required to turn a corner in the mold's runner system and the product cavity. Therefore, if one section of the product is undermodeled, it will affect the analysis of the entire product as pressure drop will not be identified and used in the fill analysis.

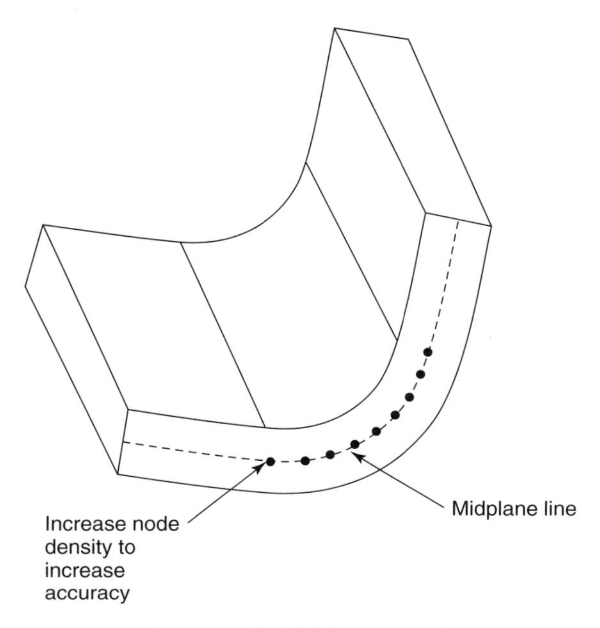

Figure 11.19 Midplane analysis.

Midplane curve lengths in the model should be exact, and the curved section must have a minimum of three nodes to describe it. If less, in a long curve, a high percentage of error results as shown in Figure 11.19. Using a higher node density in curved sections has a minimal time calculation effect, so with curves of greater radii (>1 in.) additional nodes are recommended for a more accurate analysis. This is shown in percent error versus node density in Figure 11.20, and holds true for all FEA analyses. This is predicated on desired accuracy of results and increasing processing time as ($t \propto n^4$). For a round boss, use a minimum of six nodes to describe the curve.

When describing reinforced sections, use the rib height above the surface to calculate volume. A rib should be tapered, draft for release and material conservation. The tapered wall section can be stepped, node element to node element, with the specified rib thickness the average of the rib's base and rib top thickness. This is important when shear and pressure drops are calculated in the section.

These product sections are perpendicular to the section's fill direction and often can trap air at the end of their fill length, as they may be the last areas to be filled and packed out. In any filling analysis the material will always take the path of least resistance, and areas perpendicular to the fill direction will be the last to be filled and packed out due to pressure gradients in the filling process. Therefore, the narrowing or widening of the main material flow path will have dynamic effects on the fill of a cavity. This relates to pressure variances and also when the material will solidify in the cavity. Additional elements in these areas will have a pronounced effect on filling accuracy and will affect the product's appearance, tolerances, and end-use performance.

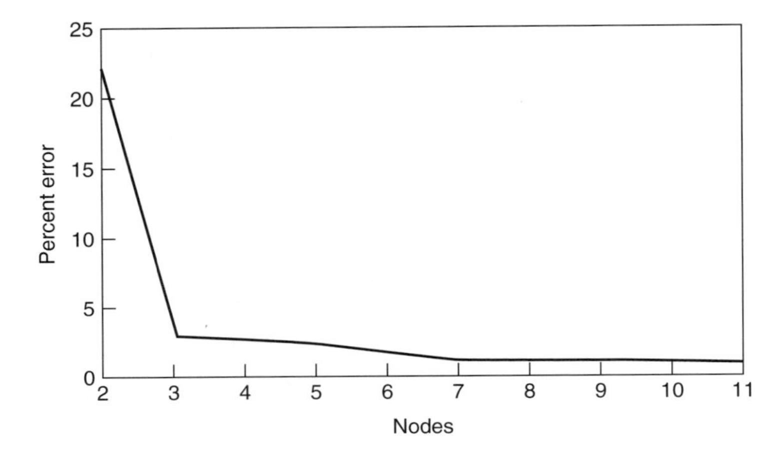

Figure 11.20 Accuracy versus node density.

CONTROL OF PRODUCT TOLERANCES

Control of the product's tolerances is defined as a function of the material, mold, and processing variables. All three have variables that will affect the product's final dimensions and quality. If low cavity pressure or resin temperature occurs in critical tolerance areas, there is a high probability that the resin will prematurely freeze off. This results in the product not being packed out to obtain the required dimensions. Plastics shrink on cooling, and variables such as pressure and material and mold temperatures will affect the product dimensions. All tolerance stackup dimensional variability affects the final product's quality level as shown in Figure 11.21. Factors to consider in defining the product tolerance limits are

1. Product tolerances
2. Processing tolerances
3. Mold cavity tolerances
4. Material shrinkage variability

Product Tolerances

The final dimensions for the product and mold have their own separate tolerances. They set the control limits for the processing variables, mold design and processing limits of the manufacturing variables. The tolerance variables for a general-purpose polystyrene for the mold and selected product tolerance are shown in Figure 11.22. The material's average mold shrinkage rate, shown in Table 11.2, must also be considered when cutting the mold steel for meeting tolerances. How the product is gated, layout in the mold cavity, and thickness selected for the product and flow and fill direction will affect the product's final dimensions.

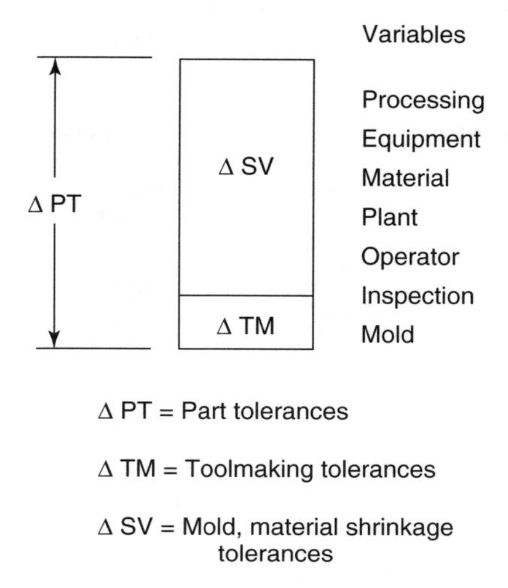

Figure 11.21 Part tolerance factors.

This is illustrated in Figures 11.23 and 11.24 for injection and fill pressure and gate location to product layout in the mold cavity. The mold designer must consider all of these variables in designing and laying out the mold.

The tightness of the product requirement—precise, technical or coarse—affects the mold tolerances. When there is a more precise demand for the product tolerances, as labeled “fine” and “commercial,” the mold cavity dimensions must be held to an even tighter tolerance. The design of the mold from steel type, number of cavities, runner and gating layout and sizing, plus the temperature control of the mold cavity can assist or hinder the molder’s efforts to produce the product to design specifications. Specify the tolerances required for the product while always maintaining the process control standards for the molding operation. Determine early in the design phase of the program what the critical dimensions are and then key in on the critical dimensions, typically only one or two, and allow for the others to vary slightly during the molding

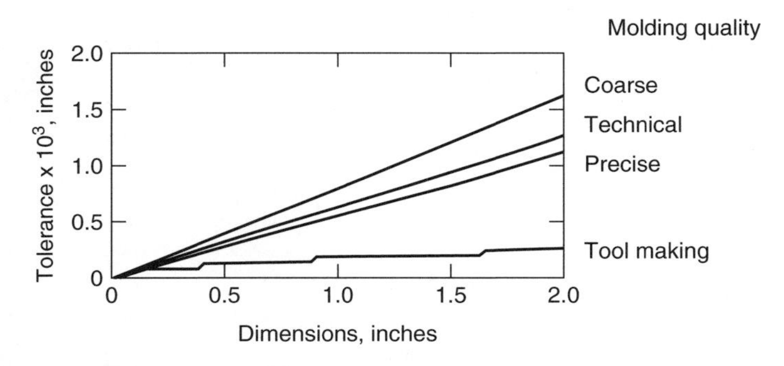

Figure 11.22 Tolerance levels for general-purpose polystyrene.

Table 11.2 Nominal mold shrinkage rates for thermoplastics in (inches\inch)

Material	Average Rate per ASTM D955		Directional Rates 0.062-in. Sample	
	0.125 in.	0.250 in.	Flow	Transverse
ABS				
Unreinforced	0.004	0.007	0.005	0.005
30% glass fiber	0.001	0.0015	0.001	0.002
Acetal, copolymer				
Unreinforced	0.017	0.021	0.022	0.018
30% glass fiber	0.003	0.008	0.003	0.016
HDPE, homopolymer				
Unreinforced	0.015	0.030	NA	NA
30% glass fiber	0.003	0.004	0.003	0.009
Nylon 6				
Unreinforced	0.013	0.016	0.014	0.014
30% glass fiber	0.0035	0.0045	0.003	0.004
Nylon 6/6				
Unreinforced	0.016	0.022	0.021	0.021
15% glass fiber +25% mineral	0.006	0.008	0.006	0.007
15% glass fiber +25% glass beads	0.006	0.008	0.006	0.008
30% glass fiber	0.005	0.0055	0.003	0.005
PBT polyester				
Unreinforced	0.012	0.018	0.018	0.015
30% glass fiber	0.003	0.0045	0.003	0.007
Polycarbonate				
Unreinforced	0.005	0.007	0.006	0.006
10% glass fiber	0.003	0.004	0.003	0.004
30% glass fiber	0.001	0.002	0.001	0.002
Polyether sulfone				
Unreinforced	0.006	0.007	0.006	0.006
30% glass fiber	0.002	0.003	0.001	0.002
Polyether ether ketone				
Unreinforced	0.011	0.013	0.009	0.011
30% glass fiber	0.002	0.003	0.002	0.004
Polyetherimide				
Unreinforced	0.005	0.007	0.006	0.006
30% glass fiber	0.002	0.004	0.001	0.002
Polyphenylene oxide/PS alloy				
Unreinforced	0.005	0.008	0.005	0.005
30% glass fiber	0.001	0.002	0.001	0.002
Polyphenylene sulfide				
Unreinforced	0.011	0.004	0.009	0.011
40% glass fiber	0.002	0.004	0.001	0.003
Polypropylene, homopolymer				
Unreinforced	0.015	0.025	NA	NA
30% glass fiber	0.0035	0.004	0.003	0.009
Polystyrene				
Unreinforced	0.004	0.006	0.005	0.005
30% glass fiber	0.0005	0.001	0.001	0.002

Source: LNP Engineering Plastics, Division of ICI Advanced Materials.

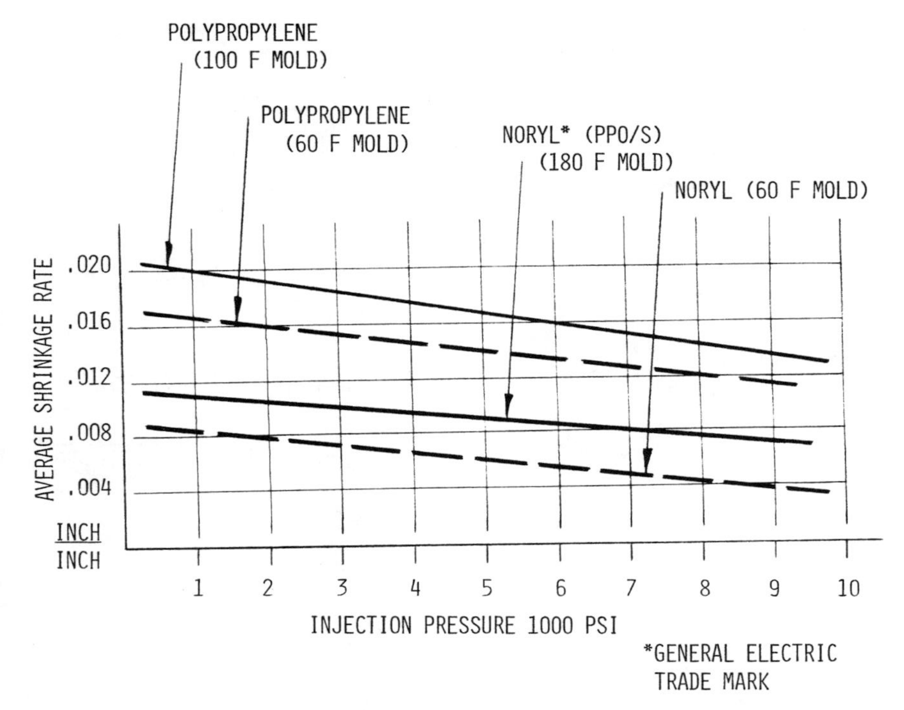

Figure 11.23 Factors affecting material shrinkage. (Adapted from Ref. 11.)

operation. By holding the key dimensions, the other dimensions will also stay within the product's dimensional requirements as long as the molding cycle is repeatable.

The mold must always be capable of repeatable producing the product, cycle to cycle within a continuously changing plant and material environment. Therefore, at the start of any program, decide on the product tolerances required, and by modeling or from experience, design the mold to meet these requirements.

Since a material's mold shrinkage is variable, the mold designer/builder, before final product cavity dimensions are cut, molds products under as near to production conditions as possible. This preproduction mold trial is to verify that the material's shrinkage dimensions in the cavity were correctly calculated. If any variance is found, the cavity dimensions can be adjusted to meet product requirements.

The product's requirements regarding mold and processing variables will establish the tolerance conditions attainable using the mold fill program and the mesh density used for the product analysis. The mold filling analysis uses the specific resin's processing variables in the analysis. This analysis assists the mold designer and design team in developing a mold layout and design to repeatably produce the product. After the mold cavity, gating, flow path, section thickness, cooling or heating channels, and tool steel are selected, the production department can select the molding process conditions for manufacture. The degrees of material tolerances that can be obtained during manufacture of a product based on dimensions are referenced in the SPI tolerance

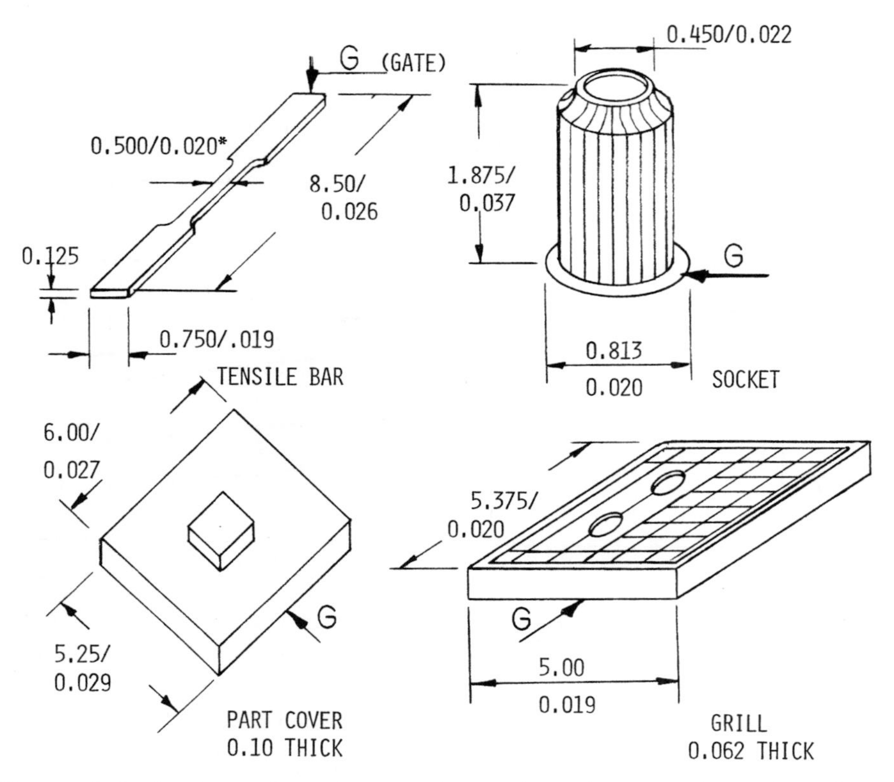

Figure 11.24 Shrinkage effects due to part geometry and gate location. Values: dimension/shrinkage in inches with acetal copolymer. (From Hoechst Celanese, adapted from Ref. 11.)

reference for individual plastic materials. The Society of the Plastics Industry, Inc. (SPI) has developed for generic plastics recommended product tolerance guidelines. Designers and mold builders and molders can use these in determining the degree of tolerances possible under commercial and fine tolerance molding conditions. This information is available in their publication *Standards and Practices of Plastics Molders,* and an example for ABS is shown in Figure 11.25. Material suppliers can also provide specific shrinkage and tolerance information on their resins and compounds.

Tolerance Factors

The product's dimensional tolerances are determined by performance requirements and dimensions required for the product's end-use performance. The product's final tolerance requirements are determined by the quality of the mold that produces the product. The more precise the product requirements, the higher the mold's tolerances and precision required. This includes the mold's temperature control and quality of the steel used to manufacture the mold. This is reflected in very precision medical and

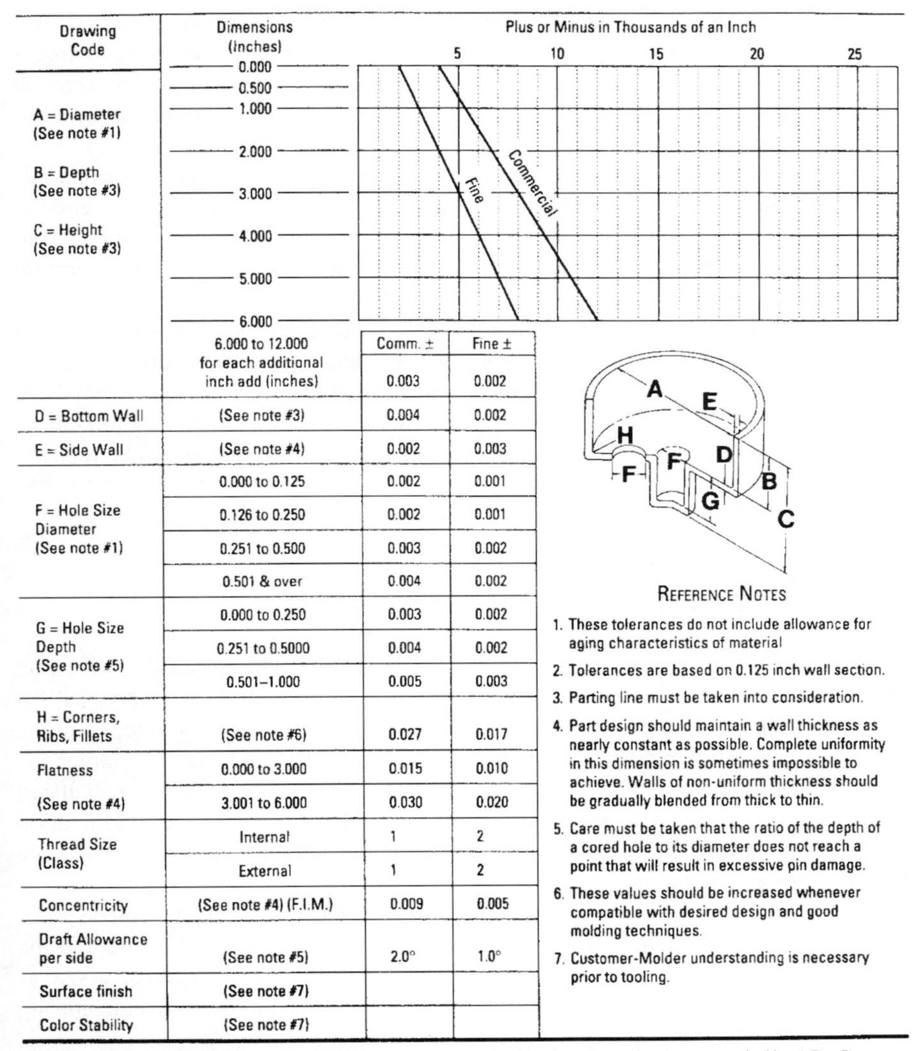

Drawing Code	Dimensions (Inches)	Comm. ±	Fine ±
A = Diameter (See note #1) B = Depth (See note #3) C = Height (See note #3)	0.000 0.500 1.000 2.000 3.000 4.000 5.000 6.000		
	6.000 to 12.000 for each additional inch add (inches)	0.003	0.002
D = Bottom Wall	(See note #3)	0.004	0.002
E = Side Wall	(See note #4)	0.002	0.003
F = Hole Size Diameter (See note #1)	0.000 to 0.125	0.002	0.001
	0.126 to 0.250	0.002	0.001
	0.251 to 0.500	0.003	0.002
	0.501 & over	0.004	0.002
G = Hole Size Depth (See note #5)	0.000 to 0.250	0.003	0.002
	0.251 to 0.5000	0.004	0.002
	0.501–1.000	0.005	0.003
H = Corners, Ribs, Fillets	(See note #6)	0.027	0.017
Flatness (See note #4)	0.000 to 3.000	0.015	0.010
	3.001 to 6.000	0.030	0.020
Thread Size (Class)	Internal	1	2
	External	1	2
Concentricity	(See note #4) (F.I.M.)	0.009	0.005
Draft Allowance per side	(See note #5)	2.0°	1.0°
Surface finish	(See note #7)		
Color Stability	(See note #7)		

REFERENCE NOTES

1. These tolerances do not include allowance for aging characteristics of material
2. Tolerances are based on 0.125 inch wall section.
3. Parting line must be taken into consideration.
4. Part design should maintain a wall thickness as nearly constant as possible. Complete uniformity in this dimension is sometimes impossible to achieve. Walls of non-uniform thickness should be gradually blended from thick to thin.
5. Care must be taken that the ratio of the depth of a cored hole to its diameter does not reach a point that will result in excessive pin damage.
6. These values should be increased whenever compatible with desired design and good molding techniques.
7. Customer-Molder understanding is necessary prior to tooling.

Note: The *Commercial* values shown below represent common production tolerances at the most economical level. The *Fine* values represent closer tolerances that can be held but at a greater cost. Any addition of fillers will compromise physical properties and alter dimensional stability. Please consult the manufacturer.

Figure 11.25 Standards and practices of plastics molders. Material: acrylonitrile butadiene styrene (ABS). (Reprinted from *Standards and Practices of Plastics Molders,* courtesy of the Society of the Plastics Industry, Inc.)

electronic molds where the cavity dimensional tolerance levels are 4–10 times tighter than the products they make. This leaves the largest possible margin for variation to occur during the molding operation for the material and process variables. The product's complexity, size, and shape, and any in-mold operations, cam pulls, unscrewing, or other features will affect the mold processing and product costs. Any in-mold operation will extend cycle time and should be minimized when possible. It should be necessary only when required for a special product function such as extracting a core to form a hidden cavity in the product, internal screw threads, or undercuts in the plane of the mold or a large undercut, which cannot be formed without an in-mold cavity operation.

Molding process and mold temperature control of the product cavity control product dimensions and tolerances. The mold cavity cooling system must be designed to maintain an entering and exiting water temperature variation of no more than $\pm 2°F$, to control product dimensions, cycle to cycle. Tight control of mold cavity temperature is essential and should never be left to last as it dramatically affects product dimensions and quality. The number of cavities in a mold, determined by the product's size, manufacturing volume, and tolerance requirements, has a major effect on piece product cost. For multicavity molds, the product's dimensional tolerances may have to be reduced, unlike those in a single-cavity mold. If the manufacturing tolerances are built into the mold, very precise product dimensional tolerances can be realized consistently. If not, then only one, at most two, critical dimensions to be specified and expected should be held in tolerance, cycle to cycle.

PROCESS CONTROL

The manufacturing cycle for the required quantity of products based on the rate of production affects the size and type of the injection-molding machine used for manufacturing the product. The injection-molding machine should be sized to fit the mold and volume of plastic melt required to produce the products. The larger the machine, the higher the machine's hourly rate. The molding cycle time is calculated according to the type of resin used, the cooling efficiency of the mold, the solidification time of the resin, and the ability of the mold, machine, and molding process to make products to the required quality level.

The design team must consider these factors. They affect the finished piece product price and have a definite effect on product quality during manufacture and the product's end-use requirement. Again, the design team analysis method can assist, as production will be present to provide information for this and other manufacturing operations

Plastic Resin Variability

Plastic resins do vary to some degree, lot to lot, in their processing characteristics. The material supplier tries to manufacture their resins uniformly with many products produced in (not continuous) batch lot manufacture. These resin batch lots are made in several-thousand-pound, multiadditive compound lots that vary within a manufacturer's process. To even out lot variability, the suppliers often blend multiple lots to achieve the material's necessary physical and processing properties to meet the product specifications. Blending helps the process technician by eliminating the need to continually adjust the molding process control variables for varying process and property values. Molders know that there is always some degree of adjustment required when a new lot of material is used or when the mold is started up. This requires startup inspection and documentation of process control specifications being performed and documented for quality and recordkeeping purposes.

Product Released for Manufacturing

When the production mold is released to manufacturing, all major design, product, and mold analyses, and any other product and program decisions, such as decoration, assembly, and packaging, must be finalized. The production personnel, in-house or custom molders, should have the best designed and best built mold to produce quality products, cycle to cycle. Manufacturing should not be forced to make products to specifications if the mold, material, and product designs are not capable.

Manufacturing input should be requested during the design team product analysis on molding variables. Manufacturing will have to fine-tune the molding cycle, but should not have to force the material and processing conditions to conform to unrealistic expectations. Their input at this stage can be used to define the molding cycle, variables, and requirements, in the early stages of the product's design program.

Manufacturing will adjust the material's melt temperature, mold temperature, pressure setting, and fill times to optimize the molding cycle to obtain product dimensions. Manufacturing should use computer-controlled, closed-loop process control feedback to control the molding process. Using output data from their process control system and product quality checks, they will know, in real time, whether the process is in control.

Closed-Loop Process Control

The manufacture of quality products, cycle to cycle, requires more than a good product design, mold, and material. The process requires the molder to have their equipment, molding machine, and auxiliary support equipment interfacing to monitor and control the manufacturing process.

The majority of new molding machines has or can be equipped with closed-loop process control systems. Older machines should be retrofitted to achieve the highest capability possible for the continued manufacture of plastic products. This can be accomplished by using closed-loop/continuous-feedback process control, a system used for continuous monitoring, adjusting, and control of the molding process. It is used to ensure that the melt is delivered to the mold at the same temperature, volume, and injection-molding pressure and is repeatable, cycle to cycle.

Determining Molding Process Capability

A closed-loop/continuous-feedback process control system is used for continuous monitoring, adjustment, and controlling of the molding process variables to ensure that the melt is delivered to the mold at the same temperature, volume, and pressure, repeatability cycle to cycle.

It is manufacturing's responsibility to be sure that their equipment—injection-molding machine, auxiliaries, and any secondary operations—are always within capability specifications to guarantee that the product meets the customer's requirements. This is accomplished by correctly designing the product and using only quality material and mold and maintaining process control during manufacture.

A correctly designed product can fail if the mold and manufacturing process do not meet manufacturing quality standards. The design team can assist the mold builder by referencing the product's critical dimensions from one reference point or product surface noted on the product design drawings.

As the number of mold cavities increases, the number of critical dimensions must be decreased on the product. The number of mold cavities that must be controlled during the molding phase of manufacture has an effect on controlling key dimensions; one cavity may be the easiest, and two or more somewhat more difficult to control, based on the number of critical dimensions selected for the control. The design team should not use typical metal dimensional tolerances on the product. The molded product's dimensions and their required tolerances should reflect the accuracy necessary for the product's function. The mold designer controls critical product dimensions using irregular product surfaces such as contour changes, cutouts, bosses, ribs, and cavity temperature to control shrinkage and tolerances. Anchor points in the products mold cavity assist in holding the product's dimensions as the resin cools and solidifies in the mold to maintain product dimensions as shown in Figure 11.26.

Production Mold Qualification

Some multicavity molds are built initially using full-size mold frames as single-cavity prototype molds for producing products for evaluation and testing. Later, when the product design is finalized, the mold is completed with additional cavities added plus completion of the cooling and ejection system, runner and gating.

If the prototype mold was designed for later incorporation into the production mold with multiple cavities, the new cavities must be duplicated exactly along with completing the ejection and cooling system. The completed production mold must then be reevaluated and products from each cavity retested if the application is critical. Also, ensure that each cavity is individually identified, should a problem later arise.

When qualifying the production tool, be sure that it is run for several days under production settings to see if dimensions drift or if any production problems occur. Sample all cavities at set times during the production run, and monitor the process variables to be sure that the process is always in real-time process control. Now is the time to optimize the cycle and make sure that it does not affect product performance.

VERIFYING MANUFACTURING CONTROL AND CAPABILITY

The major problem companies have is determining whether their manufacturing operation is truly capable of producing the product to the required dimensions, cycle to cycle. Management must evaluate the capability of their injection-molding machines

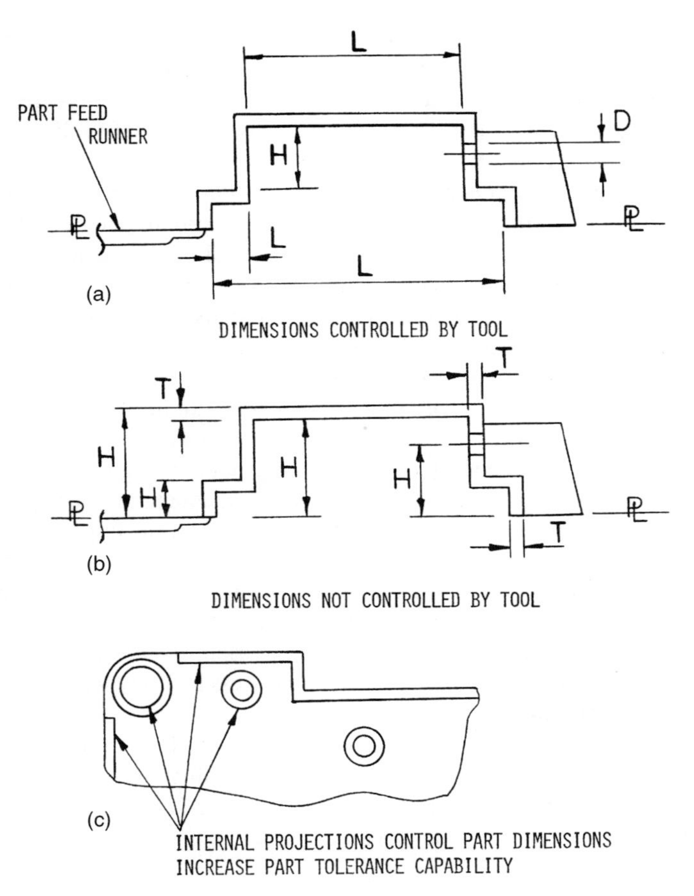

Figure 11.26 The relationship between mold cavity and part dimension control: (a) dimensions controlled by tool; (b) dimensions not controlled by tool; (c) internal projections control part dimensions increase part tolerance capability. (Adapted from Ref. 22.)

and personnel to produce the product in the most economic and profitable method. Efficiency and time management studies with economic evaluation will answer some questions. Also, an in-depth analysis of equipment and process capability can show problems in their manufacturing system so that they can be solved and prevented in the future. Equipment capability studies coupled with smart scheduling of product manufacture will assist in producing good-product quality and productivity.

No two injection-molding machines are identical, as each will have a degree of variability, and this information must be known. Also, some molds may perform better on one machine than on another because of variability of one machine's process variables, pressure, temperature, and speed controls versus those of another. These process and machine variables need to be identified and documented during the molding operation. Some resins may process better on one machine than another.

This involves the machine's size, screw design, compression ratio; melt capacity, and other properties, in order to process one resin more efficiently than that of another machine of the same class and size. Knowing this information will increase productivity and product quality when knowledgeable production scheduling using these machines is performed.

In many manufacturing operations this information is not known or used when a mold is scheduled for production. Therefore, before production begins, the manufacturing system, machine, mold, material, and auxiliary equipment should be given a capability sturdy to be sure that they can produce the product economically to specifications. By performing a process capability study, if the processing system is found not capable, it can be analyzed to determine the problem areas. If the injection-molding machine is found to be limited, then another machine can be scheduled. The new machine should also have a molding system capability study conducted. The documented information is used to write a procedure to ensure that the mold is run on machines capable of producing good repeatable product using established molding conditions to produce product to specifications. It is manufacturer's responsibility to be sure that their equipment, molding machine, auxiliaries, and any secondary equipment and operations are always within specification to guarantee that the product will meet the end-use and customer requirements. During the initial startup procedure, control and processing limits must be established as shown in Figure 11.27. These limits are necessary to establish the processing variables' range of tolerances that can be allowed to control of the manufacturing process and stay within the product's dimensional control limits.

All molding machines have a degree of variability, cycle to cycle. The amount of variability in the machine and process must be determined to establish the amount of control required to make the product. This is what is known as "determining the molding machine's and production's operational capabilities."

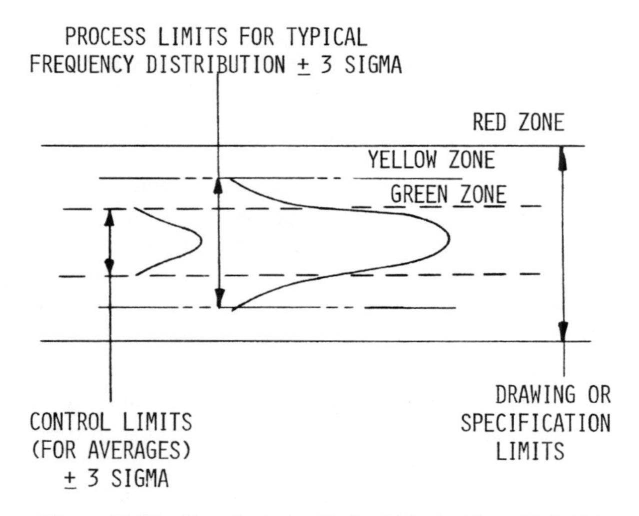

Figure 11.27 Manufacturing limits. (Adapted from Ref. 13.)

Analysis of maintenance records can reveal trends in equipment problems, and machines can be serviced before a failure occurs. It can also show areas where products need replacement or more frequent recalibration to maintain quality requirements.

Discussions with machine operators are helpful. Operators often make machine adjustments to keep the product quality within specification, and these must always be documented on the Molding Data Record Sheet Figure 11.28. If these adjustments are excessive, this indicates other problems with the equipment that will affect the products quality. Operators are responsible for the output of their machines and the product quality. They should know their equipment maintenance status and if and when repairs are required.

Establishing the capability of a machine to produce products to the customer's requirements can be performed by different methods of analysis. This is accomplished by documenting the process variables using the molding record data sheet. This requires monitoring and recording the system's, machine variable adjustments, required to keep the process in control to produce good products over a period of time. But this is seldom sufficient as conditions will change continually and only the process, snapshot, data points will be recorded. Also, minor fluctuations in the machine, plant conditions, and material variables may never be noticed and recorded.

Each lot of plastic material has a degree of product variability. The melt viscosity and other flow characteristics of the material can vary over the lot size, necessitating minor adjustments in the processing variables. As a machine's moving products wear, temperature control of the machine's hydraulic fluid can vary along with processing temperatures and plant environment, with controllers drifting or going out of calibration, which may not always be evident by merely monitoring process control settings. All molding machines have a degree of variability, cycle to cycle. The amount of variability and the amount of control required to make the product are what are needed to know in determining a true manufacturing capability.

Determining Molding Machine Capability

It is very important that all key machine and process variables be recorded during the capability analysis. There should always be a procedure or work instruction available for the operators to follow whenever a mold is run. The work instruction should describe the setup and operation of the mold and machine and become a permanent record of the program. It must be kept updated whenever the mold is run, on existing or new machines to produce products. It will also have a history of the mold and the molding machines on which it is run and continually kept up to date. This includes a startup–shutdown procedure. Procedures must be written for the startup and shutdown of a molding operation. The turnover of personnel makes it mandatory for review each time the mold is to be set in a machine for product production.

Following the procedures for system startup, the molding system is allowed to reach equilibrium or steady-state conditions. This includes the temperature variables for the mold and molding machine's barrel settings and hydraulic fluid. Variability of the machine's hydraulic fluid temperature can cause speed changes and injection and

JOB	MOLD DESCRIPTION	MACHINE SETUP INSTRUCTIONS	SPECIAL INSTRUMENTATION	LS ☐ ___ LS ☐ ___ LS ☐ ___
OPERATORS	SCREW USED		SAFETY CHECK	LS ☐ ___ LS ☐ ___ LS ☐ ___
	PRESS NO.	NOZZLE #		

DATE / TIME	Run Number	Resin	Lot Number	TEMPERATURE (°F)											PRESSURES (PSI)				Speed Control Setting	CYCLE TIMES (SEC)							Pad (Inches)	Feed (Inches)	RPM	WEIGHTS (GMS)		ALLIED EQUIPMENT				REMARKS
									MOLD																											
				Nozzle	Front	Center	Rear		Fixed	Movable	X-PLATE	THROAT		OIL	Booster	Hold	Back			Injection	Hold	Open	Overall	Booster	Ram in Motion	Screw Retraction				Full Shot	Part Only	Mold Heater No 1 °F	Mold Heater No 2 °F	Hopper Dryer °F	Dehumidifier Setting	

COMMENTS ON MOLDING OPERATION, START-UP, ETC.

Figure 11.28 Molding data record sheet.

packout pressure variances. Included in this are the plant's processing and auxiliary equipment and conditions around the molding machine, which are seldom considered. Fans are never used to cool the operator by blowing on the injection-molding machine as will cause temperature fluctuations and errors in data recorded and temperatures for the molding process of the equipment.

If the mold is equipped with cavity pressure and temperature sensors, this should be recorded as a product of the documentation. Also, for the system's support equipment, variables should be recorded to document the total process cycle variables for critical products; this includes plant temperature, humidity, and shift or startup time.

A computer with data acquisition software or terminals can be used to record the variables by connecting directly into the machine's process control output sensors. The number of process variables able to be recorded depends on the equipment used and the degree of control necessary for controlling the manufacturing process.

Process Capability Analysis. There are 19 monitoring points on an injection-molding machine. These points control the molding cycle and product quality. The molding machine process controllers control all except the mold temperature control. The manufacturing team should know their injection-molding machine's, acceptable tolerance for quality class classification (ATQCC). Knowing their processing system's ATQCC class tells them whether the machine's molding cycle variables, melt and hydraulic fluid temperature, injection fill rate, and fill and pack pressure and cycle timers can be controlled and also, whether the support equipment is capable of repeatable manufacture within the established control limits. The analysis determines when components in the system are replaced or repaired, and when control and process limits should be tightened. Analysis of the molding system when compared to the ATQCC standard molding machine values, shown in Table 11.3 for control purposes, can be conducted to evaluate the system's repeatable capability for the manufacturing process. Quality class factors are listed for the tightest class, Class 1, to just within capability Class 9. The factors were developed for extensive machinery evaluations that included over 1800 production machines' manufacturing products. The data represent the machines most capable of repeatable, cycle-to-cycle, operations to the poorest with products just within the envelope of producing acceptable quality product on molds that were classified as "highly tolerant tooling and material combinations."

The data are valid only for the machines evaluated and are used only as a guide for which all machines can be compared for capability. The data are used by molders for comparison of their current equipment against a standard that is deemed acceptable to produce quality products. The objective when comparing machine output to the tabulated data is to always improve the capability of that machine and to reach the machine's highest class level.

Mr. Denes B. Hunkar, president of Hunkar Laboratories, Inc., Cincinnati, Ohio, developed this table as a "quality yardstick comparison for machine control." Class capability varies from 1 (the highest) to 9 (just capable) in his analysis. Mr. Hunkar requires that molders use his "Gold Standard" sensors versus the sensors on the machine during the analysis. He believes that using the same quality of sensors he

Table 11.3 Acceptable tolerances for quality class classification

Parameter (Unit)	Class 1	Class 2	Class 3	Class 4	Class 5	Class 6	Class 7	Class 8	Class 9
Cycle time (s)	0.20	0.24	0.29	0.35	0.41	0.50	0.60	0.72	0.86
Hold times (s)	0.02	0.02	0.03	0.03	0.04	0.05	0.06	0.07	0.09
Inject time (s)	0.04	0.05	0.06	0.07	0.08	0.10	0.12	0.14	0.17
Clamp closed (s)	0.10	0.12	0.14	0.17	0.21	0.25	0.30	0.36	0.43
Clamp open (s)	0.10	0.12	0.14	0.17	0.21	0.25	0.30	0.36	0.43
Plasticate (s)	0.15	0.18	0.22	0.26	0.31	0.37	0.45	0.54	0.64
Cavity pressure (psi)	15.00	18.00	21.60	25.92	31.10	37.32	44.79	53.75	64.50
Peak injection pressure (psi)	20.00	24.00	28.80	34.56	41.47	49.77	59.72	71.66	86.00
Hold pressure (psi)	4.00	4.80	5.76	6.91	8.26	9.95	11.94	14.33	17.20
Backpressure (psi)	5.00	6.00	7.20	8.64	10.37	12.44	14.93	17.92	21.50
Ram stroke (in)	0.05	0.06	0.07	0.09	0.10	0.12	0.15	0.18	0.21
Mold A temperature (°F)	3.00	3.60	4.32	5.18	6.22	7.46	8.96	10.75	12.90
Mold B temperature (°F)	3.00	3.60	4.32	5.18	6.22	7.46	8.96	10.75	12.90
Oil temperature (°F)	3.00	3.60	4.32	5.18	6.22	7.46	8.96	10.75	12.90
Dew point (%)	0.01	0.01	0.01	0.02	0.02	0.02	0.03	0.04	0.04
Temp 1 (°F)	2.00	2.40	2.88	3.46	4.15	4.98	5.97	7.17	8.60
Temp 2 (°F)	2.00	2.40	2.88	3.46	4.15	4.98	5.97	7.17	8.60
Temp 3 (°F)	2.00	2.40	2.88	3.46	4.15	4.98	5.97	7.17	8.60
Temp 4 (°F)	2.00	2.40	2.88	3.46	4.15	4.98	5.97	7.17	8.60

used in developing the class ranking will ensure that the new analysis conducted on another machine will be as accurate as his original data for classifying equipment. He has found that other machine manufacturer's sensors may not be as reliable or as sensitive to minor changes as his "Gold Star" sensors. The performance comparisons of quality classifications for process control systems are shown in Table 11.4.

Using Hunkar's classification system for analysis of a machine capability can be determined according to a known class rating based on a typical molding cycle as shown in Figure 11.29. The analysis requires the recording of selected injection-molding machine variables for a minimum of 100 cycles. The data are then compared in a software program to Hunkar's standard process control class limits, from 1 to 9. The machine's output data are evaluated using an internal software evaluation formula to calculate the number of upper and lower variable values that fell in and outside the selected class specifications, with upper selected limit and lower selected limit (USL/LSL) values as shown for a Class 3 rating in Figure 11.30. For example, if a Class 3 control limit analysis were run, as shown in Figure 11.31, the out-of-control USL and LSL variable limits compared to the machine values are highlighted in the program output on the screen.

The programmed values for a Class 3 evaluation are displayed with the values obtained from the machine production evaluation. The data are statistically analyzed

Table 11.4 Performance comparisons: control design—typical short- and long-term performance capabilities (repeatability)[a]

Controllable Feature	Open-Loop Proportional	Corrected Proportional	Real-time Closed-Loop
Short-term hold pressure	±35 psi	±14 psi	±3 psi
Short-term backpressure	±27 psi	±14 psi	±3 psi
Short-term injection speed	±5%	±5%	±0.001%
Long-term hold pressure	±87 psi	±74 psi	±5 psi
Long-term backpressure	±32 psi	±28 psi	±7 psi
Long-term injection speed	±11%	±11%	±0.01%

[a]This information was obtained from a large sample monitoring of actual operating machines, under normal plant conditions. Gold-standard sensors used to assure precise comparisons. *Note:* Repeatability will have a direct relationship to part quality.

Source: Adapted from Hunkar Laboratories, Inc.

for significance versus the standard values for the class under evaluation. The variables out of tolerance for the Class 3 range are highlighted, indicating that they have significantly deviated from the standard.

The formula used for this analysis is: greater than the $\sqrt{N}$, where N is the number of samples. If a deviation count exceeds the square root of the sample, the deviating parameter is highlighted in the analysis and on the screen. This allows the deviating parameters to be brought into control, or within the Class 3 standard. This formula ignores the out-of-control spikes as adjustments or momentary cycle random out-of-control conditions that may occur during normal startup and adjusting the cycle to produce acceptable products early in the manufacturing process. The system software after the required number of cycles (100 minimum) then ranks any out-of-control

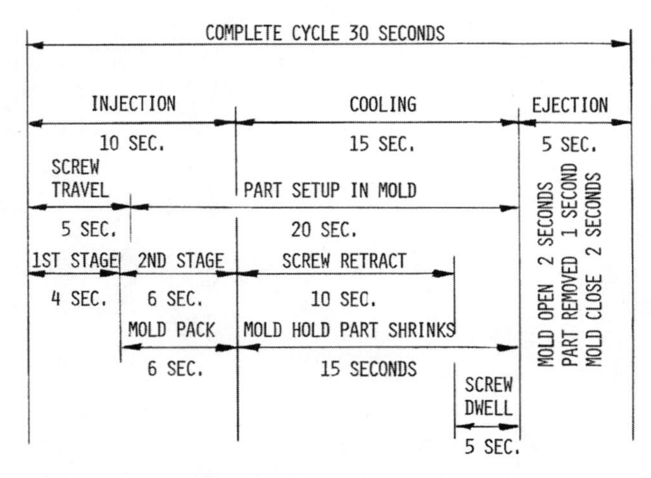

Figure 11.29 The injection-molding cycle.

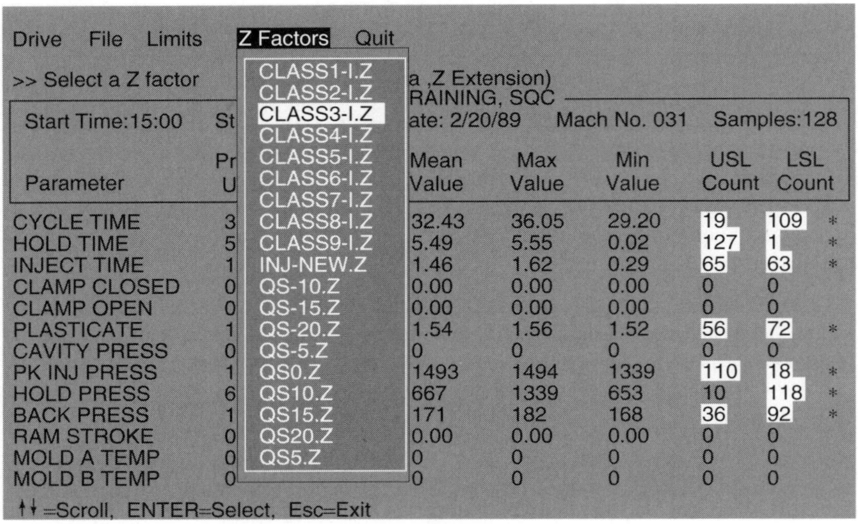

Drive File Limits **Z Factors** Quit

>> Select a Z factor … a ,Z Extension)

… RAINING, SQC

Start Time:15:00 St… ate: 2/20/89 Mach No. 031 Samples:128

Parameter	Pr U	Z Factors menu	Mean Value	Max Value	Min Value	USL Count	LSL Count	
		CLASS1-I.Z						
		CLASS2-I.Z						
		CLASS3-I.Z						
		CLASS4-I.Z						
		CLASS5-I.Z						
		CLASS6-I.Z						
		CLASS7-I.Z						
CYCLE TIME	3	CLASS8-I.Z	32.43	36.05	29.20	19	109	*
HOLD TIME	5	CLASS9-I.Z	5.49	5.55	0.02	127	1	*
INJECT TIME	1	INJ-NEW.Z	1.46	1.62	0.29	65	63	*
CLAMP CLOSED	0	QS-10.Z	0.00	0.00	0.00	0	0	
CLAMP OPEN	0	QS-15.Z	0.00	0.00	0.00	0	0	
PLASTICATE	1	QS-20.Z	1.54	1.56	1.52	56	72	*
CAVITY PRESS	0	QS-5.Z	0	0	0	0	0	
PK INJ PRESS	1	QS0.Z	1493	1494	1339	110	18	*
HOLD PRESS	6	QS10.Z	667	1339	653	10	118	*
BACK PRESS	1	QS15.Z	171	182	168	36	92	*
RAM STROKE	0	QS20.Z	0.00	0.00	0.00	0	0	
MOLD A TEMP	0	QS5.Z	0	0	0	0	0	
MOLD B TEMP	0		0	0	0	0	0	

↑↓=Scroll, ENTER=Select, Esc=Exit

Figure 11.30

machine control values for comparison in a self-contained system Pareto chart for resolution of major contributing factors as shown in Figure 11.32. The Pareto chart prioritizes by number of occurrences the variables out of the class standard for immediate corrective action. During the real-time production molding evaluation, the operator may have to make some manual adjustments for an occasional totally out-of-control data point (spikes), as shown for cycle time in Figure 11.33 as they may occur.

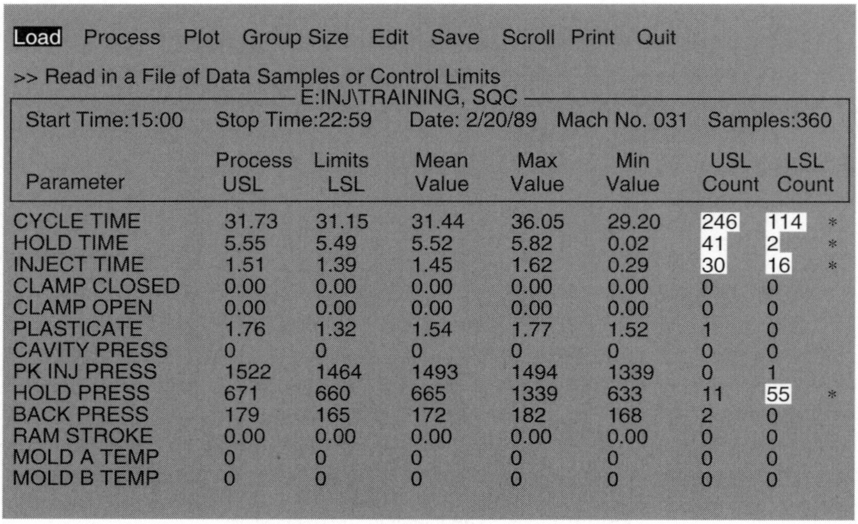

Load Process Plot Group Size Edit Save Scroll Print Quit

>> Read in a File of Data Samples or Control Limits

E:INJ\TRAINING, SQC

Start Time:15:00 Stop Time:22:59 Date: 2/20/89 Mach No. 031 Samples:360

Parameter	Process USL	Limits LSL	Mean Value	Max Value	Min Value	USL Count	LSL Count	
CYCLE TIME	31.73	31.15	31.44	36.05	29.20	246	114	*
HOLD TIME	5.55	5.49	5.52	5.82	0.02	41	2	*
INJECT TIME	1.51	1.39	1.45	1.62	0.29	30	16	*
CLAMP CLOSED	0.00	0.00	0.00	0.00	0.00	0	0	
CLAMP OPEN	0.00	0.00	0.00	0.00	0.00	0	0	
PLASTICATE	1.76	1.32	1.54	1.77	1.52	1	0	
CAVITY PRESS	0	0	0	0	0	0	0	
PK INJ PRESS	1522	1464	1493	1494	1339	0	1	
HOLD PRESS	671	660	665	1339	633	11	55	*
BACK PRESS	179	165	172	182	168	2	0	
RAM STROKE	0.00	0.00	0.00	0.00	0.00	0	0	
MOLD A TEMP	0	0	0	0	0	0	0	
MOLD B TEMP	0	0	0	0	0	0	0	

Figure 11.31

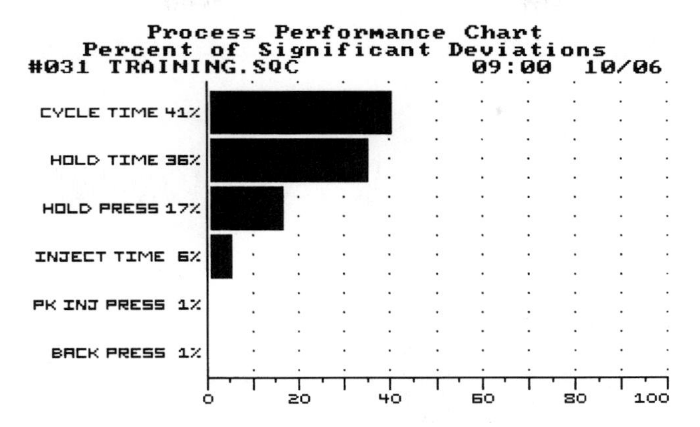

Figure 11.32 Process performance chart. Percent of significant deviations.

During the evaluation any manual cycle adjustments must be recorded on the molding record data sheet for review when the data are presented to ascertain the reason for any spikes that occurred in the analysis. When the data are evaluated, these manual or out-of-control spikes should be eliminated when analyzing the machine's true process capability. The Hunkar program has an editing feature designed to remove these manual adjustments or spikes without affecting the remaining data points (see Fig. 11.34). The rule for editing data points is if a change occurred rapidly over one or two cycles, it was a human adjustment. If it occurred gradually over 5 or 10 cycles, a natural drift trend occurred and must be considered. These spikes will be visible in all variable charts and must be eliminated when doing the full system class analysis.

Editing the data, for example, for inject time as in Figure 11.35, with the spike bracketed for removal, the data are cleansed of the spike and redisplayed in Figure 11.36. The same task is performed for "Hold Press" (pressure) in Figures 11.37 and 11.38. Figure 11.39 shows the calculated standard deviation and range for "Hold Press" with the spike removed and the data smoothed.

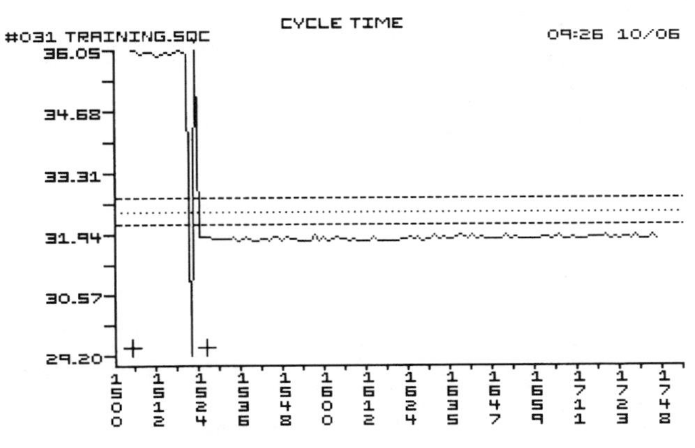

Figure 11.33 Cycle time

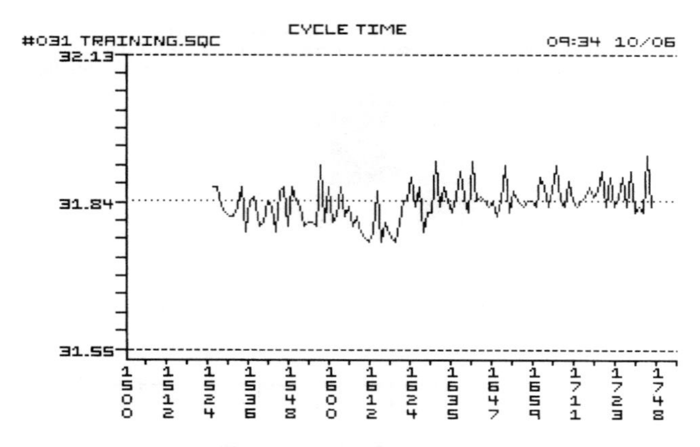

Figure 11.34 Cycle time.

In the example for the inject time (Fig. 11.35), a manual adjustment spike occurred and a process drift as shown by the points below the mean was detected following this. Editing out the spike on the graph still leaves the cycle drift, indicating that injection time was still being affected by the molding process and machine variables. It was also not adjusted correctly with 6% of the readings out of the Class 3 limits as shown in the Pareto chart in Figure 11.32.

The analysis determined that the injection time was initially long but became shorter because of a change in the resin's melt index or viscosity. The machine's hydraulic injection pressure was not changed, so the only plausible explanation is that the resins melt index changed, allowing the screw to advance faster with the same injection pressure.

After editing of the data, a new Pareto chart is generated for the out-of-control values (Fig. 11.40). The data are cleaned to show the system's significant performance

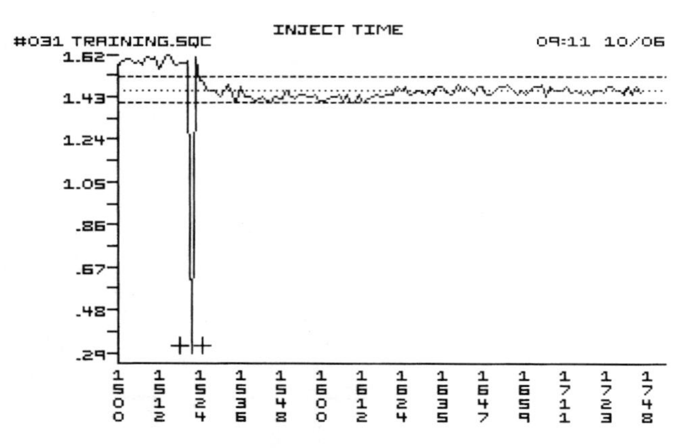

Figure 11.35 Injection time.

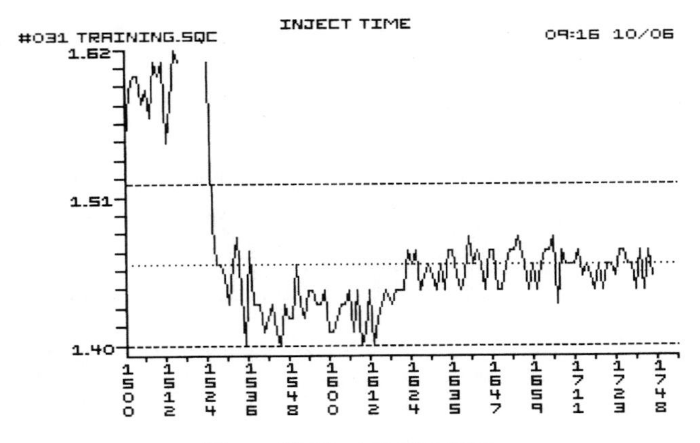

Figure 11.36 Injection time.

deviations, and variable output is then again compared to the selected Class 3 standard, (Fig. 11.41). Any remaining significant variable deviations from the class are again highlighted. The cleaning of the remaining output data is performed for the other variables of significant deviation, such as the hold pressure.

CpK CAPABILITY ANALYSIS

Because hold pressure is a critical molding variable, the CpK value for this variable for the process is further expressed in Figure 11.39 for a Class 3 standard. The variable's calculated CpK value is a measure of both dispersion and centeredness. This means that the formula for CpK takes into account both the spread of the distribution and where that distribution is in regard to the specification midpoint. CpK expresses

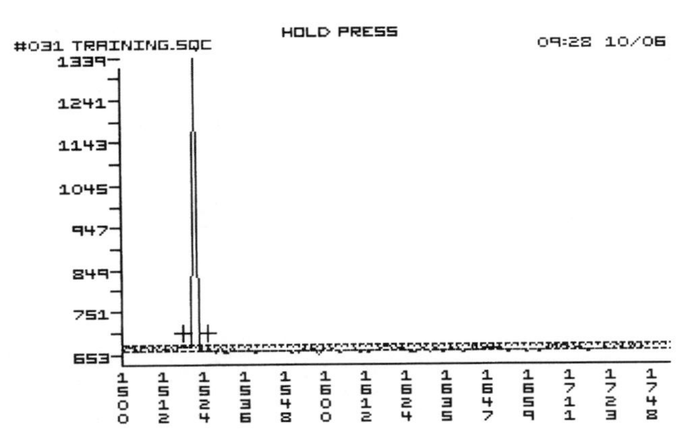

Figure 11.37 Hold pressure.

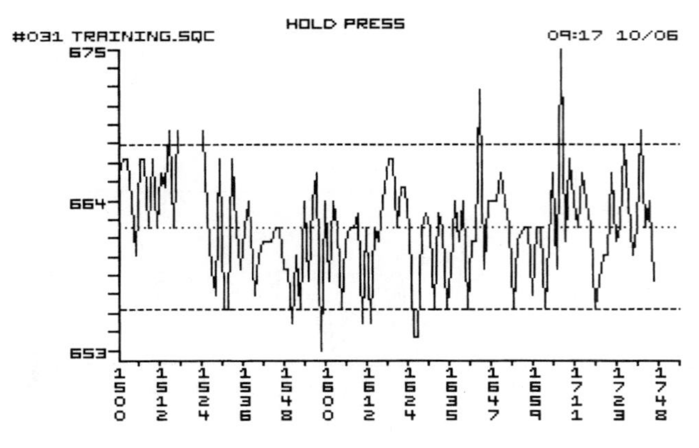

Figure 11.38 Hold pressure.

the exactness of the variable when compared to the upper and lower control limits, expressed in capability of the process to remain within the current class standard for the analysis:

$$\text{CpK} = \text{the lesser of } \frac{\text{USL} - \text{mean}}{3\sigma} \quad \text{or} \quad \frac{\text{mean} - \text{LSL}}{3\sigma}$$

Because the lesser of the two values calculated is chosen, the capability of the process is on the worst side (tail of the bell curve closest to the specification limit).

The capability of process index (CPI) measures the ratio of tolerance to 3σ (three sigma), the typical bell curve spread for 99.73% of the measurements, which lie between $\pm 3\sigma$ values as shown in Figure 11.42.

CPI measures the centeredness of the data to the mean. CpK cannot be greater than CPI for the same process. Only when the mean is exactly centered on the specification

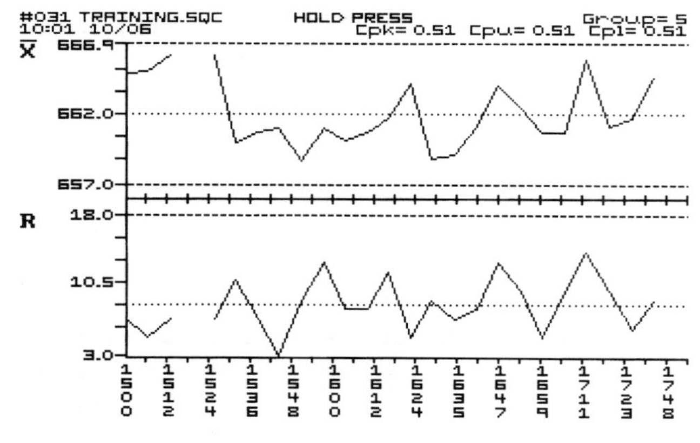

Figure 11.39 Hold pressure.

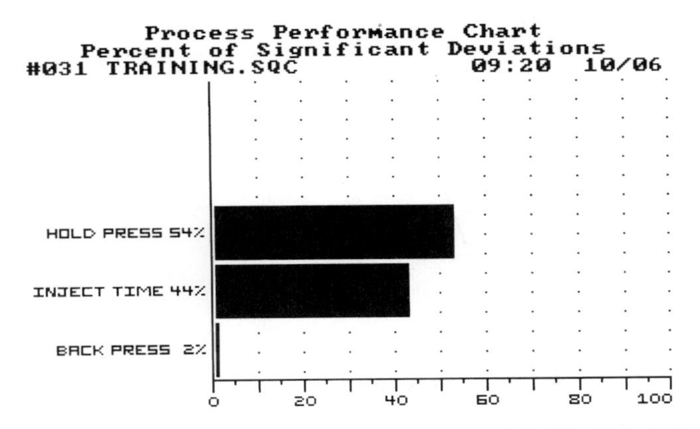

Figure 11.40 Process performance chart. Percent of significant deviations.

midpoint is CpK = CPI. Therefore, CPI is valuable as an indicator of how much better the CpK could be if the process were set up so that the center of the distribution were closer to the specification midpoint. CPI answers whether the distribution of values could fit within the tolerances (if centered). A CpK value of 1.0 indicates that 99.73% of values are within the class specification limits and when higher, it is considered to meet the required standard for a process within real-time process control. For the CpK value, the greater the value calculated above 1.0, the better, measuring both spread and centeredness as shown in Figure 11.43.

Drive File Limits **Z Factors** Quit

>> Select a Z factor

CLASS1-I.Z
CLASS2-I.Z
CLASS3-I.Z
CLASS4-I.Z
CLASS5-I.Z
CLASS6-I.Z
CLASS7-I.Z
CLASS8-I.Z
CLASS9-I.Z
INJ-NEW.Z
QS-10.Z
QS-15.Z
QS-20.Z
QS-5.Z
QS0.Z
QS10.Z
QS15.Z
QS20.Z
QS5.Z

a ,Z Extension)
RAINING, SQC

Start Time:15:00 St ate: 2/20/89 Mach No. 031 Samples:128

Parameter	Pr U	Mean Value	Max Value	Min Value	USL Count	LSL Count	
CYCLE TIME	3	31.84	31.93	31.76	0	0	
HOLD TIME	5	5.53	5.55	5.51	0	0	
INJECT TIME	1	1.46	1.62	1.40	15	4	*
CLAMP CLOSED	0	0.00	0.00	0.00	0	0	
CLAMP OPEN	0	0.00	0.00	0.00	0	0	
PLASTICATE	1	1.54	1.56	1.52	0	0	
CAVITY PRESS	0	0	0	0	0	0	
PK INJ PRESS	1	1494	1494	1492	0	0	
HOLD PRESS	6	662	675	653	6	16	*
BACK PRESS	1	171	182	168	1	0	
RAM STROKE	0	0.00	0.00	0.00	0	0	
MOLD A TEMP	0	0	0	0	0	0	
MOLD B TEMP	0	0	0	0	0	0	

↑↓ =Scroll, ENTER=Select, ESC=Exit

Figure 11.41

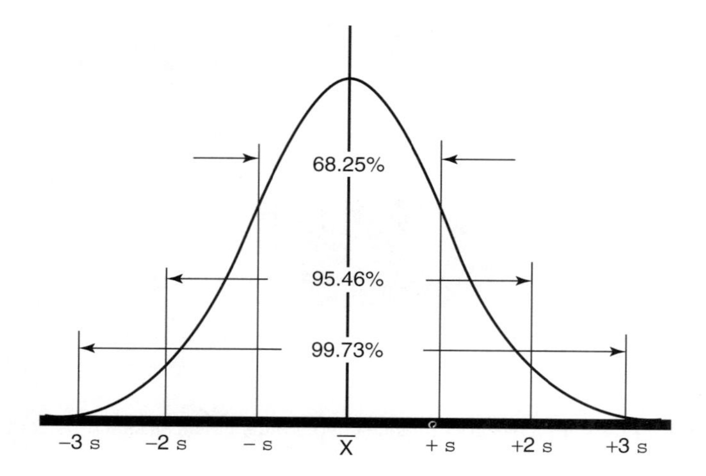

Figure 11.42 Relation of sigma to a normal distribution.

The evaluation process continues, evaluating the data to the next higher class standard, until all highlighted and out-of-control variables, to the evaluated class level, are eliminated, not highlighted, or when the last class level is reached, Class 9. As the class level increases, the standard class variable values become wider.

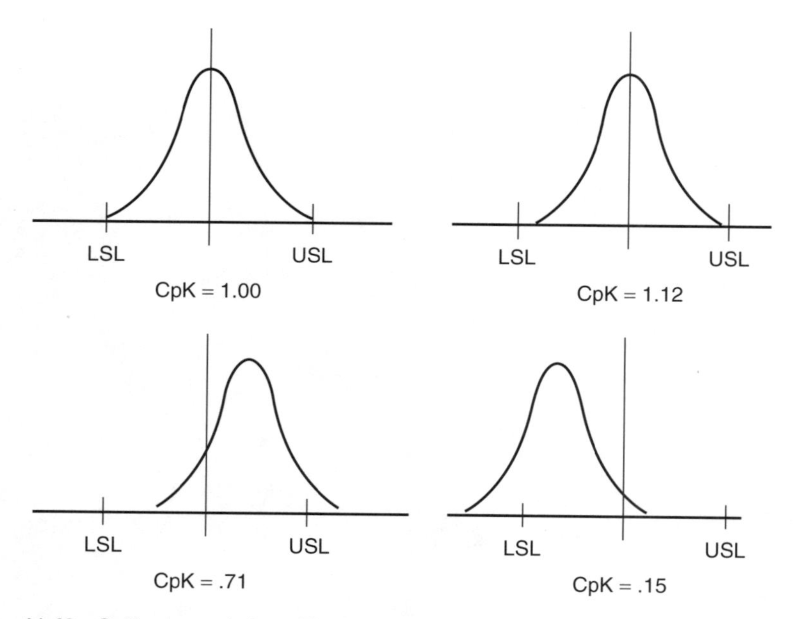

Figure 11.43 CpK value variations. The higher the CpK value, the more in control the process. The CpK is a measure of spread and centeredness. (Adapted from Ref. 15.)

Load Process Plot Group Size Edit Save Scroll Print Quit

>> Read in a File of Data Samples or Control Limits

E:\INJ\TRAINING, SQC

Start Time:15:00 Stop Time:17:48 Date: 2/20/89 Mach No. 031 Samples:128

Parameter	Process USL	Limits LSL	Mean Value	Max Value	Min Value	USL Count	LSL Count	
CYCLE TIME	32.13	31.55	31.84	31.93	31.76	0	0	
HOLD TIME	5.56	5.50	5.53	5.55	5.51	0	0	
INJECT TIME	1.52	1.40	1.46	1.62	1.40	15	4	*
CLAMP CLOSED	0.00	0.00	0.00	0.00	0.00	0	0	
CLAMP OPEN	0.00	0.00	0.00	0.00	0.00	0	0	
PLASTICATE	1.76	1.32	1.54	1.56	1.52	0	0	
CAVITY PRESS	0	0	0	0	0	0	0	
PK INJ PRESS	1523	1465	1494	1494	1492	0	0	
HOLD PRESS	668	656	662	675	653	6	16	*
BACK PRESS	178	164	171	182	168	1	0	
RAM STROKE	0.00	0.00	0.00	0.00	0.00	0	0	
MOLD A TEMP	0	0	0	0	0	0	0	
MOLD B TEMP	0	0	0	0	0	0	0	

Figure 11.44

The analysis progresses to the next-higher class level, Class 4, shown in Figure 11.44 and compared to this class level, shown in Figure 11.45. Further evaluation at higher class levels occurs, as shown in Figure 11.46 for Class 6. When all machine performance factors are within one class limit, with no out-of-tolerance values, the machine's class capability is known.

Drive File Limits Z Factors Quit

>> Select a Z factor ... a ,Z Extension)

...RAINING, SQC

Start Time:15:00 St... ...ate: 2/20/89 Mach No. 031 Samples:128

Parameter	Pr U	Z Factors	Mean Value	Max Value	Min Value	USL Count	LSL Count	
		CLASS1-I.Z						
		CLASS2-I.Z						
		CLASS3-I.Z						
		CLASS4-I.Z						
		CLASS5-I.Z						
		CLASS6-I.Z						
		CLASS7-I.Z						
CYCLE TIME	3	CLASS8-I.Z	31.84	31.93	31.76	0	0	
HOLD TIME	5	CLASS9-I.Z	5.53	5.55	5.51	0	0	
INJECT TIME	1	INJ-NEW.Z	1.46	1.62	1.40	15	0	*
CLAMP CLOSED	0	QS-10.Z	0.00	0.00	0.00	0	0	
CLAMP OPEN	0	QS-15.Z	0.00	0.00	0.00	0	0	
PLASTICATE	1	QS-20.Z	1.54	1.56	1.52	0	0	
CAVITY PRESS	0	QS-5.Z	0	0	0	0	0	
PK INJ PRESS	1	QS0.Z	1494	1494	1492	0	0	
HOLD PRESS	6	QS10.Z	662	675	653	5	6	
BACK PRESS	1	QS15.Z	171	182	168	1	0	
RAM STROKE	0	QS20.Z	0.00	0.00	0.00	0	0	
MOLD A TEMP	0	QS5.Z	0	0	0	0	0	
MOLD B TEMP	0		0	0	0	0	0	

↑↓ =Scroll, ENTER=Select, ESC=Exit

Figure 11.45

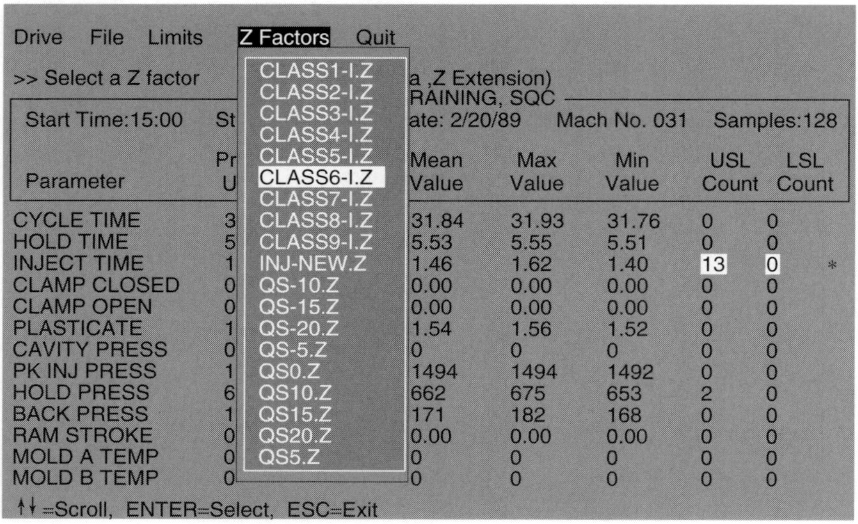

Drive File Limits **Z Factors** Quit

>> Select a Z factor ... a ,Z Extension)

...RAINING, SQC

Start Time:15:00 St... ...ate: 2/20/89 Mach No. 031 Samples:128

Parameter	Pr U		Mean Value	Max Value	Min Value	USL Count	LSL Count	
		CLASS1-I.Z						
		CLASS2-I.Z						
		CLASS3-I.Z						
		CLASS4-I.Z						
		CLASS5-I.Z						
		CLASS6-I.Z						
		CLASS7-I.Z						
CYCLE TIME	3	CLASS8-I.Z	31.84	31.93	31.76	0	0	
HOLD TIME	5	CLASS9-I.Z	5.53	5.55	5.51	0	0	
INJECT TIME	1	INJ-NEW.Z	1.46	1.62	1.40	13	0	*
CLAMP CLOSED	0	QS-10.Z	0.00	0.00	0.00	0	0	
CLAMP OPEN	0	QS-15.Z	0.00	0.00	0.00	0	0	
PLASTICATE	1	QS-20.Z	1.54	1.56	1.52	0	0	
CAVITY PRESS	0	QS-5.Z	0	0	0	0	0	
PK INJ PRESS	1	QS0.Z	1494	1494	1492	0	0	
HOLD PRESS	6	QS10.Z	662	675	653	2	0	
BACK PRESS	1	QS15.Z	171	182	168	0	0	
RAM STROKE	0	QS20.Z	0.00	0.00	0.00	0	0	
MOLD A TEMP	0	QS5.Z	0	0	0	0	0	
MOLD B TEMP	0		0	0	0	0	0	

↑↓=Scroll, ENTER=Select, ESC=Exit

Figure 11.46

For example, a Class 7 class level is reached for the molding system shown in Figure 11.47. If the product produced is acceptable at this class level of machine capability, the analysis ends. Comparing the progressive figures reveals that as higher class levels are evaluated, the out-of-tolerance variables are no longer highlighted. This indicates that the value was within that class level standard. Finally, when no

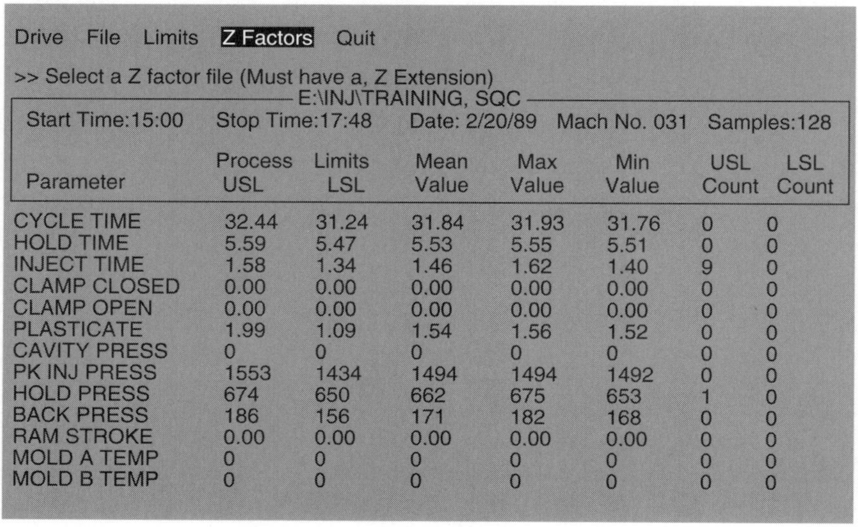

Drive File Limits **Z Factors** Quit

>> Select a Z factor file (Must have a, Z Extension)

E:\INJ\TRAINING, SQC

Start Time:15:00 Stop Time:17:48 Date: 2/20/89 Mach No. 031 Samples:128

Parameter	Process USL	Limits LSL	Mean Value	Max Value	Min Value	USL Count	LSL Count
CYCLE TIME	32.44	31.24	31.84	31.93	31.76	0	0
HOLD TIME	5.59	5.47	5.53	5.55	5.51	0	0
INJECT TIME	1.58	1.34	1.46	1.62	1.40	9	0
CLAMP CLOSED	0.00	0.00	0.00	0.00	0.00	0	0
CLAMP OPEN	0.00	0.00	0.00	0.00	0.00	0	0
PLASTICATE	1.99	1.09	1.54	1.56	1.52	0	0
CAVITY PRESS	0	0	0	0	0	0	0
PK INJ PRESS	1553	1434	1494	1494	1492	0	0
HOLD PRESS	674	650	662	675	653	1	0
BACK PRESS	186	156	171	182	168	0	0
RAM STROKE	0.00	0.00	0.00	0.00	0.00	0	0
MOLD A TEMP	0	0	0	0	0	0	0
MOLD B TEMP	0	0	0	0	0	0	0

Figure 11.47

more variables are highlighted, the process class level is reached. A new CpK value is calculated for the Class 7 level and documented. If the products produced are acceptable for the customer's requirements, then the process is deemed capable for the molding operation. But if the product is found to be off-specifications, then the process requires more analysis. The problem variables must be examined if the values cannot remain within operation parameters.

The software also has a self-diagnostic function whereby, for each variable out-of-class standard, it recommends possible solutions to fix or adjust the machine or process. This is similar to the troubleshooting guides furnished by material suppliers. Then as corrections or fixes are completed, the process is reevaluated, moving to the next lower class level for capability. This may be the repair and/or replacement of components or process control adjustments to solve the highlighted problem. Problem solving continues while working toward the next lower class level (5, 4, etc.), eliminating the cause of each problem or until the product is acceptable. Once the desired machine class system is reached, the machine's and molding system's capability values (CpK) are calculated. The software program in the system analysis performs this. This information should be communicated and documented to the manufacturing and scheduling personnel. Evaluation of the molding system can continue to monitor and assess the molding cycle to ensure that control is maintained thoroughout the production run. The evaluation system can also be tied into other molding cells to monitor their control using a 486 and higher computer system. Matching mold and product requirements to machine capability will provide the ideal machine/mold and material combinations for repeatable product and product quality and productivity.

Machine/mold/process capability analysis should be run with each new and existing mold to determine the best process and production combinations. The time spent on improving the machine's operational characteristics and evaluating which molds perform best in each machine makes basic economic and quality sense. It can aid in ensuring that a mold will be scheduled only in a machine capable of producing good product and monitored to ensure that variables remain in control.

DESIGN OF EXPERIMENTS

If the manufacturer does not have an automated analysis system and the product is not meeting specifications, there is the "design of experiments." (DOE) technique, which can be used to evaluate the manufacturing system. DOE is used to evaluate a product design, mold, and processing conditions should a problem occur with the product. DOE is used to solve a problem by concentrating on suspected problem solving variables.

DOE analyzes in a random but organized matrix, based on the number of variables to be evaluated using controlled tests and experiments. The analysis randomizes the variables to be evaluated with each variable evaluated for their high and low value effects on the product compared to the other variables. The DOE uses a minimum of test runs to arrive at a possible solution. The DOE analysis eliminates hours of evaluating all possible interacting variable combinations. DOE problem analysis is fast and very reliable in assisting a team in reaching a possible solution to a problem.

Table 11.5 Variable factors

Factor	(+) High Level	(−) Low Level
A. Mold temperature	200°F	80°F
B. Screw forward time	15 sec	5 sec
C. Injection pressure	1500 psi	500 psi
D. Screw backpressure	200 psi	50 psi
E. Injection time	3 sec	1 sec
F. Flow of material	Soft	Stiff
G. Mold cool time	30 sec	15 sec

The variables are evaluated in a systematic one-time-only analysis of their high and low values in a series of controlled experiments. The number of variables to be evaluated determines the time and number of test runs required for generating the data. After an analysis is completed, the data are evaluated and the contributing problem variables are determined. These variables are then changed, the problem corrected or eliminated, and the DOE is run again to determine the effects on the product. On the processing side this means that once a solution is reached, the operators must continue to monitor the molding system's variables to ensure that they remain in control, cycle to cycle. All the variables affecting a material or operation are listed, and for each variable all their related or associated variables are listed, until all variables, within reason, are listed, which could contribute to the cause of the problem. Variables affecting a production problem are graphically plotted using the Ishakawa or "fishbone" diagram, as shown in Figure 11.48.

The major contributing variables are then selected for analysis. These variables are then assigned a "high" value and a "low" value, which the process will use to evaluate

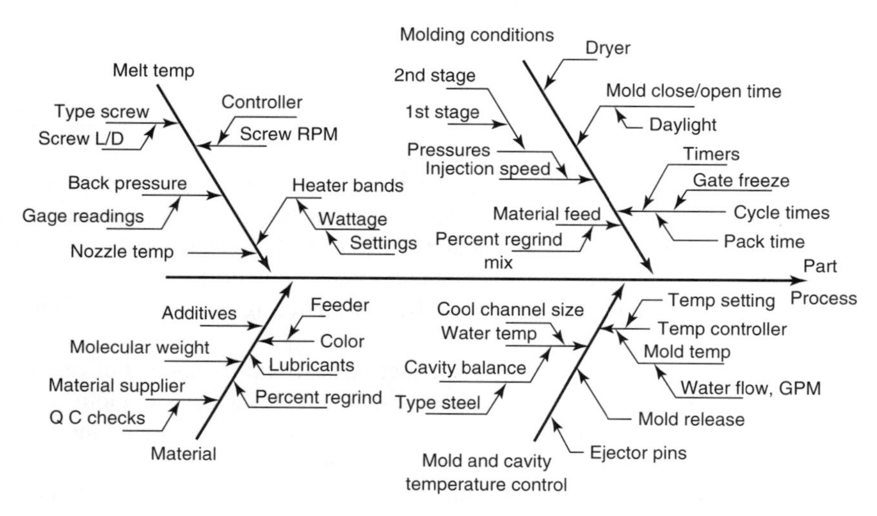

Figure 11.48 Ishakawa "fishbone" diagram.

Table 11.6 Eight run variable matrix

Run	A	B	C	D	E	F	G
1	+	+	+	−	+	−	−
2	−	+	+	+	−	+	−
3	−	−	+	+	+	−	+
4	+	−	−	+	+	+	−
5	−	+	−	−	+	+	+
6	+	−	+	−	−	+	+
7	+	+	−	+	−	−	+
8	−	−	−	−	−	−	−

Source: Adapted from Ref. 18.

the contributing influence of the variable to the problem. The number of variables to be evaluated will determine the size of the DOE matrix analysis. For example, a seven-variable matrix is listed in Table 11.5. To determine the variable's effects on the product, a key dimension or attribute is selected for analysis. This is the control used for determining the selected variable's effects on the product. An 8-run analysis for the seven variables is shown in Table 11.6. The variables, high and low values, shown as plus or minus, are used for the specific run as indicated in the table and evaluated for their effect during the trial run.

The trial is then rerun using the DOE values and the effects it has on the process. After each experimental run, the products produced are measured. The variance from the control is recorded, and the difference from the required dimension or effect is compared against the control with the difference recorded and posted in Table 11.7.

Table 11.7 Matrix for a seven-variable screening experiment factor (variable)

Run	A Mold Temperature	B Screw Forward Time	C Injection Pressure	D Screw Back- pressure	E Injection Time	F Flow of Material	G Mold Cool Time
1	+1.992	+1.992	+1.992	−1.992	+1.992	−1.992	−1.992
2	−2.001	+2.001	+2.001	+2.001	−2.001	−2.001	−2.001
3	−2.000	−2.000	+2.000	+2.000	+2.000	−2.000	+2.000
4	+1.990	−1.990	−1.990	+1.990	+1.990	+1.990	−1.990
5	−1.998	+1.998	−1.998	−1.998	+1.998	+1.998	+1.998
6	+1.995	−1.995	+1.995	−1.995	−1.995	+1.995	+1.995
7	+1.998	+1.998	−1.998	+1.998	−1.998	−1.998	+1.998
8	−1.991	−1.991	−1.991	−1.991	−1.991	−1.991	−1.991
Sum H	+7.975	+7.989	+7.988	+7.989	+7.980	+7.984	+7.991
Sum L	−7.990	−7.976	−7.977	−7.976	−7.985	−7.981	−7.974
Difference	−0.015	+0.013	+0.011	+0.013	−0.005	+0.003	+0.017
Effect	−0.004	+0.003	+0.003	+0.003	−0.001	+0.001	+0.004

Source: Adapted from Ref. 18.

The effect results, either plus or minus, are then compared for significance and other tables used to determine the variable(s) that affect the process the most. It is not necessary to use a closed-loop/continuous-feedback and control system, but this does aid in repeatable manufacture of the product. Using a process control system with machine control analysis is extremely helpful to show changes in the molding process. This is one of the first steps necessary toward computer-integrated manufacture (CIM) and total quality real-time process control.

REFERENCES

1. FDM ®1650, *Fused Deposition Model Machine,* Stratasys' Inc., SS1, Eden Prairie, MN, April 1996.
2. SLA-250/50, *Stereo Lithography Machine,* 3D Systems, Valencia, CA.
3. HyperMesh, Altair Computing, Inc., Troy, MI.
4. Figures 3, 4 and back page, *Temperature, Pressure, and Fill Profiles,* Vol. 8, No. 3, AC Technology, Ithaca, NY,
5. *Horizons,* DTM Corp., Austin, TX, DTM™, Sinterstation, winter 1998.
6. V. Wigotsky, "Some Mold Design Guidelines," *Plastics Eng.* 24–25 (Nov. 1985).
7. J. Overbeeke, "High-Speed Injection Molding," *Plastic Machinery & Engineering Magazine,* 37 (April 1993).
8. G. Engelstein, "Flow Analysis, Effects of Finite Element Meshing on Mold Filling Analysis," *Plastics Eng.* 41–43 (April 1993).
9. B. Mullins, "Finite Element Analysis: Predicted versus Actual Performance," *Plastics Design Forum,* 71–76 (March/April 1989).
10. E. Bernhardt, "Design Computer Modeling Predicts Tolerance of Molded Parts," *Plastics Eng.* 29–31 Octo. 1990.
11. B. Miller, "Predicting Part Shrinkage Is a Three-Way Street," *Plastics World,* 48–52 (Dec. 1989).
12. "Understanding Tight Tolerance Design," *Plastics Design Forum,* 61–71 (March/April 1990).
13. A.V. Feigenbaum, *Total Quality Control,* McGraw-Hill, New York, 1983.
14. D. Hunkar, *An Engineering Approach to Process Development and the Determination of Process Capability,* Document 228, Hunkar Laboratories Inc., Cincinnati, OH, 1991.
15. *"Best Introductory Book on SPC," DataMyte Handbook,* 6th ed., DataMyte Business, Allen-Bradley Comp., Inc., Dec. 1995, pp. 1–15
16. Ref. 15, pp. 5–10.
17. E. Kindlarski, "Ishikawa Diagrams for Problem Solving," *Quality Progress,* 26–30 (Dec. 1984).
18. J. Schleckser, "Troubleshooting Technique Shortens Path to Quality," *Plastics Eng.* 35–38 (July 1987).
19. *General Design Principles—Module 1*, 201742B, E. I. Du Pont de Nemours Corp., Wilmington, DE (Sept. 1992).
20. R. D. Beck, "Plastic Product Design," Van Nostrand Reinhold, New York, 1970.

CHAPTER 12

PRODUCT AND TOOLING DESIGN GUIDELINES

It is essential, after the product's design is complete, for the design team to ensure the tooling to manufacture the product is designed correctly. Tool operation, dimensional and temperature control, and surface finish correctly designed and built into the tooling are necessary if the product is to meet the customer's end-use requirements. Input from all members of the design team must continue to focus on a well-designed and temperature-controlled mold and processing system. The product design team composed of selected personnel with experience, knowledge, and talent in the design and/or manufacture of plastic products continues until the product is released for sales and accepted by the customer.

The design engineer's primary responsibility for the product is now transferred to the tooling designer. The tooling designer, with assistance from the design team, should now follow their material supplier's recommendations for the tools design and material processing guidelines. Following their material supplier's guidelines will assist the team in ensuring that the tool is capable. Then, using the material processing recommendations, their manufacturing system or supplier will always produce an acceptable product that meets their customer's requirements. Factors affecting product design are illustrated in Figure 12.1. These are the product development processes necessary to obtain a good product by considering the elements of product development.

PRODUCT MANUFACTURING ANALYSIS

The information needed to design the product began with knowing the product's end-use requirements and ends with developing manufacturing and process control procedures. The product's design; material, section thickness, radii, and methods of manufacture all affect the product's end-use performance. These items are decided

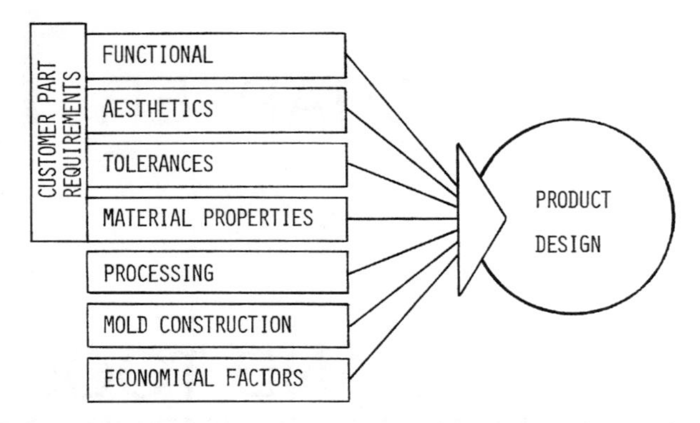

Figure 12.1 Factors affecting product design. (Adapted from Ref. 5.)

on and developed during the products design program. They have a direct influence on the tooling and manufacturing process, as plastics require additional knowledge to repeatable produce quality products.

The plastic resin's material supplier's *Molders Processing Guide* is an excellent reference for understanding any special product design, tool, and manufacturing requirements that must be followed to successfully manufacture a product in their material. The material's manufacturing requirements must always be considered early in the development program to take advantage of the material's properties and avoid making a mistake in the use of this material in your product.

MATERIAL CONSULTANTS

A plastic material's processing properties for product manufacture are melt temperature, flow length (viscosity), setup time, (time required for the molten material to solidify in the mold cavity), mold temperature control, heat transfer qualities of the mold's steels, material in-mold shrinkage plus dimensional tolerances, minimum section thickness, and gate type and location to fill the mold cavity. The material's injection-molding machine and processing variables will affect how the material flows and fills the mold cavity. The mold controls the product's final dimensions and quality using the cooling system, runner layout, and cavity gating to control the product in support of the injection-molding machine processing variables to eliminate problems during manufacture. Product material problems to control and avoid in the tool during processing are porosity, sink marks, warpage, molded-in stresses, shrinkage variances, and variable dimensional tolerances.

The majority of these problems are eliminated with a well-designed mold and an in-control processing machine and manufacturing system. The mold designer is responsible for final product tool dimensions. These dimensions are based on the following product and processing variables: material, type of runner system to feed the mold cavity, gating, weld line recognition points, and boss, rib, and internal anchor points for holding dimensional tolerances. The mold designer and team members determine the type of mold needed to manufacture the product, and how it will operate, as well as the cavity temperature and fill pressure-sensing control in the mold and product ejection system, and any other internal operations in the mold such as core pulls and unscrewing and slide operations. The mold designer selects the steel and other materials for correct heat transfer and cavity temperature control, wear, and methods of construction for the mold cavity and base. Molding problems can be minimized when the product and tooling are designed for the material to avoid and minimize manufacturing problems. The mold must be designed correctly for the material selected for manufacture using gating, venting, product ejection, and temperature control to repeatably produce the product on every molding cycle.

The manufacturing process requirements for the product's resin includes knowing the material's flow length, resistance to shear, heat stability, setup time, fiber type and percent of fill, shrinkage rate, and weld line strength, as well as the design of the tool using a balance cavity system for fill and pressure control plus temperature control, product ejection, draft, gating, venting, heat transfer, and control of the product's final dimensional quality.

UNIFORM PRODUCT SECTION

During the product design phase the design team should try to maintain as uniform a wall section or thickness for a plastic product as possible. This assists in several

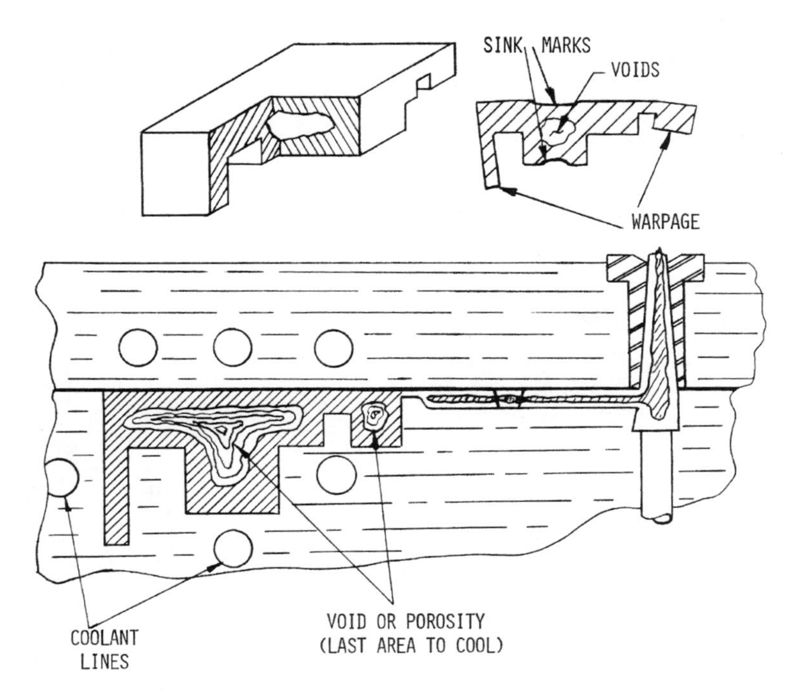

Figure 12.2 Thick/thin sections create part problems in fill and packout.

areas, such as warpage, possible processing for elimination of voids, and porosity to maintain a uniform fill and packout. During the filling of a mold cavity, the material follows the path of least resistance; the thicker product sections fill first and thin sections last. It is important to have as constant a fill and packout pressure during the molding operation as possible to avoid these types of problems in the product. It is also important to have all sections of the mold cavity fill at a uniform rate and pressure. Thick sections create a pressure drop in the mold cavity, causing a loss of injection or packing pressure to fill the thinner sections that fill only after the thick section has filled. This can lead to premature freeze-off of the melt front at the thin sections if the pressure drop is too great, resulting in a potential nonfill occurring at the thin section of the product as shown in Figure 12.2. Therefore, if a thick product section is necessary, always locate the gate at this point and fill this cavity section first. Then, as pressure in the mold cavity is increased after the thick section is filled, the thinner sections will fill. Whenever possible, all sections should be as uniform in thickness as the products design will allow, as shown in Figure 12.3.

The majority of plastic products are designed with wall sections between 0.030 to 0.187 in. in thickness. In most plastic product designs the section thickness is adequate to meet the product's performance requirements, and if not, ribs are added to increase section strength and stiffness. The use of thick sections can create process and product problems and should be avoided. The wall section or product thickness selected must be sufficient for the plastic material, being injected into the mold cavity at the gate, to

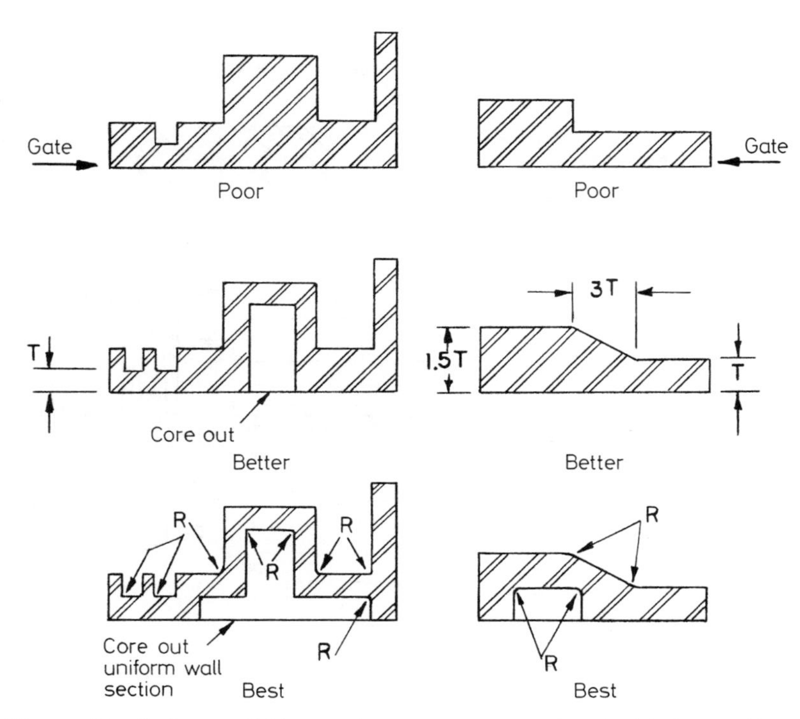

Figure 12.3 Uniform wall thickness improves part strength and minimizes problems.

flow and fill the entire length of the cavity. Each plastic material has a characteristic material flow length per section thickness, and this information is available from their material supplier. To analyze the filling of the mold cavity, use the resin's viscosity and flow characteristics with a mold fill simulation software program.

MATERIAL FLOW ANALYSIS

When using the CAD modeling system and after the FEA analysis is completed, a material flow analysis should be conducted. This analysis is usually run in conjunction with the mold temperature control and mold cooling analysis software program. This analysis will assist the designer in determining whether the products material flow length versus section thickness is adequate to fill the mold cavity under typical processing pressure and temperature. A material's flow path begins at the mold's sprue and then fills the runner system to the gate at the entrance to the mold cavity. The material must then pass through the product's restrictive gate, and then the material must flow to the furthest point in the cavity. If conditions in the mold are not controlled, or the melt and cavity temperature too cold or injection pressure is too low, the cavity will neither fill nor be correctly packed out. Product geometry, number and location of gating points on the product and the gate size, and pressure and temperature control of the molten plastic material control this condition. Also, there are pressure drops in injection pressure occurring at every point where the resin must turn a corner and at the cavity gate.

Today, injection-molding machines provide their operators with information on how the material will enter and fill the mold cavity. The operator can now adjust injection pressure or adjust the pressure and fill profile to ramp up or down the fill and pressure during the injection cycle to compensate for these pressure drops without inducing additional, undesired, shear heat into the resin. This requires the cavity orientation in the mold, type, and size of the runner system and using a balanced cavity layout that will eliminate fill and packout problems as shown in Figure 12.2. The accuracy of the material supplier's resin viscosity data in the computer simulation will determine the accuracy of the mold filling analysis.

Amorphous Resins

Amorphous resins typically require mold temperatures between 50 and 120°F. They experience relatively low mold cavity shrinkage of 0.002–0.010 in./in. These resins require cooling water to extract the heat from the resin in the mold cavity to facilitate rapid setup. Amorphous resins soften during heating and do not have an exact or sharp melting temperature as do crystalline and semicrystalline resins. Therefore, heat must be quickly extracted from the mold cavity for amorphous resins to become rigid enough for ejection from the product cavity. The more efficiently heat is removed from the molten resin, the sooner the product will set up in the mold cavity and be ready for ejection. This creates a faster cycle that can produce more products per hour as long as the product is not distorted on ejection and product requirements are achieved.

Crystalline Resins

Crystalline and semicrystalline resins have a sharp or pronounced melting temperature, unlike amorphous resins. They also experience higher material shrinkage in the range of 0.015–0.035 in./in. Each material, due to its crystalline structure, is molded at higher mold temperatures in the range of 160°F ≥400°F depending on the material. The use of higher mold temperatures allows the molder to pack out the product in the mold cavity while continuing to pack in more material to compensate for the higher material shrinkage occurring in the cavity. Higher mold temperatures also assist in the product becoming more stable when exposed to high end-use temperatures by having the maximum shrinkage taken in the mold cavity and not in service as postmold shrinkage. A hot mold also speeds crystallization of the resin, reducing the resin's setup time. Hot molds also reduce molded-in material stresses as high temperatures assist in relieving any molded-in stresses with an annealing action in the hot mold cavity. Crystalline resins molded in a cold mold (<150°F) experiences lower mold shrinkage and will exhibit greater postmold shrinkage when exposed to temperatures above the molds temperature. All materials, especially crystalline resins, want to reach their most stable state.

The more material shrinkage occurring in the mold, the less the product will experience in a high end-use temperature application. This accomplishes several purposes:

1. Maximum material shrinkage occurs in the mold, resulting in more dimensionally stable products.
2. Better product packout with uniform product weight, cycle to cycle.
3. Faster molding cycles as products become rigid more quickly as a result of their higher crystalline, sharp melting-point temperature.

Molding both types of resins in hot molds will reduce molded-in material stress. This is due to the higher temperature annealing action on the product, causing stresses to be relieved as the product slowly cools in the mold cavity. But cycle time will dramatically increase, as will manufacturing cost.

An example to consider is the material used for wire and cable ties that require a long flow path. These thin and narrow products of varying length, some many feet long, are typically made from nylon type 6/6 homopolymer resins. The molds are large with long flow lengths to economically produce the ties. The molds are often unbalanced, not designed with a runner and cavity layout to provide a uniform fill pressure at each cavity. Therefore, a material with a high flow length is required, with nylon 6/6 having this processing and flow property in thin sections. These ties are run in molds with high cavity temperatures to assist in increasing fill and flow length. It is important to avoid overheating the resin to obtain the required flow length as it can weaken the ties' physical strength properties. In some molds, just prior to opening, the mold clamps down on the ends of the wire ties and stretches them a predetermined distance. This stretching of the wire tires orients the material in their length, which results in increasing the product's tensile strength. Nylon type 6 homopolymer has

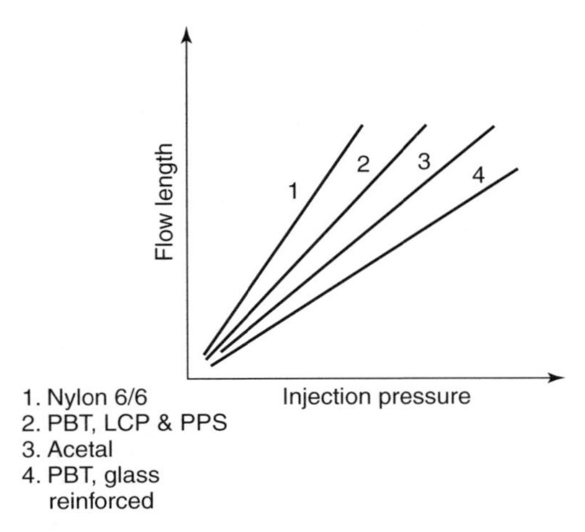

Figure 12.4 Typical spiral flow ranges. (Adapted from Ref. 14.)

tried to displace type 6/6 nylon for this application. Type 6 nylon has a higher DAM toughness and elongation, but lacks the necessary flow to fill these long sections because it has a higher melt viscosity. Therefore, type 6 nylon has not been a suitable material candidate for this high volume market.

The design team must know the limits of the products resin to fill the mold cavity. Figure 12.4 gives typical spiral flow ranges of some plastic resins. A picture of a spiral flow material flow length mold used to generate the data is shown in Figure 12.5 with the developed flow length data listed for a general purpose ABS in Table 12.1. As an example, a standard grade of ABS (acrylonitrile butadiene styrene) was evaluated in

Figure 12.5 Spiral flow test mold. (Courtesy of DSM Engineering Plastics, North America.

Table 12.1 Spiral flow for standard grade of ABS

Melt Temperature (°F)	Fill Rate (in./s)	Flow Length (in.) for Mold Temperature (°F)		Change in Percentage
		85	175	
425	0.25	12.1	13.7	13.0
425	1.75	20.5	21.6	5.4
525	0.25	22.0	25.4	15.5
525	1.75	37.3	39.6	6.2

Source: Courtesy of DSM Engineering Plastics North America.

a spiral flow mold with varying melt temperatures and fill rates or injection speeds. Another variable was mold temperature that also has an effect on flow length, but is not as significant for ABS. Spiral flow is a material flow test used to estimate a material's flow length under controlled molding conditions, pressure and temperature. The mold cavity is cut in a spiral, looping configuration from the center sprue gate outward in the mold cavity. The cavity has a specified thickness and width with a long flow path and can vary from supplier to supplier. When the material is spiral flow tested, the molding machine screw injecting the material into the mold is not allowed to bottom in the barrel. This causes the material to flow until it cools, and solidifies, the mold is opened, and the flow length is measured. Tests show that injection speed and mold temperature do affect a material's flow length.

Amorphous resins have lower flow length values than do the higher-flow engineering resins. During the design and layout of the mold the material for the product should be evaluated for flow length. Wall thickness, product geometry, and mold cavity layout must be within the material's flow length. It is too late after the mold is designed and built to realize that the material lacks the necessary flow characteristics to fill the mold and produce acceptable products.

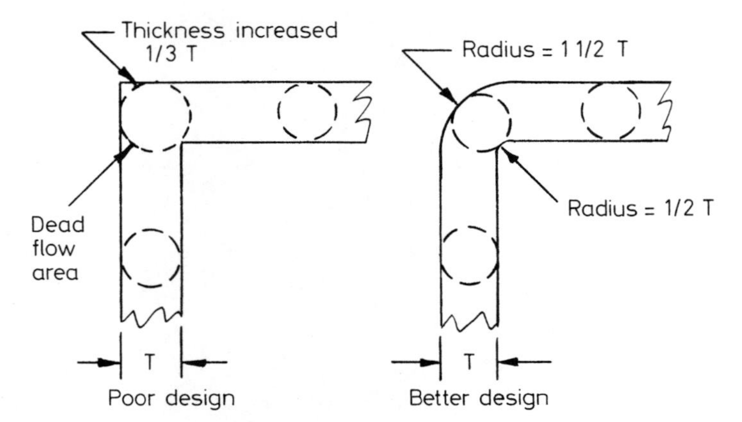

Figure 12.6 Examples of corner radii. A good design improves flow and increases part toughness

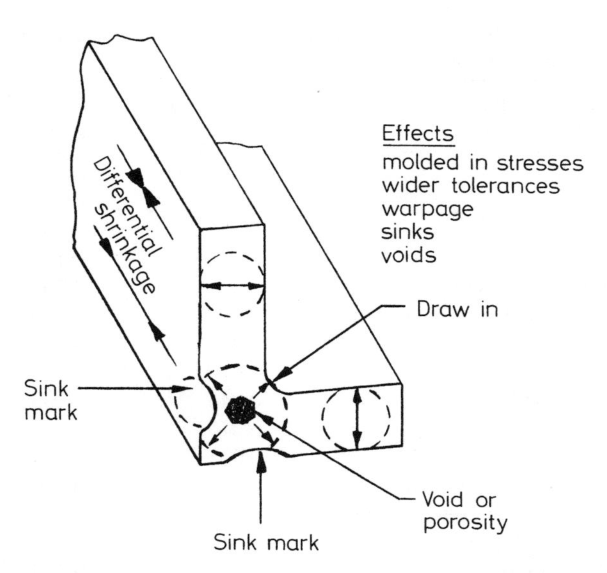

Figure 12.7 Effects of nonuniform wall thickness on molded parts. (Adapted from Ref. 1.)

Every time the resin turns a corner, in the runner system or product cavity as shown in Figure 12.6, there is a pressure drop at this section. This also occurs if the section thickness suddenly increases at a boss or thicker rib section. This is seldom a problem because the resin's flow channel remains hot and open to continue, forcing more material into these thicker sections. But if a pressure drop occurs at thin sections, the material may cool and solidify, blocking the path. When fill and flow problems occur at thin sections, local heating of the mold cavity with cartridge heaters or adding another gate or relocating the gate to a better site can often solve this problem. Poor cavity packout can cause insufficient material to fill the cavity and on cooling and shrinking, porosity in the thick section can occur, creating other product problems as shown in Figure 12.7. The rule to remember if thick sections are required in a product is to locate the gate at or as close as possible to the thick section to always fill and pack it out first to reduce pressure drops further into product filling.

These conditions must be evaluated during the product's design for the type of mold selected and the location and number of gates and temperature control to be able to produce a good product.

UNIFORM PRODUCT THICKNESS

Maintaining uniform product section thickness as shown in Figure 12.3 will eliminate fill and packout problems and assist in reducing molded-in material stresses. Recommended wall section thickness for many plastic materials are listed in Table 12.2. Ribs are used to reduce section thickness, increase product stiffness and

Table 12.2 Recommended wall thickness for thermoplastic molding materials

Thermoplastic Materials	Minimum (in.)	Maximum (in.)
ABS	0.045	0.140
Acetal	0.015	0.125
Acrylic	0.025	0.150
Cellulosics	0.025	0.187
FEP fluoroplastic	0.010	0.500
Long-strand-reinforced resin	0.075	1.000
Liquid crystal polymer (LCP)	0.008	0.120
Nylon	0.010	0.125
Polyarylate	0.045	0.160
Polycarbonate	0.040	0.375
Polyester	0.025	0.125
Polyethylene (low-density)	0.020	0.250
Polyethyelene (high-density)	0.035	0.250
Ethylene vinyl acetate	0.020	0.125
Polypropylene	0.025	0.300
Noryl® (modified PPO)[a]	0.030	0.375
Polystyrene	0.030	0.250
PVC (rigid)	0.040	0.375
Polyurethane	0.025	1.500
Surlyn	0.025	0.750

[a] Registered General Electric material trade name.

Source: Adapted from Refs. 9–11.

strength, and also provide fill paths in the product cavity. Whenever possible, position the ribbing pattern to line up in the anticipated material's flow direction in the mold cavity. This will aid in filling the mold and in reducing weld lines and pressure drops as shown Figure 12.8. Other recommended methods using ribbing and product geometry for section rigidity and at bosses are shown in Figure 12.9.

Nonuniform wall thickness will create differential shrinkage in a product. The thick sections remain hot longer, causing this increased material shrinkage, which can also lead to product warpage. This occurs more as a result of differential product shrinkage.

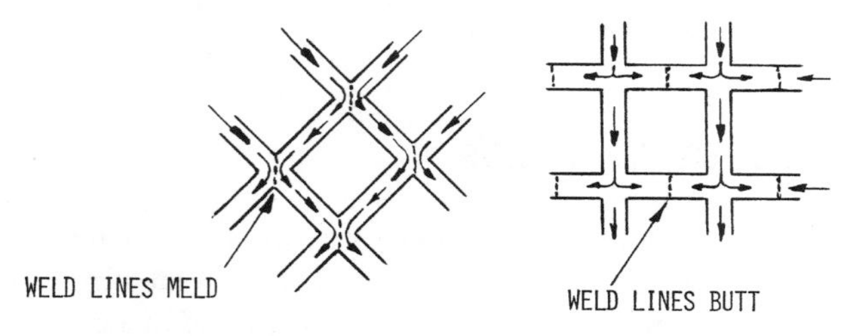

Figure 12.8 Weld line improvement. (Adapted from Ref. 2.)

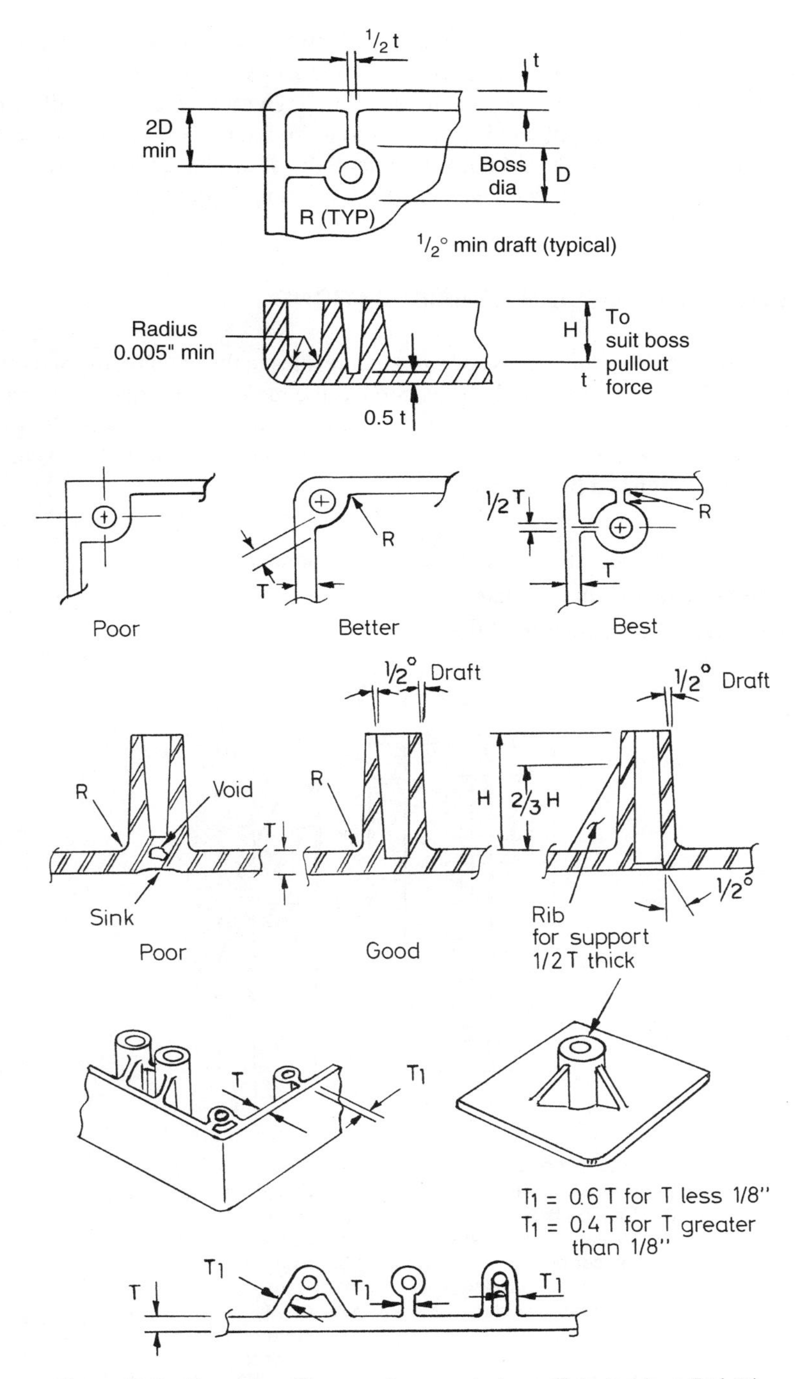

Figure 12.9 Examples of boss and corner designs. (Adapted from Ref. 8.)

This is caused by reinforcement orientation, widely varying section thickness, and differential mold temperature in the cavity. This frequently occurs with semicrystalline and crystalline materials, which have higher shrinkage rates. Therefore, all product sections must be packed out and injection-holding pressure increased to force more hot resin into the cavity to reduce porosity (bubbles and shrinkage voids). These voids cause dimensional, surface, warpage, and product strength problems.

PRODUCT MOLD DESIGN CONSIDERATIONS

Approximately 40% of the success of a product resides in the engineering and quality of the mold. The product will only be as good as the mold that produces it. A mold's design, terminology, and layout are shown for a two-plate mold in Figure 12.10.

The key factors in the design, engineering, and building of a mold to high standards required for the finished products are

1. Standard versus custom mold base and components
2. Tolerance factors in calculating shrinkage
3. Materials of construction of the mold

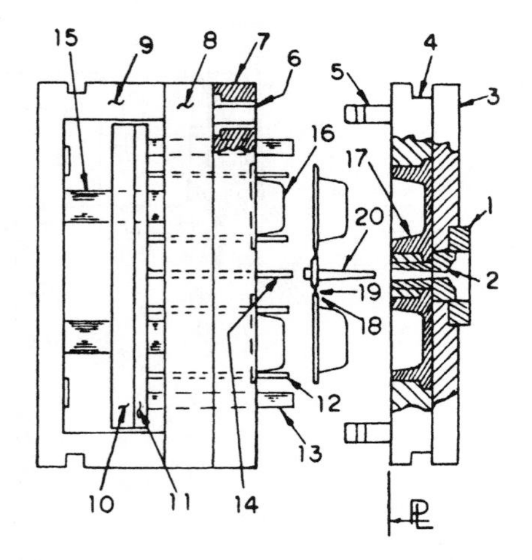

1. Locating ring
2. Sprue bushing
3. Clamp plate
4. Cavity retainer plate
5. Leader pin
6. Leader pin bushing
7. Core retainer plate
8. Support plate
9. Ejector housing
10. Ejector plate
11. Ejector retainer plate
12. Ejector pin
13. Return pin
14. Sprue puller pin
15. Support pillar
16. Core
17. Cavity
18. Gate
19. Runner
20. Sprue

Figure 12.10 Parting line operations and mold components. (Adapted from Ref. 14.)

4. Mold cavity considerations
5. Cavity-to-cavity accuracy in multicavity molds
6. Cavity temperature and mold fill pressure control
7. Runner and product gating considerations
8. Venting

The selection of a standard versus custom mold base is determined by the product to be manufactured and cost. The mold is designed to suit the product and the tolerances required to manufacture the product. Standard mold bases and their components are less expensive and adaptive to many cavity sizes. Molds can be single- or multicavity in design. Use of standard or off-the-shelf mold base components can decrease the time a mold is built and put into production. Leave this to the mold design team to assist in determining the mold base type for the program. Custom mold bases are designed for the specific product and take longer to manufacture. However, there are many standard mold plates available from suppliers to assist in supplying the steel plates and associated hardware to build the mold. Custom molds cost more and will be built specifically for the product and the cavity temperature cooling requirements to ensure that the product will meet the anticipated cycle and tolerances required for the product.

Product Mold Type Consideration

A mold flow simulation program will predict how the melt fills the mold cavity and show pressure gradients across the product's surface. The simulation should also be run with the mold cooling program to design the water channel layout to cool and control the temperature of the mold. These simulations will assist the mold designer during the mold cavity design layout in the mold. The standard two-plate mold is limited in where the gate can be located. The mold types to improve product manufacture are three-plate and hot-runner mold designs with multiple gating possibilities. These molds are more expensive but can position the product's gate away from critical areas to eliminate or reduce the effects of weld lines and obtain better mold cavity filling with fewer molded-in stresses. These molds allow placement of multiple gates almost anywhere within the product's section and also reduce the flow length of the resin that will remain hotter longer to assist in packing out the product cavity.

The design team and mold designers must always work together. Too many designers believe that after the product is designed, their portion of the program is completed. Their responsibility continues until the product is in production and successful in the market.

Mold Materials of Construction

The mold must be designed and built from materials that will ensure that it will be trouble-free as long as it is properly maintained for the anticipated life for the production of the product. Selection criteria for mold steels are corrosion and wear resistance, heat transfer qualities, plus finishing characteristics. Dissimilar mold materials are used in contact areas of high anticipated wear. This is necessary to reduce galling and allow easy replacement of high-wear mold components. Always use replaceable gate

Table 12.3 Properties to consider in selection of mold building material

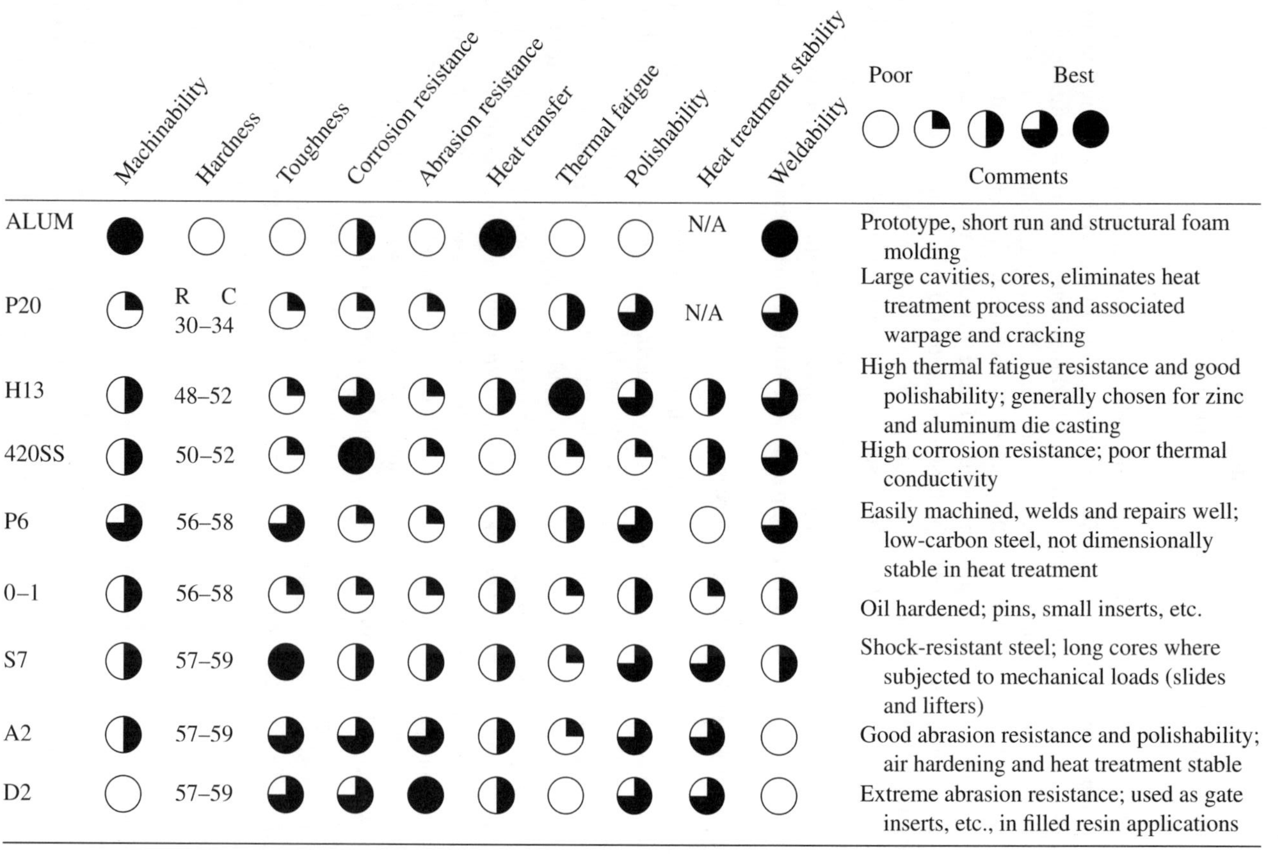

Poor ○ ◔ ◑ ◕ ● Best

	Machinability	Hardness	Toughness	Corrosion resistance	Abrasion resistance	Heat transfer	Thermal fatigue	Polishability	Heat treatment stability	Weldability	Comments
ALUM	●	○	○	◑	○	●	○	○	N/A	●	Prototype, short run and structural foam molding
P20	◔	R C 30–34	◔	◔	◔	◑	◑	◕	N/A	◕	Large cavities, cores, eliminates heat treatment process and associated warpage and cracking
H13	◑	48–52	◔	◕	◔	◑	●	◕	◑	◕	High thermal fatigue resistance and good polishability; generally chosen for zinc and aluminum die casting
420SS	◑	50–52	◔	●	◔	○	◔	◔	◑	◕	High corrosion resistance; poor thermal conductivity
P6	◕	56–58	◕	◔	◔	◑	◑	◕	○	◕	Easily machined, welds and repairs well; low-carbon steel, not dimensionally stable in heat treatment
0–1	◑	56–58	◔	◔	◔	◑	◔	◑	◔	◑	Oil hardened; pins, small inserts, etc.
S7	◑	57–59	●	◑	◑	◑	◔	◕	◕	◑	Shock-resistant steel; long cores where subjected to mechanical loads (slides and lifters)
A2	◑	57–59	◕	◕	◕	◑	◔	◕	◕	○	Good abrasion resistance and polishability; air hardening and heat treatment stable
D2	○	57–59	◕	◕	●	◑	○	◕	◕	○	Extreme abrasion resistance; used as gate inserts, etc., in filled resin applications

Source: Adapted from Ref. [16].

Table 12.4 Material heat conductivity

Material Type	Tensile Strength (psi)	Brinnell Hardness (300 UG)	Thermal Conductivity [Btu ft^{-1} h^{-1} (°F)]
S7 (annealed)	93,000	214	21
S7 (hardened)	275,000	526	21
P20	93,000	263–344	19–21
Aluminum 6000, 7000	65–70	50–100 (500 kg)	80–90
420SS	250,000	512	10–12
H13	170,000–283,000	352–530	16.4
Ampcoloy 940	95,000	210	120
Ampcoloy 18	90,000	180	40
Beryllium/copper (2.0% BE)	115,000–125,000	336–380	62
Ampcoloy 22	95,000	311	27
A10	303,000–355,000	560–680	18

Source: Adapted from Ref. [7].

blocks for each mold cavity, as this is an area of high wear due to resin flow. The gates size must be controlled to ensure that the cycle and product packout remain consistent cycle to cycle. All molds will wear, and it is best to know and control where it will occur for minimizing replacement cost, achieving a higher degree of product quality, and long mold life with minimal mold downtime. Consider mold maintenance as a prime cost in the manufacturing cycle and what mold items should be examined, measured, or replaced during each mold downtime.

The medical and electronics industries require a high degree of quality with long mold life. These molds produce millions of products yearly to very tight tolerances. Their product requirements are very critical and require very tight tolerances, with product liability a major issue. Each industry has its own definition of product quality and tolerances with varying requirements to produce acceptable products. This makes it more difficult for the mold builder and design team. They must know more about materials, design life, and tolerances for the correct mold to be built. A reference for common mold building material selectionis presented in Tables 12.3 and 12.4 and Figure 12.11. The design mold and design mold checklists in Appendix A (checklists 2 and 17) should be completed before the order for the mold is released to ensure that all questions for the product and mold are answered.

Mold Base and Components

The selection of the mold base, custom or standard to hold the mold cavity, depends on the product size, number of cavities, temperature control and molding machine size, tie-bar spacing to hold the mold in place between the platens, and cost. The design team assists in making these decisions for the mold base with direct input from production. The entire design team's input is required for the final mold design decision as the mold's final design impacts on the product's success.

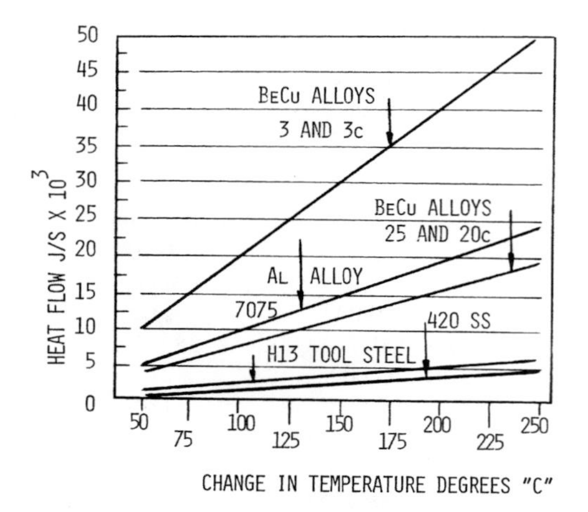

Figure 12.11 Material heat removal. (Adapted from Ref. 7.)

The type of mold selected can be a two- or three-plate, hot-runner, single- or multicavity, balanced or unbalanced mold as shown in Figures 12.12a–d and 12.13 (hot-runner mold). Other factors to consider are lead time and complexity of the mold, which includes secondary in-mold operations as slides, cams, and unscrewing required for the product. Each mold builder will have ideas and recommendations on which mold type is best for the customer's product requirements.

To specify standard (off-the-shelf) mold components is wise, as they are easily replaced if broken or worn. The use of standard mold components may eliminate the requirement for a backup mold when off-the-shelf mold component products are readily available. The requirement of having products available for sale is important for the success of a new product program. This is explained in an article in a book summarizing the competitive advantage today (R. J. Schonberger, "World Class Manufacture: The Next Decade-Building Power Strength and Value," in *Competitiveness Advantage High Quality for Low Price,* The Free Press). The competitive advantage today is shifted from quality to value for the customer.

Cavity Tolerance Control

The production mold controls the dimensional accuracy and quality of the molded product. The mold's manufacture and built-in tolerances must be specified and controlled during construction. The molding process must also be kept in control, and the higher the mold quality, the better and more in control becomes the molding process. Variables for processing conditions for the mold and material for controlling material shrinkage and tolerances are shown in Figure 12.14. The effect of injection pressure relating to packing pressure is shown in Figure 12.15. Packing pressure and injection pressure have a direct correlation. The higher the injection pressure, the higher the packing pressure. The higher pressures assists in ensuring that the cavity is filled and packed out to maximum capacity, thereby limiting the amount of shrinkage that occurs

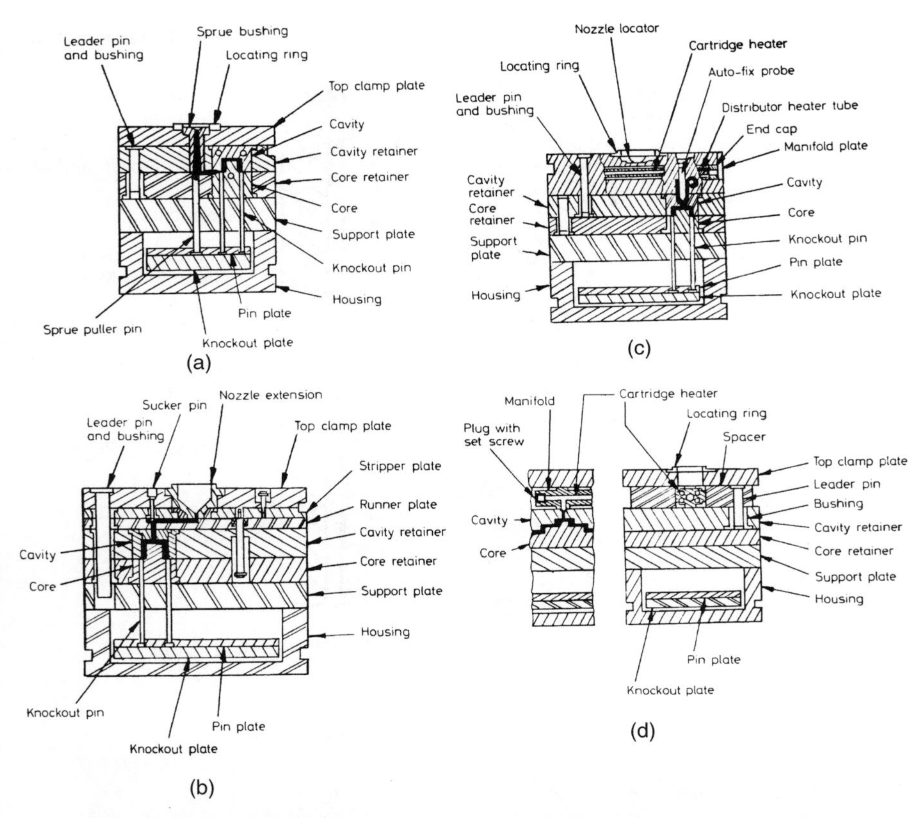

Figure 12.12 Types of molds: (a) two-plate mold; (b) three-plate mold; (c) hot-runner mold; and (d) hot-manifold mold.

in the mold cavity. The degrees of accuracy for the product are directly proportional to the design and building of the mold.

Medical items and electrical printed circuit board (PCB) connectors have tolerance control typically 4–8 times more stringent than do regular molded products. Cavity dimensions are controlled to very tight dimensions out to the fourth decimal point. The more accurate the mold, the fewer the process variable adjustments that will be required to produce repeatable good products. The degree of accuracy anticipated for commercial to fine tolerance levels for a simple plastic product versus length is listed in Table 12.5. Injection pressure, geometry, and postmold shrinkage affect the dimensions (see Figs. 12.16 and 12.17).

Production should be given a mold that does not require making processing adjustments that are out of the material's typical processing range to produce good products. If frequent or too many processing adjustments are required once the mold is in production, the mold should be reevaluated and modified to correct the problems. In many cases this occurs as a result of a poorly designed temperature controlled cavity cooling system.

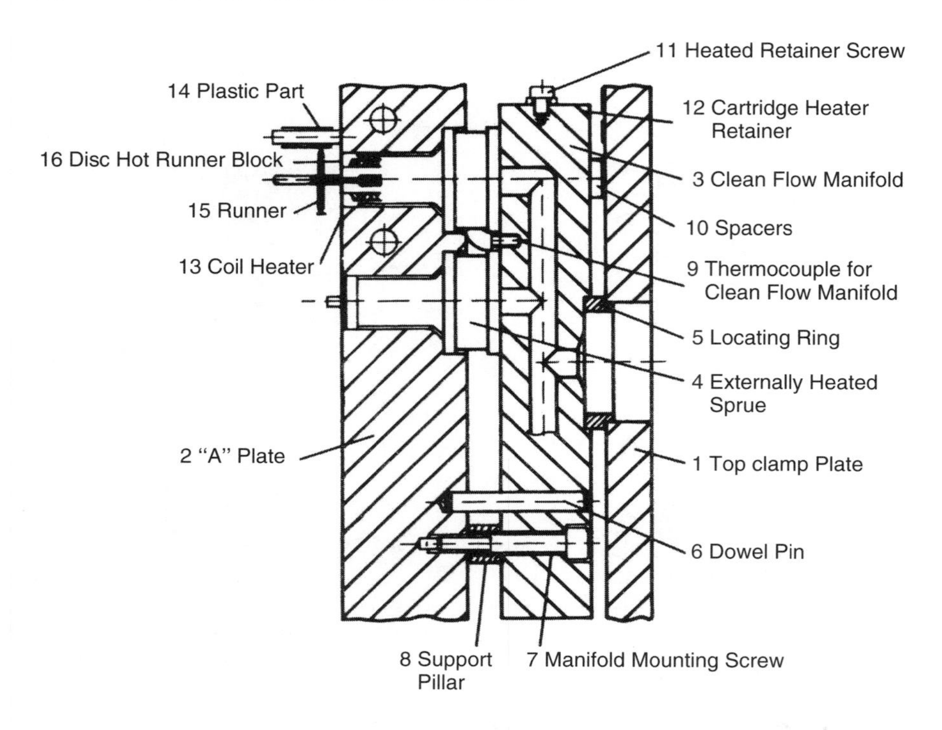

Figure 12.13 Hot-runner mold: (a) cutaway of hot runner mold system producing polycarbonate parts; (b) layout. (Courtesy of INCOE Corp.)

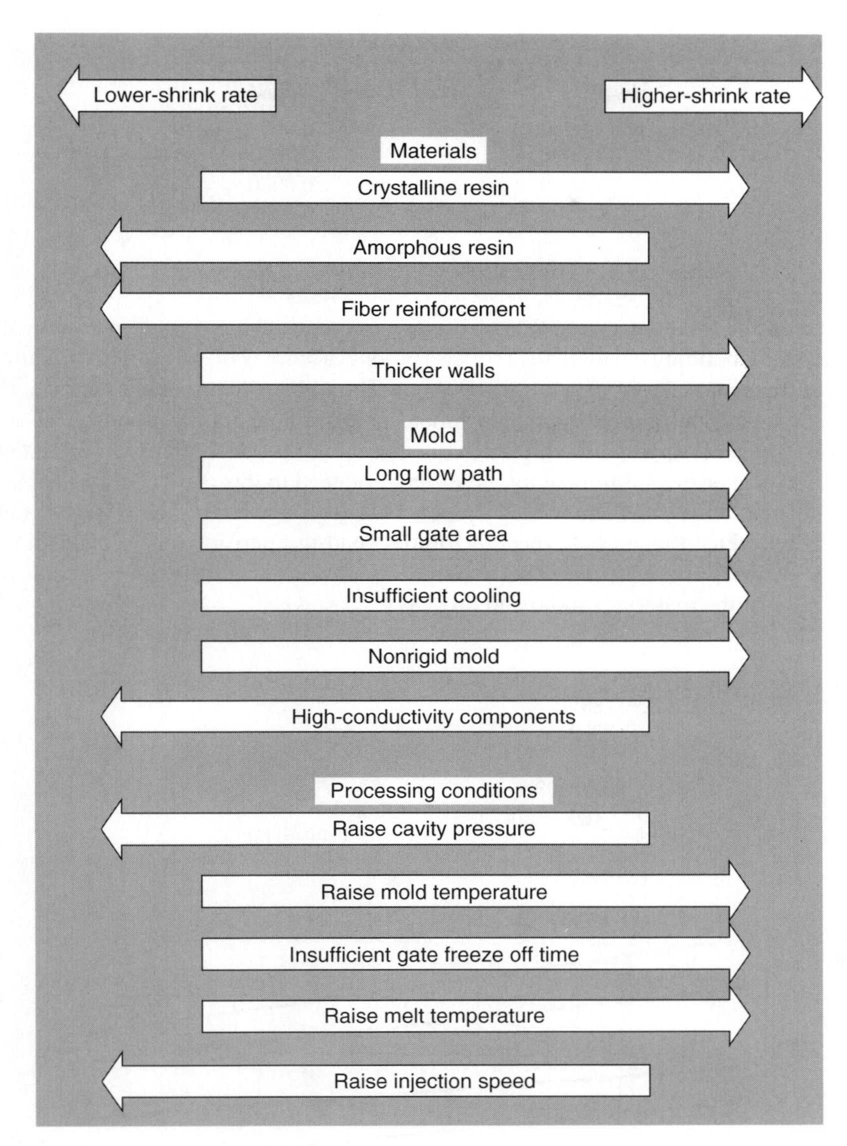

Figure 12.14 Chart illustrating effect of design and processing on shrink rates. (Adapted from Ref. 8.)

Mold Cavity Considerations

The precision in building the mold cavity is very important. When product accuracy requirements increase, the cavity dimensional tolerance requirements must also increase. For products requiring accuracy of 0.001 or 0.002 in., the maximum allowable deviation for mold dimension is ±0.0001 or 0.0002 in. This degree of mold accuracy

Table 12.5 Tolerances for simple products

Dimension (in.)		Tolerance (in.)
1.0000	±	0.0005
2.0000	±	0.0010
4.0000	±	0.0020
6.0000	±	0.0030

is not required for all products. When very precise, repeatable tolerance requirements are required, building the mold to produce a product accuracy of ±(0.004′ in.) requires a mold dimensional tolerance of ±(0.001) in. The tighter the product tolerance, the more expensive the mold to obtain the dimensional accuracy for the product. The cost of these molds is appreciably higher, requiring more time and quality in producing the cavity dimensions. Multicavity molds are produced to this precision for the medical field and other precision industry products. The design team should specify only required product tolerances to meet the product end-use performance specifications. A rule of design is to specify the required dimensions with tolerances within the capability of the material's mold and process.

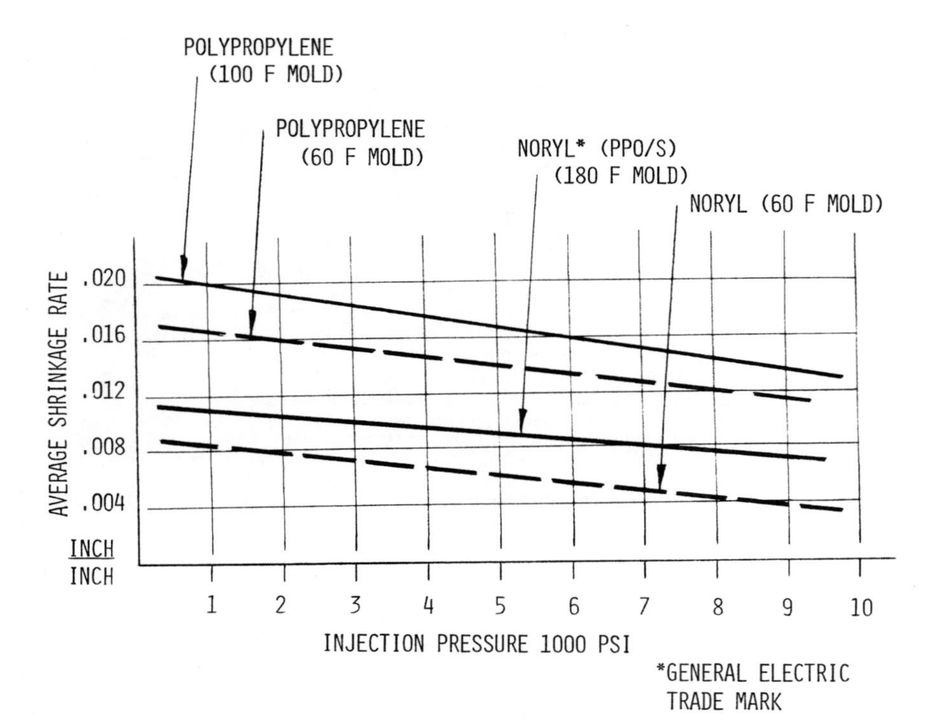

Figure 12.15 Factors affecting material shrinkage.

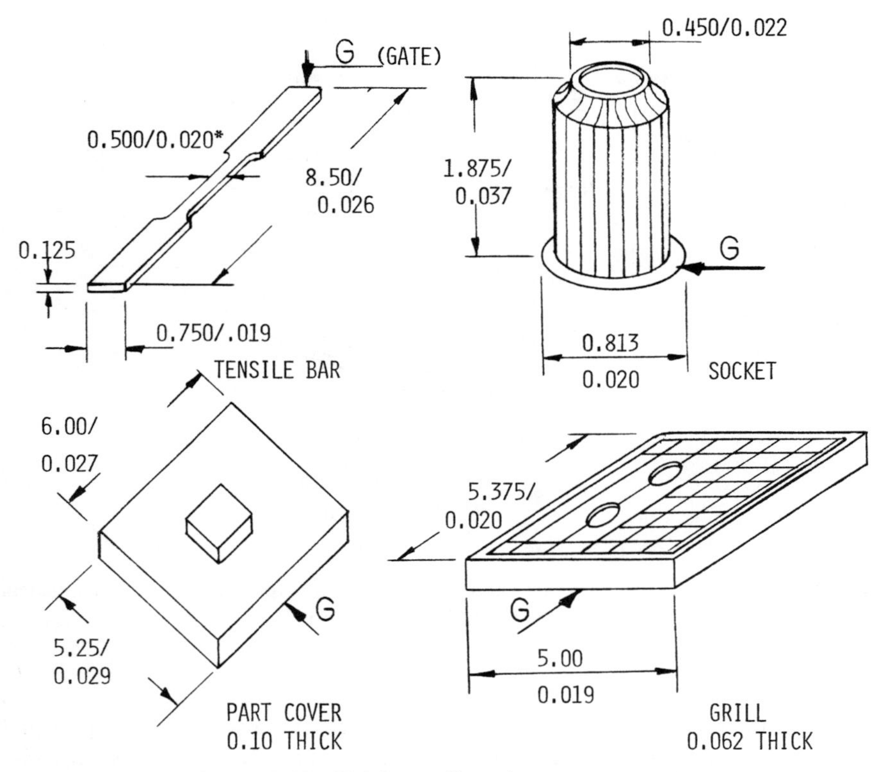

Figure 12.16 Shrinkage effects due to part geometry.

There are two options in building the mold product cavity: multiplate or as a single unit. A mold cavity made from multiple plates takes longer and may be more expensive. However, as cavity sections wear as a result of resin abrasion or moving products, replacement of worn components is easy to accomplish and less expensive. Also, where cavity plates meet, there are natural air vents to allow any trapped air to escape, which is required for a blind cavity that does not have natural venting.

Electrodischarge Machining

Electrodischarge machined (EDM) cavities is the second option. These are one-piece steel plates, machined to size and then EDM-machined to the final cavity product dimensions. The one-piece mold cavity must be designed to vent trapped air during filling. The one-piece mold cavity must be designed for good temperature control and heat transfer to the cooling system. If a standard mold base with a built-in mold cavity support system is used, the cooling layout may not adequately control the cavity temperature unless it is built into the mold cavity insert as a separate cooling system. Also, if the cavity cracks or fails, it must be replaced or repaired. This can be

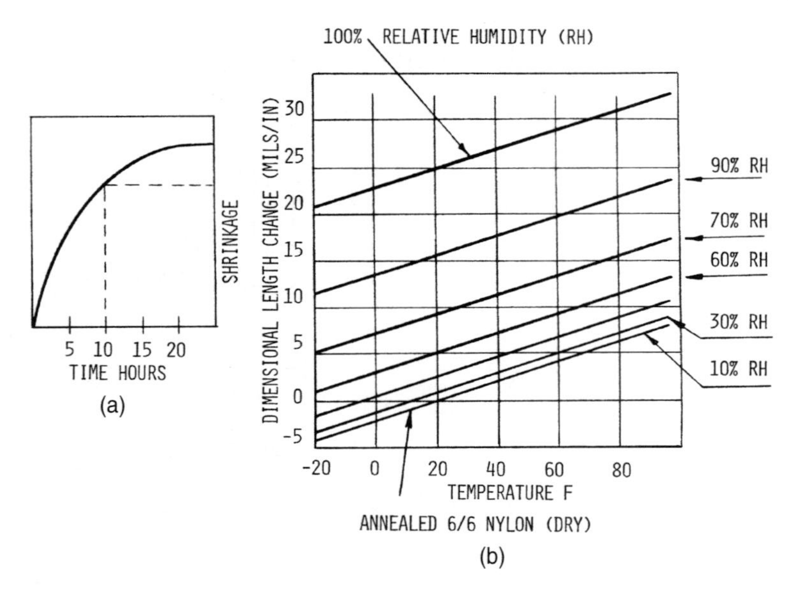

Figure 12.17 (a) Postmold shrinkage for polypropylene, and amorphous plastic; (b) dimensional growth for nylon as it absorbs moisture. (Adapted from Ref. 9.)

accomplished by having a separate spare cavity insert, which can reduce repair costs and time if the original cavity should require service. Always discuss the pros and cons with your mold designer and builder for the type of mold cavity selected.

Mold Cavity and Core Alignment

Mold cavity and core alignment is very critical for producing quality products. Any shift in a core or cavity under high injection pressure can result in mold damage or off-specification dimensions. The shutoff at the edges of the cavity must always line up and are critical to maintain packing pressure. The mold cavity shutoff, a thin ridge of mold steel, must not be damaged during processing by overclamping the mold halves together. This is a machine setup adjustment of the machine's clamping pressure to the mold's required clamp pressure that the molder makes to ensure that the cavities' steel shutoff components are not crushed or damaged during the molding operation.

Also, a product not totally ejected and free of the mold's surfaces, due to poor release or hang up in the mold, can seriously damage the mold when it closes for the next molding cycle. Any product hung up in the mold on closing can seriously damage the mold, resulting in lost production. There are failsafe systems available to ensure that the mold surface is free of products before it closes for the next cycle. Consider this option when designing the mold or during processing.

Product quality is dramatically affected if any core and cavity mismatch occurs or moving slides do not operate as designed, which can result in wear and galling of the mold. This can cause flashing at the production lines and product out-of-dimensional control.

Mold half-interlocks can be installed along with cavity interlocks and taper locks during mold manufacture to ensure that the mold halves always line up on closing. The use of these safeguards can be discussed with the mold designer. Once the mold begins to wear and product flashing begins, no amount of adjusting processing conditions will produce good products. Product cavity packing pressure, used to control final product dimensions, will be lost and so will product quality. When this occurs, the mold must be pulled and returned to original, as-built, dimensions.

Cavity-to-Cavity Accuracy

In multicavity molds each cavity must have identical dimensions. If not, the molder will have to adjust the molding conditions to the worst cavity in an attempt to product good products each cycle. In serious cases, the molder has blocked off the faulty cavity or cavities. This avoids the need to produce a rejectable product but creates an unbalance pressure condition in the mold's system, which can affect cavity product dimensions in the remaining cavities. Cavity-to-cavity product accuracy requires the mold cavity layout to be balanced. Runner and gates must be sized exactly with runner

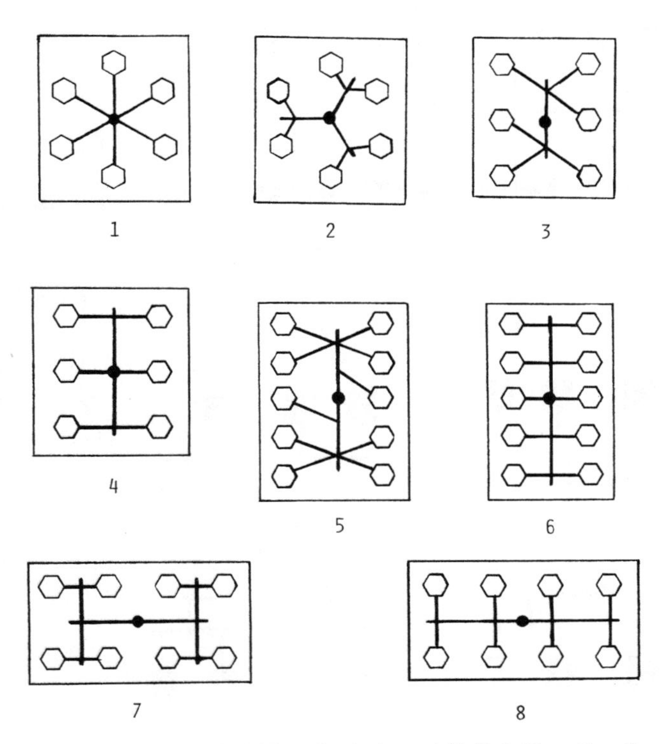

Figure 12.18 Balanced (1, 2, 3, 7) and unbalanced (4, 5, 6, 8) multicavity molds.

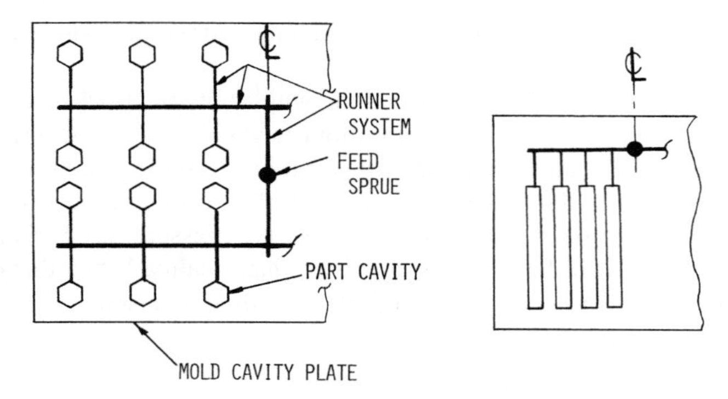

Figure 12.19 Balanced multicavity mold.

length and number of bends minimized to reduce pressure drops in the mold as shown in Figure 12.18 and seen in the balanced layouts in Figure 12.19. This is illustrated as pressures P_1 and P_2, shown in Figure 12.20. At each 90° bend in the runner and cavity there is a pressure drop for the incoming material. A pressure drop also occurs at each cavity gate. Therefore, the injection and final packing pressure indicated on

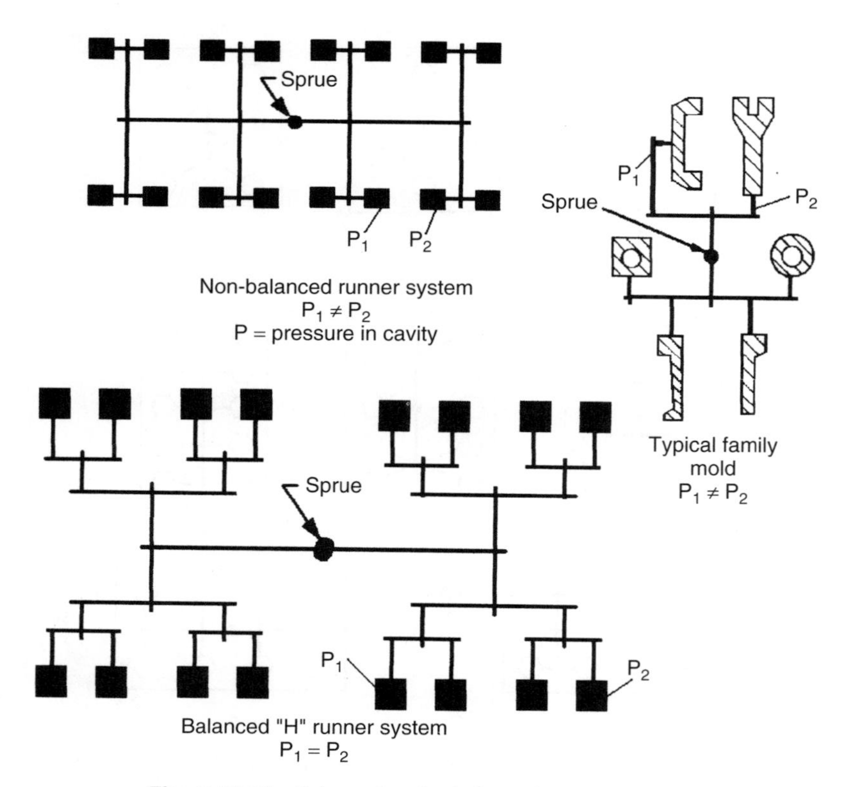

Figure 12.20 Balanced and unbalanced runner systems.

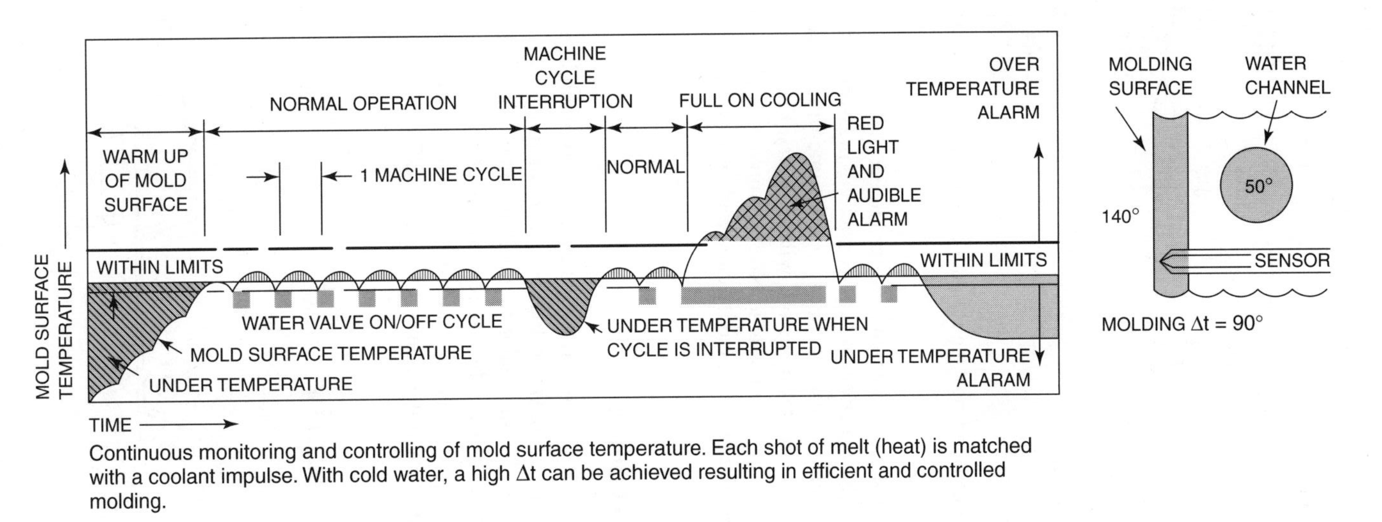

Continuous monitoring and controlling of mold surface temperature. Each shot of melt (heat) is matched with a coolant impulse. With cold water, a high Δt can be achieved resulting in efficient and controlled molding.

Figure 12.21 Relationship between mold temperature and controller functions, showing effects of unregulated cavity temperature. (Courtesy of D-M-E Co.)

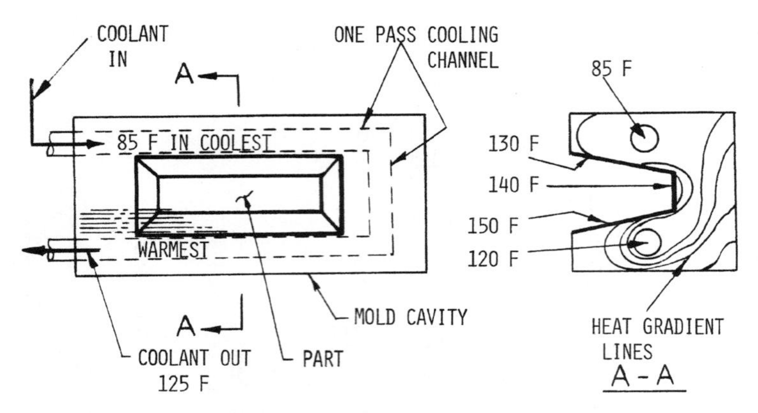

Figure 12.22 Restricted mold temperature control. (Adapted from Ref. 10.)

the machine's pressure gauges is not that produced in the mold cavity during packout. It is very important that the injection and cavity packing pressure be uniform for each product cavity and that gate freeze-off occur at the same time. When injection and packing pressure varies, product dimensions will also vary. Each cavity must also be identified. Typically, each cavity in a mold base is numbered. This is done so that if a problem should occur, the problem cavity can be identified for solving the product manufacturing problem.

Cavity Temperature Control

Controlling the product cavity temperature is essential to obtaining dimensional repeatability cycle to cycle as shown in Figure 12.21 for in-cavity and out-of-cavity temperature control. It is also the major factor in maintaining a consistent molding cycle. Cavity temperature controls the rate at which heat is removed from the molten plastic through the cavity steel and cooling system, which affects the resin's setup or solidification time as shown in Figures 12.22 and 12.23. Material setup and gate

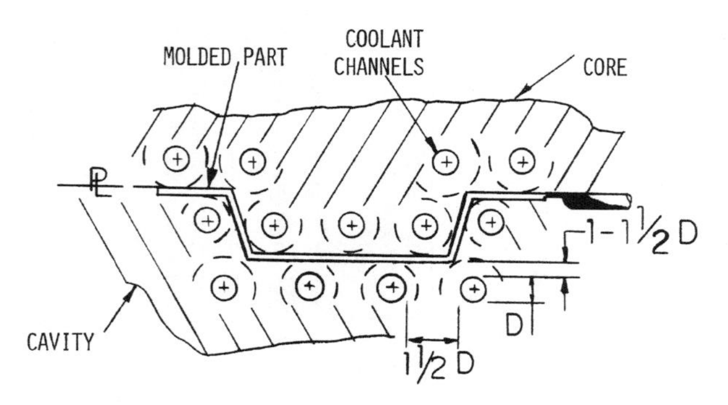

Figure 12.23 Proper location of coolant channels.

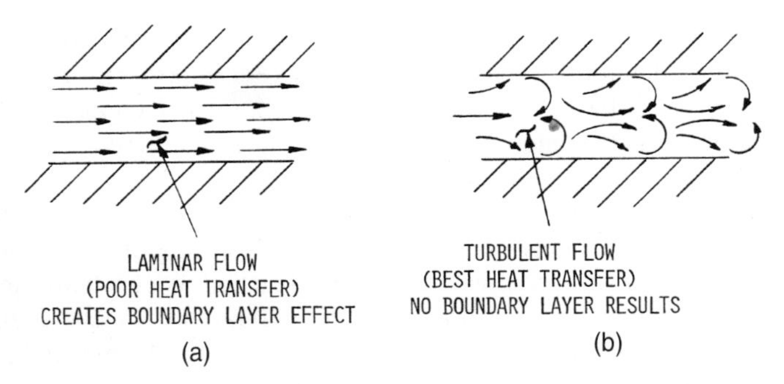

Figure 12.24 (a) Laminar and (b) turbulent flows through the cooling system.

freeze-off time determine the total cycle time of manufacture and how soon the hot product can be ejected from the mold cavity. To be successfully ejected, the product must be cooled to become rigid enough so that it will not bend or warp when ejected from the mold cavity, and also so that the ejector system does not damage the product, such as ejector pins punching through or indenting the hot material of the product.

Amorphous resins take longer to cool and solidify. Any nonsolidified product section can bend under the forces of the ejection system. When this occurs, the cycle time must be increased or the cooling system temperature lowered to remove more heat faster from the product's cavity.

Control of cavity temperature must be uniform throughout the product's section, cavity to cavity. Cavity temperature is controlled by the design of the cooling system, turbulent flow (Fig. 12.24), incoming water temperature, and flow rate of the fluid using either a series or parallel cooling channel routing (Fig. 12.25), around each

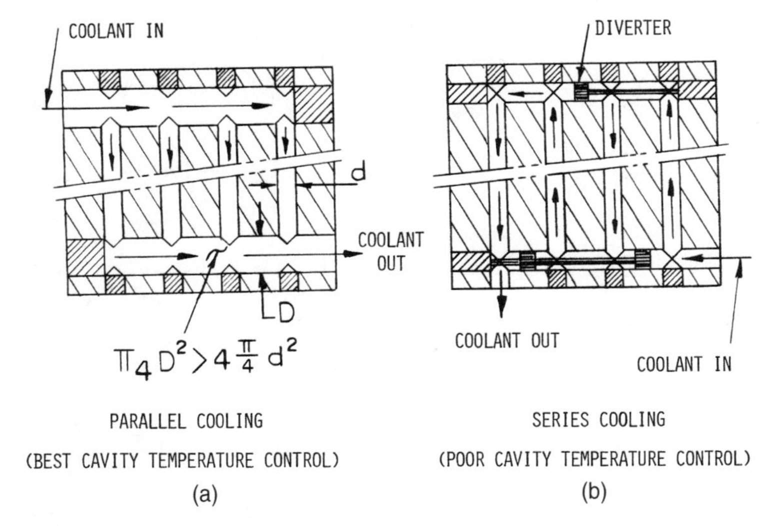

Figure 12.25 Mold cooling layouts: (a) parallel cooling (best cavity temperature control); (b) series cooling (poor cavity temperature control). (Adapted from Ref. 3.)

cavity. Individual cavity and core temperature control is recommended to maintain product dimensional control, cavity to cavity. It is very important for the control of product dimensions to have the exiting cavity cooling water temperature no more than 2°F different from the entering cavity water temperature for correct temperature control of all the individual product cavities.

Cooling feedwater temperature is accurately controlled using a central system or individual mold temperature-controlled chiller. Most molding plants use tower-cooling water that cools the equipment in their plant that is constantly maintained at 85°F (±1°) year round. This water temperature will provide adequate temperature control and heat removal for the plants operating equipment and mold cooling system.

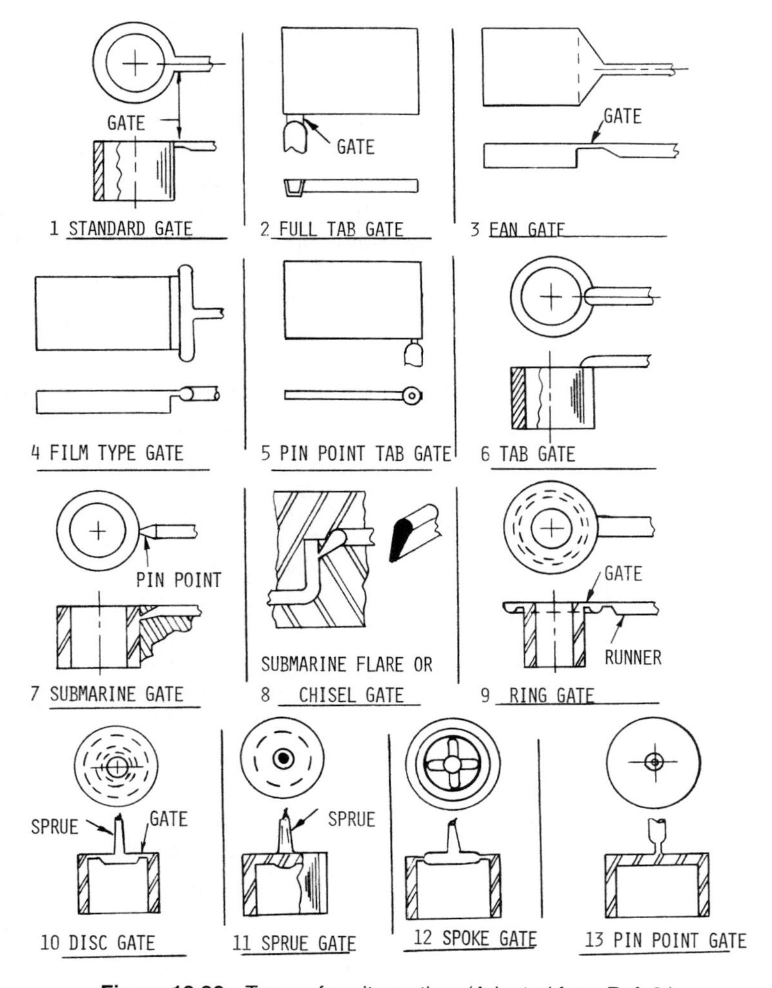

Figure 12.26 Types of cavity grating. (Adapted from Ref. 6.)

Runner and Product Gating Considerations

The product design team needs to discuss the product's gating and runner system with the mold designer and production team before the production mold's design is finalized. This involves the type of mold selected: two- or three-plate or hot-runner, as well as the volume of products, the number of cavities, and balanced or unbalanced mold cavity layout. The product's tolerance requirements are a determining factor in selecting the number of cavities allowed, the type of gating used, and the product's gates location for the cavity to control product dimensions. Examples of runner and gating designs are shown in Figures 12.26 and 12.27.

Hot-runner or hot-manifold molds are being used more today for precision products. They offer material savings, no regrind is produced, and provide precision temperature delivery of material to each cavity under uniform pressure. Once on cycle, the hot runner mold is run in automatic operation.

Large single-cavity molds (Fig. 12.28) may require multi-gating and a three-plate mold design to fill the product correctly and keep cycle time to a minimum. Multiple weld lines will be formed, and the gating must be located to reduce their effect on the products.

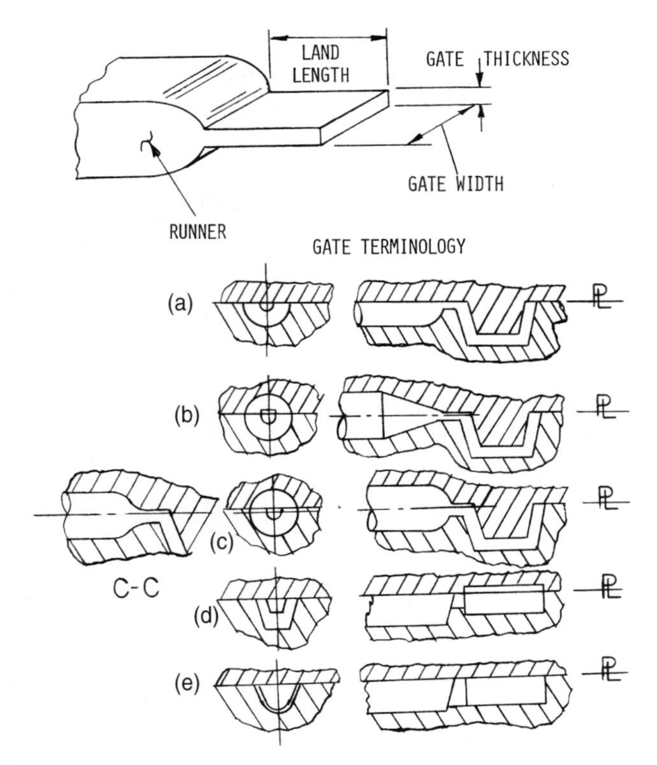

Figure 12.27 Runner and gate design combinations: (a) half round; (b) full round; (c) full round; (d) trapezoidal; and (e) U-shaped. (Adapted from Ref. 1.)

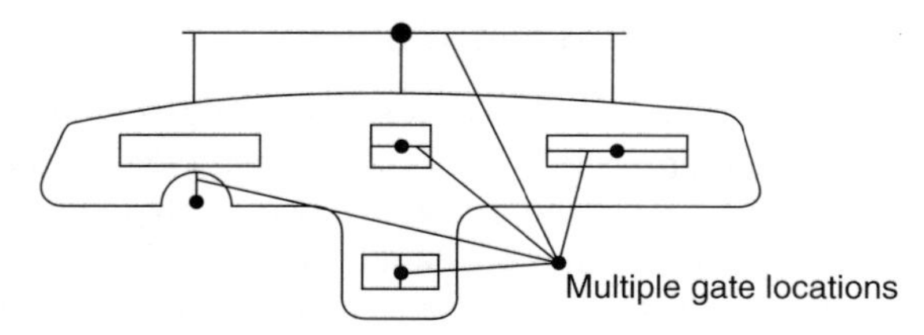

Figure 12.28 Large single-cavity multigated runner layout.

The product's dimensional tolerances control the decisions where to gate and the type of mold to use for the program. Always gate into the thickest product section to ensure that the gate remains open and the product is packed out to the required dimensions. Typical gate and runner considerations and sizes are shown for the designer's reference in Tables 12.6 and 12.7. The processing team members experience with processing different resins will assist the mold builder and design team in selecting the correct mold components and variables for producing the product to meet the customer's requirements and specifications.

Flow in the Product Cavity

The easier the material flows to fill and pack out the mold cavity to maximum product weight, as shown in Figure 12.29, the more accurate will be the product's final

Table 12.6 Recommended material runner sizes

Material	Diameter (in.)
ABS, SAN	0.187–0.375
Acetal	0.125–0.375
Acrylic	0.312–0.375
Impact acrylic	0.312–0.500
Cellulose acetate	0.312–0.500
Cellulose acetate butyrate	0.187–0.375
Cellulose propionate	0.187–0.375
Ionomer	0.090–0.375
Nylon	0.062–0.375
Polyallomers	0.187–0.375
Polycarbonate	0.187–0.375
Polyethylene	0.062–0.375
Polypropylene	0.187–0.375
Polyphenylene oxide	0.250–0.375
Polysulfone	0.250–0.375
Polystyrene	0.125–0.375
Polyvinylidene fluoride	0.125–0.312
Polyvinylchloride (plasticized)	0.125–0.375

Source: Adapted from Ref. 2.

Table 12.7 Minimum runner diameter

Part Thickness (in.)	Runner Length (in.)	Minimum Runner Diameter (in.)
0.020–0.060	≤2	$\frac{1}{16}$
0.020–0.060	>2	$\frac{1}{8}$
0.060–0.150	≤2	$\frac{1}{8}$
0.060–0.150	>2	$\frac{3}{16}$
0.150–0.250	≤2	$\frac{1}{4}$
0.150–0.250	>2	$\frac{5}{16}$

dimensions. The injection pressure and fill rate, how fast the material flows into the mold cavity are functions of the molding machine's injection pressure. The hydraulic fluid and pump pressure must be controlled over the entire manufacturing time to obtain uniform and equal product dimensions. The machine's hydraulic system controls the molding machine's forward screw speed as it pushes material into the mold cavity during the injection stroke. Molding the product to maximum product weight is a quality tool used to control product quality. Maximum product weight ensures that

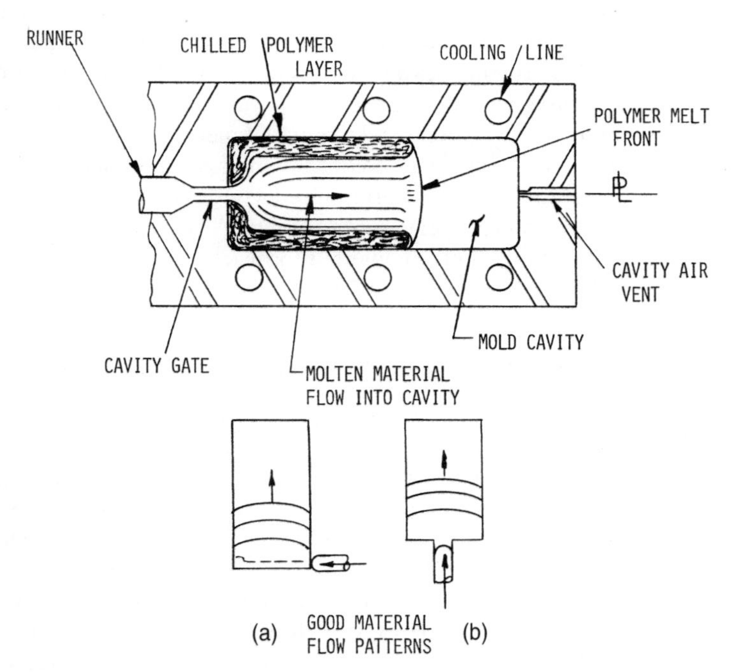

Figure 12.29 Correct resin flow into mold cavity: (a) slow even fill to establish melt front; (b) gate design and placement to obtain uniform melt flow.

the product is always packed out uniformly on each molding cycle. Product weight can be measured at the molding machine by the operator or by robotic measuring equipment. Any product found to be underweight can be segregated and later examined dimensionally to see if it is in tolerance. This quality test is used to monitor the consistency of the molding cycle. Any change in product weight indicates that a molding machine or material variable has changed. The operator can then troubleshoot the manufacturing system and get the process back in control.

Cavity Pressure Sensors

Cavity pressure sensing is used to control mold filling and packout and is becoming more widely used with computer process control molding. Sensors are inserted into specific areas of the mold cavity to record pressure directly and are located behind or under cavity knockout ejector pins. Pressure sensors are located in the runner, gate, and typically in the furthest fill point in the mold cavity.

Pressure drops in the runner and cavity during filling make it difficult to actually know the fill and packout pressure obtained in the mold during the fill cycle. When the molder can repeatably use pressure values to fill and packout the mold cavity, the product will be repeatable, cycle to cycle. Using the sensors, pressures in the system are monitored and used to control the pressure profile during fill and packout. Sensors in the back of a cavity transmit to the process control unit the actual fill and then the pack pressure developed within the product cavity. A sensor located just inside the gate area coupled with the back of the cavity will transmit actual pressure values that are used to control the pressure to pack out the product cavity. Using these sensors, packout pressure on each cycle can be repeatable to ensure that the product has attained maximum product weight and required dimensions on each cycle.

Pressure sensors also can signal gate freeze-off and assist in reducing cycle time. Sensors can also signal the process control unit of viscosity variance and aid in attaining complete process control over the entire molding cycle. Contact your material and mold equipment suppliers for complete information on this assistance to the manufacturing cycle.

Controlling Molded-in Stresses in the Product

The amount of molded-in material stress is harder to control and determine when it has occurred. More viscous materials and harder-flow resins that require high injection and packing pressures will usually have more molded-in residual material stress. Molded-in stresses are observed when the product cools down and warps. If this warpage cannot be corrected by adjusting molding process variables, fixtures can be used to control warpage as the hot product cools. This is to be avoided whenever possible, but some products, because of their design, must be fixtured. The problem with this is if the product is exposed to temperatures higher than the as-molded temperature, the stresses induced into the product on cooling are relieved. This may cause the product to warp, and if not firmly anchored to a mating product, stronger than itself, can result in a failure or an assembly or major in-service problem. To determine whether this will be a problem, the molded product can be oven-aged at elevated or end-use temperatures to see if this will become a problem in service.

Reinforced materials can warp as a result of both molded-in stresses and fiber orientation. Reinforcement and fillers will line up in the material flow direction in the mold cavity and with this orientation, differential shrinkage can cause the product to warp in the transverse, higher-shrinkage direction, if not restrained. Fiber orientation controls the reinforced material's shrinkage direction. Products with section thickness <0.100 in. have minimal shrinkage in the fiber-oriented direction and more in the transverse direction. The fibers in the products skin, usually estimated at 0.020 in. thick, are in the flow direction. Thicker products (>0.100 in.) have more uniform shrinkage due to a more random orientation of the fibers caused by flow turbulence during filling at the mold cavity. Most reinforced materials use short glass fibers that are less than 0.100 in. in length and will line up in the flow direction. This creates some strength and shrinkage variances in the product's cross-flow directions. Shrinkage increases, and material strength is lower in the direction transverse to fiber orientation in the product. The mold designer considers wall thickness and cavity filling orientation when calculating product cavity dimensions for thin/thick-walled reinforced and nonreinforced products. It is important before mold cavity design is finalized that a computer simulation be run to ensure that all problem areas are considered before finalizing the mold's dimensions, runner system, and gating.

If the product design indicates a problem may exist, a higher-flow resin may be evaluated, but the shrinkage may be different, resulting in different product dimensions. Also, if reinforced, a different type of reinforcement such as a mineral or mineral/glass combination may be a solution that may yield better flow with the least amount of sacrifice in physical properties for the product. But this requires a reevaluation of the product's physical properties and mold cavity shrinkage. Always run the product and resin simulation before the first mold steel product is cut.

Interior corners of all products should be radii to promote flow; lower pressures drops in the cavity, and reduce stress concentration. Examples of recommended corner and right-angled corner designs are shown in Figure 12.30. The use of the falling-dart impact test is an excellent guide in evaluating the effects of radii and corner design strength for different product sections as illustrated for the molded polysulfone resin in Figure 12.31.

Control of Weld Lines

Weld lines in a product can be a potential failure point as shown in Figure 12.32. Weld lines are always suspected weak points in a product, and their fusion point should always be tested to destruction. Never design the mold gating system so that a weld line is formed at a critical design point. It is not always possible to eliminate weld lines. If you are unable to eliminate a weld line situation, then position it so

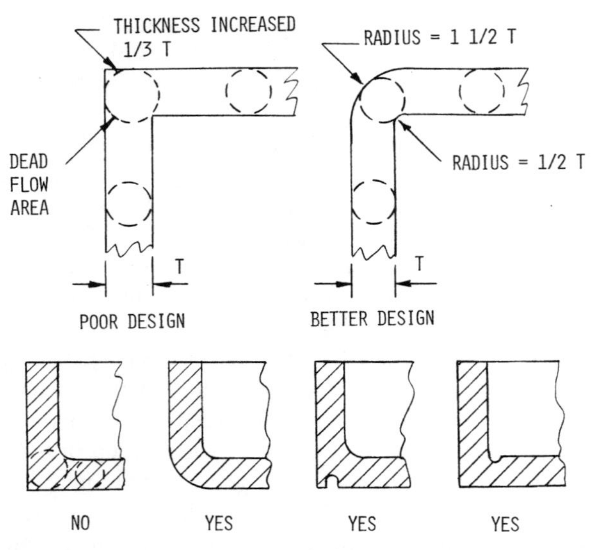

Figure 12.30 Methods of reducing corner stress with a radius. A good design improves flow and increases part toughness.

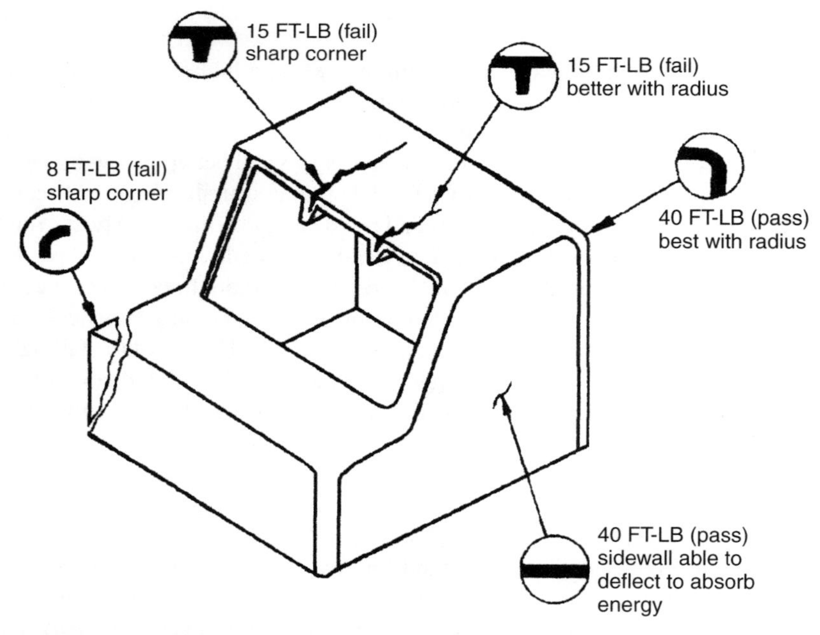

Figure 12.31 Effects of radii during a falling art impact test. Material: polysulfone. (Adapted from Ref. 3.)

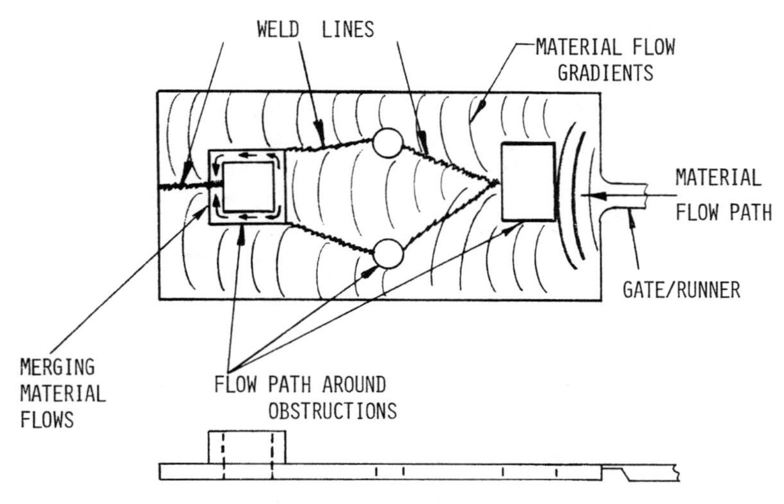

Figure 12.32 Weld line formation in molded parts.

that it is located in a noncritical section of the product. This is one reason why the design team should be composed of both tool designers and manufacturing personnel who have experience with these types of problems and can work to eliminate or minimize them for the product. Weld lines are formed when a melt front is divided by an in-mold obstruction or from multiple gating when two or more melt fronts meet as shown in Figures 12.33 and 12.34. An obstruction in the mold, such as a core pin for a hole, divides the melt flow that must meet on the opposite side of the pin and fuse. The degree of melt fusion depends on the melt's temperature, air entrapment at their meeting point, velocity of meeting melt flows, pressure of the melt entering the mold cavity, and the angle of meeting for the converging melt fronts. When tests show a problem, then a change in the location of the gating is recommended, to move the weld line to a less critical area. If this is not possible, localized cavity heating, using an in-mold cartridge heater to assist in improving weld line strength, is recommended.

If a weak, non-venting-related, weld line forms at the edge of the product in a critical section, consider the use of an overflow tab. The overflow tab extends beyond the cavity production line to allow the melt to bypass an obstruction or poor angle of meeting melt fronts at the mold edge. Flow tabs help the material flow around the obstruction, control hot fuses better, and have air exhausted away from the critical section. The overflow tab allows the melt fronts to meet beyond the product's parting line to obtain a better fusion bond within the product. The tab can often assist in improving the joint strength and later be trimmed from the product. It is impossible to eliminate all weld lines; it is only possible to reduce their effect on the product.

For internal product weld line problems such as at a boss, ribs can be used to help with poor weld line strength. Ribs properly located give more material area for strength and remove trapped air from the product's section.

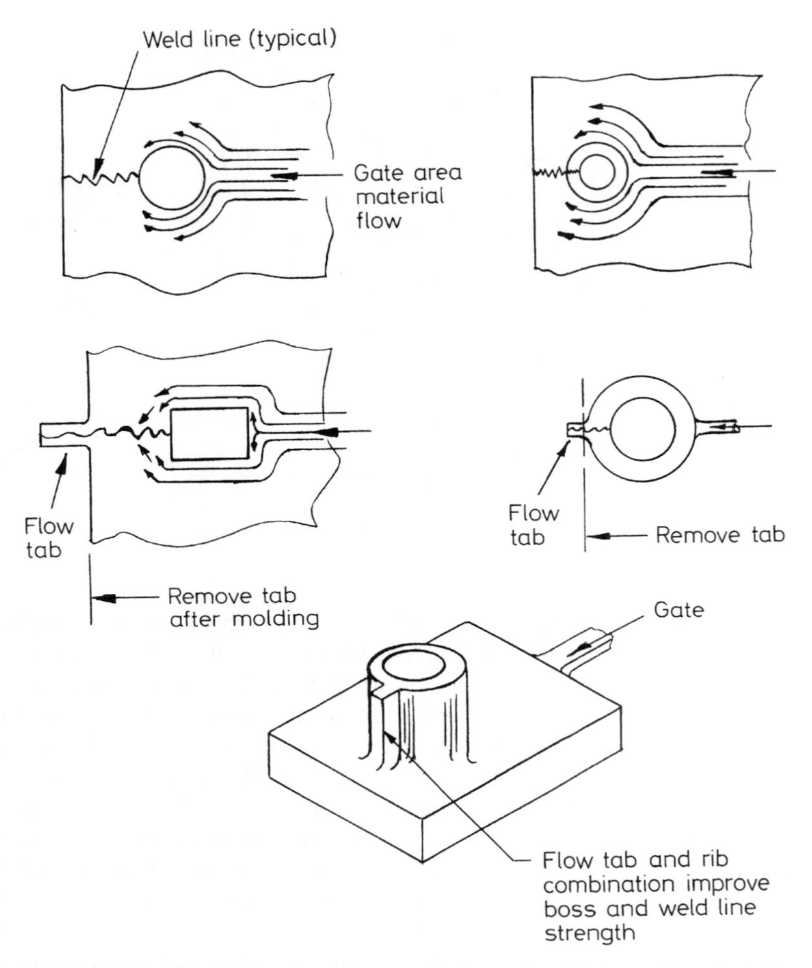

Figure 12.33 Material flow patterns around obstructions. (Adapted from Ref. 2.)

Weld lines in fiber-reinforced resins are more critical as the reinforcement does not cross or intermix at the weld line boundary, which results in lower joint strength. Also, when calculating material strength at a weld line, only the parent material's physical property strength should be used.

Placing the product's gate close to where the weld line will form, if unavoidable, as shown in Figure 12.35, can improve joint strength by increasing turbulence in the melt's flow as they converge and meet. High melt temperatures will assist in improving the joint strength. Weld lines are often visible if a poor fusion bond is formed. Localized heating of the mold cavity surface, using a cartridge heater, in some instances has improved joint strength and surface appearance. But this usually requires a mold revision and if space is not available, can cause other mold and product problems if not installed correctly.

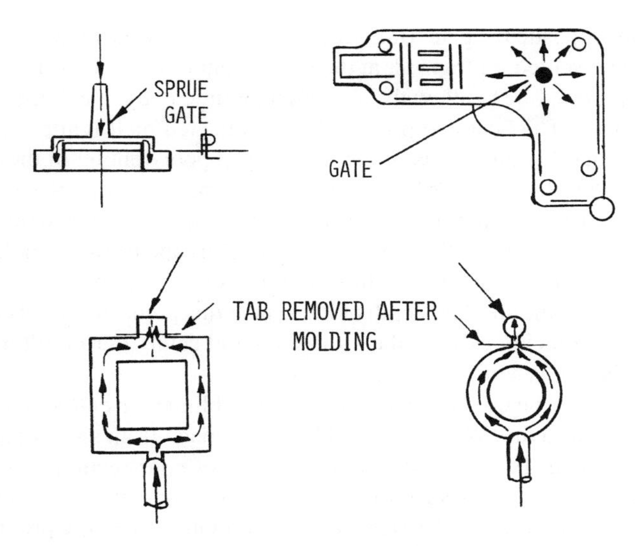

Figure 12.34 Gate location that minimizes weld lines and improves dimensions.

Venting the Mold Cavity

Air entrapment in the mold cavity is a result of poorly designed cooling and mold venting. If a weld line or burn mark is visible at the edge of a product, poor cavity venting caused it. Trapped air at a weld line will slow down and cool the melt flow, causing a poor fusion bond. This is evident in the molded product if the weld line joint is very visible and the product fails frequently at this point. A burn mark on the product, usually noticed as a black burn charred mark, is an indication of poor venting. Burn marks occur at the edge of a product or at a blind cavity trapping air

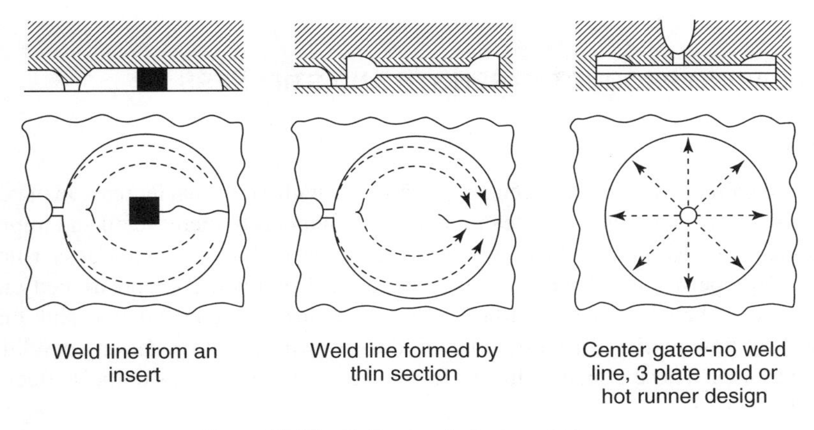

Figure 12.35 Gating to eliminate weld lines.

and are usually opposite the gate. Burn marks are created by the melt compressing the air not able to be vented, or trapped, at the cavity vent point that has too small a vent at the mold's parting line to allow the air sufficient time to escape during injection of the molten plastic. The heat and pressure of the injected melt ignite the trapped air as is done in a diesel engine. This is one reason why poor venting of the mold cavity is often called *dieseling*. To verify a venting problem, place thin pieces of masking tape on the mold's production line at the burn mark. If air was trapped during the next injection cycle, the tape will allow the trapped air to escape from the mold cavity and the burn mark will be eliminated. If this eliminates the burn mark, have your mold repair shop increase the size and depth of the vent at this point. All mold cavities must be well vented to quickly exhaust the air in the mold cavity during filling. A typical mold-venting layout is shown in Figure 12.36.

An ejector pin can be used to vent trapped air in a blind nonvented cavity section of the mold's product cavity. With more mold cavities made from solid steel by electric discharge machining (EDM), based on the method of construction, which does not provide natural venting between assembled mold sections, the necessity of having adequate venting in the cavity becomes more important. When the product cavity is made with multiple plates, more venting sites are available for trapped air to escape at the joints of the plates. A dummy ejector pin can also be used to vent blind mold cavities as long as it operates to clean the vent during each cycle. If it does not operate, it will eventually fill with resin and become useless.

There is also a porous steel that can be used locally to vent air from a nonventing mold cavity section that can be machined and finished to match the products finish. This material is porous enough for air to escape but not allow melt to plug up the surface. It is important to keep this section and other areas of the mold cavity free of any mold deposits that the resin may deposit during typical molding operations. When this porous metal insert is used, be sure that the steel insert can bleed the trapped air into a waterline or open space in the mold frame where the vent is located. Contact your mold material supplier for the name and type of this material to meet your requirements.

INCREASING PRODUCT STRENGTH AND STIFFNESS

Ribs

Ribs will increase a product's strength and stiffness with a minimal increase in material and mold costs. Ribs also provide in-cavity flow channels to increase fill and improve cavity packout. Ribs can reduce a product's section wall thickness without sacrificing physical properties and lead to a more economical product. Ribs positioned under a product's "show" or visible surface may create surface effects such as sink mark, blush, and flow marks caused by poor rib design or incorrect processing conditions by the molder. These effects will vary from material to material as to their effects on a product's surface.

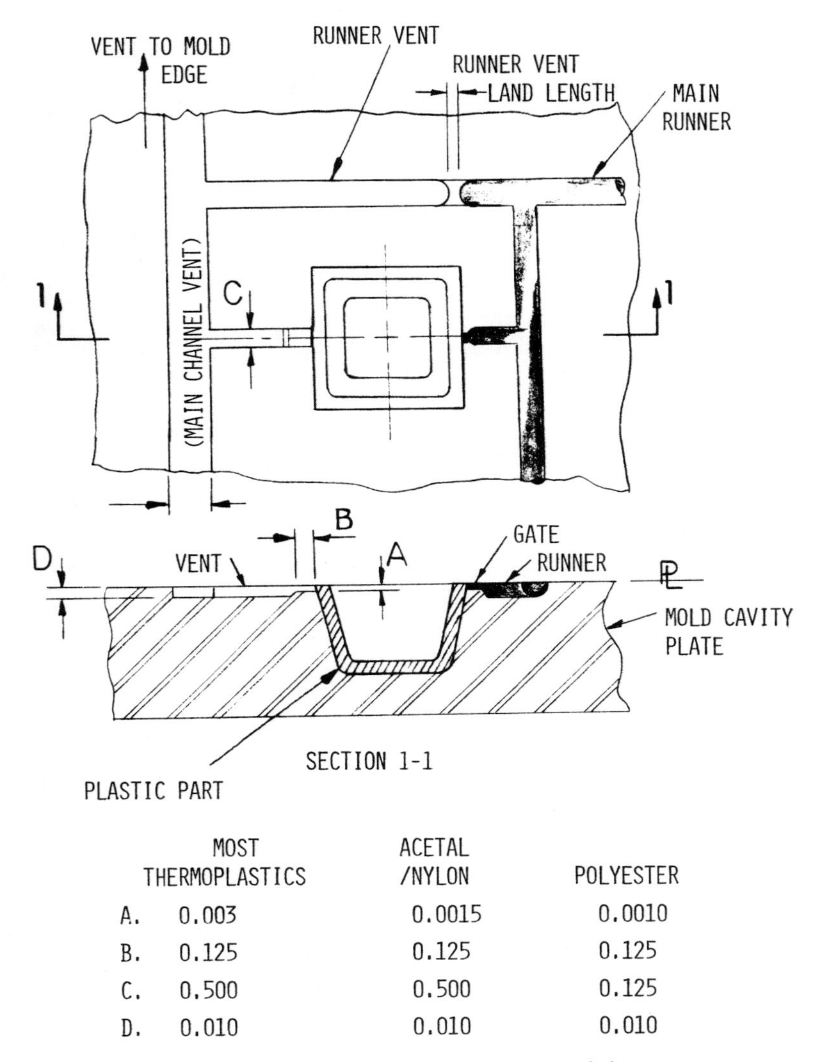

	MOST THERMOPLASTICS	ACETAL /NYLON	POLYESTER
A.	0.003	0.0015	0.0010
B.	0.125	0.125	0.125
C.	0.500	0.500	0.125
D.	0.010	0.010	0.010

Figure 12.36 Simple mold edge vents and sizes.

Therefore, a rib's size and thickness should be minimized. If the material selected is susceptible to having the rib visible on a show surface, the effects can be minimized. This is done by controlling the size of the ribs and by possibly modifying the show surface. The smaller the rib, the less the effect, and if this effect is still visible using a textured finish on the show surface, this procedure will mask the blush and rib marks. A break in the product's contour lines at a rib can also reduce these surface affects as shown in Figure 12.37.

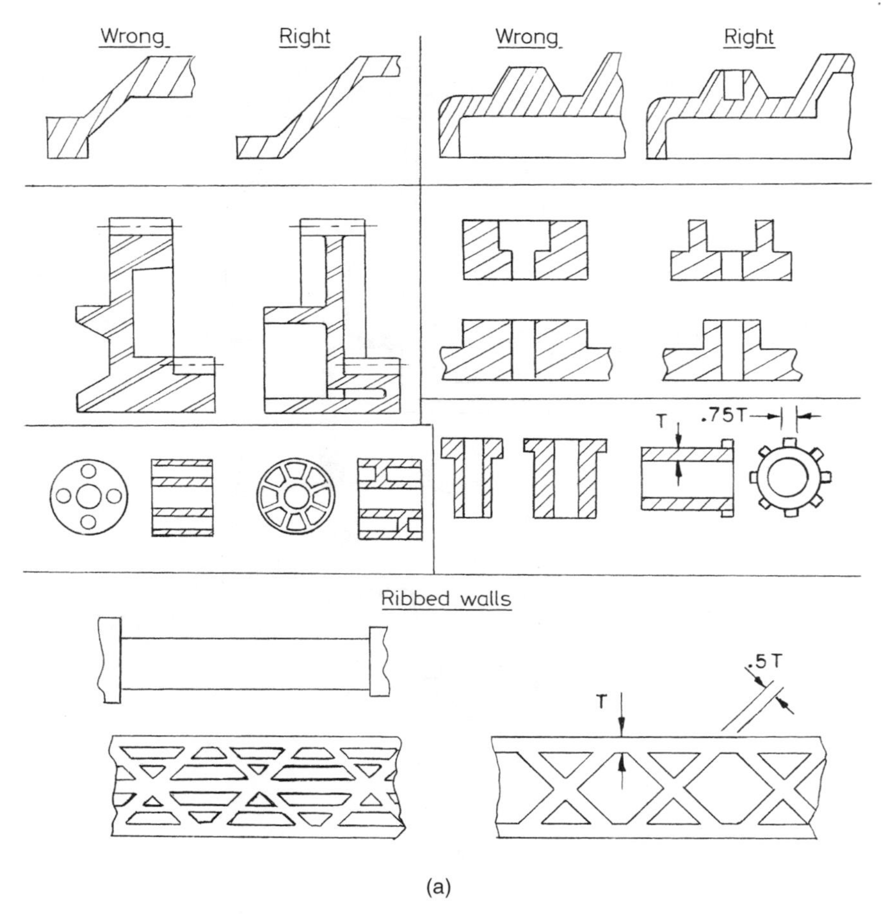

Figure 12.37 (a) Examples of designs for wall thickness. (Adapted from Ref. 4.) (b) The use of ribbing and part geometry to enhance strength and ridgidity. (Adapted from Ref. 5.)

Radii

Always use a minimum radius of 0.032 in. on all interior corners and ribs to aid flow and reduce stress concentration or risers. Sharp corners are high stress points in the product. A radius will increase the product's impact loading capability by a factor of ≥ 5 over a sharp corner as shown in Figure 12.38, presenting test results of different corner conditions.

Ensure that the mold drawings always specify radii at all interior corners and they are adequate for the specific section. Never rely on a general note in the drawing information block for specifying radii. Radii are too important and should be specified where required and verified as cut in the mold cavity. The note may be missed, misunderstood, or simply ignored by the machinist, and the necessary radii may be omitted.

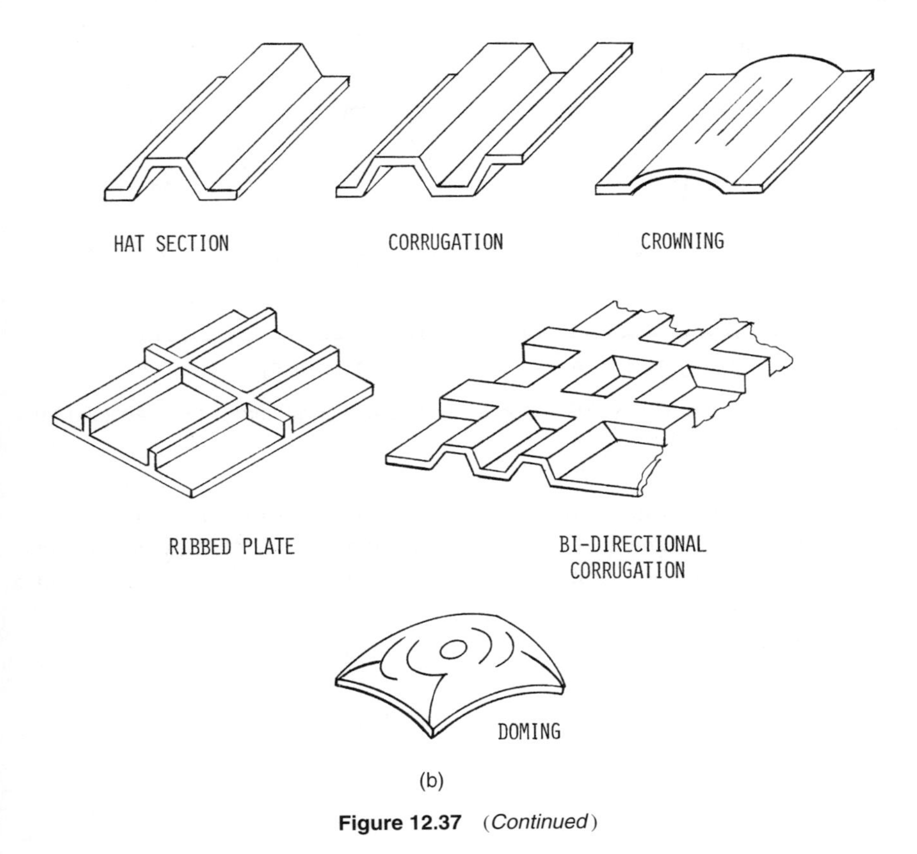

(b)

Figure 12.37 (*Continued*)

Radii reduce stress concentrations at critical product sections with the stress concentration factor (radius/thickness ratio) reduced to an acceptable value of ≤ 1.5 as shown in Figure 12.39. The design team must always ensure that the mold builder puts in the radii specified on the product design drawings. Never permit the mold builder to omit radii as they greatly increase the product strength. The effect of an adequate radius shows as the stress is more evenly distributed in the material at the corner.

Visual Measurement of Stresses

Internal stress values can be calculated using finite-element analysis (FEA) techniques with a high mesh density when located in a critical section. Opaque products can be evaluated for internal stress points and patterns when molded in a clear material. By varying molding conditions and using a material with similar flow properties, conclusions can be reached on appropriate packing and injection pressures, as well as other process variables. Residual molded-in-stresses will then be visible and in

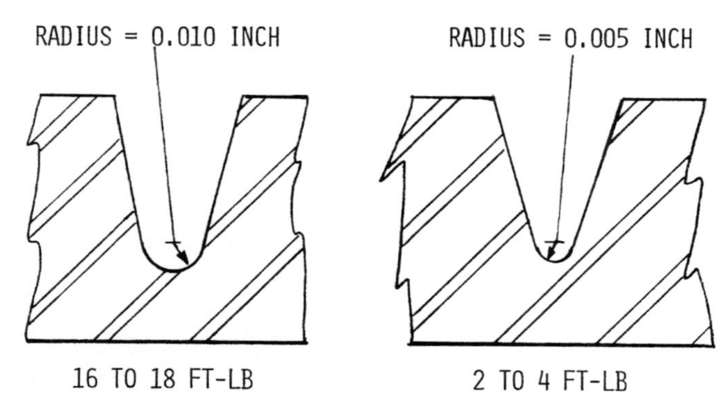

Figure 12.38 Effect of radius size for polycarbonate during an impact test. (Adapted from Ref. 9.)

some cases can be measured. Any end-use loads on the clear product will show the effects of stresses in simulated service. Because the analysis of stresses in opaque products is more difficult, finite product analysis may be necessary to determine the actual stresses. Since this is only an analytical technique, end-use testing may also be required.

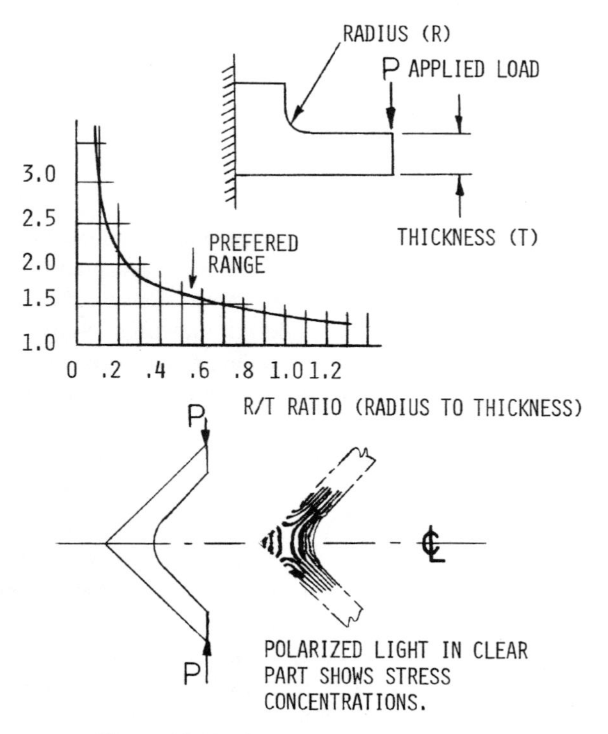

Figure 12.39 Stress concentration factor.

Strain patterns can be seen but not measured

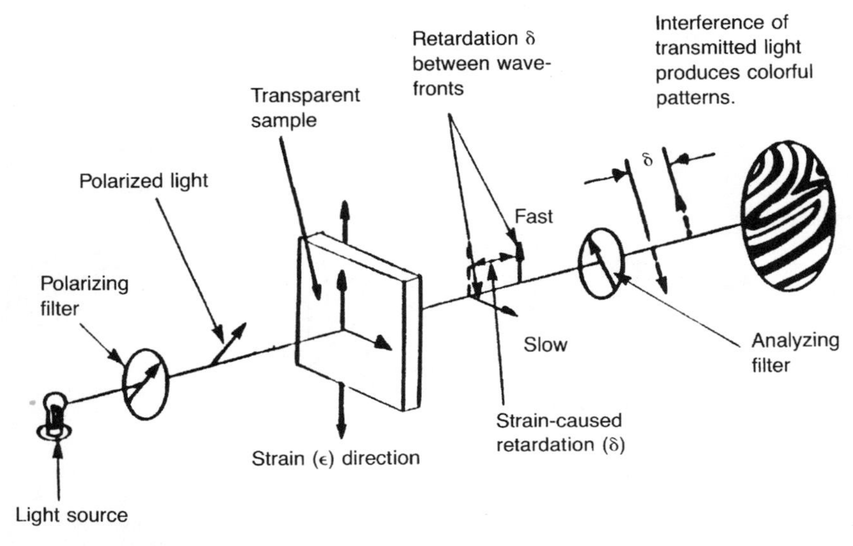

Figure 12.40 Polarized light shows strain patterns in transparent parts. (Courtesy of Strainoptic Technologies, Inc., North Wales, PA.)

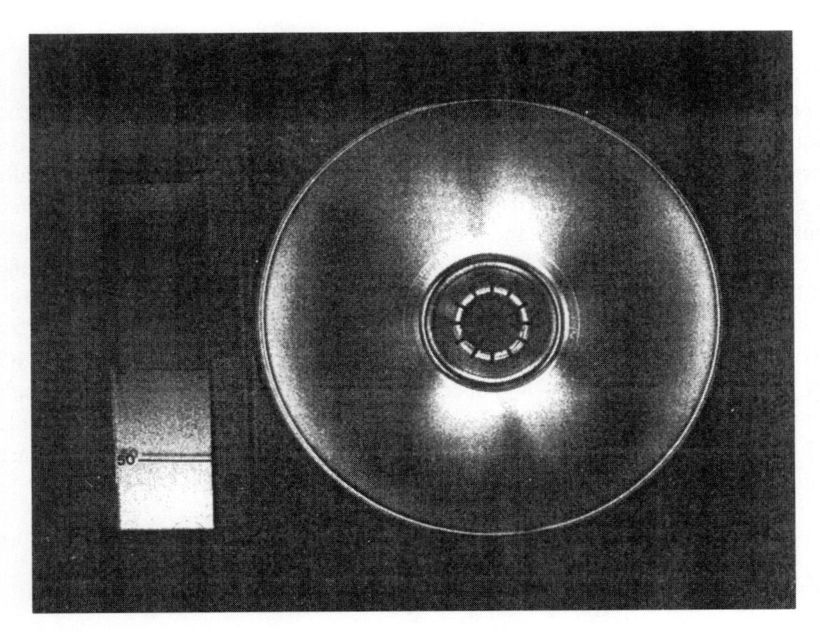

Figure 12.41 QC of strain pattern in a compact disk. (Courtesy of Strainoptic Technologies, Inc., North Wales, PA.)

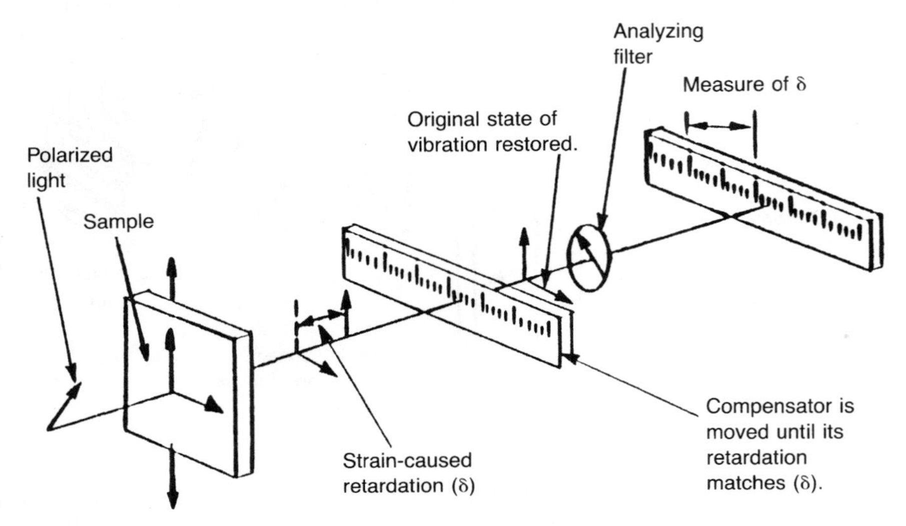

Figure 12.42 Measurement of strains in transparent parts. (Courtesy of Strainoptic Technologies, Inc., North Wales, PA.

To visually see the effects of stress in a product and their reduction using radii, mold the product in clear acrylic, polycarbonate, or polystyrene. Then view the product under polarized light before, unstressed, and after it is subjected to end-use forces, stressed, in a static position. Observing through regular polarized sunglasses allows this visual look at stress patterns in the product. The stress patterns will be visible as curved lines at stress points and the greater density of lines will indicate points of higher stress. The stress patterns will be visible with high stress points shown by the visible lines, indicating the severity of stress in the clear plastic at key points by the stress patterns seen in the product as shown in Figures 12.40 and 12.41.

The stresses can also be measured optically using spectral contents analysis (SCA). SCA uses personal computer and special software to produce a spectral signature. When digitized and compared with a standard photoelastic response, the strain level can be calculated as shown in Figure 12.42. The equipment setup for these measurements is shown in Figure 12.43.

DRAFT AND PRODUCT EJECTION IN THE MOLD

Draft is required to assist in breaking the suction of the material from the cavity and core when the product is ejected from the mold. Draft or taper on the mold cavity

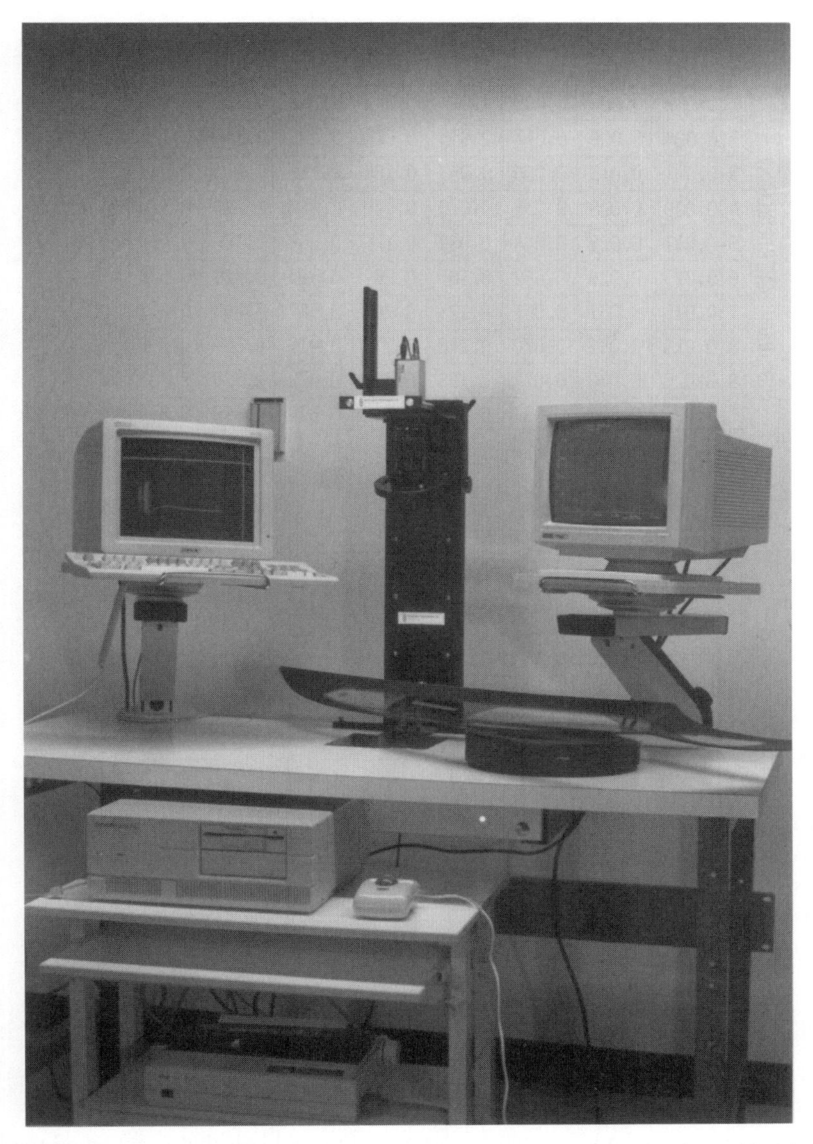

Figure 12.43 Digital image analysis system for birefringence inspection in molded lenses. (Courtesy of Strainoptic Technologies, Inc., North Wales, PA.)

sidewalls must be specified on the mold drawings. It is required on both mold cavity and core sections. The angle of draft specified will be greater on the mold side, cavity or core, where the designer wants the product to release from first when the mold opens. Since all plastics shrink on cooling, mold designers specify less draft on the mold half on which they want the product to remain when the mold opens. The mold's

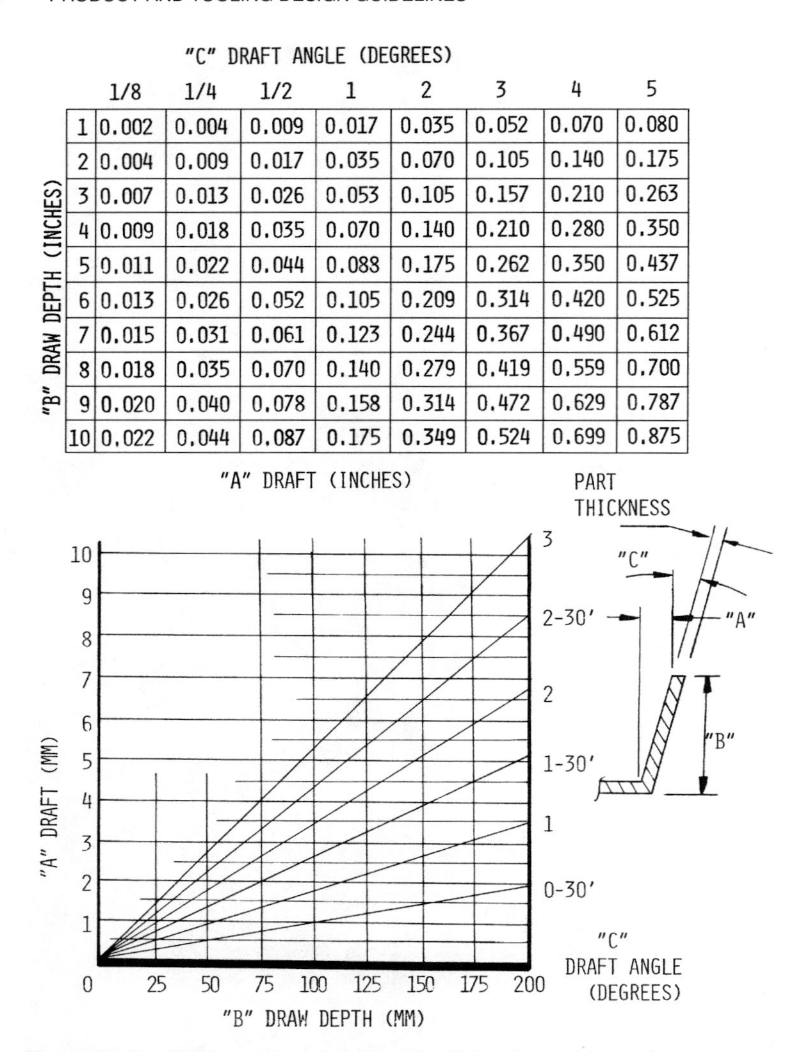

"C" DRAFT ANGLE (DEGREES); "B" DRAW DEPTH (INCHES); "A" DRAFT (INCHES)

"B"	1/8	1/4	1/2	1	2	3	4	5
1	0.002	0.004	0.009	0.017	0.035	0.052	0.070	0.080
2	0.004	0.009	0.017	0.035	0.070	0.105	0.140	0.175
3	0.007	0.013	0.026	0.053	0.105	0.157	0.210	0.263
4	0.009	0.018	0.035	0.070	0.140	0.210	0.280	0.350
5	0.011	0.022	0.044	0.088	0.175	0.262	0.350	0.437
6	0.013	0.026	0.052	0.105	0.209	0.314	0.420	0.525
7	0.015	0.031	0.061	0.123	0.244	0.367	0.490	0.612
8	0.018	0.035	0.070	0.140	0.279	0.419	0.559	0.700
9	0.020	0.040	0.078	0.158	0.314	0.472	0.629	0.787
10	0.022	0.044	0.087	0.175	0.349	0.524	0.699	0.875

Figure 12.44 Draft angle graph: (a) metric; (b) English. (Adapted from Red. 11.)

product ejection system is located on the moving mold half that is activated to eject the product from the mold as the mold opens. The mold's core is often designed for the product to shrink onto and then be ejected by the mold's ejector system. This is done to assist removal of the product should a problem occur in the ejection system, but this is product-dependent.

The amount of draft specified depends on the material and the surface finish on the mold. The mold designer must consider the finish on the product, material lubricity and/or tackiness and its material shrinkage, and any molded-in undercuts. A smooth, highly polished mold finish may not always guarantee a product's easy release if

Table 12.8 Draft angle in degrees

Unreinforced Resins	Draw <1 in. Deep	Draw >1 in. Deep
Acetal	0–0.25	0.5
Acrylic	0.75	1.5
Cellulosic	0.50	1.0
HDPE	0.75	2.0
LDPE	0.75	2.0
Nylon	0–0.125	.25–0.50
Polycarbonate	0.50	1.50
PBT polyester	0.75	1.75
Polyetherimide	0.25	1.50
Polypropylene	0.50	1.50
PPE	0.50	2.00
Polyphenylene oxide	0.50	2.00
Polystyrene	0.50	1.00
PVC	1.00	1.50
SAN	1.50	2.00
TPPE	0.25	0.50

the draw (length of the core) is very long. Sufficient draft is required to break the material's attraction or suction of the plastic to the mold surface. Critical bearing surfaces usually require some draft, and the experienced mold designer can recommend the minimum for the specific resin used for the product. The minimum recommended draft angle for basic plastic resins is listed in Table 12.8 and Figure 12.44, which list draft angles for calculating the draft dimension for the mold cavity.

Textured Cavity Surfaces

When specifying draft, be aware that textured surfaces will require a larger draft angle because of the hill/valley effect of a textured surface. More draft is required when the mold opens to ensure that the moving steel, releasing from the textured side, clears the plastic without smearing or rubbing against the textured surface.

The textured surface of a mold core or cavity pulling first should have a texture depth less than the texture height, which would rub against the mold cavity surface during the ejection stroke of the mold's operation. This is necessary to ensure that the textured surface is not smeared against the mold and to facilitate a smooth nonabrasive pull against the hot textured surface of the product.

For some textured surface patterns, shown in Figure 12.45, the designer should, when possible, orient the deeper textured sections of the pattern in the direction of the mold opening direction. Be sure to discuss any deep-textured pattern with the mold designer to ensure that it can be used without damaging the product's finish on ejection.

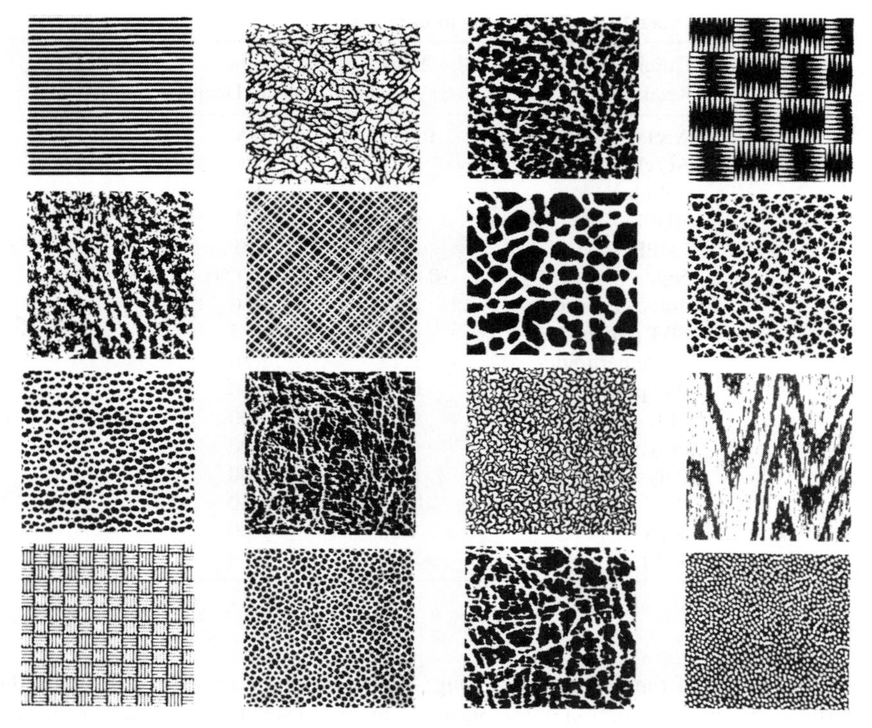

Figure 12.45 Texturing patterns. (Courtesy of the Akron Metal Etching Co., Akron, OH.)

Mold Finish and Polishing

The final polishing of the cavity and core also affects the product's appearance and release from the mold's core and cavity. The final polishing direction for a high-gloss surface should always be in the direction of the opening mold. Problems as product sticking and nonrelease from the mold cavity surface, have resulted when the final polish direction was not in the mold opening direction, especially for very tacky resins.

A problem has developed that illustrates this effect. A molded hammer handle's urethane handgrip on a long polished core was difficult to almost impossible to obtain release from the core. The product, specifically the urethane handgrips, would not release from the core, and mold release was not successful in obtaining release from the polished core. To solve the problem, the core surface was slightly modified. A very light shot peening was performed on the mold core using small glass beads to just interrupt the polished surface.

The shot peening slightly modified the mold's core surface by breaking up the smoothness of the polished core. The slight indentions on the surface of the core, not visible to the eye, broke up the suction tension effect of the urethane, which resulted in an effortless release of the hammer handle material, a very tacky, pliable, and still hot urethane resin, without the use of mold release.

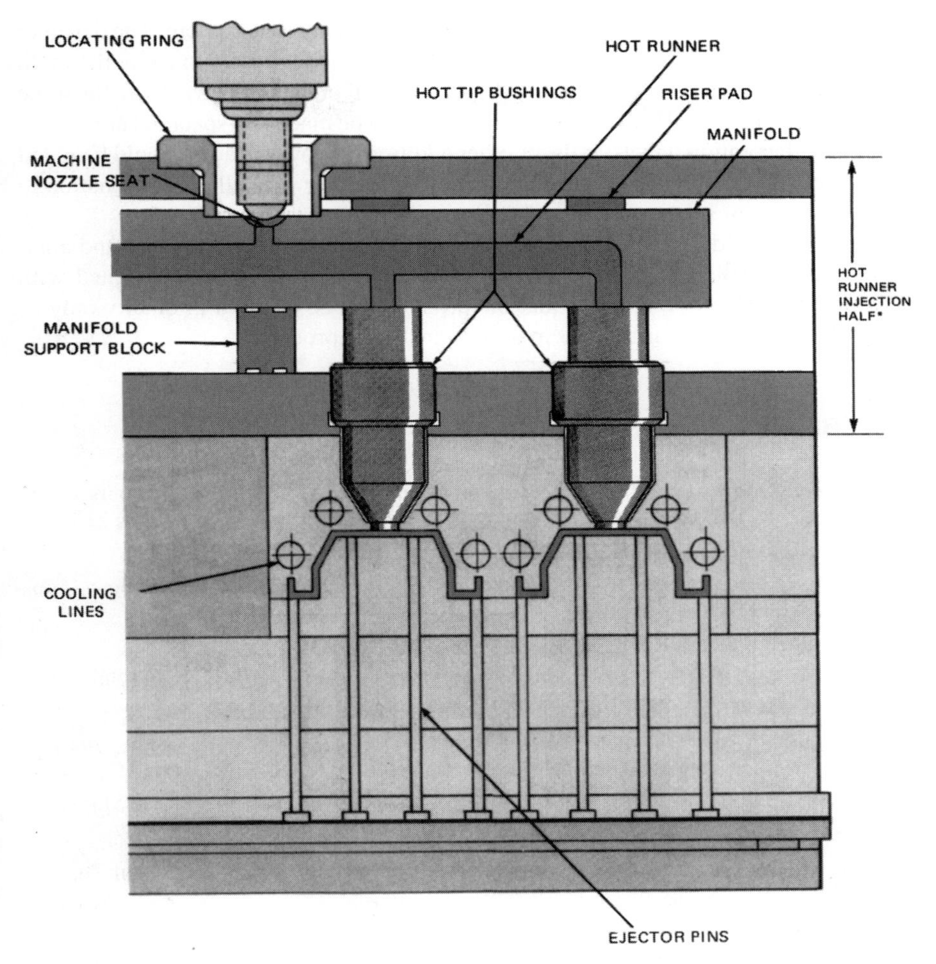

Figure 12.46 Hot-runner multicavity mold. (Courtesy of INCOE Corp.)

Mold Ejector System

The mold's cavity ejector system pushes the product out of the mold cavity at the end of the molding cycle as the mold opens. The mold designer will specify the product ejection system using knockout pins, stripper plates, or rings. The specific ejector system used depends on the product's configuration and shape in the mold cavity. Typically, knockout pins, located in the cavity surface and activated by the mold stripper plate on opening are used as shown in Figure 12.46, for a hot-runner system mold. The number and their position in the mold cavity are located to uniformly apply pressure on the product to remove it from the core or cavity when the mold opens. Stripper plates and rings are used to assist or replace the knockout pin system if distortion, puncture, warping, or marking on the product is a problem. The designer should be aware of this and, depending on how the product is positioned in the mold cavity, may specify no ejector pin marks are allowed on a show surface. If for some reason this is impossible; ejection pin positions may be placed on specified areas to be hidden by decoration, labels, foils, or other additions. Show surfaces should be noted on the decoration drawings to provide the mold designer with all the information to design the ejector system correctly for the mold cavity.

Following the design and manufacturing guidelines and using the mold and manufacturing checklists will greatly assist in reducing any problems associated with manufacture. Always remember that the product will be as good in quality only as the mold and manufacturing system used to make the product.

REFERENCES

1. *A Guide for Designing with Engineering Plastics,* E. I. Du Pont de Nemours Corp., Wilmington, DE, 1990.
2. R. D. Beck, *Plastic Product Design,* Van Nostrand Reinhold, New York, 1970.
3. *General Design Principles—Module 1,* 201742 B, E. I. Du Pont de Nemours Corp., Wilmington, DE (Sept. 1992).
4. "Planning for Screw-Holding Bosses," *Plastics World* 96 (April 1990).
5. "Rigid-Vinyl Design Considerations," *Plastics Design Forum* 82 (Nov./Dec. 1990).
6. *Modern Plastics Encyclopedia,* McGraw-Hill, New York, 1985–1986.
7. C. Mobery, "Cycle Time Reduction through High-Thermal-Conductivity Metals," *Plastics, Engineering* 42–44 (Feb. 1988).
8. B. Miller, "Predicting Part Shrinkage Is a Three-Way Street," *Plastics World* 48–52 (Dec. 1989).
9. *Zytel® Molding Guide,* E-23285, E. I. Du Pont de Nemours Corp., Wilmington, DE,
10. W. J. Tobin, "Predicting Tooling Delivery," *Plastics Machinery & Engineering* 55 (Feb. 1990).
11. *Injection Molds for Thermoplastics,* DST Engineering Plastic North America, Reading, PA (August, 1992).
12. "Designing for Impact Resistance," *Plastics Design Forum* 104 (Nov./Dec. 1989).
13. *Delrin® Molding Guide,* E-49702, E. I. Du Pont de Nemours Corp., Wilmington, DE,
14. *Designing with Plastics: The Fundamentals,* Engineering Plastics Division, Hoechst Celanese, 46–93 15M/490 (1992).
15. K. S. Mehta, "Identifying and Correcting Part-Design Problems," *Plastics Design Forum* 35–46 (Nov./Dec. 1986).
16. R. Noller, "Understanding Tight Tolerance Design," *Plastics Design Forum* 61–72 (March/April 1990).

CHAPTER 13

PLASTIC PRODUCT DESIGN AND DEVELOPMENT PROGRAM ANALYSIS

Developing a new product or replacing an existing product with a different material is a team effort. When a new material is considered, there are often many opportunities to add or develop new functions or enhancements to the existing product to make it more valuable to the customer. The major problem most companies experience during the design or redesign of a product, such as converting an existing metal or other material to plastic, is lack of internal company and customer communication.

Using the design team method, specific selected team members meet at required intervals to discuss and decide on the best methods to use in bringing the product to commercialization. The team members do not necessarily have to be experts in the use of each material or product design function, but need to know when and with whom they should seek out information when decisions must be made affecting the success and profitability of the program. The team members must also have the ability to obtain any information they lack when developing a product. Using the experience and support of your suppliers is an accepted and recommended procedure.

With the use of plastics, which are carbon-based molecular materials, the design team may observe a nonisotropic material whose property values are continuously changing under temperature, environment, and end-use conditions. Therefore, their design assumptions and analysis must be used to ensure that the product will be a success in its application. Today there are many successful product case histories that designers can draw on for obtaining the information necessary to complete a successful design program. These case histories are available from their material suppliers whose products are used in the case study.

The important factor in beginning a new design program is "teaming" with the right service and support companies. These are the material and equipment suppliers and, when necessary, you may employ a qualified outside consultant with the knowledge and experience to assist you in obtaining answers necessary to successfully complete the program. Remember to always obtain references when hiring a consultant, and be sure that that person's experience is in the area that your team members require their assistance. The success of a program begins with developing the necessary information to make accurate and quality design, material, and manufacturing decisions. The best possible product design must be coupled with a tool designed to produce the volume, part tolerances, and quality demanded by the customer. The tool controls the products final dimensions and quality coupled with a good reliable material and processing repeatability. Also, using a manufacturing facility with ISO 9000 certification will assist the customer confidence in achieving each lot of products to a specified quality standard to which they have contracted with their supplier to provide. As the design program ends and production begins, document what was agreed to and learned to guide future programs in your database. Collect information from all team members, as no single person will be aware of everything that occurred to make the program successful. Remember that there are no right or wrong conclusions made when these items were discussed.

Information that should be documented is

1. What surprises were encountered, why were they encountered, and what were their effects and solutions?
2. Did the product's performance meet, exceed, or fall short of what was anticipated?
3. Did the failure mode and effects analysis (FMEA) design analysis, tooling processing analysis, mold fill, and cool, predict the final results? Details in these areas can save future setup time, meshing density, element type, and size for future programs.
4. Were tolerances realistic, attainable, and needed for the program?
5. Did the original material selection meet the objectives, or was an alternate required?
6. Were there any quality or process problems with the resin, and was regrind used?
7. Was the tool designed to meet product expectations and revisions required for manufacturing?
8. Was an FMEA performed during the design analysis phase and prior to manufacture for the program to ensure that all variables were analyzed?

The final consideration is how the design team interacted, made decisions, and kept on schedule. Bring out positive methods that aid decisionmaking and those that may have hindered final results, leaving out personalities unless an asset or serious drawback to the program occurs.

For new programs and team members, a program review and analysis can prove beneficial and aid a new team in developing new or converting old metal product to plastic in the future.

Producing a product with the lowest possible liability risk factors is a necessity and is essential for doing business. Consumer and industrial product safety are required to avoid the possibility of product liability that is first and foremost in product design requirements.

SUCCESSFUL PRODUCT DEVELOPMENT PLAN

The final consideration is that the program must be profitable, stay on schedule, within cost and projections, and meet end-use product requirements and specifications. In simplistic terms, the part development plan outlined below, if followed, should lead to the successful introduction of a plastic part or assembly to the marketplace:

1. Determine the market for the product with estimated sales volume requirements, material, and cost.
2. Use "checklists" to develop information for the products design and manufacture.
3. Organize the "design team" to meet the program's requirements.
4. Initiate preliminary design analysis and cost studies.
5. Complete part analysis using FMEA, mold fill/cool analysis, and model part analysis to evaluate end-use requirements.
6. Test the product for meeting end-use design and life requirements.
7. Evaluate manufacturing requirements—method, tooling, tolerances, and scheduling—and perform a FMEA before manufacture.
8. Evaluate information and commit to program if all indications look successful.
9. Finalize product design, procure tooling, and schedule manufacturing to meet program requirements.
10. Evaluate tooling and manufacturing with capability studies, complete procedures for manufacture, set quality limits, and process control to meet specifications.
11. Begin production and monitor in real-time manufacturing process.

These 11 items for developing a successful product program are few, but a lot of planning and work is involved in these steps to develop a program for "doing it right the first time."

APPENDIX A

CHECKLIST FORMS

Permission for use of copyright checklists is granted by the author.

CHECKLIST 1

ASSEMBLY CHECKLIST

DATE:

CUSTOMER:
ADDRESS:

PART NAME: JOB NUMBER:
MANUFACTURING START DATE:
PRODUCTION SUPERVISOR:
ASSEMBLY START DATE: ASSEMBLY SUPERVISOR:

ASSEMBLY DRAWING NO.:
PART NUMBERS:
TYPE OF ASSEMBLY REQUIRED:
PROCEDURE AVAILABLE: PROCEDURE NUMBER:
SPECIAL INSTRUCTIONS:
SPECIAL EQUIPMENT REQUIRED:
EQUIPMENT:
SPECIAL MATERIALS REQUIRED:
MATERIALS:
OSHA REQUIREMENTS:
OPERATOR TRAINING REQUIRED: WHAT: PROVIDED BY WHOM:

COLOR OR TEXTURE MATCH REQUIRED: TYPE:
WHO APPROVES:
PROCEDURE: PROCEDURE NUMBER:

PURCHASED PARTS REQUIRED: PURCHASE ORDER NO.:

RECEIVED IN-HOUSE: WHERE IN INVENTORY:
PART NUMBERS:
QUANTITY REQUIRED:
QUALITY APPROVED FOR ASSEMBLY:

TOTAL ASSEMBLY IN HOUSE: OUTSIDE SUPPLIER:
SUPPLIER:
CONTACT: PHONE: FAX: E-MAIL:

FINISHED TESTING OF ASSEMBLY REQUIRED:
REQUIREMENTS:
WHO APPROVES:
TEST PROCEDURE NO.:
WHO DOES TESTING:
REJECTS SALVAGEABLE: HOW: PROCEDURE NO.:
WHO APPROVES:
DISPOSITION OF REJECTS:

PACKAGING REQUIREMENTS:
PROCEDURE: PROCEDURE NO.:

JUST-IN-TIME PRODUCT:
SPECIAL INSTRUCTIONS:
DOCUMENTATION REQUIRED: WHAT:
DOCUMENTATION TO WHOM:
SHIPPING CONTACT:

CHECKLIST 2

PRODUCT DESIGN CHECKLIST

DATE:

CUSTOMER:
ADDRESS:

CUSTOMER CONTACT: ALTERNATE:
PHONE: FAX: E-MAIL:
PROGRAM START DATE: EST. COMPLETION DATE:
JUST-IN-TIME PROGRAM: ANTICIPATED QUANTITY/SHIP FREQUENCY: _______/______

PART NAME: JOB NO.:
DRAWING AVAILABLE: DRAWING NO.:
PART AVAILABLE: MODEL/PROTOTYPE:

PART DESCRIPTION NEW: EXISTING: COMPETITORS:
FUNCTION OF PART:

EXISTING MATERIAL: PROPOSED MATERIAL:
PART WEIGHT & SPECIFIC GRAVITY OF MATERIAL:_________/_________

NUMBER PARTS IN ASSEMBLY, EXISTING: PROPOSED:
FUNCTION OF ADJACENT PARTS:
ABLE TO COMBINE FUNCTIONS:
WHAT:

PURCHASED PARTS USED: TYPE:
TYPE:
TYPE:

CUSTOMER INCENTIVE FOR PROJECT: COMPENSATION:
PERFORMANCE IMPROVEMENT: REDESIGN:
WEIGHT SAVINGS: COST REDUCTION:
MEET NEW REQUIREMENTS: ALTERNATE SUPPLIER NEEDED:
OTHER CONSIDERATIONS:

PRODUCTION INFORMATION:
MANUFACTURING METHOD:
VOLUME (PARTS/YEAR):
SUPPLIER, IN-HOUSE: OUTSIDE: WHO:

DESIGN CONSIDERATIONS (OBTAIN SKETCH OF FORCES ACTING ON PART)
PART FUNCTION:
CUSTOMER LIABILITY IF FAILURE OCCURS:

OPERATING CONDITIONS:	NORMAL	MAXIMUM	MINIMUM
TEMPERATURE:	:	:	:
SERVICE LIFE (HOURS)	:	:	:
FORCES (LBS/TORQUE/ETC.)	:	:	:
DURATAION OF FORCES			
TIME ON:	:	:	:
TIME OFF:	:	:	:
MAXIMUM DEFLECTION ALLOWED:	:	:	:

LOAD BEARING APPLICATION:
BUCKLING CONSIDERED:

IMPACT FORCES: WHERE ON PART (SKETCH): TYPE:
REPEATED: ONE TIME: FREQUENCY:
DROP HEIGHT: LOAD: IMPACT ENERGY:

VIBRATION EFFECTS CONSIDERED:
VIBRATION INPUT:
WEIGHT OF ASSEMBLY: WEIGHT OF INTERNAL COMPONENTS:
MOUNTING OF COMPONENTS (METHOD):
ATTACHES TO ANOTHER ASSEMBLY OR PART: PART:
FUNCTION OF THIS PART:
OPERATING SPEED (RPM):
EXCITING FREQUENCY (CPS):

DISPLACEMENT (INCHES/MM):
ACCELERATION (G FORCES):

ENVIRONMENTAL CONDITIONS:
CHEMICAL EXPOSURE: TYPE: CONCENTRATION:
CHEMICAL MAKEUP:
MOISTURE (HUMIDITY): PERCENT: TEMPERATURE:
WATER EXPOSURE TYPE (FRESH/SALT/BOILING/STEAM):
TEMPERATURE:
RADIATION: TYPE: EXPOSURE LEVEL: TIME:
SUNLIGHT: EXPOSURE TYPE (DIRECT/INDIRECT): TIME:
UV PROTECTION REQUIRED: COLOR FADING A FACTOR:
AMBIENT TEMPERATURE NOT OPERATING: OPERATING:

ELECTRICAL REQUIREMENTS:
MAXIMUM CURRENT SUPPORTED:
MAXIMUM VOLTAGE:
INSULATION PROPERTIES REQUIRED: TYPE:
EMI/EMP PROTECTION REQUIRED: VALUES REQUIRED:

FLAMMABILITY REQUIREMENTS
REQUIRED: TYPE:
SMOKE GENERATION LIMITS REQ'D.: OXYGEN LEVEL LIMITS:
MUST MEET UL FLAME REQUIREMENTS: VO/V1/V2/HB/V5

AGENCY REQUIREMENTS (UL/CSA/NSF/FDA/ETC): AGENCY:
REQUIREMENTS:
CUSTOMER REQUIREMENTS:
OTHER REQUIREMENTS:

ABRASION & WEAR REQUIRED:
TYPE:
MATING PART MATERIAL:
COEFF. FRICTION REQ'D.: STATIC: DYNAMIC:
LUBRICATION REQ'D.: TYPE: CHEMICAL COMPOSITION:
SELF LUBRICATING: INTERNAL: EXTERNAL:
TYPE LUBRICATION ALLOWED:

SAFETY FACTOR REQUIREMENTS:
PART LIABILITY: SEVERITY IF FAILURE OCCURS:
DEGREE OF LIABILITY TO SUPPLIER:
PART FAILURE IMPACT ON PRODUCT/APPLICATION:
CONSUMER/INDUSTRY APPLICATION:
INSTRUCTIONS ON PRODUCT TO OPERATE:

WARNING LABELS REQUIRED:
WHAT INFORMATION ON LABEL:
WHO SUPPLIES:
AGENCY REQUIREMENTS/TESTING REQUIRED: WHAT: WHO:
QUALITY REQUIREMENTS:
CRITICAL TOLERANCES:
HOW MANY: WHERE: TOLERANCE:
IMPACT ON PART FUNCTION IF NOT MET:
PART REQUIREMENTS (FLASH/WARPAGE/SINK/POROSITY):
SPECIFICATIONS ESTABLISHED: WITHIN SUPPLIER CAPABILITY:
MATERIAL/PARTS INCOMING TESTING REQ'D.:
TESTING SPECIFIED:
TESTING REQUIREMENTS:
CUSTOMER REQUIREMENTS ESTABLISHED:
REQUIREMENTS:

WILL CUSTOMER TEST INCOMING PRODUCT: HOW: WHERE:
TESTS REQUIRED TO MEET:
TEST EQUIPMENT:
PROCEDURE ESTABLISHED:
PERSONNEL TRAINED:
SPECIAL EQUIPMENT REQ'D.:
CUSTOMER CONTACT: PHONE:
SUPPLIER WITNESS TESTS: MUST SCHEDULE:
IF FAILURE OCCURS, HOW ARE DISPUTES SETTLED:
BY WHOM:

INTEGRATION OF COMBINING PART FUNCTIONS CONSIDERED:
WHAT OPERATIONS CAN BE COMBINED:
MATERIAL CAPABLE:
ASSEMBLY METHODS CONSIDERED:

PART REQUIRES ASSEMBLY:
SNAP/PRESS FIT:
MECHANICAL FASTENERS:
WELDING (THERMAL/SONIC):
ADHESIVES/SOLVENTS:
OTHER:
PLANT HAS EQUIPMENT TO DO ASSEMBLY: WHAT IS REQUIRED:
PERSONNEL TRAINING REQUIRED: WHAT:
IF NO, USE OUTSIDE COMPANY:
EQUIPMENT REQUIRED:
COST FACTOR ON PRODUCT

DECORATION REQUIREMENTS:
COLORED: COMPOUNDED: S & P: CONCENTRATES:
COLOR: REQUIREMENTS: COLOR SAMPLE:
MUST MATCH MATING PART: COLOR/MATERIAL/PAINT:
PIGMENT TYPE ALLOWED: WILL IT AFFECT MATERIALS PROPERTIES:
TESTING REQUIREMENTS: WHAT:
SPECIAL CLEANING REQUIRED: WHAT: AFFECTS ON MATERIAL:
PAINTED: PAINT TYPE: PRIMER: NUMBER COATS:
OSHA REQUIREMENTS INVOLVED: WHAT:

METALIZED: TYPE OF METAL: THICKNESS:
PLATED: TYPE OF METAL: THICKNESS:
ETCHING OF MATERIAL REQUIRED: WHAT: CONDITIONS:
IN-HOUSE: OUTSIDE: WHO:

PRINTING: TYPE: MOLDED IN: ON SURFACE:
SURFACE PREPARATION REQUIRED: WHAT:
FOILS/DECALS: TYPE: SUPPLIER:

PRINTING:
INFORMATION REQUIRED:
WHO FURNISHES:
FINISH ON PART: CLASS A: TEXTURED: DEPTH:
FINISH TYPE: HIGH POLISH: SAMPLE AVAILABLE:
MUST MATCH MATING PART: SAMPLE AVAILABLE:

TESTING REQUIREMENTS:
TESTS TO MEET:
PROCEDURE:
PROTOTYPE TESTED:
PRODUCTION TESTS:
END USE REQUIREMENTS:

PROTOTYPE:
MOLDED:
MODELED:
SLA/SLS/OTHER:

CUSTOMER INFORMATION:
DESIGN DEADLINE: EXTENSION TIME AVAILABLE:
ARE ALL DESIGN REQUIREMENTS/END USE INFORMATION AVAILABLE:
IF NOT, WHAT IS MISSING: WHEN AVAILABLE:

END USE TEST AVAILABLE: WHO TESTS:
WHERE: HOW MANY CYCLES:
CONDITION OF PART: CONDITIONING REQUIRED: WHAT:
CONTACT: PHONE:
WHO GIVES FINAL APPROVAL OF DESIGN:

PRELIMINARY COST FIGURES COMPLETED:
ANTICIPATED PIECE PART COST:
QUOTE NUMBER:

HAVE ALL DEPARTMENTS BEEN CONTACTED FOR THEIR DESIGN INPUT:
SALES:
ENGINEERING:
PURCHASING:
MANUFACTURING:
QUALITY:
TOOLING:
ASSEMBLY:
DECORATION:
PACKAGING:
SHIPPING
MATERIAL SUPPLIERS:
OUTSIDE SOURCES REQUIRED:
UPPER MANAGEMENT APPROVAL:

ADDITIONAL INFORMATION REQUIRED, NOT LISTED TO ASSIST IN UNDERSTANDING COMPLETELY THE FUNCTION, MANUFACTURING, QUALITY, ASSEMBLY, AND ANY OTHER ABUSE OR REQUIREMENTS THE PART MUST WITHSTAND OR ENVIRONMENTAL STRESSES NOT LISTED:

DESIGNER:
DESIGN TEAM SIGNOFF:
COMPANY REPRESENTATIVE: APPROVAL DATE:

CHECKLIST 3

DECORATING CHECKLIST

DATE:

CUSTOMER:
ADDRESS:

PART NAME: JOB NUMBER:
MANUFACTURING START DATE:
PRODUCTION SUPERVISOR:
DECORATING START DATE: DECORATION SUPERVISOR:

DECORATION REQUIRED: TYPE:
DRAWING NUMBER:
BEFORE OR AFTER ASSEMBLY:
PROCEDURE NO.:
COLOR MATCH/TEXTURE REQUIRED:
APPROVAL BY WHOM:
PROCEDURE NO.:
REJECT HANDLING PROCEDURE:
SALVAGEABLE:
PROCEDURE NO.:

PART SURFACE PREPARATION REQUIRED: TYPE:
PART SURFACE TESTING REQUIRED:
TEST REQUIREMENTS:
EQUIPMENT REQUIRED:
EQUIPMENT
PROCEDURE NO.:

IN-HOUSE DECORATION: OUTSIDE: WHO:
TRAINING REQUIRED: WHAT:
FIXTURES REQUIRED: WHAT:
SPECIAL: ORDERED: PURCHASE ORDER NO.: RECEIVED:

DECORATING MATERIALS ORDERED:
SUPPLIER:
MATERIALS:
PURCHASE ORDER NO.:
SPECIAL REQUIREMENTS:
CERTIFICATION REQUIRED: WHAT: BY WHOM:

OSHA REQUIREMENTS TO BE MET:
REQUIREMENTS:
SPECIAL EQUIPMENT REQUIRED: WHAT: ORDERED:
PURCHASE ORDER NO.: RECEIVED:

DECORATED PARTS TO:
SPECIAL HANDLING REQUIRED:
PARTS TO STORAGE/STATION:

JUST-IN-TIME PRODUCT:
SPECIAL INSTRUCTIONS:
DOCUMENT:
PACKAGING CONTACT:

CHECKLIST 4

MATERIALS CHECKLIST

DATE:

CUSTOMER:
ADDRESS:

PART NAME: JOB NUMBER:
PRODUCTION START DATE: PRODUCTION SUPERVISOR:
MATERIAL: PRODUCT CODE: VOLUME:
ALTERNATE SUPPLIER: PRODUCT CODE:
CRITICAL PARTS REQUIRING USE OF SAME LOT NUMBER OF MATERIAL;
DUE TO COLOR, DIMENSIONS:
PART NUMBERS:
ALTERNATE PARTS PRODUCTION START DATE:

PRODUCT VOLUME: PRODUCT WEIGHT: MATERIAL REQ'D. LBS.:
ORDER SIZE, LBS.: ALL ONE LOT NUMBER OR MIXED: YES/NO

CONFIRMED:
MATERIAL CERTIFICATION REQUIRED*: YES/NO
CERTIFICATION TO:
SPECIAL REQUIREMENTS, MATERIAL VALUES, COLOR, PROPERTIES, SPECIFICATION:
VALUES REQUIRED:
CERTIFICATION REQUIRED WITH EACH SHIPMENT: YES/NO
PRIOR TO RECEIPT OF MATERIAL: YES/NO
TEST VALUES ON MATERIAL REQUIRED:
VALUES REQUIRED:

PACKAGE TYPE: BAGS-DRUMS-GAYLORDS-BULK:
PRICE PER POUND/KILO: VOLUME DISCOUNT:
COLORED MATERIAL: YES/NO METHOD: COMPOUNDED-S/P-CONCENTRATE (TYPE)
CONCENTRATE SOURCE:
COLOR SAMPLE REQUIRED**: YES/NO TYPE: COLOR CHIP-SURFACE TYPE-RESIN
PIGMENT CHANGES PERMITTED: YES/NO MUST NOTIFY IF REQUIRED: YES/NO
NOTIFY WHO:

SPECIAL INCOMING MATERIAL TESTING REQUIRED: YES/NO
CONTACT:
TESTS:

QC CONTACT AT RECEIVING:
PRODUCTION CONTACT:
MATERIAL ROUTING ON RECEIPT: WAREHOUSE-SILO-PRODUCTION-OUTSIDE MOLDER
HOLD TILL TESTING COMPLETED: YES/NO CONTACT:
DISPOSITION IF MATERIAL FAILS INCOMING TESTS:
NOTIFY CONTACTS IN: QC:
PRODUCTION:
SALES:
PURCHASING:

OTHER PARTS REQUIRED FOR PRODUCT SALE:
PRODUCT NAME: PART NUMBER:
SUPPLIER: CONTACT:
DATE REQUIRED:

* SEE INSPECTION & MATERIAL FLOW SHEET NO.:
** SEE COLOR MATCH REQUEST FOR VERIFICATION:

PURCHASING REPRESENTATIVE:

CHECKLIST 5

MANUFACTURING CHECKLIST

DATE:

CUSTOMER:
ADDRESS:

PART NAME: JOB NUMBER:
MANUFACTURING START DATE:
PRODUCTION SUPERVISOR:
SET UP TIME & DATE: SET UP TECH.:

MACHINE NUMBER/SIZE SCREW TYPE: CHECK RING:
NOZZLE TYPE: MOLD INSULATED FROM PLATENS:
PROCESS PROCEDURE: PROCESS CONTROL CHART NO.:
PRODUCT SPECIFICATION SET: DOCUMENT NO.:
SPECIFICATION LIMITS ESTABLISHED: MEAN: UCL: LCL:

MOLD NUMBER/SIZE:
OWNERSHIP:
SPRUE BUSHING FITS NOZZLE: SPECIAL BUSHING REQUIRED:
PART NO.:
MAINTENANCE COMPLETED: SIGNED OFF BY: DATE:
SPECIAL REQUIREMENTS:
INSTALLATION PROCEDURE: NUMBER:
QUICK CHANGE: MOLD PREHEAT REQ'D.: TEMPERATURE:
TIME:
MOLD RELEASE ALLOWED: TYPE:

PRODUCTION EQUIPMENT:
DRYER: DRY TO % MOISTURE: TEMPERATURE:
DRY TIME:
HOPPER/CENTRAL/SIDE DRYER TYPE: FILTERS CLEAN:
DESICCANT GOOD: CLEAN:
DESICCANT BED DRIED: START BEFORE MATERIAL ADDED:
TEMP.: TIME:

MOLD CHILLER NO.: TEMPERATURE SETTING: FLOW GP:
PRESSURE:
COOLING MEDIUM: SPECIAL HOSES REQ'D.: TYPE:
FITTINGS:
SETUP PROCEDURE AVAILABLE: DOCUMENT NO.:

GRINDER NO.: LAST INSPECTED: FILTER & UNIT VACUUMED CLEAN:
BLADE SHARPENED: LAST MATERIAL GROUND:
SCREEN SIZE IN HOPPER:

ROBOT NO.: SET UP PROCEDURE: DOCUMENT NO.:
SET UP BY:
SPECIAL INSTRUCTIONS REQUIRED: DOCUMENT NO.:

PART HANDLING:
OPERATOR: GLOVES REQUIRED: TYPE:
PROTECT PART SURFACE: HOW: ELECTRO-STATIC:
CONVEYOR: MACHINE NO.:
SPRUE PICKER: MACHINE NO.:
MOLD SWEEP: MACHINE NO.:

SPECIAL OPERATIONS AT PRESS: WHAT: EQUIPMENT REQ'D.:
SPECIAL OPERATOR TRAINING: WHAT:

PACKAGE PRODUCT AT PRESS: HOW:
SPECIAL REQUIREMENTS: PACKAGING SUPERVISOR:
MATERIAL:
SUPPLIER:
PACKAGE TYPE: HOPPER LOADING METHOD:
HOPPER CAPACITY:
DRYING TIME REQUIRED: SAMPLE TEST PRIOR TO START:
% MOISTURE ALLOWED:
REGRIND ALLOWED: PERCENT: HOW BLENDED AT HOPPER:
PROCEDURE NO.: WHO ADDS TO HOPPER: FREQUENCY:

PROCESS CONTROL LIMITS ESTABLISHED: DOCUMENT NO.:

QUALITY CHECKS AT PRESS: WHO APPROVES:
PROCEDURE NO.:
TEST EQUIPMENT:
VERIFY WHAT VARIABLES AT START UP:
PROCEDURE NO.:
WHO APPROVES SAVING FIRST PRODUCTION PARTS:
SAMPLES SAVED: HOW MANY: QUALITY CHECKS:

ANY SECONDARY OPERATIONS AT PRESS:
WHAT:
SPECIAL PART HANDLING REQUIRED: WHAT:

PRODUCTION PROBLEMS CONTACT:
QUALITY ASSURANCE CONTACT:
MAINTENANCE CONTACT:
MATERIAL HANDLING CONTACT:

PARTS BOXED/PALLETIZED/COUNTED/WEIGH COUNTED/WHAT

PARTS TO STORAGE/STATION/HOLDING POINT:
PARTS PROTECTED: HOW & WITH WHAT:

ANY SPECIAL INSTRUCTIONS:

CHECKLIST 6

PACKAGING CHECKLIST

DATE:

CUSTOMER:
CONTACT: PHONE: FAX: E-MAIL:
ADDRESS:

PART NAME: JOB NO.:
PRODUCTION SUPERVISOR:
MANUFACTURING START DATE: QUANTITY:
PACKAGING REQUIRED: TYPE: SPECIAL ORDER:
PURCHASE ORDER ISSUED: P.O. NO.: WHEN:
LEAD TIME TO ORDER PACKAGING:
PACKING DUE IN: NOTIFY WHOM:

PART PROTECTION REQ'D. BEFORE PACKING: WHAT:

JUST-IN-TIME MANUFACTURE USED:
SPECIAL PACKAGING REQUIRED FOR SHIPMENT: IF NOT, DUNNAGE AVAILABLE:
WHAT TYPE: WHO FURNISHES: SPECIAL REQUIREMENT:
SUPPLIER:
REQUIREMENTS:
REUSABLE: HOW RETURNED TO SUPPLIER: BY WHOM:

PART PACKAGING PERFORMED WHERE: BY WHOM:
SPECIAL TRAINING REQ'D.:
WHAT:

NUMBER OF PARTS PER PACKAGE:
NUMBER OF PARTS PER CARTON:
NUMBER OF CARTONS PER PALLET:
ARE PALLETS STACKABLE: HOW MANY PALLETS HIGH ALLOWED:

STORAGE REQUIRED BEFORE SHIPPING:
SECURED AREA:
QS9000 INSPECTION REQ'D.: WHAT:
PROCEDURE NUMBER:
WHO PAYS: HOW LONG:

BAR CODING REQUIRED: WHO SUPPLIES:
BAR CODE SPECIFIED:
SPECIAL INSTRUCTIONS REQUIRED: WHAT:
LOT NO.:
DATE OF MANUFACTURE:
PRODUCT NAME:
PRODUCT CODE:
OTHER INFORMATION:
SPECIAL PACKAGING REQUIRED: WHAT:
WHO SUPPLIES:

PACKAGING PROCEDURE DOCUMENTED:
DOCUMENT NUMBER:

SPECIAL TRUCKING REQUIRED FOR SHIPMENT: WHAT:
SHIPPER: CONTACT: PHONE:

CHECKLIST 7

PRODUCT DEVELOPMENT CHECKLIST

DATE:

CUSTOMER:
ADDRESS:
CONTACT: PHONE: FAX: E-MAIL:
ALTERNATE:

CUSTOMER-/-IN-HOUSE PART DEVELOPMENT NUMBER:
MARKET ESTABLISHED: BENEFITS TO MARKET:
MARKET SIZE: ESTIMATED VOLUME/YEAR:
USERS OF PRODUCT:

ANTICIPATED SALES PRICE: HOW CALCULATED: HOW SOLD:
MANUFACTURE IN-HOUSE: OUTSIDE SUPPLIER: JOINT:
OUTSIDE SUPPLIER(s):

PURCHASED PARTS REQUIRED:
PARTS:
SUPPLIERS:
COST:

NEW PART: EXISTING: REDESIGN REQ'D.:
METAL REPLACEMENT:
COMPETITION: WHO: SALE PRICE:
MARKET SIZE: SHARE DESIRED: ESTIMATED SELL PRICE:

PATENTABLE: APPLIED FOR: DATE:
PATENT NO.:

PROGRAM ASSETS AVAILABLE: REQUIRED:
MARKET INTRODUCTION DATE ANTICIPATED:
ASSISTANCE REQ'D.: TYPE:
WHOM: WHAT AREAS:
PROBABILITY OF PROGRAM SUCCESS:
ESTIMATED COMPLETION DATE:

PROJECT TEAM LEADER: ALTERNATE:
PROJECT START DATE: ____________________
DECISION DATES: ALL REQ'D. INFO. AVAILABLE
____________________ ____________
ASSETS AVAILABLE START DATE

DEVELOPMENT TEAM MEMBERS:
CUSTOMER IF DESIGNATED:
SALES:
ENGINEERING:
DESIGN:
PRODUCTION:
TOOLING:
QUALITY:
PURCHASING
FINANCE:
MANAGEMENT:
SUPPLIERS:

PART DEVELOPMENT (TEAM ANALYSIS OF PRODUCT)
PART REQUIREMENTS (GENERAL, SPECIFIC, LIABILITY ITEMS), BE SPECIFIC:
1.
2.
3.
BENEFITS TO USER:
LIMITATIONS OF CURRENT PRODUCT:
COMPETITIONS PRODUCT EVALUATION:
QUALITY REQUIREMENTS:
IMPROVEMENTS POSSIBLE:
POSSIBLE TO COMBINE FUNCTIONS:
POSSIBLE TO CHANGE MATERIAL:

CHECKLIST ANALYSIS:
CUSTOMER REQUIREMENTS: DATE:

SALES/CONTRACT: DATE:
ENGINEERING: DATE:
PROBLEM: DATE:
DESIGN: DATE:
MATERIAL: DATE:
PROGRAM SCHEDULING: DATE:
MANUFACTURING: DATE:
TOOLING: DATE:
PURCHASING: DATE:
SUPPLIERS: DATE:
PRICE ESTIMATION: DATE:
DEVELOPMENT: DATE:
ASSEMBLY: DATE:
DECORATION: DATE:
PACKAGING: DATE:
SHIPPING: DATE:

AGENCY AND CODE REQUIREMENTS:
WHO: WHAT:
CUSTOMER REQUIREMENTS:
WHAT:
PROGRAM STATUS, CONTINUE: NEED MORE INFORMATION:
WHAT:
TERMINATE: REASONS:

CONTINUE: APPROVED BY: DATE:

PART DESIGN & MATERIAL SELECTION:
DESIGN CHECKLIST COMPLETED TO SUIT REQUIREMENTS:
ADDITIONAL INFORMATION REQUIRED:
WHAT:
FROM WHOM: REQUIRED BY:
AVAILABLE: NEEDS TO BE DEVELOPED: BY WHEN:
HOW:
BY WHOM:

PART/DESIGN ANALYSIS:
DESIGNER:
TYPE-CALCULATIONS: FEA: SLA/SLS: MODEL:
PROTOTYPE:
TIME ESTIMATED TO COMPLETE: COST:
COMPLETION DATE:

MATERIAL CANDIDATES:

A: WHY:
B: WHY:
C: WHY:

SUPPLIER A:
SUPPLIER B:
SUPPLIER C:

SUPPLIER PROPERTY DATA AVAILABLE: REQUESTED: DATE:
REQUIRED DATA:
IF NOT AVALIABLE, CAN IT BE DEVELOPED: BY WHOM:

BY WHEN:
SUPPLIER CONTACT: PHONE: FAX: E-MAIL:
PROTOTYPE: TESTABLE: TYPE:

WHO PROVIDES: WHEN:
SPECIFIED IN CONTRACT:
COST ESTIMATE:
FULL SIZE: MATERIAL:
PROTOTYPE TESTABLE: TYPE OF TESTS REQUIRED:

ACTUAL END USE CONDITIONS: SIMULATED:
CONDITIONS:
REQUIREMENTS TO PASS:
WHAT IDENTIFIES FAILURE/PASS:
WHO DETERMINES:
AGENCY/CODE REQUIREMENTS TESTABLE:
WHAT:
TESTING TIME:
TEST COST:
NUMBER OF TESTS: SAMPLES REQUIRED:
SUPPLIER TEST DATA REQUIRED:
PROCEDURE DEFINED: PROCEDURE NUMBER:
WHO DOES TESTING:
CONTACT: PHONE: FAX: E-MAIL:
WHO EVALUATES DATA:
DOCUMENTATION REQ'D.:
CERTIFICATION REQ'D.:
TEST RESULTS IN WHAT FORM:

PASS/FAIL:
COMMENTS:

PROJECT STATUS CHECK POINT:
CONTINUE: TERMINATE: NEED MORE DATA:

WHAT:
BY WHOM:

DESIGN FINALIZED:
CUSTOMER APPROVED: BY WHOM: DATE: TITLE:

MATERIAL SELECTED:
SUPPLIER: PRODUCT CODE:
ALT. SUPPLIER: PRODUCT CODE:
CAN EITHER BE SUBSITUTED AT WILL:
DECISION AUTHORIZED BY ONLY:

EACH MATERIAL MUST BE END USE TESTED BEFORE FINAL APPROVAL:
SUPPLIER ON CERTIFIED SUPPLIER LIST:
IF NOT, WHEN:
WHO APPROVED:

QA APPROVAL STATUS OF SUPPLIERS:
SUPPLIER CERTIFICATION TYPE REQUIRED:
SPECIFIC LOT DATA:
TYPICAL LOT DATA:
SPECIAL REQUIREMENTS:
APPROVAL STATUS:

PURCHASED PARTS REQUIRED: WHAT:
SUPPLIERS:
CERTIFICATION REQ'D.:
VENDOR AUDITED FOR QUALITY: WHEN: STATUS OF AUDIT:

CRITICAL DIMENSIONS:
DRAWING AVAILABLE FOR DISCUSSION: DRAWING NO.:
DIMENSIONS ATTAINABLE: PLASTIC TOLERANCES:
NUMBER OF CRITICAL TOLERANCES:
WHERE 1: 2. 3.

INSERTS USED: TYPE:
IN MOLD: AT ASSEMBLY:
SCREWS USED: TYPE:

OTHER ASSEMBLY METHODS:
SNAP/PRESS FIT:
SONIC:
THERMAL:
SOLVENT/ADHESIVES:

QUALITY REQUIREMENTS: (SEE QUALITY CHECKLIST):
WHAT MAJOR REQUIREMENTS:
WHO DETERMINES:
WHEN VERIFIED:
BY WHOM:
TEST EQUIPMENT REQUIRED:
WHAT:
AVAILABLE:
SUPPLIED BY WHOM:
COST OF TESTING:
CUSTOMER TO VERIFY TESTS: ONLY DATA: BOTH:
PROCEDURE NUMBER:
DOCUMENTATION REQUIRED:
WHAT:
HOW TO REPORT:

MANUFACTURING METHOD (SEE CHECKLIST)
METHOD:
TOOLING:
SPECIAL REQM'TS.:
WHERE: BY WHOM:
CONTACT: PHONE: FAX: E-MAIL:
CAPABILITY OF EQUIPMENT EVALUATED: CP: CpK: CR:
PERSONNEL TRAINING REQ'D.:

PROCESS CONTROL USED: CLOSED LOOP FEEDBACK: OTHER:
REAL TIME PROCESS CONTROL USED:
MANUFACTURING PROCEDURE DOCUMENTED:
PROCEDURE NUMBER:
WORK INSTRUCTIONS:

MOLD DESIGN (SEE CHECKLIST)
COMPLETED:
ALL TEAM MEMBERS APPROVED DESIGN:
IF NOT, WHO DISAGREES:
WHY:
HOW RESOLVED: BY WHOM: DATE:

SPECIAL REQUIREMENTS: WHAT:
MOLD TYPE: NUMBER OF CAVITIES:
BALANCED RUNNER SYSTEM: REPLACEABLE GATE BLOCK:

SUPPLIER:
CONTACT: ALTERNATE:
PHONE: FAX: E-MAIL:

ESTIMATED PRICE: DELIVERED WHEN:
MOLD SPECIAL FEATURES: WHAT:
CORE PULLS/UNSCREWING:

MOLD FLOW ANALYSIS: BY WHOM: RESULTS:
MOLD COOL ANALYSIS: BY WHOM: RESULTS:

MOLD TRYOUT: WHERE: BY WHOM:
PRESS SIZE: OUNCES: TONS OF CLAMP:
PLATEN SIZE: MOLD FIT WITHIN:
PROCESS CONTROL USED FOR TRYOUT:

MATERIAL: GRADE: LOT NUMBER:
REGRIND ALLOWED: USED: PERCENTAGE: HOW BLENDED:
TRIAL DATE: LENGTH OF TRIAL:
GOOD PARTS PRODUCED:
IF NOT, WHAT WAS PROBLEM:
HOW WILL IT BE CORRECTED:
BY WHOM:
WHEN:
RETRIAL OF MOLD SCHEDULED: WHEN: WHERE:
MOLD TRIAL RESULTS:

FINAL MOLD TRIAL DATE:
LENGTH OF TRIAL: TIME: CYCLES: GOOD PARTS PRODUCED:
PARTS MEET CUSTOMER REQUIREMENTS:
IF NOT, WHAT WAS LACKING:
FIXABLE: PROCESSING: MOLD: MATERIAL:
TOOL APPROVAL: DATE: BY WHOM:

EXISTING TOOLING: LAST MOLDED AT:
MAINTENANCE PERFORMED TO MEET PRODUCT REQUIREMENTS:
QUALITY ASSURANCE APPROVED ALL DIMENSIONS:
DATE:
CONTACT:
REASON FOR TRANSFER:
TOOL DRAWINGS AVAILABLE:
PARTS LIST AVAILABLE:
KNOWN PROBLEMS WITH TOOL, DOCUMENTED:
CORRECTED:
BY WHOM:
VERIFIED: BY WHOM: TITLE: DATE:
MOLD APPROVED FOR PRODUCTION:
BY WHOM: TITLE: DATE:

ASSEMBLY METHOD (SEE CHECKLIST)
COMPLETED:
ASSEMBLY REQUIRED: FIXTURES REQ'D.: AVAILABLE:
MUST BE DEVELOPED: BY WHEN: BY WHOM: WHO PAYS:
ASSEMBLY DRAWING AVAILABLE: DRAWING NUMBER:
TYPE OF ASSEMBLY:
PRESS FIT: SNAP FIT: SONIC: THERMAL: SOLVENTS:
ADHESIVES:
SCREWS: OTHER OR COMBINATION OF METHODS:
REPAIRABLE: TYPE ALLOWED:
SEALED UNIT: TYPE:
HAND/MACHINE ASSEMBLY: REQUIRED ASSEMBLY RATE:
PROCESS/SPECIFICATIONS DEFINED: DOCUMENT NUMBER:
PART CLEANING REQUIRED: HOW: WITH WHAT:
MUST KEEP PART DRY AS MOLDED: HOW: WITH WHAT:
STORED WHERE:

ASSEMBLY TESTING REQUIRED: WHAT:
PROCEDURE NUMBER: TESTING SPECIFICATIONS:
SPECIFICATION NUMBER:

ASSEMBLY EQUIPMENT REQUIRED: IS IT CAPABLE:
MUST IT BE CALIBRATED: TO WHAT:
SPECIFICATIONS:

ADHESIVE/SOLVENTS USED: SYSTEM:
SUPPLIER: CONTACT: PHONE:
MSDA SHEETS REQUIRED WITH ORDER:
OSHA REQUIREMENTS:

PURCHASED PARTS IN ASSEMBLY:
WHAT:
SUPPLIER:
CONTACT: PHONE: FAX: E-MAIL:
QUALITY RATING: APPROVED SUPPLIER:
INSPECT BEFORE ASSEMBLY: SPECIFICATION:
BY WHOM: WHEN: WHERE:

DECORATION METHOD (SEE CHECKLIST)
COMPLETED:

PACKAGING METHOD (SEE CHECKLIST)
COMPLETED:

PIECE PART COST ANALYSIS (SEE PRICE CHECK SHEET)
COMPLETED: BY WHOM: DATE:
APPROVED BY MANAGEMENT: DATE:
APPROVED BY PRODUCTION: DATE:

MATERIAL COST:

MANUFACTURING COST:
MOLD COST:
HOW PAID FOR:
BY WHOM:
AMORTIZED OVER PRODUCTION RUN:

ASSEMBLY COST:
PURCHASED PART COSTS:
DECORATION COST:
PACKAGING COST:

TOTAL COST OF PROGRAM:

FINAL PROGRAM ANALYSIS:
CONTINUE:
TERMINATED:
COMMENTS:

APPROVED BY:
DATE:
CUSTOMER REPRESENTATIVE:
DATE:

PROGRAM START DATE ANTICIPATED:

CHECKLIST 8

PROGRAM SCHEDULING CHECKLIST

CUSTOMER: DATE:
ADDRESS:
CONTACT: PHONE: FAX: E-MAIL:

QUOTE NUMBER: COMPLETION DATE: REVIEW DATE:
QUOTED BY:
SUBMITTED TO CUSTOMER:
ACCEPT/REJECT: REASON:
CONTRACT SIGNED: BY WHOM: TITLE: . DATE:

PROGRAM CHECKLISTS COMPLETED, DATE:
SALES & CONTRACT: MANUFACTURING:
PART DEVELOPMENT: QUALITY:
PART DESIGN: ASSEMBLY:
MATERIAL: DECORATION:
PURCHASING: PACKING & SHIPPING:
MOLD: WARRANTY PROBLEM SOLVING:
PRICING:

MANUFACTURING DOCUMENTS COMPLETED:
JOB TRAVELER:
JOB NUMBER FOR TRACKING:

BAR CODING USED: LABELS SPECIFIED: TYPE:
INFORMATION REQUIRED:

MOLD EXISTING: STATUS:
REPAIR REQ'D.: WHAT:
BY WHOM: WHO PAYS:
MOLD TRIAL DATE: WHERE:
BY WHOM: MATERIAL:
ACCEPT/REJECT MOLD: REASON:
CAN MOLD BE MODIFIED TO MAKE PARTS: BY WHOM:
WHEN:
MODIFICATIONS REQUIRED:
MOLD ACCEPTED BY PRODUCTION: BY WHOM: DATE:

NEW MOLD: WHO DESIGNS:
WHO PAYS: HOW:
START: FINISH:
SCHEDULE DETERMINED FOR MANUFACTURE: WHEN:
MOLD TRIAL DATE: WHERE: BY WHOM:
MATERIAL:
RESULTS:
MODIFICATIONS REQUIRED:
WHEN COMPLETED:
MOLD ACCEPTED BY PRODUCTION: BY WHOM: DATE:

PURCHASING:
RESIN:
PO/DATE: DUE IN: PACKAGE: LBS:
STORED WHERE:
BAR CODED INTERNALLY:

FINISHED PARTS: WHAT:
PO/DATE: DUE IN: INSPECTION REQ'D.:
WHAT:
PROCEDURE NO.:
STORED WHERE:
SPECIAL STORAGE REQUIRED:
WHAT:

SPECIAL EQUIPMENT: WHAT:
PO/DATE: DUE IN: INSPECTION REQ'D.:
WHAT:
PROCEDURE NO.:

MANUFACTURING:
START DATE: FINISH DATE:
QUANTITY TO PRODUCE: TO ORDER: MONTHLY:
QUARTERLY:

MOLD NUMBER:
MACHINE NUMBER:
PROCEDURE DOCUMENTED: DOCUMENT NUMBER:

NEW MOLD SETUP: DOCUMENT DATA REQUIRED:
BY WHOM:
COMPLETED: DOCUMENT NUMBER:

AUXILIARIES:
DRYER: CHILLER: FEEDER:
CONVEYOR: PART SEPARATOR: ROBOT:
GRINDER: WEIGH SCALE: PACKER:
BLENDER: OTHER:

ASSEMBLY:
START DATE: FINISH DATE:
SPECIAL EQM'T.: WHAT:
PO/DATE: DUE IN:
SPECIAL PARTS:
WHAT:
PO/DATE: DUE IN: QUANTITY: INSPECTION REQ'D.:

DECORATION:
START DATE: FINISH DATE:
SPECIAL EQM'T.: WHAT:
PO/DATE: DUE IN:
SPECIAL PARTS:
WHAT:
PO/DATE: DUE IN: QUANTITY: INSPECTION REQ'D.:

PACKING:
START DATE: FINISH DATE:
SPECIAL PACKAGING REQ'D.:
WHAT:
PO/DATE: DUE IN: QUANTITY:

PART/MATERIAL TESTING REQUIREMENTS:
WHAT:
WHERE:
WHEN:
BY WHOM:
TEST PROCEDUE:
CUSTOMER TO VERIFY:
CERTIFICATIONS REQUIRED:

WHAT:
INCOMING:
MATERIAL: TEST:
MATERIAL: TEST:

IN PROCESS:
PART: TEST:
PART: TEST:

ASSEMBLY:
PART: TEST:
PART: TEST:

FINAL:
TEST: TYPE: INSPECTION:

SPECIAL EQM'T. REQUIRED:
WHAT:
CUSTOMER SUPPLIED/PURCHASED:
PO/DATE: DUE IN:

PRODUCT CERTIFICATION REQ'D.:
WHAT:
DUE TO CUSTOMER: HOW: WHEN:
TO WHOM:

INVOICING: DATE:
INVOICE NUMBER:
AMOUNT INVOICED:
QUANTITY SHIPPED:
QUANTITY ORDERED:
OVER/UNDER % ALLOWED: PERCENT:
DISCOUNTS:
TERMS:

REORDER ANTICIPATED:
WHEN:
QUANTITY:
PRICE:

CHECKLIST 9

PRICE ESTIMATING CHECKLIST

CUSTOMER: DATE:
ADDRESS:
CONTACT: PHONE: FAX: E-MAIL:

PART NAME: JOB NUMBER:
DRAWING NO.:

PIECE PART COST ESTIMATING PER 1000 PARTS

A. MATERIAL	:	:	:	:
B. RESIN COST ($/LB)	:	:	:	:
C. SPECIFIC GRAVITY (Sg)	:	:	:	:
D. PART WEIGHT (lbs)	:	:	:	:
E. PART WEIGHT (D × 1000)	:	:	:	:
F. MATERIAL COST (B × E)/0.95	:	:	:	:
G. CYCLE TIME (CT)	:	:	:	:
H. NUMBER OF CAVITIES (NC)[a]	:	:	:	:
I. PARTS/HOUR (H/G × 3600)	:	:	:	:

J. CAVITY AREA (PROJECTED)[b]	:	:	:	:
K. CLAMP FORCE[c] (CF) TONS × (J × MATERIAL FACTOR)	:	:	:	:
L. SHOT WEIGHT(oz) (D × H × W[d] × 16 oz/lb)	:	:	:	:
M. MACHINE HOUR COST (RATE × (MC[e])	:	:	:	:
N. PROCESSING COST ($/1000 PARTS) M/I × 1000	:	:	:	:
O. ADJUSTED PROCESSING COSTS[f] [N/(0.95)(0.80)]	:	:	:	:
TOTAL COST (PROCESSING PER 1000 PARTS)	:	:	:	:

[a] Assumed three shifts/day, 6 days/week (*f), one years production produced
[b] Projected cavity area and runner/sprue, mold cavity in square inches x number of cavities, plus runner and sprue area of mold surface in square inches
[c] 80% to 20% maximum shot weight of resin, use material clamp factor to estimate tons of clamp required
[d] Use reference chart for shot weight Figure A.
[e] Use machine hour rate chart Figure B, adjust for current machine rates
[f] Assumes 95% yield and 80% utility of molding process

CHECKLIST 10

PURCHASING CHECKLIST

DATE:

CUSTOMER:
ADDRESS:
CONTACT: PHONE: FAX: E-MAIL:

CONTRACT NUMBER:
JOB NUMBER:
PRODUCTION START DATE:
JOB SCHEDULE COMPLETED:
APPROVED BY: DATE:

PURCHASING REQUIREMENTS:
BUYER: PHONE:
BUYER: PHONE:

PROTOTYPES: TYPE:
SUPPLIER:
CONTACT: PHONE: FAX: E-MAIL:
PURCHASE ORDER REQ'D.: P.O. NO.: DATE:
DUE BY: RECEIVING CONTACT:

NOTIFY DEPARTMENT MANAGER(s) WHEN SPECIFIC MATERIALS ARE RECEIVED:
PURCHASING: PHONE:
PRODUCTION: PHONE:
ENGINEERING: PHONE:

ASSEMBLY: PHONE:
DECORATION: PHONE:
QUALITY: PHONE:
PACKAGING: PHONE:

MATERIAL & FINISHED GOODS AND PARTS:
PRIME MATERIAL SUPPLIER:
GRADE:
CERTIFICATION REQUIRED: WHAT: SPECIFIC LOT DATA:
REQUIRED PRIOR TO OR WITH RECEIVING DOCUMENTS:
TEST RESULTS REQUIRED: WHAT: SEND TO:
MSDS REQUESTED WITH ORDER:
QUANTITY ORDERED: POUNDS: PACKAGE TYPE:
PURCHASE ORDER NUMBER: ORDERED DATE:
CONTACT: PHONE:
DUE TO ARRIVE ON OR BEFORE:
SHIPPER PRO NUMBER REQUIRED: WHAT IS NUMBER:
TRUCKING COMPANY:
RECEIVING NOTIFY ON RECEIPT:
PURCHASING: PHONE:
PRODUCTION: PHONE:
QUALITY: PHONE:

INCOMING TESTING REQUIRED:
QUALITY CONTACT: PHONE:
PRODUCTION: PHONE:

TESTING RESULTS:
QUALITY APPROVED BY: DATE:
INVENTORY PLACEMENT:
WHERE: BAR CODED TO INVENTORY: HEATED AREA REQ'D.:
TEMP. REQ'D.:

IF REJECTED, REASON: BY WHOM: DATE:
SUPPLIER QUALITY CONTACT: PHONE: E-MAIL:
DISPOSITION:
SEGREGATED FROM CURRENT INVENTORY: WHERE:
BY WHOM:
SPECIAL REJECTION LABEL ON PACKAGING:

REORDER REQUIRED:
PURCHASING NOTIFIED: WHEN: BY WHOM:
NEW P.O. NUMBER:
ABLE TO MEET PRODUCTION START DATE:
PRODUCTION NOTIFIED: WHO: WHEN:

SCHEDULING REQUIRED TO BE ADJUSTED:
SALES NOTIFIED: WHO: WHEN:
CUSTOMER CONTACTED BY SALES: WHEN:
COMMENTS:

TOOLING:
SUPPLIER:
CONTACT: PHONE: FAX: E-MAIL:
CONTRACT/P.O. NO.: DATE ENTERED:
DUE BY DATE: RECEIVED ON DATE:
PRODUCTION NOTIFIED: WHOM: DATE:
TOOLING NOTIFIED: WHOM: DATE:
QUALITY: WHOM: DATE:
ENGINEERING: WHOM: DATE:

PURCHASED PARTS FOR MOLDING PRODUCT:
PARTS:
PURCHASE ORDER NUMBER: ENTERED:
QUANTITY ORDERED:
DUE IN BY:
CERTIFICATION REQUIRED:
WHAT:
SUPPLIER:
ON APPROVAL LIST: NEEDS APPROVAL: WHO/WHEN APPROVED:
CONTACT: PHONE: FAX: E-MAIL:
INCOMING TESTING REQUIRED:
WHAT:
QUALITY CONTACT: ALTERNATE:
ACCEPTED/REJECTED: BY WHOM: DATE:
REASON:
REORDER: P.O. NUMBER: REQUIRED BY:
DUE IN:

PURCHASED PARTS FOR ASSEMBLY:
PARTS:
QUANTITY ORDERED:
PURCHASE ORDER NUMBER: DATE:
CERTIFICATION REQUIRED:
WHAT:
SUPPLIER:
ON APPROVAL LIST: NEEDS APPROVAL: WHO/WHEN APPROVED:
CONTACT: PHONE: FAX: E-MAIL:

INCOMING TESTING REQUIRED:
WHAT:
QUALITY CONTACT: ALTERNATE:
ACCEPTED/REJECTED? DATE:
REASON:
REORDER: P.O. NUMBER: REQUIRED BY:
DUE BY:

PARTS:
QUANTITY ORDERED: DATE:
PURCHASE ORDER NUMBER:
CERTIFICATION REQUIRED:
WHAT:
SUPPLIER:
ON APPROVAL LIST: NEEDS APPROVAL: WHO/WHEN APPROVED:
CONTACT: PHONE: FAX: E-MAIL:

INCOMING TESTING REQUIRED:
WHAT:
QUALITY CONTACT:
ACCEPTED/REJECTED: DATE:
REASON:
REORDER: P.O. NUMBER REQUIRED BY:
DUE BY:

PURCHASED PARTS FOR DECORATION:
PARTS:
QUANTITY ORDERED:
PURCHASE ORDER: DATE:

PACKAGING:
PURCHASE ORDER:
QUANTITY:
SPECIAL REQUIREMENTS:
REQUIRED BY:

RECEIVING TO NOTIFY:
PURCHASING: PHONE:
PRODUCTION: PHONE:

SPECIAL MANUFACTURING EQUIPMENT REQUIRED:
WHAT:
SUPPLIER:
CONTACT: PHONE: FAX: E-MAIL:

P.O. NUMBER:	ORDER DATE:	DUE DATE:
NOTIFY:		
PURCHASING:	PHONE:	
PRODUCTION:	PHONE:	

CHECKLIST 11

QUALITY CHECKLIST

DATE:

CUSTOMER:
ADDRESS:
CONTACT: PHONE: FAX: E-MAIL:

PART NAME: JOB NUMBER:
MANUFACTURING START DATE:
PRODUCTION SUPERVISOR:
MATERIAL: PRODUCT CODE:
SUPPLIER:

QUALITY PROCEDURES PER ISO9000/QS9000/OTHER
QUALITY INSPECTOR:

CUSTOMER QUALITY REQUIREMENTS KNOWN:
DOCUMENT: REVISION:
ENGINEERING CHANGE ORDERS RECEIVED: WHAT:
INCORPORATED INTO PRODUCTION: WHEN: BY WHOM:
ANY DEVIATIONS ALLOWED: WHAT:
WHO APPROVED AT CUSTOMER: WHEN: TITLE:

PART REQUIREMENTS:
PHYSICAL: DOCUMENT: DRAWING:
CHEMICAL: DOCUMENT:
ELECTRICAL: DOCUMENT:
AGENCY REQUIREMENTS: WHAT:

CODE REQUIREMENTS: WHAT:

PART DESIGN DOCUMENTED: DRAWING NO.:
MATERIAL DOCUMENTED: CERTIFICATION RECEIVED:
RESULTS:
INCOMING INSP/TEST RESULTS:
CONFIRMED BY: DEPT.: TITLE:

REVIEW OF PROCEDURES BY: ALL CURRENT:
MANUFACTURING: REVIEWED BY: DATE:
DECORATING: REVIEWED BY: DATE:
ASSEMBLY: REVIEWED BY: DATE:
FINAL TESTING: REVIEWED BY: DATE:
PACKAGING: REVIEWED BY: DATE:
SHIPPING: REVIEWED BY: DATE:

MATERIAL SAFETY DATA SHEETS AVAILABLE & CURRENT:
MANUFACTURING EQUIPMENT MAINTENANCE CURRENT:
TOOLING MAINTENANCE CURRENT:
AUXILIARY EQUIPMENT MAINTENANCE CURRENT:

PROCESS CONTROL LIMITS ESTABLISHED: BY WHOM:
PART QUALITY LIMITS DOCUMENTED: BY WHOM:
DOCUMENT:
TEST & INSPECTION EQUIPMENT DOCUMENTED: BY WHOM:
PROCEDURE:
STATISTICAL PROCESS CONTROL DATA REVIEWED:
PROCESS CONTROL:
DOCUMENTED FOR RECORDS:

QFD, ANALYSIS COMPLETED WITH CUSTOMER: DATE:
FMEA, ANALYSIS COMPLETED: DATE:
FISH BONE ANALYSIS COMPLETED: DATE:
MEASUREMENT TOOL ANALYSIS: DATE:
METRIC REQUIREMENTS COMPLETED: DATE:
SPC, REQUIREMENTS ESTABLISHED: DATE:
SIX SIGMA ANALYSIS COMPLETED: DATE:
ALL MEASUREMENT ITEMS IN CERTIFICATION: DATE:

TEST EQUIPMENT AVALIABLE:
WHAT IS REQUIRED:
INSPECTOR: ALTERNATE:

CUSTOMER ON SITE INSPECTION REQUIRED:
DURING MANUFACTURE: FINAL:

CUSTOMER INSPECTOR: ALTERNATE:
PHONE: FAX: E-MAIL:
SAME INSPECTION EQUIPMENT USED BY CUSTOMER:
WHAT IF NOT:
WHO SUPPLIES:
AGENCY TESTING REQUIRED: CONTACT:
ADDRESS:
NUMBER OF PARTS TO SEND: SENT:
INFORMATION DUE BACK WHEN:
INFORMATION/FORMS REQUIRED:

WAS "REAL TIME" PROCESS CONTROL USED DURING MANUFACTURE:
COMPUTER OUTPUT SAVED AND FILED:
FILE NAME:

QUALITY RECORDS REVIEWED AND SIGNED OFF FOR SHIPMENT:
BY WHOM: DATE:

CHECKLIST 12

SALES CONTRACT CHECKLIST

DATE:

CUSTOMER:
ADDRESS:
CONTACT: PHONE: FAX: E-MAIL:

APPLICATION:
VOLUME/YEAR: ANTICIPATED PART PRICE:
RELEASE QUANTITY: FREQUENCY:
PART SIZE (SQ. IN.)
DRAWING/SKETCH/PROTOTYPE AVAILABLE: TYPE: DWG. NO.
WHO DESIGNS PRODUCT:

REQUIREMENTS:
PART DEVELOPMENT CHECKLIST USED:
PART IS NEW/EXISTING/REDESIGN:
USERS:
AGENCY/CODE APPROVAL REQ'D.:
SPECIAL SITUATION:
SUPPLIER CERTIFICATION REQ'D.: WHAT:

TYPE OF MANUFACTURE:
ANTICIPATED MATERIAL: TYPE:
SUPPLIER:
IS COMPANY CAPABLE OF SUPPLYING:

PURCHASED PARTS USED: WHAT:
WHO FURNISHES: INVENTORY REQM'TS.:
ASSEMBLY REQ'D.: TYPE: CHECKLIST USED:
DECORATION REQ'D.: TYPE: CHECKLIST USED:
PACKAGING & SHIPPING REQM'TS.:

MOLD/TOOLING:
WHO SUPPLIES:
WHO DESIGNS:
TYPE: NUMBER OF CAVITIES: BALANCED:
SPECIAL IN MOLD REQM'TS.: WHAT:

EXISTING MOLD:
CONDITION:
REASON FOR TRANSFER:
LAST MOLDER:
CONTACT: PHONE: FAX: E-MAIL:
IN-HOUSE TRIAL TO ACCESS CONDITION:
WHEN: WHERE:
MOLD DRAWINGS AVAILABLE: BILL OF MATERIALS:
WHO BUILT CURRENT MOLD:
CONTACT: PHONE: FAX: E-MAIL:
SPECIAL EQUIPMENT REQ'D. TO RUN MOLD: WHAT:
WHO FURNISHES:
PROCESS CONDITION RECORDS AVAILABLE: WHERE;
PARTS AVAILABLE:
MATERIAL USED:
SUPPLIER:
GRADE: AMOUNT ON HAND:

NEW TOOL:
WHO DESIGNS:
WHO OWNS:
ANTICIPATED COST:
CUSTOMER PAYMENT METHOD, DIRECT ON APPROVAL, WITH PARTIAL PAYMENTS:
AMORTIZED OVER PRODUCTION RUN AS PARTS ARE DELIVERED:
WHO APPROVES TOOLING:
WHO APPROVES FIRST PARTS OFF TOOLING:
CONTACT: PHONE: FAX: E-MAIL:

MOLD CHECKLIST USED:
MAINTENANCE REQUIREMENTS:
WHO APPROVES REPAIRS:
WHO PAYS: APPROVAL REQ'D. BEFORE REPAIR:

CONTACT: PHONE: FAX: E-MAIL:

QUALITY REQUIREMENTS:
QUALITY CHECKLIST USED:
INCOMING MATERIAL TESTS REQ'D.:
PURCHASED PARTS TESTS REQ'D.:
REQUIREMENTS SPECIFIED:

PROTOTYPE TESTING REQUIRED:
TYPE, MODEL/MOLDED PART/SLA/SLS MODEL, OTHER:
WHO FURNISHES:
TIME REQUIREMENTS TO PROVIDE:
COST ANTICIPATED:
IN-PROCESS TESTING REQ'D.: WHAT:
REQUIREMENTS:
EQUIPMENT REQUIRED:
WHO FURNISHES: WHAT:

END USE TESTING REQUIRED:
WHAT:
WHO PERFORMS:
REQUIREMENTS:
TEST EQUIPMENT REQUIRED: WHO FURNISHES:

TESTING DOCUMENTATION (SPC) REQUIRED:
INCOMING:
PRODUCTION:
ASSEMBLY:
DECORATION:
SPECIAL DOCUMENTATION REQ'D.:
WHAT:
CUSTOMER REQUIRED DOCUMENTATION PRIOR TO OR AT TIME OF SHIPMENT:
WHAT:
PROBLEM RESOLUTION:
CONTACT: PHONE: FAX: E-MAIL:

CUSTOMER TESTS AT INCOMING: WHAT TESTING:
PROCEDURES DOCUMENTED: WHAT:
WHO FURNISHES:
CONTACT: PHONE: FAX: E-MAIL:

PRODUCTION:
FIRST ARTICLE WHO APPROVES:
REQUIREMENTS (QUALITY CHECK LIST OR OTHER USED):

FORM/FIT/FUNCTION:
AESTHETICS:
DIMENSIONAL:
COLOR APPROVAL:
SPECIAL SPECIFICATIONS:

ANTICIPATED RELEASE QUANTITIES PER ORDER:
JUST-IN-TIME PRODUCTION REQ'D.: FREQUENCY:
SHIPMENT DISTANCE:
INVENTORY REQUIREMENTS: WHO PAYS:
PAYMENT TERMS/METHOD:
DIRECT: AMORTIZED: OTHER:

CONTRACT TERMS SPECIFIED:
RELEASE ON PURCHASE ORDERS: TIMED RELEASES: WHO SPECIFIES:
CUSTOMER APPROVAL TO SHIP: WHO:
SHIPPER SPECIFIED:
SHIPPING PAID BY: TERMS:
ON RELEASE:
CONTRACT TERMS USED: WHAT TERMS:

CUSTOMER QUOTE TO:
ADDRESS:
QUOTE DUE DATE: TIME:
QUOTE DELIVERED BY: HOW:

QUOTE ITEMS REQUIRED PIECE PART PRICE:
SCHEDULING:
MATERIAL:
TOOLING:
ASSEMBLY:
DECORATION:
PACKING:

SPECIAL TESTING:
SPECIAL REQM'TS.:
WHAT:

QUOTE NUMBER:
QUOTED BY:
APPROVED BY:
SALES CONTACT: PHONE: FAX: E-MAIL:
SUBMITTED DATE:
CUSTOMER RESPONSE DATE:

ACCEPTED/REJECTED BY: WHY:
ANY SPECIAL TERMS REQUIRED:
WHAT:
REQUOTE ALLOWED IF REJECTED:
TIME LIMITS:
DUE BY DATE:

CONTRACT SUBMITTED DATE:
CUSTOMER APPROVED DATE:
CUSTOMER OFFICER APPROVAL:
CONTRACT SUPPLIER APPROVED BY: DATE:

CHECKLIST 13

WARRANTY PROBLEM CHECKLIST

DATE:

CUSTOMER:
ADDRESS:
CONTACT: PHONE: FAX: E-MAIL:
INDUSTRY/MARKET OF USE:

PROBLEM:

PROBLEM REPORTED BY:
REPORT OF PROBLEM SUBMITTED: PROBLEM.:

OCCURRED AT DEVELOPMENT: PROTOTYPE: FINAL DESIGN:
PRODUCTION:
ASSEMBLY: DECORATION: END USE: OTHER:

FAILURE DEFINED AS:

DESIGN-MATERIAL-PURCHASED PARTS-ASSEMBLY-DECORATION-PACKAGING:
SHIPPING-OTHER AREA, DESCRIBE IN DETAIL:
FAILURE OCCURRENCE–ONCE: SEVERAL TIMES: REPEATABLE:
NO. TIMES:
SAME POINT OR AREA ON PART: VARIABLE: WHERE:
SKETCH SHOWING LOCATIONS:

SAMPLE OF FAILED PARTS AVAILABLE:
SENT TO: WHEN: BY WHOM:

FAILURE OCCURRED AT:
MANUFACTURE: ASSEMBLY: WAREHOUSE: SHIPPING:
END USE:

OCCURRED IN WINTER: SUMMER: SPRING: FALL:
TROPICAL: DRY AREA: OTHER CONDITIONS, DESCRIBE:
SECTION OF COUNTRY:

SERIOUSNESS:
LIABILITY INVOLVED: WHAT EXTENT:
STATUS OF FAILURE: KNOWN: MUST BE INVESTIGATED:

MOLDED PART FAILURE ANALYSIS:
MATERIAL: SUPPLIER: PRODUCT NO.: LOT NO.:
CERTIFIED BY SUPPLIER: WHAT CERTIFICATIONS:
INCOMING TEST RECORD: DATE TESTED: TEST RESULTS:
REGRIND USED: PERCENTAGE: NUMBER OF PASSES
ALLOWED:

CHEMICAL DATA AVAILABLE:
PHYSICAL DATA AVAILABLE:
SAMPLE OF PART RETAINED:
ENGINEERING CHANGE ORDER: NUMBER: DATE:
APPROVED BY:
CUSTOMER APPROVAL REQUIRED: WHOM:
GRANTED BY: ON DATE:
ALL COMPANY DEPARTMENTS NOTIFIED OF CHANGE ORDER AND
THEIR SIGNATURE ON APPROVAL SIGN-OFF SHEET:
CONFIRMED BY: TITLE: DATE:

INCORPORATED INTO PART: WHEN:

AGENCY/CODE APPROVAL GRANTED: BY WHOM: DATE:

PART REQUIRED AGENCY/CODE CERTIFICATION:
WHAT:
CONTACT: PHONE: FAX: E-MAIL:

COLORED MATERIAL: BLENDED WHERE: CONCENTRATE:
SUPPLIER:
LOT/P.O. NUMBER:

CONTACT: PHONE: FAX: E-MAIL:
ANY CHANGES IN PIGMENT SYSTEM INGREDIENTS DURING MANUFACTURE:
WHAT:
LOT SAMPLES AVAILABLE: TEST RESULTS FROM SUPPLIER:

MOLDED PART: PART NUMBER: DATE MFG'D.:
PURCHASED PART: PART NUMBER: DATE MFG'D.:
SUPPLIER:
MOLD NUMBER:
CAVITY NUMBER: CONSISTENT WITH FAILED PARTS:
MOLD NUMBER: DRAWING OF MOLD AVAILABLE:
SUPPLIER INCOMING INSPECTION RECORD: DATE: TEST NO.:

PART FAILURE ANALYSIS:

MECHANICAL FAILURE:
FAILURE TYPE:
DESCRIBE TYPE OF FAILURE AND IF DURING USE:
CUSTOMER:
SEVERITY:
REPAIRABLE:
HOW:

ELECTRICAL FAILURE:
USED AS:
FAILURE TYPE:
CUSTOMER:
FREQUENCY OF OCCURRENCE:
SEVERITY:
REPAIRABLE:
HOW:

QUALITY ASSURANCE:
ANY TESTING SHOWED PROBLEMS:
METHOD OF TESTING BASED ON FAILURE:
ANY REPORTS OF PRIOR FAILURES OF THIS TYPE:
CUSTOMER REACTION:
SEVERITY:
MATERIAL PROBLEM:
MOLDING PROBLEM:
END USE APPLICATION TOO SEVERE FOR PART/MATERIAL:
ANALYSIS OF FAILURE:

MANUFACTURED IN HOUSE: OUTSIDE MOLDER: WHO:
CONTACT: PHONE: FAX: E-MAIL:
PRODUCTION DATE: LOT NO.: SHIFT:
PROCEDURE FOR MANUFACTURE FOLLOWED:
PROBLEMS NOTED DURING PRODUCTION: WHAT:
PROCESS CONTROL RECORDS REVIEWED: IN REAL TIME:
BY WHOM:
MOLDING PRESS NO.:
MOLD NUMBER:
MAINTENANCE PERFORMED LAST:
MAINTENANCE RECORDS AVAILABLE:
SPC PROCESS DATA AVAILABLE:

PART ANALYSIS/TEST RESULTS:
VISUAL INSPECTION:
ANALYTICAL RESULTS:
PHYSICAL:
CHEMICAL:
DSC:
TGA:
IR:
OTHER:
MATERIAL SUPPLIER ANALYSIS/INPUT:

SOLUTION TO PROBLEM:

CORRECTIVE ACTION RESPONSE ASSIGNED TO:
DATE:
TIME ESTIMATED TO RESOLVE:
ESTIMATED COST TO COMPANY:

CHECKLIST 14

ENGINEERING CHANGE REQUEST CHECK SHEET

ECR NO. ____________

1. DRAWING NO. ____________________________ ECN NO. ____________

   TITLE: ____________________________

REQ. BY. ____________

DATE: ____________

ENG. SER MGE ______

2. DESCRIPTION OF CHANGE (DETAILED): ____________________________

3. INFORMED OR NEED APPROVAL OF CHANGE REQUEST (I = INFORMED A = APPROVAL

(I) (A)

_ _ MFG ENG MGR _ _ I.E. MGR _ _ Q.C. MGR _ _ PROD___

_ _ MFG MGR _ _ MKT MGR _ _ CONTROLLER _ _ CONT MGR_

4. ENG. ASSIGNED__________ PROD. PLANNER________ EST. DATE OF RELEASE_____

5. ITEMS TO BE INVESTIGATED (INCLUDE W.O., P.O., & Q.C. TEST NOS. USED FOR INVESTIGATIONS

PRELIMINARY ECN REQUIRED_ _ _ _ YES_ _ NO_ _

6. APPROVALE FOR ECN RELEASE:

PUR MGR ____

MFG ENG MGR ____ I.E. MGR ____ Q.C. MGR____

MFG MGR ____ MKT MGR ____ CONTROLER ____

COST IMPACT

STD $/UNIT

MATERIAL ______

LABOR ______

BURDEN ______

TOTAL ______

7. NOTE:

ANNUALIZED

X VOLUME ______

TOOLING ______

SCRAP/

OBSOLETE ______

TOTAL COST______

CHECKLIST 15

SUPPLIER QUALITY SELF-SURVEY REPORT

Supplier Name: ______________________ Audit Date: __________
Address: ______________________
______________________ Auditor(s): ______________
Phone: ______________________ ______________
Fax: ______________________ ______________
Product Type: ______________________ ______________
Contact: ______________________ ______________

1. **Organization** **YES** **NO** **N/A**

1.1 Quality Manager: ______________________
Reports to: ______________________
Title: ______________________
1.2 Total number of employees: ___________
1.3 Total number of Quality Personnel: ___________
1.4 Is there a Quality Manual? (attach copy)
1.5 Quality System conforms to the following specification: _____ _____ _____
____
____
1.6 Years in business: __________ Average Annual Sales: __________

2. **Operation**

2.1 Current Contracts: ______________________
2.2 Major Customers: ______________________
2.3 __________%MIL __________ % Commercial

2.4 Facilities: List (sq. ft.)______________________

____

2.5 Equipment List (attach) _____ _____ _____
Tool List (attach) _____ _____ _____

3. **Procurement/P.O. Review**

3.1 Does Quality review P.O. requirements? _____ _____ _____
3.2 Is there a system to inform customers of order status? _____ _____ _____

4. **Technical Document Control**

4.1 Do the drawings and other technical documents used for production and inspection purposes reflect the revision on the P.O.? _____ _____ _____
4.2 Are obsolete drawings promptly removed from all points of issue or use? _____ _____ _____

5. **Purchasing**

5.1 Are documented procedures utilized for the purchase of items which ensure they meet the specified requirements? _____ _____ _____
5.2 Are suppliers qualified via a documented evaluation procedure and qualification criteria? _____ _____ _____
5.3 Is a qualified supplier list maintained? _____ _____ _____

6. **Customer Supplied Items**

6.1 Are there adequate controls for customer supplied items which minimize its loss, damage, or misuse? _____ _____ _____
6.2 Are customer supplied items inspected to ensure their suitability for use? _____ _____ _____

7. **Process Control**

7.1 Are the appropriate procedures available at locations where operations essential to the effective functioning of the quality system are performed? _____ _____ _____
7.2 Are items provided traceability when required? _____ _____ _____
7.3 Are documented work orders/procedures in use? _____ _____ _____
7.4 Are detailed procedures, training, and personnel qualifications available for special processes? _____ _____ _____
7.5 List general workmanship standards used:

____

____

____

____

7.6 Is good housekeeping control in place? _____ _____ _____

8. **Inspection and Testing**

8.1 Are adequate inspection instructions or criteria available to material, in-process, and final inspection personnel? _____ _____ _____
8.2 Are incoming materials adequately inspected and results documented? _____ _____ _____
8.3 Are accepted, rejected, and materials awaiting inspection adequately identified and segregated? _____ _____ _____
8.4 Are in-process and final inspections and tests performed by adequately trained/qualified personnel? _____ _____ _____
8.5 Are required in-process and final inspections and tests adequately identified including their acceptance criteria? _____ _____ _____
8.6 Are results of in-process and final inspections adequately documented? _____ _____ _____
8.7 Are products failing inspections and tests identified and segregated? _____ _____ _____

9. **Inspection, Measuring, and Test Equipment**

9.1 Does the supplier control, calibrate, and maintain gauges and measuring equipment to demonstrate conformance to the specified requirements? _____ _____ _____
9.2 List Standards:

____

____

____

10. **Control of Nonconforming Product**

10.1 Does the supplier maintain procedures to ensure that items which do not conform to specified requirements are prevented from inadvertent release to the customer? _____ _____ _____
10.2 Is responsibility for review, documentation, and authority for disposition of nonconforming items defined? _____ _____ _____

11. **Corrective Action**

11.1 Does the supplier document and maintain procedures for investigating causes of nonconforming items and the corrective action needed to prevent recurrence? _____ _____ _____

12. **Quality Records**

12.1 Does the supplier maintain pertinent documented quality records? _____ _____ _____

12.2 Are supplier quality records maintained? _____ _____ _____

12.3 Are quality records stored such as to minimize loss or deterioration and are records readily retrievable? _____ _____ _____

13. Training

13.1 Does the supplier maintain procedures for identifying training needs and provide the training to applicable personnel? _____ _____ _____

14. Handling, Storage, Packaging, Preservation, and Delivery

14.1 Are written procedures in use to control the quality of items during handling, storage, and delivery? _____ _____ _____

CHECKLIST 16

SUPPLIER SURVEY REPORT

TYPE OF SURVEY: DATE OF SURVEY:________

1.__ PRE-SURVEY

2.__ INITIAL

3.__ FOLLOW-UP

PRODUCT SERVICE: ____

COMPANY NAME: ____

ADDRESS: ____

CITY, STATE & ZIP: ____

TELEPHONE NO.: (   ) ______________________ FAX NO.: ________

APPROXIMATE NUMBER OF PEOPLE IN WORKFORCE:

MANUFACTURING:_______ ENGINEERING:_______ QUALITY:_______

OTHER:_______ TOTAL NUMBER OF EMPLOYEES:________

UNION?: _________ NAME: ____________________________

LENGTH OF CONTRACT:______ EXPIRATION DATE:_______________

PRODUCTS/SERVICES PROVIDED TO: (INCLUDE APPLICABLE PART NUMBERS AND ACTIVE PURCHASE ORDER NUMBER) ATTACH ADDITIONAL PAGES AS REQUIRED.

____

____

____

PERSONNEL CONTACTED:____________ TITLE OR FUNCTION:________

____

____

____

____

DISTANCE TO PLANT MILES/TIME:______ PLANT SQUARE FEET:________
GENERAL CONDITION OF FACILITIES/EQUIPMENT:____________________

____

____

HOUSEKEEPING OBSERVATIONS:____

____

____

AUDIT RATING:_____ CLASSIFICATION:__ ACCEPTABLE
__CONDITIONAL
__MARGINAL
__UNACCEPTABLE

AUDIT PERFORMED BY:_________________

____

____

____

SUMMARY OF SURVEY

COMPANY:________________ DATE OF SURVEY:__________ TYPE:____________

SUPPLIER SYSTEM ELEMENTS:	RATING	
	POTENTIAL	ACTUAL
1. ORGANIZATION AND MANAGEMENT POLICIES	100	
2. SPECIFICATION REVIEW AND DESIGN ASSURANCE	70	
3. MANUFACTURING PLANNING AND CONTROLS	320	
4. STATISTICAL PROCESS CONTROL	50	
5. MEASUREMENT AND TEST EQUIPMENT	60	
6. FIRST-ARTICLE INSPECTION	30	
7. CONSIGNED MATERIAL	30	
8. HANDLING, PRESERVATION, PACKAGING AND SHIPMENT	40	
9. SUPPLIER PURCHASED MATERIAL CONTROL	90	
10. PERSONNEL TRAINING, CERTIFICATION, EDUCATION AND MOTIVATION	40	
11. RECORDS AND CHANGE CONTROL	55	
12. CONTROL OF NONCONFORMANCE	50	
13. COSTS OF QUALITY	20	
14. CORRECTIVE ACTION AND RECURRENCE PREVENTION	90	
15. QUALITY SYSTEM AUDIT	70	
TOTAL:	815	

SUPPLIER AUDIT RATING (ACTUAL/POTENTIAL × 100) ________ %

SUPPLIER SYSTEM EVALUATION

EVALUATION CRITERIA:

1. Organization and Management Policies	Potential	Audit
a. Is there a documented and approved company quality policy?	10	
b. Are functional responsibilities for quality defined?	4	
b.1 Is there a company organization chart showing the relationship of the quality organization to management and other departments?	3	
b.2 Does the quality organization have the independent reporting authority required to be effective?	3	
c. Is the quality function adequately staffed to maintain effective control and assurance?	10	
d. Is the focus of the quality system "prevention" versus to "detection" oriented?	10	
e. Is the machine operator held accountable for rejects?	10	
f. Is the quality system documented in the form of a Quality Manual?	20	
g. Are detailed procedures and work instructions available for use by quality personnel?	20	
h. Is there a system for the review, approval, control, and maintenance of procedures and work instructions?	10	
TOTAL:	100	

Comments: ____

2. Specification Review and Design Assurance	Potential	Audit
a. Are contracts and purchase orders reviewed for quality requirements and manufacturability?	20	
b. Are formal design reviews held?	5	
b.1 Are unique part requirements identified?	5	
b.2 Are the critical tolerances and characteristics of product designated?	5	
c. If applicable, are reliability prediction and failure mode and effect analysis performed for new products?	5	
c.1 Is action taken to minimize probability and effect of failure?	5	
d. If applicable, is a safety analysis performed for all new products and/or ASE code and agency approval required?	5	
d.1 Do procedures exist for eliminating the probability of safety related failures?	5	
e. If applicable, are test procedures prepared for qualification tests of pre-production, engineering and production of first articles?	5	
e.1 Are qualification tests witnessed and verified by quality control personnel?	5	
e.2 Are records of qualification tests maintained including date and results of tests?	5	
TOTAL:	70	

Comments: ______________________________________

____

____

3. Manufacturing Planning and Controls	Potential points	Audit points
a. Are routing sheets, operation sheets, and work instructions utilized and checked for compatibility with drawing requirements?	20	
b. Are special workmanship requirements designated on the applicable work instructions?	10	
c. Are process capability studies performed for all new products?	20	
d. Are the instruments used to measure product conformance of adequate precision?	20	
e. Are processes analyzed for trends and possible future corrective action?	10	
f. Does equipment have built-in process control correction for process drift? (or are alternate methods and equipment used to perform the same function?)	10	
g. Are traceability procedures in effect that identify sources of raw material or component parts?	10	
h. Do environmental controls give consideration to temperature, humidity, vibration and other controllable factors affecting product quality?	10	
i. Does supplier furnish any special processes (e.g., heat treating, plating, welding, etc.) which require certified personnel?	10	
j. When required, are certified personnel adequately trained?		
k. Are there any special processes that require periodic recertification of equipment?	10	
l. Are production flow charts and control plans developed for all production parts?	5	
l.1 Have adequate inspection stations been established throughout the manufacturing process?	5	
m. Do "route sheets" (e.g. shop travelers or move tickets) accompany parts through the manufacturing process?	10	
n. Do production workers sign off route sheets from operation to operation?	10	
o. Are inspection stations identified on the route sheets?	10	
o.1 If so, are inspection operations stamped, indicating product status at all stages of production?	10	
p. Are written instructions for all manufacturing, assembly, inspection and test operations available at the work stations?	10	

3. Manufacturing Planning and Controls (*Cont.*)	Potential points	Audit points
q. Are the instructions clear and easily understood by the operators?	10	
r. Does shop documentation enable traceability to the responsible production department, machine and operator?	10	
s. Is first piece inspection performed?	10	
t. Is roving in-process inspection performed?	10	
u. Is final acceptance inspection performed?	10	
v. Are all inspections performed to written inspection procedures?	10	
w. Do inspection plans include acceptance criteria?	10	
x. Are results of all inspections recorded?	10	
y. Are inspection records used for trend analysis and corrective action?	10	
z. Are current drawings, specifications and purchase orders available to and used by inspection?	10	
aa. Is adequate test and inspection equipment available when needed?	10	
ab. Are the test and inspection facilities adequate?	2	
ac. Are visual aids used to define workmanship standards for manufacturing and inspection personnel?	4	
ac.1 Are they available at the work stations?	4	
TOTAL:	320	

Comments: ____

____

____

4. Statistical Process Control	Potential points	Audit points
a. Will statistical process control (SPC) methods be used for ongoing control of the process in "real time"?	10	
b. Have supplier personnel been adequately trained in SPC methods?	10	
c. Are control charts being used for process control?	10	
d. If control charts are used, is corrective action taken when the process shows lack of control?	10	
e. Are valid acceptance sampling plans specified and properly used in inspection and production?	10	
TOTAL:	50	

Comments: ____________________________________

____

____

5. Measurement and Test Equipment	Potential points	Audit points
a. Does the supplier have a written system for calibration or measuring and test equipment?	10	
b. Do written procedures exist for recall and maintenance of measuring and test equipment?	10	
c. Are employee-owned tools and gages subject to the same controls as those owned by the company?	5	
d. Does the supplier have detailed written procedures for each calibration?	5	
e. Are adequate records of calibration kept on file? in inspection and production?	5	
f. Where possible, are labels affixed to measuring and test equipment indicating: date of last calibration, next due date, and by whom calibrated?	5	
g. Is calibration performed under adequate environmental control?	5	
h. If any production tooling is used for inspection or testing, is it included in the calibration system?	5	

5. Measurement and Test Equipment (*Cont.*)	Potential points	Audit points
i. Are calibrations made against certified higher accuracy standards which have known valid relationships to national standards?	5	
j. Are adequate facilities used for storage of measuring and test equipment?	5	
TOTAL:	60	

Comments: ____________________________________
____
____

6. First-Article Inspection	Potential points	Audit points
a. Is there a procedure to assure that initial pre-production sample are submitted to customer for approval?	10	
b. Is there a written procedure for performing first-article verification to drawings and specifications?	10	
c. Are all required production gages and test equipment available at the time of first-article submissions?	5	
d. Are "control plans" completed by the Supplier prior to first-article submission?	5	
TOTAL:	30	

Comments: ____________________________________
____
____
____

7. Consigned Material

	Potential points	Audit points
a. Does the Supplier perform incoming examination upon receipt of any consigned material?	10	
b. Is consigned material uniquely identified and segregated for storage, control and proper use in production?	5	
c. If tests are required, will personnel be qualified to perform such tests?	5	
d. Are suitable records maintained for the control, inventory, and use of consigned material?	10	
TOTAL:	30	

Comments: ____

____

____

8. Handling, Preservation, Packaging and Shipment

	Potential points	Audit points
a. Are special handling requirements and procedures available to production?	5	
b. Are parts and materials handled correctly to prevent damage?	5	
c. Are materials and parts correctly identified to prevent intermixing?	10	
d. Are inventory materials and parts protected from damage, corrosion, contamination and age limit requirements?	5	
e. Are procedures available for the control of handling, preservation, packaging and shipping to assure conformance to contractual requirements?	5	
f. Are outgoing shipments checked for: verification of acceptance; damage in handling; conformance to customer requirements and inclusion of documentation?	10	
TOTAL:	40	

Comments: ____

____

____

9. Supplier Purchased Material Control

	Potential points	Audit points
a. Does the Supplier have a formal purchasing function?	10	
b. Are written purchasing operation procedures available?	10	
c. Does the Supplier have formally approved procurement sources?	10	
d. Is source approval based on pre-award survey of the Suppliers and quality history?	10	
e. Do Supplier's purchase orders contain applicable quality provisions such as: chemical and physical analysis, certification of test results, special treatment, source inspection, and other quality data or evidence of acceptability?	10	
f. Does the Supplier's quality system provide for the control of procured items prior to release to inventory?	10	
g. Are all such incoming inspections and tests performed against written inspection plans?	10	
h. Does the Supplier perform source surveillance?	5	
i. Does the quality system provide for early information feedback to Supplier?	5	
j. Is corrective action required of the Supplier on nonconforming supplies?	10	
TOTAL:	90	

Comments: ______________________________

____

____

10. Personnel Training, Certification, Education and Motivation

	Potential points	Audit points
a. Does the Supplier have a formal documented education and training program for personnel responsible for the determination of quality?	10	
b. Is there a certification program for persons performing or inspecting special processes such as: welding, decorating, part assembly and nondestructive testing?	10	
c. Are records of proficiency tests maintained?	5	

10. Personnel Training, Certification, Education and Motivation (*Cont.*)

	Potential points	Audit points
d. Are personnel performing the work required to show evidence of periodic certification?	5	
e. Is there an employee participation program for quality or product improvement, such as "Zero Defects" or "Quality Awareness"?	5	
f. Does the Supplier sponsor and promote participation in technical and professional societies, such as SME and ASQO?	5	
TOTAL:	40	

Comments: ______________________________________

____

____

11. Records and Change Control

	Potential points	Audit points
a. Does the Supplier have a formal release system for drawings and specifications?	10	
b. Is there a formal system to assure that latest applicable documentation is available to purchasing, manufacturing, inspection and testing functions?	5	
c. Is a change order system set up to assure that changes required by purchase order revisions are incorporated?	5	
d. Is adequate control of the distribution and replacement of drawings maintained to assure the removal of obsolete information from production use?	10	
e. Is there a configuration management system which records the configuration status of all delivered items?	5	
f. Are adequate records maintained and analysis performed for inspections and test operations?	10	
g. Does management receive and use quality status reports?	10	
TOTAL:	55	

Comments: ______________________________________

____

____

12. Control of Nonconformance

Question	Potential points	Audit points
a. Is there a positive system with written procedures for identification, segregation, and disposition of nonconforming material to prevent inadvertent entry into production?	20	
b. Is there a formal Material Review Board?	5	
c. Are material review actions studied for corrective actions?	5	
d. Are records of material disposition used for trend analysis and corrective action?	10	
e. Are formal methods required for any rework or repair actions?	10	
TOTAL:	50	

Comments: ____

____

____

13. Costs of Quality

Question	Potential points	Audit points
a. Are there procedures for the collection of quality costs?	10	
b. Are detailed quality cost reports analyzed by management?	5	
c. Does evidence exist that quality costs are being effectively collected, analyzed and used for management action?	5	
TOTAL:	20	

Comments: ____

____

____

14. Corrective Action and Recurrence Prevention

	Potential points	Audit points
a. Do written procedures define the Supplier's corrective action system?	10	
b. Is evidence of corrective action documented?	10	
c. Is corrective action extended to second-tier Suppliers?	10	
d. Does the corrective action system include analysis of nonconformance trends?	10	
e. Does the corrective action system include analysis of scrap and rework to determine cause?	10	
f. Is timely corrective action promptly taken on root causes of nonconformances?	10	
g. Does the Supplier's management review the effectiveness of the corrective action system?	10	
h. Is customer data put into the corrective system for formal analysis and follow-up?	10	
i. Is the corrective action system used as a basis for product improvements?	10	
TOTAL:	90	

Comments: ______________________________

____

____

15. Quality System Audit

	Potential points	Audit points
a. Does the Supplier have formal internal quality audit procedures?	20	
b. Do the periodic audits address each of the Supplier System Elements outlined in this requirements document?	10	
c. Are audit findings reported to appropriate levels of management?	10	
d. Are records of audits maintained, reviewed, and used as a basis for follow-up audits?	10	

15. Quality System Audit (*Cont.*)	Potential points	Audit points
e. Are reports on audit deficiencies maintained and used for corrective action?	10	
f. Are audit trends analyzed and used by management as a basis for corrective action?	10	
TOTAL:	70	

Comments: ______________________________________

____

____

CHECKLIST 17

MOLD DESIGN CHECKLIST

DATE________________
CUSTOMER____________________________ CONTACT____________
PART____________ DRAWING NO.________ REV. NO.____________
MOLD NO.____________ CONTRACT NO.________
PHONE NO.____________

A.	PART DRAWING	MOLD HALF		COMMENTS
		(A) FH	(B) MH	
___	Utility of part checked	____	____	____
___	Material selected (Vendor and No.)	____	____	____
___	Shrinkage calculated correct?	____	____	____
___	Gate type and location selected	____	____	____
___	Runner sized for flow distance	____	____	____
___	Matching contours for assembly	____	____	____
___	Tolerances attainable (Major/Minor)	____	____	____
___	Decorating required	____	____	____
___	Inspection equipment noted	____	____	____
___	Cooling fixture required	____	____	____
___	Assembly operations required	____	____	____
___	Draft and part finish noted	____	____	____
___	Quality Level requirement	____	____	____
___		____	____	____
___		____	____	____
___		____	____	____

B. INJECTION MOLDING MACHINE	MOLD HALF		COMMENTS
	(A) FH	(B) MH	
___ Press No.______ oz______ tons________	___	___	___
___ Screw type_______	___	___	___
___ Process setup sheets available	___	___	___
___ Press available	___	___	___
PRESS DATA			
___ Clamp force required ________	___	___	___
___ Injection press (Max) ________	___	___	___
___ Part surface area ________	___	___	___
___ Nozzle type ________	___	___	___
___ Nozzle angle and diameter ________	___	___	___
___ Sprue bushing matches nozzle ________	___	___	___
___ Tiebar "daylight" (opening) ________	___	___	___
___ Melt capacity of press ________	___	___	___
___ Ejection (hydraulic/mechanical)	___	___	___
___ Cycle parameters ________	___	___	___
___ Inject/pack pressures ________	___	___	___
___ Cycle times ________	___	___	___
___ Barrel temperature profile ________	___	___	___
___ Back pressure ________	___	___	___
___ Shot volume ________	___	___	___
___ Pad length ________	___	___	___
___ Suckback required ________	___	___	___
___ Screw RPMs ________	___	___	___
___ Screw recovery time ________	___	___	___
___ External connections ________	___	___	___
___ Hydraulic ________	___	___	___
___ Air________	___	___	___
___ Electrical ________	___	___	___
___ Water ________	___	___	___
___ Locating ring diameter ________	___	___	___
___ Mold clamping ________	___	___	___
___ Core puller and control ________	___	___	___
___ Ejector release mechanism ________	___	___	___
___ Ejector coupling ________	___	___	___
___ Limit switch ________	___	___	___
___ Heating connections ________	___	___	___
___ Regrind allowed______%________	___	___	___
________	___	___	___
________	___	___	___
________	___	___	___
C. AUXILIARY EQUIPMENT			
___ Material drying required	___	___	___
___ Dryer type____Temperature____Time ____	___	___	___
___ Mold temperature control (unit no.)______	___	___	___
___ Cool/heat________	___	___	___

		MOLD HALF		COMMENTS
		(A) FH	(B) MH	
___	Separate circuits in mold___	___	___	___
___	Degating of part___	___	___	___
___	In tool___	___	___	___
___	Hand___	___	___	___
___	Part separator___	___	___	___
___	Robot___	___	___	___
___	Special mold support equipment___	___	___	___
___	Operation___Type___	___	___	___
___	Process control limits (estimate)___	___	___	___
___	Feed to hopper___	___	___	___
D.	MOLD (SUMMARY) MOLD NO.			
___	Type___	___	___	___
___	Dimensions___	___	___	___
___	Daylight required___	___	___	___
___	Centering diameter (+/– in.)___	___	___	___
___	Sprue bushing radius/diameter___	___	___	___
___	Parting line marked___	___	___	___
___	Ejector stroke dimensioned___	___	___	___
___	Heat treatment noted___	___	___	___
___	Venting noted___	___	___	___
___	Cavity finish noted___	___	___	___
___	Draft ___	___	___	___
___	Cooling circuits (individual) ___	___	___	___
___	Special cores for heat removal ___	___	___	___
___	Cores separate cooling circuits ___	___	___	___
___	Ejector system adequate ___	___	___	___
___	Sprue puller (type) ___	___	___	___
___	Balanced cavity layout ___	___	___	___
___	Tolerances anticipated ___	___	___	___
___	Gating into part ___	___	___	___
___	Press for sampling ___	___	___	___
___	Press for production___	___	___	___
___	Materials for part ___	___	___	___
___	Shrinkage ___	___	___	___
___	Cavities numbered ___	___	___	___
___	Thermocouples in mold for temperature control ___	___	___	___
___	Uniform texture ___	___	___	___
___	Inserts required ___	___	___	___
E.	HOT RUNNER TYPE___VENDOR___			
___	Heat capacity/No. units___	___	___	___
___	Gate type___	___	___	___
___	No. of plugs___	___	___	___
___	Electrical load___	___	___	___

	MOLD HALF		COMMENTS
	(A) FH	(B) MH	
___ No. thermocouples (type)________	________	________	________
___ Temperature settings________	________	________	________
________	________	________	________
________	________	________	________
________	________	________	________
F. UNSCREWING MOLDS	________	________	________
___ Method/connections dimensioned________	________	________	________
___ Mechanical________	________	________	________
___ Electrical________	________	________	________
___ Hydraulic________	________	________	________
___ Air________	________	________	________
___ Internal/external to mold________	________	________	________
___ Core stripper plate travel________	________	________	________
G. QUICK MOLD CHANGE			
___ Tool built for QMC________	________	________	________
___ Service conn. in back plate________	________	________	________
___ Machine adapted for QMC________	________	________	________
___ Installation method (equipment available)________	________	________	________
___ Fittings/type________	________	________	________
___ Temperature control prior to mounting________	________	________	________
H. PARTS LIST			
___ All items recorded________	________	________	________
___ Quantities correct________	________	________	________
___ Spare parts noted________	________	________	________
___ Vendors listed________	________	________	________
___ Heat treatment noted & process________	________	________	________
___ Surface treatment noted and specified________	________	________	________
________	________	________	________
________	________	________	________

COMMENTS

____

____

____

____

Date________________________ Reviewed by________________(Design)

________________(Purchasing)

________________(Production)

________________(Quality)

________________(GM)

Injection mold specification sheet.

Date________

Customer:______________________________ Contact:____________________

Part:______________________________ Dwg. No.________ Rev. No.________

Mold No.______________________________ Contract No.____________________

Phone No.______________________________

Material______________________________ Supplier________ Shrinkage________

A. MOLD TYPE

- ______ 2 Plate
- ______ 3 Plate
- ______ MUD
- ______ Hot runner
- ______ Quick change
- ______ Surface area
- ______ Production life
- ______ (parts/year)

B. CAVITIES

- ______ Number
- ______ Balanced
- ______ Inserted
- ______ Fired half (FH)
- ______ Move half (MH)
- ______ Internal radii
- ______ External radii

RUNNER

- ______ Full round
- ______ Semi round
- ______ Quarter round
- ______ Trapezoidal
- ______ U-shaped
- ______ Cold slug traps

C. GATING

- ___ Sprue gate ___ Outside ___ Inside
- ___ Film
- ___ Submarine ___ Flare ___ Pinpoint
- ___ Tab
- ___ Pin
- ___ Fan
- ___ Hot runner: Type ____________________
  Supplier ____________________

- ___ Standard
- ___ Ring
- ___ Disc
- ___ Full edge
- ___ Spider
- ___ Pinpoint center
- ___ Degating type

D. COOLING (TEMPERATURE) ____________________

- ______ Optimum cooling
- ______ Cooled cores
- ______ Cooled core pins
- ______ Each cavity separate
- ______ Cooled cavity plates
- ______ Individual circuits
- ______ Surge cooling
- ______ Turbulent flow
- ______ Water line diameter
  - FH ________
  - MH ________
  - Cores ________
- ______ Connections/quick release

E. HEATING (TEMPERATURE) ____________________

- ______ Hot water
- ______ Hot oil
- ______ Cartridge watts________________
- ______ Coil watts________________
- ______ Band watts________________
- ______ Plugs watts________________
- ______ Cavity
- ______ Core
  - FC Temp________
  - MH Temp________
- ______ Connection quick release

F. MATERIAL

	P-20	H-13	A-2	D-2	42OSS	P-6	O-1	S-7	DERYL CU	SIL BRON2				CASE HRD	HARDN'D	QUENCHED/	TEMPERED		NITRITED	CHROMED		
______ Mold set																						
______ Cavity plates FH																						
______ Inserts FH																						
______ Cavity Plates MH																						
______ Inserts MH																						
______ Slides																						
______ Back-up plates																						
______ Ejector plate																						
______ ______________																						
______ ______________																						
______ ______________																						

G. DEMOLDING

______ Stripper plate
______ Round ejectors
______ Stripper sleeve
______ Two stage strip
______ Flat ejectors
______ Angled
______ Attainable w/gating selected
______ Time checked after molding
______ Part end-use environment temperature__________________, __________________

______ Unscrewing timed by__________
______ Hydraulic/air/mechanical
______ Sprue puller plate
______ Slide FH______
______ Slide MH______
______ Sprue puller FH/MH Type______

M. SPECIFIC INFORMATION

______ Receive all detail drawings with mold and parts list
______ Receive sample parts with measured points
______ Receive one set drawings with parts list with mold
______ No corrections for distortion on hardening
______ ____
______ ____
______ Price ___________ Cavity
______ Estimated Price ___________ Cavity
______ Fixed Price ___________ Cavity
______ Delivery time: Weeks ___________ Penalty if late___________

______ Mold acceptance based on
______ Parts to print
______ Part (Form/Fit/Function)
______ End-use testing
______ Cavity/mold revisions negotiable
______ Cost
______ By you
______ By us

Design ______ By us ______ By you
Material ______ By us ______ By you
Sampling ______ By us ______ By you

______ Part drawings available
______ Mold drawings available
______ Parts list available

Mold accepted by______________________
____
____
____

Date ______________________________Signature

APPENDIX B

GLOSSARY OF TERMS USED IN DESIGN AND DEVELOPMENT OF PLASTIC PRODUCTS

Abrasion The wearing away of some surface area by its contact with another material.

Absorption *See* **Moisture**.

Accelerated aging Aging by artificial means to obtain an indication on how a material will behave under normal conditions over a prolonged period.

Accelerated weathering Duplicating or reproducing weather conditions by machine-made means.

Acceptable quality level The minimum quality level at which a product will be accepted or rejected

Acetal resin A crystalline thermoplastic material made from formaldehyde. Trade names: Delrin and Celcon.

Acrylics The name given to plastics produced by the polymerization of acrylic acid derivatives, usually including methyl methacrylate. An amorphous thermoplastic material that is clear.

Acrylonitrile, butadiene, styrene (ABS) A thermoplastic classified as an elastomers-modified styrene.

Additive A material added to resin prior to molding or forming to add a desired property or characteristic to the finished product or to assist in the processing of the material.

Adsorption *See* **Moisture**.

Aesthetics Referring to the external surface appearance of a plastic product.

Aging The change of a material over time under defined natural or synthetic environmental conditions, leading to improvement or deterioration of properties. *See also* **Accelerated aging**; **Artificial aging**.

Air vent A small outlet or area around the periphery of a mold cavity used to prevent entrapment of gases within the mold cavity.

Algorithms A mathematical procedure and series of equations used in a computer with support software to produce a result based on input data.

Ambient temperature Temperature of the medium surrounding an object. Used to denote prevailing room temperature.

Amorphous Plastic materials that have no definite order of crystallinity.

Amortized The cost of an item spread out equally over time or a specified number of parts. Frequently used when estimating finished product costs if the cost of a mold or capital equipment is spread out and added into the piece part cost.

Analog Refers to a needle on a scale readout that is almost instantaneous from the input signal to the output readout. Determined by the design of the circuitry. Gives an approximate reading based on the detail of the readout scale.

Angular welding See **Ultrasonic sealing**.

Anneal (1) To head a molded plastic article to a predetermined temperature and slowly cool it to relieve stresses. (Annealing of molded or machined parts may be done dry, as in an oven, or; wet, as in a heated tank of mineral oil.) Often done with the part in a holding fixture. (2) To heat steel to a predetermined temperature above the critical range and slowly cool it to relieve stresses and reduce hardness.

Antioxidant A substance added to a material to inhibit oxidation.

Antistatic agents (antistats) Agents when added to the molding material or applied on the surface of the molded part, make it less able to conduct electricity (thus hindering the fixation of dust).

Approved supplier A product supplier who has been rated satisfactorily on previous jobs. May involve a detailed analysis of manufacturing and quality capability to be sure that it meets customer requirements.

Arc resistance The time required for a given electric current to render the plastic surface of a material conductive because of carbonization by the arc flame.

Artificial aging The accelerated testing of plastic specimens to determine their changes in properties. Carried out over a short period of time, such tests are indicative of what may be expected of a material under service conditions over extended periods. Typical investigations include those for dimensional stability; the effect of immersion in water, chemicals, and solvents; light stability; and resistance to fatigue, among others.

Ashing The reduction of a polymer by high heat to yield any inorganic fillers or reinforcements. Used to verify the percentage of nonorganic content in the material.

ASTM Abbreviation for the American Society for Testing and Materials.

Automatic mold A mold or die in injection or compression molding that repeatedly goes through the entire cycle without human assistance.

Auxiliary equipment Refers to equipment, other than the injection-molding machine and mold, required to ensure the manufactured product would be made correctly, including dryers, chillers, material and part conveyors, and robots.

Balanced mold A mold laid out with runner and mold cavities spaced and sized for uniform flow, fill, and packing pressure throughout the system.

Ball mill A crushing/abrading machine with balls in a fixed race used to crush materials fed into the crushing chamber.

Barcoding The electronic/optical bar recognition system for identification, storage, printout, and retrieval of specified data and information.

Barrel (extruder) In injection molding, extrusion, or bottle-blowing equipment. It is the hollow tube in which the plastic material is gradually heated and melted and from which it is extruded into a die or rammed into the mold cavity under pressure.

Bezel A grooved rim or flange. An example is a television bezel.

Blanket purchase order (BPO) A purchase order placed with a supplier for materials over a set time period. The customer then releases material as required or as specified.

Blend Any combination of mixtures of a base resin with additives or modifiers. The base resin has been modified.

Blister A raised area on the surface of a molded part caused by the pressure of gases or air inside the material.

Bloom A visible exudation or efflorescence on the surface of a plastic. Bloom can be caused by lubricant, plasticizer, or other means.

Blow molding (1) A molding process used primarily to produce hollow objects. (2) A molding process in which a hollow tube (a parison) is forced into a shape of the mold cavity using internal air pressure. The two primary types are injection blow molding and extrusion blow molding.

Blow pin A hollow pin inserted or made to contact the blowing mold so that the blowing medium can be introduced into the parison or hollow form and expanded to conform to the mold cavity.

Blowing agent (forming agents) An additive capable of producing a cellular structure in a plastic or rubber mass.

Blush The tendency of a plastic to turn white or chalky in areas that are highly stressed, such as gate blush.

Boss Projection on a plastic product designed to add strength, facilitate alignment during assembly, and provide for a point to fasten or screw the product together.

Branched chains In polymer chemistry, sidechains attached to the main original polymer chain.

Bubble A spherical, internal void; globule or air or other gas trapped within a plastic. *See* **Void**.

Bubbler A cooling tube in a blind mold cavity that allows cooling medium to flow to control the temperature of a core or section of the mold.

Burned A carbonized condition showing evidence of thermal decomposition through some discoloration, distortion, or localized destruction of the surface of the plastic. Usually caused by poor venting of the mold cavity.

Burning (1) Overheating the resin in the barrel causing discoloration and, if long enough, charring the material. (2) Caused by trapped gasses in a poor or nonvented area of the mold. The gasses may ignite, due to pressure and temperature, as in a diesel engine, and discolor or char the product.

Buttress thread A type of screw thread used for transmitting power in only one direction. It has efficiency of the square thread and the strength of the V thread.

CC, certificate of compliance A letter or form furnished by a supplier who states that the material meets company or predetermined customer requirements.

Cadmium A heavy-metal element used as a pigment in plastics. Now being replaced because of its hazardous nature.

Caprolactam A cyclic amide compound containing six carbon atoms. When the ring is opened, caprolactam is polymerizable into a nylon resin known as *type six* (6) *nylon*[11] or *polycaprolactam*.

Carbon black A black pigment produced by the incomplete burning of natural gas or oil. It is widely used as filler, particularly in the rubber industry and wire/cable applications. Because it possesses useful ultraviolet protective properties, it is also use in molding compounds intended for outside weathering applications.

Cavity Depression in the mold that usually forms the outer surface of the product. Depending on number of such depressions, molds are designated as a single cavity mold, a multicavity mold, or a family cavity mold.

Cavity number A sequential number engraved in a mold cavity and reproduced on the molded part for later reference in case a problem ever occurs with the part. Used in multicavity molds of similar parts.

Cellular plastics Foamed plastic products.

Cementing A process of joining two similar plastic materials to themselves or to dissimilar materials by means of solvents.

Center gated mold An injection or transfer mold in which the cavity is filled with molding material through a sprue or gate directly into the center of the part.

Chalking Dry, chalklike appearance or deposit on the surface of a plastic. *See* **Haze; Bloom**.

Change request A written request, often called an *engineering change order* (ECO) to modify or alter the dimensions, material, tolerances, or manufacture or a part now in or soon to be in production. Use to ensure all interested and involved department personnel are informed and can comment and approve or disapprove of the pending change.

Charpy A type of pendulum test for toughness. Typically known as the European method for measuring material toughness. *See* **Impact test**.

Chemical resistance Ability of a material to retain utility and appearance following contact with chemical agents.

Chiller A refrigeration unit used to supply cooling water in a closed-loop system for a mold or equipment used to regulate temperature.

Chromium plating An electrolytic process that deposits a hard film of chromium metal onto working surfaces of other materials. Used when resistance to corrosion, abrasion, and/or erosion is required.

CIM Computer-integrated manufacture, the use of computer technology to manufacture and control a product.

Clamping area The largest rate molding area an injection or transfer press can hold closed under full molding pressure.

Clamping force (clamping pressure) In injection molding, the pressure is applied to the mold to keep it closed despite the fluid pressure of the compressed molding material within the cavity and runner system.

Clarity Material clearness or lack of haze.

Closed loop System used with microprocessor for control of a machine's cycle. See **Feedback**.

Closed-loop/ continuous-feedback process control A system collecting operation variable manufacturing data, analyzing the data in "real time" and using software to adjust the machine variables for the next cycle when required. Used to adjust and perform in a continuous operation, cycle to cycle.

Coefficient of expansion The fractional change in a specified dimension (sometimes volume) of a material for a unit change in temperature. Values for plastics range from 0.01 to 0.2 mils/in. · C (ASTM D696).

Coefficient of friction The value calculated under a known set of conditions, such as pressure, surface, speed, temperature, and material to develop a number—either static or dynamic—of the resistance of the material to slide or roll. The lower the value, the higher the material's lubricity.

Coining The peening over or compressing of a material to change its original shape or form.

Cold flow A plastic exhibits cold flow when it does not return to its original dimensions after being subjected to stress. *See also* **Creep**.

Collapsing core A mold core that which collapses as a result of a center pin or other operation that makes the core smaller, usually in diameter, while or just before the mold opens so that the part can be ejected.

Color concentrate A mixture of a measured amount of dye or pigment and a specific plastic material base. A more precise color can be obtained using concentrates than using raw colors. *Note:* Care should be taken to verify that the color concentrate base is compatible with the plastic it is to color. Color concentrate is normally used at 1–4% of the plastic material to be colored.

Colorfast The ability to resist change in color.

Color standard The exact color a plastic resin or product must match to be acceptable. Resin suppliers often submit color chip samples of the matched resin color to be compared to the molded part. The color chip, or standard, is usually 2 × 3 in. With one polished surface and various textured surfaces on the opposite side. Suppliers use similar standards to verify the color of each lot of resin shipped to their customers.

Combination mold *See* **Family mold**.

Commodity resin Usually associated with the higher-volume lower-priced plastics, with low to medium physical properties. Examples are PE, PP, PS, and acrylic, PVC, EVA, and ABS. Used for less critical applications.

Compound A mixture of polymer(s) with all materials necessary for the finished product.

Compression ratio In the extruder of an injection/blow-molder screw, the ratio of volume available in the first flight at the hopper to the last flight at the end of the screw.

Compressive strength Crushing load at the failure of a specimen divided by the original sectional area of the specimen (ASTM D695).

Concentricity (1) The relationship of all circular surfaces with the same center. (2) Relationship of all inside dimensions to all outside dimensions. Usually, as with diameter, expressed in thousandths of an inch (FIM=full indicator movement). Deviation from concentricity is often referred to as *runout*.

Conditioning The subjection of a material to a stipulated treatment so that it will respond in a uniform way to subsequent testing or processing. The term is frequently used to refer to the treatment given before testing. ASTM standard conditions for a plastic testing laboratory are 23°C (73.4°F + 3.6°F) and 50 + 5% relative humidity.

Conditioning chamber An enclosure used to prepare parts for their next step in the assembly or decoration process. Parts can be stress-relieved, humidity- or moisture-conditioned. or impregnated with another element.

Consigned material Material given over to another supplier for care and use in manufacturing a customer's product.

Contamination Any foreign body in a material that affects or detracts from the part's quality.

Control plan A written plan that lists step-by-step procedures describing how a specific operation will be conducted and followed.

Controllers The instruments, timers, and pressure controls used to control and regulate the molding cycle or any manufacturing cycle.

Cooling channels Channels or passageways within the body of a mold through which a cooling or heating medium can be circulated to control temperature on the mold surface. May also be used for heating a mold by circulating hot water, steam, hot oil, or other heated fluid through channels as in molding thermoplastic materials.

Cooling factor The amount of time estimated for the molds cooling system to solidify the parts material in the mold so that it can be ejected without distortion.

Cooling fixture A block of steel, wood, or composite material that is similar to the shape of the molded piece. The hot molded part is taken from the mold, placed on it, and allowed to cool without distorting. Also known as a *shrink fixture*.

Cooling time The time period required after the gate freezes for the part to solidify and become rigid enough for ejection from the mold cavity.

Copolymer A polymer produced by polymerization of two or more monomers. Can also be done as a secondary compounding operation on an extruder.

Core (1) Male element in die that produces a hole or recess in a part. (2) Part of a complex mold that molds undercut parts. Cores are usually withdrawn to one side before the main sections of the mold open. (3) A channel in a mold for circulation of a heat transfer medium. (4) The central member of a laminate.

Core pin A pin used for forming a hole or opening in a plastic mold.

Core pin plate Plate holding core pin.

Coring The removal of excess material from the cross section of a molded part to attain a more uniform wall thickness.

Corona treatment Exposing a plastic part to a corona discharge increasing receptivity to inks, lacquers, paints, and adhesives. *See also* Surface treatment.

Corrosion Material that is eaten away by chemical reactions at the surface area.

Crazing Fine cracks that may extend in a network on or under the surface or through a layer of plastic material.

Creep The dimensional change with time of a material under load, following the initial instantaneous elastic deformation. *See also* Cold flow (ASTM D 674).

Critical path A method of charting/scheduling for identifying key elements in the path to completion of a program.

Crosslinking The chemical combination of molecules to form thermally stable bonds within a polymer, not broken by heating.

Crystallinity A state of molecular structure in some resins that denotes uniformity and compactness of the molecular chains forming the polymer. Normally attributed to the formation of solid crystals with a definite geometric form. High crystallinity causes a polymer to be less transparent or opaque.

Cure That portion of the molding cycle during which the plastic material in the mold becomes sufficiently rigid or hard to permit ejection.

Curing time The time between the end of injection pressure and the opening of the mold.

Cycle The complete, repetitive sequence of operations in a process or part of a process. In molding, the cycle time is the period, or elapsed time, between a certain point in one cycle and the same point in the next.

Daylight opening Clearance between two platens of a press in the open position. Mold daylight describes the opening distance of mold halves for part removal.

Degating The removal of the part from the runner system.

Degradation A deleterious change in the chemical structure, physical properties, and/or appearance of a plastic, usually caused by exposure to heat.

Delaminate To split or separate a laminated plastic material along the plane of its layers.

Density Weight per unit volume of a substance, expressed in grams per cubic centimeter or pounds per cubic foot.

Design of experiments (DOE) A problem-solving technique developed by Taguchi using a testing process with an orthogonal array to analyze data and determine the main contributing factors in the solution to the problem.

Design stress A long-term stress, including creep factors and safety factors that is used in designing structural fabrication.

Destaticization Treating plastic materials to minimize their accumulation of static electricity and subsequently the amount of dust picked up by the plastics because of such charges. *See* **Antistatic agents**.

Destructive test Any test performed on a part in an attempt to destroy it, often performed to see how much abuse the part can tolerate without failing.

Deterioration A permanent change in the physical properties of a plastic as evident by impairment of these properties.

Diaphragm gate Gate used in molding annular or tubular articles. Gate forms a solid web across the opening of the part.

Die A metal form in making or punch in plastic products. It is used interchangeably with mold. In extrusion it refers to the tooling forming the plastic shape that the molten plastic is forced through.

Dielectric constant Normally the relative dielectric constant. For practical purposes, the ratio of the capacitance of an assembly of two electrodes, separated solely by plastic insulating material, to its capacitance when the electrodes are separated by air. A relative measure of nonconductance.

Dielectric heating (electronic heating or radiofrequency heating) The plastic to be heated forms the dielectric of a condenser to which a high frequency voltage is applied. Dielectric loss in the material is the basis of the process that is used for sealing vinyl films.

Dielectric strength The maximum electrical voltage a material can sustain before it is broken down, or "arced through," in volts per mil of thickness.

Dieseling *See* **Burning**.

Differential shrinkage Nonuniform material shrinkage in a product.

Digital Numerical output device that must index, number by number, from the initial output reading to the final output reading. More accurate than a similar analog device, but slower. Gives exact reading.

Dimensional stability Ability of a plastic part to retain the precise shape in which it was molded

Discoloration Any change from the original color; often caused by overheating, light exposure, irradiation, and high stress at local points on the part, or chemical attack.

Dished Showing a symmetric distortion of a flat or curved section of a plastic object so that, as normally viewed, it appears concave or more concave than intended. *See* **Warpage**.

Disk gate *See* **Diaphragm gate**.

Dispersion Finely divided particles of material in suspension in another substance.

Domed Showing a symmetric distortion of a flat or curved section of a plastic object so that, as normally viewed, it appears convex or more convex than intended. *See* **Warpage.**

Dowel Pin used to maintain alignment between two or more parts of a mold.

Draft A taper or slope in a mold required to facilitate removal of the molded piece. The opposite of this is called *backdraft,* and this is on the opposite mating part of the mold.

Drop test *See* **Impact test**.

Dry as molded (DAM) Term used to describe a part immediately after it is removed from a mold and allowed to cool down. All physical, chemical, and electrical property tests are performed on nonconditioned test bars and results recorded on the data sheets. Parts and test bars in this state (DAM) are felt to be their weakest in some properties, as they have not had time to condition or relieve any molded-in stresses.

Dry coloring Method commonly used to color plastic by tumble blending uncolored particles of the plastic materials with selected dyes and pigments.

Dryers Auxiliary equipment used to dry resins before processing to ensure that surface properties are within manufactured specifications. There are several styles of dryers, including ovens, microwave, and hot-air desiccant bed and refrigeration types.

Ductility The extent to which a solid material can be drawn into a thinner cross section without breaking.

Dunnage A product specifically designed to convey, hold, transport, or protect a product during manufacture or shipment to a customer.

Durometer hardness The hardness of a material as measured by the Shore durometer.

Dyes Intensely colored synthetics or natural organic chemicals that are soluble in most common solvents and dissolve in the plastic substrate while imparting color. Characterized by good transparency, high tincturial strength, and low specific gravity.

Economic order quantity Ordering a product in a quantity for cost savings and for projected use in a predetermined time period.

Ejection The removal of the finished part from the mold cavity by mechanical means.

Ejection time The time in the cycle when the mold opens, the part is ejected, the mold closes, and clamping pressure is applied.

Ejector pin (ejector sleeve) A rod, pin, or sleeve that pushes a molding off of a core or out of a cavity. It is attached to an ejector bar or plate that can be activated by the ejector rod(s) or the press or by auxiliary hydraulic or air cylinders.

Ejector pin retainer plate Retainer plate onto which ejector pins are assembled.

Ejector return pins ProjectioOns that push the ejector assembly back as the mold closes. Also called safety pins and position pushbacks.

Ejector rod or bar A bar that activates the ejector assembly when the mold is open.

Elastic deformation A deformation in which a substance returns to its original dimensions on release or the deforming stress.

Elasticity That property of a material by virtue of which it tends to recover to its original size and shape after deformation. If the strain is proportional to the applied stress, the material is said to exhibit Hookean or ideal elasticity.

Elastomer A material that at room temperature can be stretched repeatedly under low stress to at least twice its length and snaps back to the original length on release of the stress.

Electric discharge machining (EDM) A metalworking process applicable to mold construction in which controlled sparking is used to erode the workpiece.

Electroplating Deposition of metals on certain plastics and mold for finish.

Elongation The increase in length of a material under test, expressed as a percentage difference between the original length and the length at the moment of the break.

EMI/EMP Electromotive interference and electromotive protection, terminology used to describe electronics vulnerable to electrical radiation interference, which can affect and damage solid-state devices.

Embossing Techniques used to create depressions of a specific pattern in plastics film and sheeting. Such embossing is in the form of surface patterns on the molded part by photoengraving or a similar process.

Endothermic An action or reaction that absorbs heat.

End use The function the part or assembly was originally designed and manufactured to perform.

Engineering resin Associated with plastics having medium to high physical properties used for structural and demanding applications. Examples are nylon, acetal, PBT, PET, PC, PPS, and LCP.

Environmental stress cracking (ESC) The susceptibility of a thermoplastic article to crack or craze under the influence of certain chemicals or aging, weather, and stress. Standard ASTM test methods that include requirements for environmental stress cracking are indexed to ASTM standards.

Etch To treat a mold with an acid, leaving parts of the mold in relief to form the desired design. To prepare the surface of certain plastics for a secondary process, as nylon for electroplating.

Ethylene–vinyl acetate A plastic copolymer made from the two monomers, ethylene and vinyl acetate. This copolymer is similar to polyethylene, but has considerable increased flexibility.

Exothermic Pertaining to an action or reaction that gives off heat.

Extrusion The plasticizing of a material in an extruder (barrel-and-screw or plunger assembly) and forcing of the molten material or extrudate through a die or into a mold. The initial part of the molding process.

Extrusion plastomer An instrument used to determine melt flow index (MFI). *See* **Melt index**.

Fabricate To work a material into a finished form by machining, forming, or other operations. In the broadest sense, it means *to manufacture*.

Failure mode and effects analysis A quality assurance tool analyzing all manufacturing operations in a continuous step-by-step manner to determine any variables in an operation that can affect the operation. Once these are determined, ways should be developed to control the variability and selection of control methods for the control of these variables to produce a repeatable good product, cycle to cycle.

Family mold (1) A multicavity mold in which each cavity forms a part that often has a direct relationship in usage to the other parts in the mold. Family molds can have more than one cavity making the same part, but they will still always have that same direct relationship to usage. (2) A multicavity mold in which each cavity forms one of the component parts of the assembled object. The term often applied to molds in which parts from different customers are grouped together in one mold for economy of production. Sometimes called a *combination mold*.

Fan gate A shallow gate somewhat wider than the runner from which it extends.

Feathered thread A thread that is thin at its end (comes out sharp) and does not end abruptly. Usually found in screw-machined parts.

Feedback Information returned to a system or process to maintain the output within specified limits.

Fiber Thin strands of glass used to reinforce both thermoplastic and thermosetting materials. One-inch-long fibers are occasionally used, but the more common lengths are 0.25–0.50 in. long and often less than 0.100 in. long.

Fill rate The pressure–tie relationship used to described the filling of the mold cavity.

Filler An inert substance added to plastics for the purpose of improving physical properties or processability or reducing the cost of the material.

Fillet A rounded inside corner of a plastic piece. The rounded outside corner is called a *bevel*.

Fines Very small particles mixed in with larger particles. Often generated during the regrinding of plastic parts for reprocessing again with virgin material.

Finish (1) To complete the secondary work on a molded part so that it is ready for use. Operations such as filling, deflashing, buffing, drilling, tapping, and degating are commonly called *finishing operations*. (2) The plastic forming the opening of a bottle, shaped to accommodate a specific closure. The ultimate surface structure of a part. (3) *See* **Surface finish**.

Finite-element analysis (FEA) A stress analysis technique of a part using a computer-generated model that can take finite sections of the part for analysis of the forces and loads the part will experience in service. It generates a part section analysis that shows the force concentrations in the section and determines if the material selected will be suitable for the part by calculating the stresses in the material.

First surface The front surface of a plastic part, nearest the eye.

Fishbone diagram A problem analysis technique used to list all the variables and steps in the solution to a problem. All contributing elements are associated with each factor and taken back to their starting points to ensure that all variable elements are considered.

Fissure A narrow opening crack in a material.

Fixture Means of holding a part during a machine or other operation.

Flakes Resin residue formed on the inside of pipes during material transfer. Created by the friction of the pellets against the surface of the transfer piping. With time, they build up, flake off, and can cause feed problems at the throat of the injection-molding machine.

Flame-retarded A resin modified by flame-inhibiting additives so that exposure to a flame will not burn or will self-extinguish. Some resins will not burn as thermosets; others can be modified to meet agency flame/burning specifications; and others, depending on their base materials, may not be capable of being modified.

Flammability Measure of the extent to which a material will support combustion.

Flash Extra plastic attached to a molding along the parting line. Under most conditions it is objectionable and must be removed before parts are judged acceptable. Requires reworking of the mold halves' parting line or the venting evaluated.

Flash gate Usually a long gate extending from a runner parallel to an edge of a molded part along the flash or parting line of the mold.

Flash limits The predetermined melt overflow at or in a fusion joint area that does not affect the integrity or quality of a joint.

Flash line A raised line appearing on the surface of a mold and formed at the junction of mold faces.

Flash trap A molded-in lip or blind recess on a part that is used for trapping excess molten material (flash) during an assembly operation as with sonic or spin welding. Negates a flash trimming operation.

Flex bar An ASTM-specified test bar used to develop physical property data for plastic materials. Usually $\geq 4 \times \frac{1}{2} \times \frac{1}{8}$ in. thick, depending on the ASTM specification.

Flexural strength Ability of a material to flex without permanent distortion or breaking (ASTM D790).

Flow (1) A qualitative description of the fluidity of a plastic material during the process of molding. (2) A quantitative value of fluidity when expressing a melt flow index. *See* **Melt index**.

Flowchart A line chart that traces a process from start to finish.

Flow length The actual distance a material will flow under a set of molding machine conditions. Influenced by the processing and mold design variables, the composition of the polymer, and any additives in the polymer.

Flow line A mark on a molded piece made by the meeting of two flow fronts during molding. Also called *weld line* or *weld mark*. *See* **Weld line**.

Flow marks Wavy surface appearance on a molded object caused by improper flow of the material into the mold. *See* **Splay marks**.

Fluoropolymer A generic name given to fluorine based plastic, trade-named Teflon, plastic material.

Foamed plastics Resins in sponge form. The sponge may be flexible or rigid, the cells closed or interconnected. The density anything from that of the solid parent resin down to, in some cases, 2 lb/ft^3.

Force (1) (Physics) that which changes the state of rest or motion in matter, measured by the rate of change of momentum. (2) That portion of the mold that forms the inside of the molded part. *See also* **Core**; **Plunger**.

Freeze off Refers to the gate area when it solidifies as well as any area in the resin flow system when the melt becomes too cool to flow and solidifies.

Friction welding A means of assembling thermoplastic parts by melting them along their line of contact through friction. *See also* **Spin welding**.

Full indicator movement (FIM) A term in current use to identify tolerance with respect to concentricity. "Former practices" terms are *full indicator reading* (FIR) and *total indicator reading* (TIR) runout.

Fusion bond (1) The joining of two melt fronts that meet and solidify in a mold cavity. (2) The bond formed during the assembly operation where the joint line is melted prior to assembly. *See* **Hot-plate welding**; **Induction welding**.

Galling A surface area that is worn away by another by rubbing against it.

Gardner A type of drop-weight impact test. *See* **Impact test**.

Gardner test *See* **Impact test**.

Gas-assisted injection molding (GAM) An injection molding process that introduces a gas (usually nitrogen) into the plasticized material, to form voids in strategic locations.

Gaseous blowing agent A compressed gas such as compressed air or nitrogen used to create a cellular structure or controlled voids in a rubber or plastic material.

Gate In injection and transfer molding the orifice through which the melt enters the mold cavity.

Gauges The measuring instruments used to determine if the part meets customer specification, including go/no-go plugs, micrometers, and vernier calipers.

Gaylord A plastics term used to identify a box of resin versus a bag or drum. Box size and weight of resin can vary depending on the density of the resin and the supplier's box size. Box size usually conform to the size of a standard pallet on which it is shipped.

Gel-permeating chromatograph (GPC) Used to determine molecular weight distribution.

Generic Descriptive of an entire type or class of plastic resins. The base resin is one of a family of polymers, but there may be hundreds of product combinations.

Glass transition point The temperature range indicated by the change from a viscous or rubbery state to a hard, brittle state in a polymer.

Grinder (granulator) Machine with a series of knifeblades and a sizing screen to chop up parts, sprue and runner, and other plastic materials for reuse or resale. Available in many sixes, styles, and capacities. Used to make regrind.

Guide pins Devices that maintain proper alignment of mold sections and cavity as the mold closes.

Gusset An angular piece of material used to support or strengthen two adjoining walls.

Hardness The resistance of a material to compression and indentation. Among the most important methods of testing this property are Brinell hardness, Rockwell hardness, and Shore hardness, the latter for more rubbery materials.

Haze The degree of cloudiness in a plastic material.

Heading The mechanical, thermal, or ultrasonic deformation of a pin to form a locking attachment to retain whatever is under the deformed material.

Heat distortion point An arbitrary value of deformation under a given set of test conditions. In ASTM test D648, it is defined as a total deflection of 0.010″ in a rectangular bar supported at both ends under a load of 66 or 264 psi while submersed in oil. The temperature is raised 2°C per minute until this deflection is reached.

Heat sealing A process of joining two or more thermoplastic films or sheets by heat and pressure.

Heat stability The resistance of a plastic material to chemical deterioration during processing.

Heat stabilizer An ingredient added to a polymer to improve its processing or end-use resistance to elevated temperatures. The term is used in different contexts depending on the benefit to be derived from the additive. For processing, it retards changes in resin color. For end use, it protects the surface of the part exposed to elevated temperature from oxidation. It does not imply that a resin can be use beyond its recommended end-use temperature rating if it is heat-stabilized.

Heater bands The only heat source for the barrel and nozzle temperature control divided usually, into read, middle, and front, and nozzle temperature control sections. They are very accurate resistance heaters with high heat output.

Heating chamber Injection molding, that part of the machine in which the cold feed is reduced to a hot melt. Also called *heating cylinder* or *barrel.*

Hermetic As in seal, to form a bond that is pressuretight, so that air or gasses cannot enter or escape.

Holding pressure The pressure maintained on the melt after the cavity is filled and until the gate freezes off. See **Packing pressure**.

Holdup time See **Residence time**.

Homopolymer The product of the polymerization of a single monomer.

Hoop stress The circular stresses referred to in round, usually pressure-type, containers.

Hopper A conical reservoir from which the molding powder or pellets feed into the extruder screw.

Hopper feeder Usually part of the resin drying system, but can be an independent system, to convey material to the machine's feed hopper using vacuum or positive air pressure.

Hot/heated manifold mold A thermoplastic injection mold in which the portion of the mold that contains the runner system has its own heating elements to keep the molding material in a plastic state ready for injection into the cavities, from which the manifold is insulated.

Hot-plate welding The use of a heated tool to cause surface melting of a plastic part at the joint are. It is then removed prior to the joint surfaces being pressed together to form a fusion bond.

Hot-runner mold A thermoplastic injection mold in which the runners are insulated from the chilled cavities and remain hot so that the center of the runner never cools in a normal cycle operation. Runners are not usually ejected with the molded pieces. Called *insulated runner molds* when heating elements are not used in the mold. *Note:* A heated manifold mold is a hot-runner mold that is both heated and insulated; an insulated mold is a hot-runner mold that does not contain heaters.

Hot stamping Engraving operation in which roll leaf is dyed or (metalized foil) stamped with heated metal dies onto the face of the plastics. Also called *branding*.

Hot tip The precise controller and gating mechanism of a hot-runner mold.

Hydrolysis Chemical decomposition of a material involving the addition of water.

Impact strength (1) The ability of a material to withstand shock loading. (2) The work done in fracturing, under shock loading, a specified test specimen in a specified manner. (3) The relative susceptibility of plastic articles to fracture under stress applied at high speeds.

Impact test Often associated with the Gardner (ball or falling dart) test, with a known weight falling at a known distance and hitting a part, thereby subjecting it to an instantaneous high load. Could also be a pendulum-type impact test. ASTM impact tests for material properties are the Izod, Charpy, and tensile impact tests.

Induction welding The use of radio, magnetic, or electrical energy to form a melt through the application of a foreign medium at the joint line to form a fusion bond.

Inert pigment A pigment that does not react with any components of a paint.

Initiator Any foreign additive mixed in a material to cause a chemical or physical reaction in the melt or liquid stage.

Injection molding A molding procedure whereby a heat-softened plastic material is fed into a cavity (mold), which gives the article the desired shape using a screw and ram. Used with both thermoplastic and thermosetting materials.

Injection pressure The pressure in the mold during the injection of plasticized material into the mold cavity. Expressed in psi, with the hydraulic system pressure being used to indicate changes, when there are no sensors in the mold.

Injection time The time it takes for the screw's forward motion to fill the mold cavity with melt.

Inorganic A mineral compound not composed of carbon atoms.

Insert An integral part of plastics molding. It consists of metal or other material, which may be molded into position or may be pressed into the molding after the molding is completed. Also, a removable or interchangeable component of the mold.

Ishakawa Developed the "fishbone diagram" method of analysis.

ISO 9000 International Organization of Standardization, the current world-class recognized quality standard for all businesses in the world.

Izod A type of pendulum impact. *See* **Impact test**.

Izod impact test An impact test in which a notched sample bar is held at one end and broken by a blow. This is a test for shock loading. *See* **Impact test**.

Jetting Turbulent flow of plastic from an undersized gate or thin section into a thicker mold section, as opposed to laminar flow of material progressing radially from a gate to the extremities of the cavity. May also result from shooting material into a mold cavity where there is no core or immediate cavity wall to break up the flow of the material coming through the gate.

Just-in-time (JIT) Refers to practice developed to minimize customer inventory by the Japanese. The supplier provides the product, at predetermined intervals, so that it can proceed directly to the customer's assembly line. This practice demands excellent quality control and production schedules. Customers who use JIT must demand the same care and treatment from their own suppliers. Suppliers and customers are usually located within a few hours shipping time of each other to make it work effectively.

Kirksite An alloy of aluminum and zinc used for the construction of prototype molds. It imparts a high degree of heat conductivity to the mold.

Knit line A line on a part where opposing melt fronts meet. Treated by material flow around obstructions or multiple gating. *See* **Weld line**.

Knockout pin A pin that pushes a cured molded article out of a mold. Sometimes called an *ejector pin*.

Laminar flow Laminar flow of thermoplastic resins in a mold is accompanied by solidification of the layer in contact with the mold surface that acts as an insulating tube through which material follows to fill the remainder of the cavity. This type of flow is essential to duplication of the mold surface.

Land (1) The horizontal bearing surface of a semipositive or flash mold by which excess material escapes. (2) The bearing surface along the top of the flights of a screw in an extruder. (3) The surface of an extrusion die parallel to the direction of the melt flow. (4) The bearing surfaces of any mold. (5) The gate, when entering a part, has either one or two dimensions. There is always one more dimension involved, which is the length of the gate itself. This would be called the "land." On a round gate, it is the second dimension. On a rectangular or square gate, it is the third dimension.

Letdown ratio Quantification of the quantity of one ingredient to be mixed with a base material to obtain the desired results.

Lifters *See* **Slides**.

Light resistance The ability of a plastics material to resist fading after exposure to sunlight or ultraviolet light (ASTM D1501). Light stability is the measure of this resistance.

Locked-in stress *See* **Residual stress**.

Lot number A number assigned to a specific lot of material or parts. Used for traceability and accountability by the supplier and customers on all paperwork for the product.

Lubricants (1) A processing aid to assist material flow in the barrel of an injection-molding machine or extruder. Can be a solid, such as sodium or zinc styrate, or a liquid usually compounded into the base material. (2) Internally lubricated resins that use oils, Teflon, molybdenum disulfide, or other materials to give the molded part a lower coefficient of friction.

Macbeth A lighting system used for checking color.

Manifold A pipe channel, or mold, with several inlets or outlets.

Marriages Referring to poorly cut pellets that are too hot and bond together in strings or clumps that are not trapped and removed by the screen prior to packaging.

Master curve The acceptable or required curve that all subsequent test curves must match.

Matched metal die molding Method of molding reinforced plastics between two close-fitting metal molds mounted in a press.

Material review board (MRB) A panel of representatives from departments of the company who are involved with a product. It decides if the material and/ or product meet customer requirements if a question or problem about quality arises.

Matrix Refers to the base resin or material used for a molded product.

Matte finish A type of dull, nonreflective finish. *See* **Surface finish.**

Melt fracture An elastic strain set up in a molten polymer as the polymer flows through the die. It shows up in irregularities on the surface of the plastic product.

Melt front The exposed surface of molten resin as it flows into a mold. The melt front advances as the molten resin is continuously pushed through its center section.

Melt generation capacity Ability of the injection-molding machine's barrel and screw to produce the required melt quantity for the size of the barrel-and-screw combination required for the molding cycle. Used to size the molding machine on the basis of polystyrene melt generation capacity and listed as ounces of melt generation capacity.

Melt index (MI) or melt flow index (MFI) The amount, in grams, of a thermoplastic resin that can be forced through a 0.0825-in. orifice when subjected to the prescribed force (grams) in 10 min at the prescribed temperature (°C) using an extrusion plastomer (ASTM D1238).

Melt strength The strength of the plastic while in the molten state.

Melt temperature (1) The temperature at which a resin melts or softens and begins to have flow tendencies. (2) The recommended processing temperature of resin melt for correct processing. (3) The temperature of the melt when taken with a pyrometer melt probe.

Metal plating The process of plating a plastic part by chemically etching the surface to accept a base metal on which the subsequent layers of metal are deposited. Usually a multistep process. Not all plastics can be metallized.

Metalizing A general term used to cover all processes by which plastics are coated with metal.

Meter SI (International System of Units) length unit equal cm to 100 or 39.37 in.

Metering equipment A machine or system to accurately meter additives or regrind to the machine's hopper or feed throat. Comes in many sizes and types to suit each particular application, including augers, shuttle plates, photoelectric eyes, and positive or negative weight loss belt feeders.

Metering screw An extrusion or injection molding screw that has constant shallow depth and pitch section, usually over the last three to four flights.

Methyl methacrylate An amorphous thermoplastic resin. A common name is *acrylic resin*.

Microprocessor Computer system that stores, analyzes, and adjusts the controls of a machine per the parameters established during the operation of the machine it is controlling. Only operates within preset limits. Continuously analyzes output data to adjust and maintain the machine's cycle within programmed limits. Can also store data and output them as directed by programming.

Migration of plasticizer Loss of plasticizer from an elastomer's plastic compound with subsequent absorption by an adjacent medium or lower plasticizer concentration.

Mil English unit of length equal to 0.001 in. or 0.00254 cm.

Milestone chart Usually identified as a go/no-go decision point in a critical path flowchart or schedule.

Modifiers Any additive to a resin that improves the processing or end-use properties of the polymer. An example would be plasticizers added to PVC resin to make it soft and pliable and improve its impact strength. All PVC resins use different modifiers to meet desired product requirements. This is true of almost all plastic resins currently manufactured.

Modulus of elasticity The ratio of stress to strain in a material that is elastically deformed (ASTM D790)

Moisture (1) *Absorption,* the pickup of moisture from the atmosphere by a material that penetrates the interior. (2) *Adsorption,* surface retention of moisture from the atmosphere.

Moisture conditioning A method used to ensure that a product has a predetermined amount of moisture absorbed by the product. Typically done by placing parts in a moisture-conditioned/maintained enclosure at ambient or elevated temperature to drive the action of absorption. The moisture level is maintained at the required moisture level for absorption. The operation is timed, and parts are weighed to determine the correct amount of moisture absorption, a percent of part weight.

Moisture vapor transmission rate (MTR) The rate at which water vapor permeates through a plastic film or wall at a specified temperature and relative humidity (ASTM E96).

Mold (1) A medium or tool designed to hold a cavity form to make a desired shape and/or size of product. (2) To process a plastic material using an injection-molding process.

Mold deposits Material buildup on a cavity's surface due to plateout of resin, usually in a gaseous state. Can also be attributed to additives in a resin adhering to the mold's surface.

Mold open time *See* **Ejection time**.

Mold release (1) A lubricant used to coat a mold cavity to prevent the molded piece from sticking, thereby facilitating its removal from the mold. (2) Additives put into a material to serve as a mold release. Also called a *release agent.*

Molding A group of plastics processes using molds.

Molding cycle The period of time required to complete a cycle and produce a product.

Molding material Plastic material in varying stages of granulation often comprising plastic or resin filler, pigments, plasticizers, and other ingredients, ready for use in the molding operation. Also called *molding compound* or *powder.*

Molding pressure (1) The pressure applied directly or indirectly on the compound to allow the complete transformation to a solid dense part. (2) The pressure developed by a ram or screw to push molten plastic into a mold cavity. *See* **Injection pressure**.

Molding shrinkage The difference in dimensions, expressed in inches per inch, between a part and the mold cavity in which it was molded. Both the part and the mold cavity are at normal room temperature when measured. Also called *mold shrinkage* and *contraction.*

Molecular weight (MW) (average molecular weight) The sum of the atomic masses of the elements forming the molecule, indicating the relative size typical chain length of the polymer molecule.

Monomer A low-molecular-weight-reactive chemical that polymerizes to form a polymer.

Morphology The study of the physical form and structure of a material.

Mottle A mixture of colors or shades giving a complicated pattern of specks, spots, or streaks.

Movable platen The moving platen of an injection- or compression-molding machine in which half of the mold is secured during operation. This platen is move by either a hydraulic ram or a toggle mechanism.

Multi-cavity mold A mold having more than one cavity or impression for forming finished items during one machine cycle.

Necking The localized reduction of the cross-sectional area of an object.

Node A single point on an FEA model. A node is the starting and connection points of a mesh. All nodes connect to each other in a 2D or 3D geometric analysis.

Nonpolar Incapable of having a significant dielectric loss. Polystyrene and polyethylene are nonpolar.

Nonrigid plastic A plastic that has a modulus of elasticity (either in flexure or in tension) of $\leq$10,000 psi at 25°C and $>$505 relative humidity (ASTM D747).

Notch-sensitive A plastic material is said to be notch-sensitive if it will break when it has been scratched, notched, or cracked. Glass is considered to be highly notch-sensitive.

Nozzle The hollow cored metal nose screwed into the extrusion end of (1) the heating cylinder of an injection machine or (2) a transfer chamber (where this is a separate structure).

Nucleation (nucleator) With crystalline polymer, any foreign additive that assists or acts as a starting site for crystallinity within the resin. These initiators can reduce cycle time by speeding up the crystalline formations, thereby causing the part to solidify to hasten its ejection from the mold.

Nylon A generic term for polyamides. A crystalline thermoplastic.

Olefin plastics Plastics produced from olefins (polyolefins). Examples are polyethylene and polypropylene.

Opaque A material that will not transmit light and is not transparent.

Optical comparator An inspection machine using optics to compare the outline of a part to its required dimensions on a graphic screen.

Orange peel An unintentional rugged surface that gives an appearance resembling the skin of an orange.

Organic Refers to the chemistry of carbon compounds.

Orientation The alignment of the crystalline structure in polymeric materials so as to produce a highly uniform structure. Can be accomplished by cold drawing or stretching during fabrication.

Orifice An opening in a die or other metal piece used to meter (control the flow of) fluid material.

Out-of-round Referring to nonuniform radius or diameter.

Overflow tab A small, localized extension of a part at a weld line junction to allow a longer material flow path for; the purpose of obtaining a better fusion bond of the meeting melt fronts.

Oxidation (1) Degradation of a material through contact with air. (2) A chemical reaction involving a combination with oxygen to form new compounds.

Oxygen index An indication of flammability that states the necessary amount of oxygen required to sustain combustion.

Pack time The amount of time that packing pressure is kept on the screw until the gate freezes off. Occurs immediately after the initial injection stroke ends.

Packing pressure The pressure applied just before the part cavity fills, and is maintained to keep melt flowing into the mold cavity to compensate for in-mold material shrinkage and until the gate freezes off. Packing pressure must be the same as the injection pressure so that the mold cavity is not depressurized during the final packing period.

Parallel to the draw The axis of the cored position (hole) or insert parallel to the up-and-down (vertical) movement of the mold as it opens and closes.

Parallels The support spacers between the mold and press platen and clamping plate. Also called *risers* or *support fillers*.

Pareto analysis An analytical and statistical technique used to determine part defect type and quantity. Ranks each type of defect as a percentage of the total number of defects found, based on the quantity of each type of defect.

Parison Term given the extruded molten material, usually hollow, to form a product in an extrusion/blow-molded operation.

Part separator A machine or system used to separate parts from the runner system automatically after molding. Separated parts go to their net station, and the runner moves to a granulator for reuse if permitted. System may use blades, rigid pins, or a degating station with parts placed by a robot for separation.

Parting agent *See* **Mold release**.

Parting line (1) The point in the mold where two or more metal surfaces meet, creating a shutoff. (2) Mark on a molding or casting where halves of a mold meet in closing.

Partitioned mold cooling *See* **Bubbler**.

Pastel A tint. Mass tone to which white has been added.

Permeability (1) The passage of diffusion of a gas, vapor, liquid, or solid through a material without chemically or physically affecting it. (2) The rate of the passage in (1).

Perpendicular to the draw At a 90° angle from parallel to the draw of the mold opening.

Piece part price The calculated finished part cost based on material, processing, assembly, decoration, and packaging, including productivity and overhead costs.

Pigment Imparts color to plastic while remaining a dispersion of undissolved particles.

Pigmented Color pigments are added to a resin to produce a desired color in the plastic resin after molding. Pigments can be either organic or inorganic material. The inorganic pigments are usually heavy metals that are carcinogenic and no longer used.

Pinpoint gate A restricted orifice through which molten plastic flows into a mold cavity. Also called *restricted gate*.

Pitch With respect to extruder or injection molding, the distance from any point on the flight of a screw line to the corresponding point on an adjacent flight, measured parallel to the axis of the screw line or threading.

Plastic (1) One of the high polymeric materials, either natural or synthetic, exclusive of rubbers, which either melt and flow with heat and pressure, as with a thermoplastic, or chemically "set," as with a thermoset material. (2) Capable of flow under pressure or tensile stress (verb).

Plastic deformation The deformation of a material under load that is not recoverable after the load is removed. Opposite of elastic deformation.

Plastic memory A phenomenon of a plastic to return, in some degree, to its original form upon heating.

Plasticate To soften by heating or kneading.

Plasticity A property of plastics that allows the material to be deformed continuously and permanently without rupture upon the application of a force that exceeds the yield value of the material.

Plasticize To make a material soft and moldable with the addition of heat and/or pressure or a plasticizer.

Plastisol Name given to liquid polymer compounds, usually in the elastomer family used to make different products.

Platens The mounting plates of an injection or compression-molding machine to which the entire mold assembly is bolted.

Plating *See* **Metalizing**.

Plunger The part of a transfer or injection press that applies pressure to the unmelted plastic material to push it into the chamber. This, in turn, forces plastic melt at the front of the chamber out through the nozzle. *See* **Ram**.

Polyallomers Crystalline thermoplastic polymers made from two or more differed monomers, usually ethylene and propylene.

Polyamides A group of crystalline thermoplastics, of which nylon is typical.

Polycarbonate resin An amorphous thermoplastic material. It is transparent and can be injection molded, extruded, thermoformed, and blow-molded. It is known for its high impact force retention capabilities but is solvent-sensitive. The material is amorphous.

Polyethylene A crystalline thermoplastic material made by polymerizing ethylene gas.

Polyimide Classified as a thermoplastic, it cannot be processed by conventional molding methods. The polymer has rings of four carbon atoms tightly bound together. It has excellent resistance to heat.

Polyliner (1) A perforated, longitudinally ribbed sleeve that fits inside the cylinder of an injection-molding machine. Used as a replacement for conventional injection cylinder torpedoes (older machines). (2) A plastic bag placed inside a carton or box to prevent moisture and foreign material contamination during shipment of resin to a customer.

Polymer A high-molecular-weight organic compound—natural or synthetic—whose structure a repeated small unit, the MER, can represent. Examples are polyethylene, rubber, and cellulose. Synthetic polymers are formed by addition of condensation polymerization of monomers. Some polymers are elastomers and some are plastics.

Polymerization A chemical reaction in which the molecules or a monomer are linked together to form large molecules whose molecular weight is a multiple of the original substance. When two or more monomers are involved, the process is called *copolymerization*. Addition and condensation are the two major types of reaction.

Polyphenylene oxide (PPO) An amorphous thermoplastic. This material is noted for its useful temperature range from −275 to 375°F.

Polystyrene An amorphous thermoplastic made by polymerizing styrene.

Polysulfone An amorphous thermoplastic noted for its high strength.

Polyvinyl chloride (PVC) A thermoplastic material made by the polymerization of vinyl chloride with peroxide catalysts. The pure polymer is brittle and difficult to process. It yields a flexible material when compounded with plasticizers.

Postannealing Stress relieving of molded parts by external means, hot air, or oil, humidity chambers, or submersion in a fluid.

Postforming A process used to impart a shape to a previously molded article.

Postmold shrinkage The shrinkage occurring after a part has been removed from the mold. Influenced by the material and chemical properties of the resin and its molding conditions. Also influenced by end-use conditions and environmental conditions.

Potentiometer An electrical control device that senses changes in voltage or a potential difference by comparison to a standard voltage and can transmit a signal to a control switch.

Preplastication Technique of premelting injection-molding powders in a separate chamber, then transferring the melt to the injection cylinder. Device used for preplastication is commonly known as a *preplasticizer*.

Press fit An interference assembly between two mating parts, with friction holding the parts together. Parts assembled are under considerable stress.

Pressure drop The decrease in pressure on a fluid attributed to the number of turns it has to make and the distance it must flow to fill a cavity.

Pressure gradient lines A hypothetical set of pressure lines in a part created by the material's pressure drop as the part is filled. The farther the material flows from the gate, the lower the packout pressure.

Process control procedures A separate document, often included as an attachment to the quality control manual, which is a detailed description of the methods to be followed in the manufacture of a product. A copy may be attached to the work order for reference and revised as required should changes in the product occur.

Processing aid A resin additive that improves processing characteristics.

Product certification The certificate or letter stating that the material or product meets or exceeds customer requirements. Values are often listed for the tested or measured results. Signed by a key representative of the company to verify accuracy.

Projected surface area The exposed resin area of a mold on the parting line that transmits the injection pressure on the closed mold halves. Includes part, runner, and sprue surfaces expressed in inches squared of the surface area.

Prototype mold A simplified mold construction often made from a light-metal casting alloy or from an epoxy resin in order to obtain information for the final mold and/or part design.

Purging Cleaning one color or type of material from the cylinder of an injection molding machine or extruder by forcing it out with the new color or material to be used in subsequent production. Purging materials are also available.

Pyrometer An electrical thermometer for measuring high temperatures. Unit comes with two probes to measure melt and surface temperatures.

QS 9000 Automotive harmonization of Dammler/Chrysler, Ford, and General Motors with input from the trucking manufacturers. An add on to ISO 9000 requiring documentation and verification in greater depth and detail to the automotive suppliers specifications and requirements. Soon to be combined with a world wide automotive specification in or about 2002.

Quality assurance A separate department established to direct the quality function of the business and systems responsibility areas. Major concentration is directed at assisting and auditing the activities of the quality control department in their efforts to ensure that quality products are produced.

Quality circles A quality analysis group consisting of employees with specific departmental knowledge used to provide suggestions and ways to solve a procedural or manufacturing quality problem. If found acceptable, the group's findings and solutions are then passed on to upper management for implementation.

Quality control A department set up to be technically involved in the control of product quality. Involved in the principal inspection and testing of a product, with limited systems responsibility.

Quality control manual A document that states the company's quality objectives and how they will be implemented, documented, and followed in the manufacture and conducting of business with their customers.

Quality rated *See* **Approved supplier**.

Quench A method of rapidly cooling thermoplastic molded parts as soon as they are removed from the mold. Submerging the parts in water generally does this.

Quick mold change An efficient method of quickly changing over to a new molding program, often staging mold, equipment and tools at the machine to reduce setup charges and program cost.

RFQ, request for quote A request for a supplier to furnish a price and delivery quote to a customer with in a specified time period as defined by the instructions in the quote.

Ram The press member that enters the cavity block and exerts pressure on the molding compound designated as the "top force" or "bottom force" by position in the assembly. *See* **Plunger**.

Real time An action or operation occurring in present time.

Reciprocating screw A combination injection and plasticizing unit in which an extrusion device with a reciprocation screw is used to plasticize the material. Injection of material into a mold can take place by direct extrusion into the mold, by reciprocation the screw as an injection plunger, or by a combination of the two. When the screw serves as an injection plunger, this unit—the screw and the barrel—acts as a holding measuring and injection changer.

Recycled plastics A plastic material prepared from previously used or processed plastic materials that have been cleaned and reground.

Regrind (1) Waste plastics that are recovered and processed for reuse. (2) Plastics that have been ground or palletized at least twice, nonvirgin resin pellets.

Reinforced molding compound A material reinforced with special fillers to meet specific requirements, such as glass fibers, mineral, or other reinforcing modified medium.

Release agent *See* **Mold release**.

Relief angle (1) The angle of the cutaway portion of the pinchoff blade measured from a line parallel to the pinchoff land. (2) In a mold the angle between the narrow pinchoff land and the cutaway portion adjacent to the pinchoff land. (3) A rounding of a 90° internal or external corner to provide for lower stress concentrations at this point on the product.

Residence time The amount of time a resin is subjected to heat in the barrel of an injection-molding machine.

Residual stress The stresses remaining in a plastic part as a result of thermal or mechanical treatment.

Resin (1) Any of a class of solid or semisolid organic products of natural or synthetic origin, generally of high molecular weight with no definite melting point. (2) In a broad sense any polymer that is a basic material for plastics. *See* **Polymer**.

Resin pocket An apparent accumulation of excess resin in a small localized section that is visible on cut edges of molded surfaces. Also called *resin segregation.*

Restricted gate Sometimes referred to as *pinpoint gate.* A small opening between the runner and the cavity in an injection or transfer mold.

Retainer plate The plate on which demountable pieces, such as mold cavities, ejector pins, guide pins, and bushings are mounted during molding; usually drilled for steam or water.

Retaining pin A pin on which an insert is placed and located prior to molding.

Rib An object designed into a plastic part to provide lateral, longitudinal. or horizontal support and additional strength to the section to which it is added.

Ring gate A gate or annular opening that circles around a core pin or molded part through which melt is injected into a mold cavity.

Rockwell hardness A common method of testing materials for resistance to indentation in which a diamond or steel ball, under pressure, is used to pierce the test specimen (ASTN D785).

Runner In an injection or transfer mold, the channel that connects the sprue with the gate to the cavity. The channel through which the molten plastic flows into the mold cavity.

Runner system With plastics, the sprue, runners, and gates that lead the material from the nozzle of an injection-molding machine to the mold cavity.

Salt-and-pepper blends Resin blends of different concentrate additives, in pellet form, mixed with virgin resin to make a different product. Usually associated with color concentrate blends, that, when melted and mixed by the injection-molding machine's screw, yield a uniform colored melt for a product.

SAN An abbreviation for *styrene acrylonitrile copolymers.*

Scrap A product or material that is out of specification to the point of being unusable.

Screw The main component of the "reciprocation screw" injection-molding machine. Has various sizes, lengths, and compression ratios to feed, compress, melt, and meter for injection into the mold cavity. Basically divided into three major sections but there can be more. *Feed section*—deep screw depths to convey the resin into the next screw's section. *Transition section*—gradually decreasing screw depths when resin is compressed, forced against the barrel's surface, and melts. *Metering section*—the molten melt is further compressed in a shallow, uniform screw depth conveying forward as the screw turns past the check ring at the front of the screw.

Screw flights The circular groves cut into the screw whose size, depth, and shape convey the pellets down the barrel, compressing and melting them while preparing the melt for the next molding cycle.

Screw plasticating injection molding *See* **Injection molding**.

Scuff mark An imperfection on a part's show surface caused by dragging the part against the mold's surface during ejection from the mold cavity.

Sealing diameter That portion of a metal insert that is free of knurl and is allowed to enter the mold to prevent the flow of plastic material.

Second-surface decorating A method of decorating a transparent plastic part from the back or reverse side. The decoration is visible through the part, but is not exposed.

Secondary finishing operations Operations performed on a product after it has achieved its primary form required to complete the manufacturing of the product, such as decorating assembly or packaging, etc.

Semiautomatic molding machine A molding machine in which only part of the operation is controlled by direct human action. The machine according to a predetermined program controls the automatic part of the operation.

Setup charge A monetary amount calculated to cover the expense of preparing a machine for the next molding operation, time, material, equipment, and labor. A set fee, actual cost or percent of overhead.

Shear Stress developed because of the action of the layers in the material attempting to slide against or separate in a parallel direction.

Shear heat The rise in temperature created by the compression and longitudinal pressure on the resin in the barrel by the screw's pumping and turning action.

Shelf life The that time a material, such as an additive for a molding compound, can be stored without losing any of its original physical or functional properties.

Shore hardness A method of determining the hardness of a plastic material. This device consists of a small conical hammer fitted with a diamond point and acting in a glass tube. The hammer is made to strike the material under test and the degree of rebound is noted on a graduated scale. Generally, the harder the material, the greater the rebound (ASTM D2240).

Short or short shot A molded part produced when the mold has not been filled completely. Often done intentionally during startup to determine the correct amount of material to initially inject during the injection and fill stroke.

Shot The yield from one complete molding cycle, including sprue, runner, and flash.

Shot capacity The maximum volume of material that a machine can produce from one forward motion of the plunger or screw. All machines are rated using polystyrene as the melt standard and presented in ounces or pounds of melt per cycle.

Shot peening Impacting the surface of the material with hard, small, round beads of materials to disrupt the surface flatness. Used to stress-relieve welds and to improve the release of plastic resins on smooth core surfaces.

Shot volume *See* **Shot capacity**.

Shot weight The amount of molten resin generated in the barrel and injected into the mold cavity on a typical molding cycle to fill and pack out the mold cavity to the correct part weight.

Shrink fixture *See* **Cooling fixture**.

Shrinkage In a plastic, the reduction in dimensions after cooling.

Shrinkage allowance The additional dimensions that must be added to a mold to compensate for shrinkage of a plastic material on cooling.

SI units International System of Units.

Side actions (side coring or side draw pins) (1) An action built into a mold that operates at an angle to the normal open-and-close action and facilitates the removal of parts that would not clear a cavity or core on the normal press action. (2) Projections used to core a hole in a direction other than the line of closing of a mold and that must be withdrawn before the part is ejected.

Silicone (1) Chemical derived from silica used in molding as a release agent and general lubricant. (2) A silicon-based thermoset plastic material.

Sink mark A depression or dimple on the surface of an injection-molded part due to collapsing of the surface following local internal shrinkage after the gate seals. May also be an incipient short shot.

Six sigma The new quality term and methodology for identifying a process control technique to control a process within six (6) sigma limits that reduces defects to 3.4 defects per million, a reduction of 20,000 times.

Skins *See* **Flakes**.

Slides Caming sections of a mold cavity that form complex three-dimensional part sections that must operate and move before the part can be ejected from the mold. Used to form openings and sections of parts 90° to the part's release from the mold cavity.

Snap fit An assembly of two mating parts, with one or both parts deflecting until the mating parts are together. They then return to their as-molded condition or nearly so, depending on the design of the attachment. Parts can be under high to low stress after assembly.

Solvent Any substance, but usually a liquid, that dissolves other substances.

Solvent welding (solvent cementing, solvent bonding) A method of bonding thermoplastic articles of like materials to each other by using a solvent capable of softening the surfaces to be bonded. Thermoplastic materials that can be bonded by this method are ABS, acrylics, cellulosics, nylons, polycarbonate, polystyrene, and vinyls.

Sonic bonding High frequency vibrations generated by a transducer and transmitted in a tuned horn that contacts a part. The vibration energy generates heat and while pressure is applied to form a seal/connection or shape.

Specific gravity The density (mass per unit volume) of a liquid or solid material divided by that of water (ASTM D792).

Specification A written statement that dictates the material, dimensions, and workmanship of a manufactured product.

Specular gloss The relative reflective appearance of a material as judged visually.

Spider gate Multigating of a part through a system of radial runners from the sprue.

Spin welding The process of fusing two objects by forcing them together while one of the pair is spinning, until frictional heat melts the interface. Spinning is then stopped and pressure held until the objects are fused together.

Spiral flow test A method of determining the flow properties of a thermoplastic or thermoset material, in which the resin flows along the path of the spiral cavity that is circular in design from the sprue. The length of the material that flows into the cavity and its weight gives a relative indication of the flow properties of the resin.

Splay marks or splay Marks or lines found on the surface of the part after molding that may be caused by overheating the material, moisture in the material, or flow paths in the part. Usually white, silver, or gold in color. Also called "silver streaking."

Split cavity A cavity or a mold that has been made in sections.

Split ring mold A mold in which a split cavity block is assembled in a chase to permit the forming of undercuts in a molded part. These parts are ejected from the mold and then separated from the molded part.

Spot welding The localized fusion bonding of two adjacent plastic parts. Does not require a molded protrusion or hole in the parts. To be effective, used where two parallel and flat surfaces meet.

Spreader/torpedo A streamlined metal block placed in the path or flow of the plastic material in the heating cylinder of the extruder and injection-molding machines to spread it into intimate contact with the heating areas.

Sprue Feed opening provided in the injection or transfer mold. Also, a slug formed at this hole. "Spur" is a shop term for the sprue slug.

Sprue bushing A hardened steel insert in an injection mold that contains the tapered sprue hole and has a suitable seat for the nozzle of the injection cylinder. Must form a pressure/materialtight seal with the machine nozzle at the end of the barrel and the sprue bushing.

Sprue gate A passageway through which molten plastic flows from the nozzle into the mold cavity.

Stabilizer An ingredient used in the formulation of some plastics to assist in maintaining the physical and chemical properties of the compounded materials at their initial values throughout the processing and service life of the material

Staking A term used in fastening. The forming of a head on a protruding stud for the purpose of holding component parts together. Cold staking, hot staking, or ultrasonic heating may perform stacking. *See* **Heading**.

Static discharge A method to eliminate static electrical charge on a molded part so that it will not attract foreign particles to its surface and contaminate the part surface.

Stationary platen The plate of an injection/compression-molding machine to which the front plate of the mold is secured during operation. This platen does not move during the normal operation. Typically the platen at the end of the barrel nozzle.

Statistical process control The gathering of variable data using quality control methodology and charting the results to monitor and control a process.

Stereolithography A three-dimensional modeling process that produces copies of solid or surface models in special plastic resins. This process uses a moving laser beam, directed by computer, to copy or draw sections of the computer generated drawing or model onto the surface of photocurable liquid plastic. After each pass the model indexes down into the resin for the next layer to be developed.

Storage life *See* **Shelf life**.

Strain The dimensionless numbers (or units of length/length, i.e., inch per inch) that characterize the change of dimensions of a test specimen during controlled deformation. In tensile testing, the elongation divided by the original gauge length of the test specimen.

Strength of material Refers to the structural engineering analysis of a product to determine its strength properties.

Stress The force applied to produce a deformation in the material. The ratio of applied load to the original cross-sectional area of a test specimen (psi).

Stress concentration Sections or areas in a part where the molded-in or physical forces are very high or magnified by a force or action.

Stress crack External or internal cracks in a plastic caused by tensile stresses less than its short-term mechanical strength can withstand.

Striation (1) A separation of colors resulting in a linear effect or color variation. (2) In blow molding, the rippling of thick parisons. (3) A longitudinal line in a plastic due to a disturbance in the melt path.

Stripper plate A plate that strips a molded part from the core pins or cores.

Styrenic Indicates a group of plastics materials that are polymers, either whole or partially polymerized from styrene monomer.

Submarine gate A type of edge gate where the opening from the runner into the mold is located below the parting line or mold surface, as opposed to conventional edge gating, where the opening is machined into the surface of the mold on the parting line of the mold halves. With submarine gates, the part is broken off from the runner system on opening of the mold or ejection from the mold.

Surface finish Finish of a molded product. Refer to the SPI-SPE Mold Finishes Comparison Kit, available from DME Corporation, Detroit, MI.

Surface treatment Any method of treating a material so as to alter the surface and render it receptive to inks, paints, lacquers, and adhesives such as chemical, flame, and electronic treatments.

Swaging An assembly technique, similar to heading, where the plastic material is deformed to a specific shape to assemble one or more parts.

Tab-gated A small removable tab of approximately the same thickness as the mold item, usually located perpendicular to the item. The tab is used as a site for edge gate location, usually on items with large flat areas.

Taber mill *See* **Ball mill**.

Taguchi *See* **Design of experiments**.

Tapping Cutting threads in the walls of a circular hole.

Temperature gradient The slope of a graphed temperature curve. An increasing or decreasing temperature profile on the barrel of he molding machine is an example.

Tensile impact test A test whereby the sample is clamped in a fixture attached to a swinging pendulum. The swinging pendulum strikes a stationary anvil, causing the test sample to rupture. This is similar to the Izod test. See **Impact test**.

Texturizing The etching or cutting of a pattern on a mold surface to be reproduced on the molded part.

Thermal conductivity Ability of a material to conduct heat.

Thermal expansion The linear rate at which a material expands or contracts due to a rise or fall in temperature. Each material is unique and has its own rate of expansion and contraction. Expressed in in./in. · °F (mm/mm · °C).

Thermal stress cracking (TSC) Crazing and cracking of some thermoplastic resins that results from overexposure to elevated temperatures.

Thermocouple A thermoelectric heat-sensing element mounted in or on machinery and the mold to transmit accurate temperature signals to a control and readout unit.

Thermoelastomers *See* **Elastomer**.

Thermoplastic (TP) (1) Capable of being repeatedly softened by heat and hardened by cooling. (2) A material that will repeatedly soften when heated and harden when cooled. Typical of the thermoplastic family are the styrenic polymers and copolymers, acrylics, cellulosics, polyethylene, polypropylene, vinyls, nylons, and the various fluorocarbon materials.

Thermoset (TS) A material that undergoes or has undergone a chemical reaction by the action of heat and pressure, catalysts, ultraviolet light, or other agent, leading to a relatively infusible state. Typical of the plastics in the thermosetting family are the aminos (melamine and urea), unsaturated polyesters, alkyds, epoxies, and phenolics. A common thermoset goes through three stages. *A stage*—an early stage when the material is soluble in certain liquids, fusible, and will flow. *B stage*—an intermediate stage at which the material softens when heated and swells in contact with certain liquids, but does not dissolve or fuse. Molding compounds resins are in this state. *C stage*—the final stage is the TS reaction when the material is insoluble, infusible, and cured.

Thinner A liquid that can extend a solution, but not reduce the power of the solvent.

Thread plug, ring, or core A part of the mold that shapes a thread.

Tie bars Bars that provide structural rigidity to the clamping mechanism of a press and usually guide platen movement.

Timers Analog or digital timers used to accurately control the molding cycle operations of occurrences.

TIR (total indicator reading) An abbreviation used to identify tolerances with respect to concentricity. *Note:* The term TIR is a "former practices" term; the more acceptable current term is FIM (full indicator movement).

Toggle or toggle action A mechanism that exerts pressure developed by the application of force on a knee joint. It is used as a method of closing presses and also serves to apply pressure at the same time.

Tolerance A specified allowance for deviation in weighing and measuring or for deviations from the standard dimensions of weight (*SPI Guidelines of Plastic Custom Molders*).

Tool *See* **Mold**.

Torpedo *See* **Spreader/torpedo**.

Torsional The twisting or turning motion of a part. Torsional stress is created when one end of a part is twisted in one direction while the other is held rigid or twisted in the other direction.

Translucent The quality of transmitting light without being transparent.

Transparent A material with a high degree of light transmission that can be easily seen through.

Tumbling (1) Finishing operation for small plastic article by which gates, flash, and fines are removed and/or surfaces are polished by rotating them in a barrel together with wooden pegs, sawdust, and polishing compounds. (2) Adding color to a material through tumble blending.

Tunnel gate *See* **Submarine gate**.

Ultimate strength Strength (stress in psi) at the breakpoint in a tensile test.

Ultrasonic insertion The inserting of metal into a thermoplastic part by the application of vibratory mechanical pressure at ultrasonic frequencies.

Ultrasonic sealing or bonding A method in which sealing is accomplished through the application of vibratory mechanical pressure at ultrasonic frequencies (20–40 kC). Electrical energy is converted to ultrasonic vibrations through the use of either a magnetostrictive or piezoelectric transducer. The vibratory pressures at the interface in the sealing area develop localized heat losses that melt the plastic surfaces effecting the seal.

Unbalanced mold A nonuniform layout of mold cavities and runner system, fill rate, packing pressure, and part quality will vary from cavity to cavity. Used only for noncritical, standalone parts.

Undercut (1) Having a protuberance or indentation that impedes withdrawal from a mold in its normal open/closed movement. Flexible materials can be ejected intact even with slight undercuts. (2) Any such protuberance or indentation; depends also on design of mold.

Unit mold (1) Mold designed for quick-changing interchangeable cavity parts. (2) A mold that comprises only a single cavity, frequently a pilot for the production set of molds.

Universal testing machine A machine used to determine tensile, flexural, or compressive properties of a material in test bar form.

Unscrewing mold equipment Operations requiring cores and thread forming operations requiring that they be withdrawn prior to part ejection from the mold, which form a blind or cavity in the part, often in the plane of the mold, face.

UV (ultraviolet) stabilizer Any chemical compound that when added to thermoplastic material, selectively absorbs UV rays. Carbon black is a natural UV absorber and used extensively in plastic materials.

Vacuum metalizing Process in which surfaces are thinly coated with metal by exposing them to the vapor of metal that has been evaporated under vacuum (one millionth of normal atmospheric pressure).

Vendor A company or person who sells or supplies a part or service to another for a price.

Vent In a mold, a shallow channel or minute hole cut in the cavity at the parting line to allow air to escape as the material enters.

Vibration welding *See* **Ultrasonic sealing**.

Vicat softening temperature The temperature at which a plastic is penetrated to 1 mm depth by a flat-ended circular metal pin, while in an increasing temperature-controlled silicone fluid bath (ASTM D1525).

Vinyl Usually polyvinyl chloride, but may be used to identify other polyvinyl plastics.

Virgin plastics or virgin material Material not previously used or processed and meeting manufacturer's specifications.

Viscosisty A measurement of resistance of a material to flow.

Void A void or bubble occurring in the center of a heavy thermoplastic part, usually caused by excessive shrinkage, moisture, or trapped air in the melt.

Volume Synonym for capacity or displacement.

Volume resistivity The electrical resistance between opposite faces of a 1-cm cube of insulating material. It is measured under prescribed conditions using a direct-current potential after a specified time of electrification. It is commonly expressed in ohm-centimeters. Also called specific insulation resistance (ASTM D257).

Warpage Dimensional distortion in a plastic object after molding.

Web gate *See* **Diaphragm gate**.

Weigh packing A method, often automated, in which product is packed in a container, based on individual part weight or combination thereof. Weight is often compared to part count for small parts.

Weld line *See* **Flow line**.

Welding Joining thermoplastic parts by one of several heat-softening processes. Butt fusion; spin welding, ultrasonic, and hot gas or plate welding. Each is different and unique but accomplishes the same end result.

Welding horn The ultrasonic energy transmission and pressure-transmitting tool used for ultrasonic welding. Each welding horn is tuned to specific amplitudes to efficiently perform the welding operation.

Wetting agent An ingredient or solution used to lower the surface tension between two materials, so that good coverage and bonding occur.

Witness line Lines left on a molded part by poor mating and fit of side action cores.

Yield value (1) *Yield strength*—in tensile testing, the stress, usually in psi at which there is no increase in stress with a corresponding increase in strain: usually the first peak on the curve. (2) *Yield point*—the specific limiting deviation from the proportional stress–strain curve.

Young's modulus See **Modulus of elasticity**.

Zero defects A quality control method where anyone in the production cycle who discovers a quality problem can stop the assembly line or manufacturing process until the problem is corrected. The problem associated with this method is that upper management is seldom informed that a problem occurred. This lack of knowledge may prevent a complete repair from being initiated, and the problem may continue, to occur.

INDEX